Engineering Hydrology Handbook

Engineering Hydrology Handbook

Editor: Stacy Keach

STATES
ACADEMIC PRESS
www.statesacademicpress.com

States Academic Press,
109 South 5th Street,
Brooklyn, NY 11249, USA

Visit us on the World Wide Web at:
www.statesacademicpress.com

ISBN: 978-1-63989-178-8 (Hardback)

Cataloging-in-Publication Data

Engineering hydrology handbook / edited by Stacy Keach.
p. cm.
Includes bibliographical references and index.
ISBN 978-1-63989-178-8
1. Hydraulic engineering. 2. Hydrology. 3. Hydraulics. 4. Water-supply engineering.
5. Fluid mechanics. I. Keach, Stacy.
TC145 .H36 2022
627--dc23

Table of Contents

Preface

This book has been a concerted effort by a group of academicians, researchers and scientists, who have contributed their research works for the realization of the book. This book has materialized in the wake of emerging advancements and innovations in this field. Therefore, the need of the hour was to compile all the required researches and disseminate the knowledge to a broad spectrum of people comprising of students, researchers and specialists of the field.

Engineering hydrology is the field of engineering that deals with the study of movement, occurrence, distribution, and the properties of water on the Earth or beneath its surface and in the atmosphere. The primary applications of engineering hydrology include the calculating of rainfall, surface runoff and precipitation, determining water balance, enabling real-time flood forecasting and warning, etc. It also aims to establish a relationship between ground and surface water observed in catchments, design and operation of hydraulic structures and generation of hydropower. This book provides significant information of engineering hydrology to help develop a good understanding of this field and its related fields. It outlines the processes and applications of engineering hydrology in detail. In this book, using case studies and examples, constant effort has been made to make the understanding of the difficult concepts of engineering hydrology as easy and informative as possible, for the readers.

At the end of the preface, I would like to thank the authors for their brilliant chapters and the publisher for guiding us all-through the making of the book till its final stage. Also, I would like to thank my family for providing the support and encouragement throughout my academic career and research projects.

Editor

1

Coupling X-ray microtomography and macroscopic soil measurements: a method to enhance near-saturation functions?

E. Beckers[1], **E. Plougonven**[2,*], **N. Gigot**[1], **A. Léonard**[2], **C. Roisin**[3], **Y. Brostaux**[4], **and A. Degré**[1]

[1]Univ. Liège, GxABT, Soil–Water Systems, 2 Passage des Déportés, 5030 Gembloux, Belgium
[2]Univ. Liège, Department of Applied Chemistry, Laboratory of Chemical Engineering, Sart-Tilman, 4000 Liège, Belgium
[3]Walloon Agricultural Research Centre of Gembloux (CRA-W), Department of Agriculture and Natural Environment, Soil Fertility and Water Protection Unit, 4 Rue du Bordia, 5030 Gembloux, Belgium
[4]Univ. Liège, GxABT, Applied Statistics, Computer Science and Mathematics Unit, 2 Passage des Déportés, 5030 Gembloux, Belgium
[*]now at: ICMCB-CNRS/Group 4, 87 Avenue du Docteur Albert Schweitzer, 33608 Pessac, France

Correspondence to: E. Beckers (eleonore.beckers@ulg.ac.be)

Abstract. Agricultural management practices influence soil structure, but the characterization of these modifications and consequences are still not completely understood. In this study, we combine X-ray microtomography with retention and hydraulic conductivity measurements in the context of tillage simplification. First, this association is used to validate microtomography information with a quick scan method. Secondly, X-ray microtomography is used to increase our knowledge of soil structural differences. Notably, we show a good match for retention and conductivity functions between macroscopic measurements and microtomographic information. Microtomography refines the shape of the retention function, highlighting the presence of a secondary pore system in our soils. Analysis of structural parameters for these pores appears to be of interest and offers additional clues for soil structure differentiation, through – among others – connectivity and tortuosity parameters. These elements make microtomography a highly competitive instrument for routine soil characterization.

1 Introduction

Tillage simplification has become a popular practice in recent years, mainly in order to save energy and reduce erosion. Agricultural management practices influence soil structure, as stated in numerous papers which have addressed the effects of tillage intensity on soil (see Strudley et al., 2008 for a review on the subject). However, changes in soil hydrodynamic behaviour are not fully understood and are still studied (Schlüter et al., 2011; Eden et al., 2012; Holthusen et al., 2012 for example). Furthermore, researchers have drawn divergent conclusions over the impact of tillage practices on soil hydraulic properties (e.g. Green et al., 2003; Cousin et al., 2004; Bhattacharyya et al., 2006; Strudley et al., 2008), though for most of them, pore size distribution, connectivity and orientation are involved. These changes in porosity suggest that the related hydrodynamic properties will be affected, as well as water fluxes and their partition. But the characterization of these modifications and consequences remains a challenge.

Plot-scale measurements, on the one hand, allow characterization of the global behaviour, but do not provide mechanistic explanations of structural modifications. The divergence in the literature with regard to agricultural management impacts shows the inability of these measurements to comprehend them completely. On the other hand, microscale characterization, involving small soil samples (with accuracy to within a micron or less), can offer helpful insight on the pore structure, but might not be representative at the plot scale. Notably, X-ray tomography has been recently used in order to characterize changes in soil pore distribution in different contexts, and attempts to link microtomography information to hydrodynamic measurement.

Already in 1997 Olsen and Borresen (1997) were studying pore characteristics depending on tillage intensity with computed tomography. However, with a pixel size of 0.5 mm, they only concluded about the macroporosity distribution in the soil profile. Since then, soil porosity has been analysed many times using X-ray tomography (see Taina et al., 2008 for the state of the art), but research on the link between macroscopic measurements and microscopic investigation of the soil structure remains scarce. Wiermann et al. (2000) showed the interest of this technique by combining water retention, hydraulic conductivity and tomography analyses to compare soil reaction to precompression stress depending on management practices. Kumar et al. (2010) and Kim et al. (2010) tried to explain saturated hydraulic conductivity (K_{sat}) differences by pore parameter measurements with computed tomography, and found that most of these parameters were correlated with K_{sat}. Rachman et al. (2005) and Quinton et al. (2009) studied macroporosity through X-ray tomography and water retention curves; they concluded that these methods lead to comparable results for porosity distributions. Dal Ferro et al. (2012), for their part, analysed soil porosity with mercury intrusion porosimetry and X-ray microtomography. They highlighted the fact that microtomography does not take into account all the microporosity, and therefore advised combining microtomography analyses with other techniques. Cousin et al. (2004) conducted a two-scale study in order to determine more parameters: qualitative observation of macroporosity through stained infiltration combined with laboratory hydraulic conductivity measurements and tomographic analyses. They reported a better plot-scale conductivity in no tillage due to the presence of earthworm tunnels. The scarcity of these tunnels leads to the need for macroscopic measurements while tomographic observations allow a quantitative characterization of the bulk soil pore network. However, the achieved resolution in their study (i.e. 0.4 mm pixel size) was not sufficient to confirm a link between the pore network and in situ hydraulic conductivity measurements. Finally, Bayer et al. (2004) tested the ability of microtomography to provide water retention curves through a dynamic setup. Their results were in good agreement with a classical multistep outflow experiment. It can be seen that microtomography has become an interesting tool for the study of soil pore networks, as it provides a 3-D visualization of the internal soil structure. Furthermore, microtomography offers the possibility to extract pore size distribution as well as structural parameters. Among them, pore connectivity, which influences hydrodynamics (Vogel and Roth, 1998), or specific surface (surface area/volume) (Gerke, 2012) can be estimated.

In this study, we test X-ray microtomography as a tool to help differentiate soil structure modification due to tillage intensity through multiscale measurements. This association could help highlight the most influential microstructural factors affecting macroscopic transport phenomena. Water retention and hydraulic conductivity measurements on the one hand, and X-ray microtomography (34 μm pixel size) on the other hand, compose the experimentation campaign. Water retention and hydraulic conductivity measurements are used frequently in other studies to show the effect of tillage intensity on soils. Furthermore, they are linked together and can lead to different estimations of the same parameter. This choice enables us to validate microtomographic results and to use the latter as an explanatory element of the fundamental processes highlighted by macroscopic measurements. However, the consistency of the results depends on the quality of the tomographic reconstructions. Quality is, among other factors, correlated with acquisition time. As a result microtomography as a hydraulic measurement technique is often considered as time-consuming in comparison with other techniques. In our study, we show that relatively fast scans supply not only a good match with other measurements, but also improve retention and hydraulic curves modelling near saturation ($|h| < 10$ mbar). Microstructural parameters calculated on the pore network appear to be of interest to characterize tillage differences. These elements make microtomography a competitive instrument for routine soil characterization.

2 Material and methods

In this paper, we test X-ray microtomography as a tool to help quantification of soil structure modification due to tillage intensity. To do this, we couple what we call "micro" and "macroscopic" measurements: "microscopic" measurements refer to measurements that supply characterization at the pore scale; in opposition, the expression "macroscopic measurements" gathers all the measurements that are not at the pore scale. Establishment of 3-D soil strength profiles serves as a first visualization of soil structural differences and as a basis to organize sampling and measurement campaigns. Microscopic investigations consist in measuring 3-D morphological parameters using X-ray microtomography (μCT). Macroscopic investigations include retention data with the pressure plate method (Richards' apparatus), saturated and unsaturated soil hydraulic conductivity.

Retention and hydraulic curves are derived from both microscopic and macroscopic investigations, and compared. Morphological parameters obtained with X-ray microtomography are analysed with principal component analysis.

Variability at the plot scale was assessed through the use of at least seven repetitions for each soil measurement, which seems sufficient to obtain a good approximation of the global behaviour for each object. For reasons of space, statistical analyses concerning soils comparison are not presented here. It has to be noted, however, that when the term significant is used in the following paragraphs it refers to a p value of maximum 0.05.

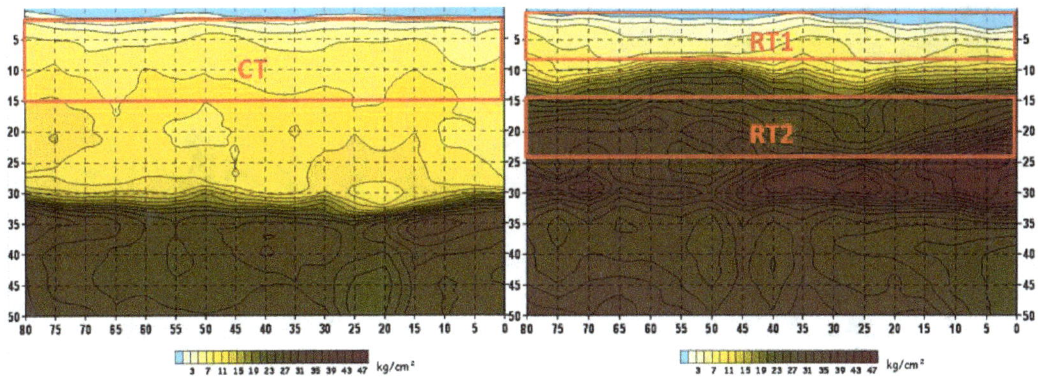

Figure 1. 3-D soil strength profiles. Left panel: CT profile; right panel: RT profile.

2.1 Location

Our field experiment takes place in Gentinnes (Walloon Brabant, Belgium), on a field organized in a Latin square scheme. Since 2004, plots (one plot is used here) have been cultivated in conventional tillage (CT), deep loosening (not studied here), or in reduced tillage (RT). The latter consists in sowing after stubble ploughing of about 10 cm. The crop rotation is sugar beet, winter wheat, flax. The soil is mainly composed of silt loam and can be classified as a Luvisol.

A fully automated penetrometer mounted on a small vehicle was used to illustrate soil resistance to penetration (see Fig. 1). The surface area of $160 \times 80\,\text{cm}^2$ and its position were chosen considering the characteristics of the agricultural management practices (more information can be found in Roisin, 2007). On this figure, we can observe two different soil horizons for both practices. In the CT profile, the second horizon appears at approximately 30 cm. The upper layer, from 0 to 30 cm, is quite homogeneous with a slight gradient along the depth. In the RT profile, the second horizon appears between 10 and 15 cm. The old plough pan is still observed at 30 cm depth. Because of these observations and the variation in tillage depth between management practices (10 cm for RT vs. 25 cm for CT), two horizons are investigated for RT – RT1 between 0 and 10 cm and RT2 between 12 and 25 cm (see Fig. 1) – while CT is considered as homogeneous.

2.2 Measurement techniques

2.2.1 X-ray microtomography

X-ray microtomography consists of performing series of X-ray radiograms at different angles, producing enough information to algorithmically reconstruct a 3-D X-ray attenuation map of the sample. The transmitted X-ray intensity depends on the attenuation coefficient of each material located along the X-ray path, in a cumulative way. The attenuation coefficient is related to the material properties, i.e. its density and atomic number (Attix and Roesch, 1968) and to the energy of the incident beam.

Sampling

Soil sample dimensions have been chosen according to the tomograph characteristics. The cylindrical samples are 5 cm in height and 3 cm in diameter, allowing for a good compromise between resolution and time acquisition. Eight soil samples ($35\,\text{cm}^3$) are removed from each horizon (CT, RT1 and RT2) and are exposed to a 15 bar pressure according to Richards' procedure (Richards, 1948; DIN ISO 11274, 2012). Samples are therefore around the wilting point and pores with a radius above $0.1\,\mu$m are meant to be free of water. Samples were collected very carefully. However, they are prone to cracks close to the sample container. As a result, the radius of image investigation is reduced a little to eliminate these effects. Final radii are the same for all our samples. Figure 2 gives a 3-D representation of pores for each horizon.

X-ray microtomograph

Samples are scanned using a Skyscan-1172 high-resolution desktop micro-CT system (Skyscan, Kontich, Belgium). The cone beam source operates at 100 kV, and an aluminium filter is used. The detector configuration (16 bit X-ray camera with 2x2 binning, creating 2048×1024 pixel radiograms) and the distances between source–object–camera are adjusted to produce images with a pixel size of $34\,\mu$m. The rotation step is 0.4° over 180°. We perform here what can be called a fast scan, in a total of about 2 h per sample. Since the objects are larger than the field of view of the detector, several sets of radiograms are taken and stitched together. The final projections are actually mosaics of 3 by 2 radiograms, meaning that one set of radiograms is acquired in about 20 min, which is relatively fast. In fact, the aim of this procedure is to have a good compromise between the acquisition quality, time and number of samples.

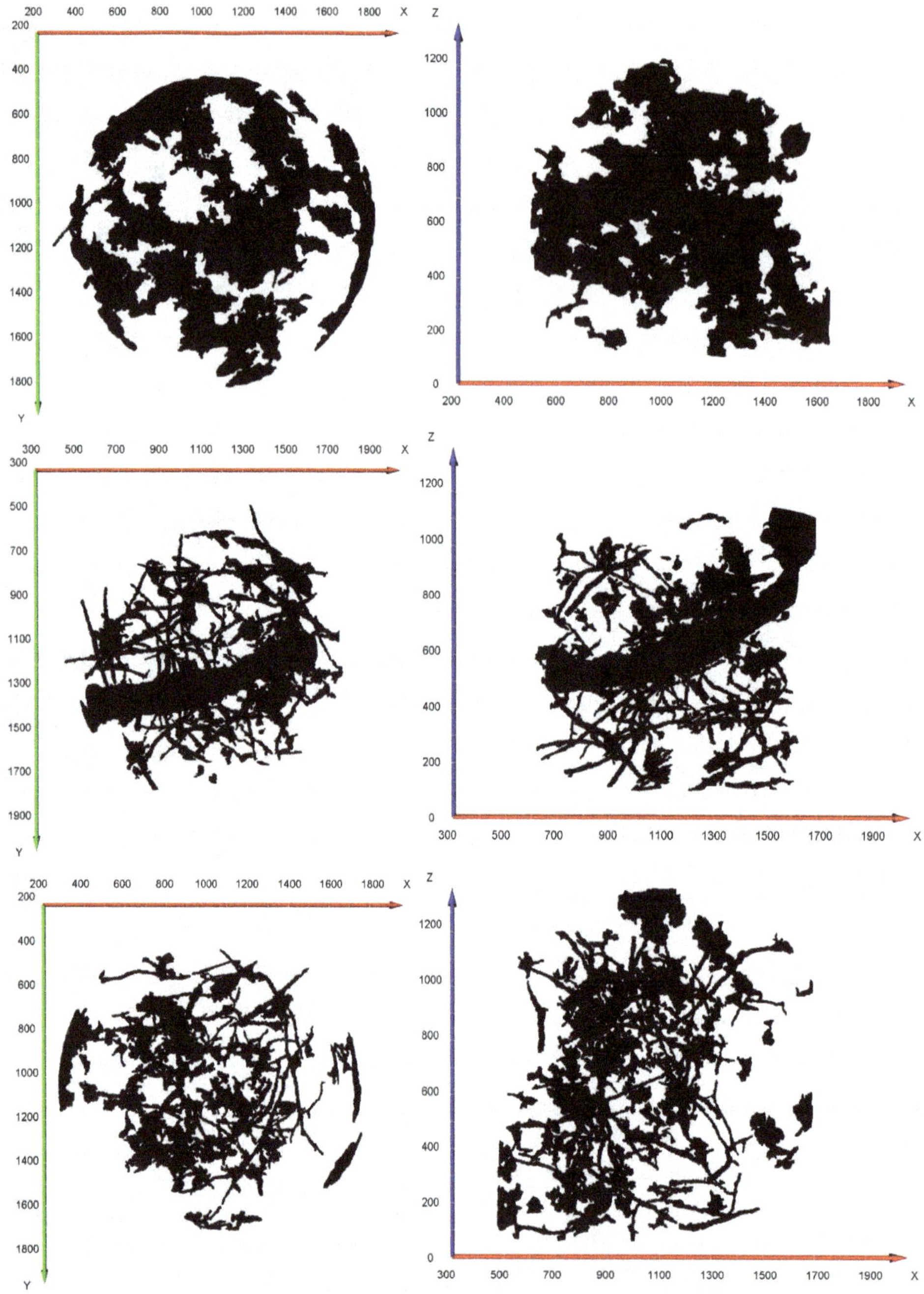

Figure 2. 3-D visualization of pores for each soil object for half samples (i.e. 2.5 cm height and 3 cm in diameter); CT, RT1, RT2, from top to bottom panels; XY view at the left panels; XZ view at the right panels.

Image reconstruction

Reconstruction is performed with the NRecon software® supplied with the Skyscan Micro-CT system. Ring artefact correction and rotation axis misalignment compensation is used. The resulting cross sections are then stacked and imported into Avizo® to be processed.

Image processing

In order to process our microtomograms, we use an algorithm developed by Plougonven (2009; Plougonven and Bernard, 2011) and integrated into the Avizo® software. This algorithm is organized in two steps, a pre-processing of the images (noise reduction thanks to a greyscale opening, followed by a Gaussian filtering) and a post-processing to decompose the porosity into individual pores, and calculate morphological parameters. Between these two steps, a threshold value needs to be chosen.

Segmentation

The choice of the segmentation method is not so obvious. Numerous methods exist and these methods can be global (i.e. a same threshold value for the entire domain) as well as local (i.e. the value is adapted regarding local characteristics). Review papers or research articles comparing segmentation methods (for example Iassonov et al., 2009; Wang et al., 2011) show that there is no ideal segmentation method in the context of soil analyses. Following Iassonov et al. (2009), the method has to be chosen considering the context of the study, and factors to be taken into account are principally "the reliability and consistency of the method, as well as the computational efficiency and the automated character of it". Local vs. global methods show different advantages considering these factors. Local methods dedicated to soils are very promising (e.g. Schlüter et al., 2010; Hapca et al., 2013; Houston et al., 2013), but they are more demanding in terms of time and computational resource. In the present study, we decide to apply a single mean threshold value (based on the soil porosity). This procedure has been validated and shows results nearly as good as the local methods cited before (see Beckers et al., 2014 for more details). Furthermore, the studied objects are soils with structural differences only. Our approach is comparative and our choices' impact can be supposed equivalent for the different studied objects.

Post-process

Once the threshold value is chosen, the part of the algorithm calculating morphological parameters can be applied in Avizo®. It provides local 3-D quantification based on pore space decomposition. The decomposition method is a watershed-based approach on the Euclidean distance transform of the porous phase, given specific pre-processing (Plougonven and Bernard, 2011), with the basins defined from both topological information (i.e. the intersection of branches in the skeleton) and geometrical information (post-merging of basins given certain geometric conditions, for more details see Plougonven, 2009). For each pore (following the definition of the current pore space decomposition), parameters are computed as illustrated in Fig. 3: volume (Vol), surface (determined from the number of voxel faces), barycentre, inertia tensor (defined as if the pore was an object with uniform mass, and on which the inertia matrix is computed), number of neighbouring pores (N_c), surface area of the connections (Sf_c) and equivalent radius of the connections. Additionally, we compute the specific surface (SS) for each pore, and the pore deformation (Def), defined as the ratio between minimum and maximum components of the inertia tensor. Using this deformation, an elliptic cylinder was fitted to the pores in order to calculate a mean radius (R). Finally, we calculate a variable related to specific connectivity (SC):

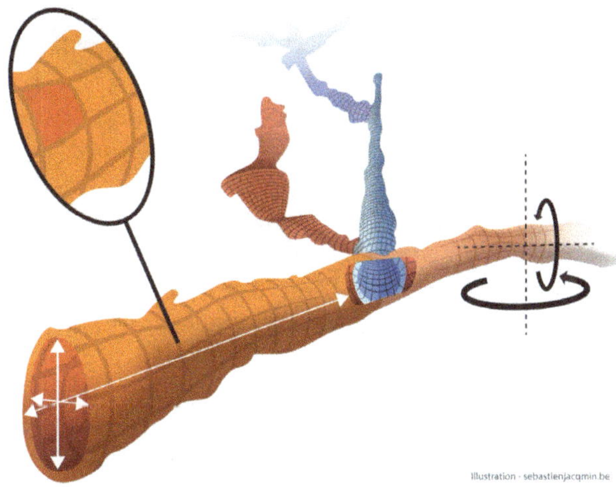

Figure 3. Representation of a theoretical pore network. The following morphological measurements are illustrated: on the left, dimensions of the pore as an elliptic cylinder (double white arrows); on the left, pore surface measurement determined from the number of voxel faces (see magnifying glass); in the middle, connection characterization (pore hole); on the right, inertia tensors measurements (black arrows).

$$SC = \frac{N_c A_c}{V_p} \tag{1}$$

where N_c is the number of connections, A_c the mean surface area of the connections (L^2) and V_p the pore volume (L^3).

2.2.2 Water retention

Retention points are obtained thanks to the Richards procedure on one hand (Richards, 1948; Dane and Hopmans, 2002 cited by Solone et al., 2012; DIN ISO 11274, 2012), and thanks to X-ray microtomography on the other hand.

With Richards' procedure, samples are saturated by upward moisturizing during 48 h, exposed to increasing pressures between 10 mbar to 15 bar (pF 1 to 4.2), and weighed between each stage. Seven soil samples ($100\,cm^3$) are taken from each horizon (CT, RT1 and RT2).

Pore size distribution derived from tomographic results allows calculating retention data points between pF1 (10 mbar) and saturation. Samples are around the wilting point during scanning procedure (see Sect. 2.2.1).

Pore size distribution is obtained from microtomographic measurements using the following relationship (capillary theory, Jurin's law):

$$r = \frac{2\sigma \cos(\alpha)}{\rho g h} \tag{2}$$

where r is the pore radius (L), h is the pressure head (L), σ is the liquid surface tension ($M\,T^{-2}$), α is the contact angle between the liquid and the soil, ρ is the liquid density ($M\,L^{-3}$) and g is the gravitational acceleration ($L\,T^{-2}$). The radius is determined with structural parameters measured thanks to X-ray microtomography (see Sect. 2.2.1, Post-process).

2.2.3 Hydraulic conductivity

In situ measurements

In situ infiltration measurements are performed by a 20 cm diameter tension-infiltrometer (TI) (Eijkelkamp Agrisearch Equipment). Eight repetitions are made for each management practice. For each location, three infiltration measurements for tensions of −9, −6 and −3 mbar are performed. Measurements are long enough to have at least 18 min of steady-state infiltration. We only use the unsaturated flows, saturated hydraulic conductivity being measured directly in the laboratory. Concerning measurements for RT1, this layer presents a shallow thickness (max 10 cm, see Fig. 1). Moreover, to obtain the mandatory plane surface for TI measurements, a part of it had to be removed. As a result, these measurements are likely to be attributed to RT2.

Laboratory measurements

We use a permeameter (Laboratory-Permeameter, Eijkelkamp, Giesbeek, Netherlands) to measure saturated hydraulic conductivity on 100 cm^3 soil samples. The basic principle is to create a pressure gradient between the sides of the sample and to measure the flow coming out. The constant head method is used (Klute, 1986 cited by Bayer et al., 2004). Considering the possible change in pore orientation, we measure this conductivity in the two main orientations: vertical conductivity vs. horizontal conductivity (parallel to the slope). Measurements are performed on eight samples for each object (CT, RT1 and RT2) and each orientation, and the mean value for K_{sat} is calculated in each direction. In this paper we use the highest value between the mean horizontal and the mean vertical one, considering that the 3-D pore network characteristics are more correlated with this highest value.

2.3 Measurements analysis

2.3.1 Retention and hydraulic functions

Continuous functions can be adjusted on our retention data points coming from Richards' measurements and microtomography. Using K_{sat} measured with the permeameter and parameters from these fitted retention data, $K(h)$ curves can be compared with TI measurements for each object. The agreement between these points serves as an indicator of the relevance of microtomography measurements.

However, numerous models exist. Since we study the near-saturation behaviour, this part of the curve will be of great importance. According to Durner (1994), the largest differences in retention and hydraulic predictions are not caused by the choice of the single porosity model, but by taking into account – or not – supplementary pore systems. Indeed, Durner (1994) states that errors in hydraulic predictions can come from the fact that some pore distributions are bimodal (or multimodal) instead of unimodal, each mode being considered as one pore system. In such a context, one model of each type has been adjusted (using the Levenberg–Marquardt algorithm, Levenberg, 1944; Marquardt, 1963): the unimodal from van Genuchten (1980) and the bimodal from Durner (1994). The associated hydraulic model is of Mualem (1976) for both cases.

The van Genuchten formulation is

$$\theta(h) = \theta_r + \frac{\theta_s - \theta_r}{[1 + |\alpha h|^n]^m} \tag{3}$$

with θ_r the residual water content, θ_s the saturated water content, n a pore size distribution parameter, α the inverse of the bubbling pressure (L^{-1}) and m a function of n ($m = 1 - 1/n$). The associated hydraulic conductivity is expressed as follows (Mualem, 1976):

$$K(h) = K_s S_e^l \left[1 - \left(1 - S_e^{1/m}\right)^m\right]^2 \tag{4}$$

with K_s the saturated hydraulic conductivity ($L T^{-1}$), l a pore connectivity parameter, and S_e the effective saturation:

$$S_e = \frac{\theta - \theta_r}{\theta_s - \theta_r}. \tag{5}$$

With the dual porosity model (Durner, 1994) the effective saturation becomes

$$S_e = w_1 \left[1 + (\alpha_1 h)^{n_1}\right]^{-m_1} + w_2 \left[1 + (\alpha_2 h)^{n_2}\right]^{-m_2} \tag{6}$$

with w the weighing factor, and subscripts 1 and 2 referring to each part of the porosity. The hydraulic function is

$$K(S_e) = K_s \frac{(w_1 S_{e1} + w_2 S_{e2})^l \left(w_1 \alpha_1 \left[1 - \left(1 - S_{e1}^{\frac{1}{m1}}\right)^{m1}\right] + w_2 \alpha_2 \left[1 - \left(1 - S_{e2}^{\frac{1}{m2}}\right)^{m2}\right]\right)^2}{(w_1 \alpha_1 + w_2 \alpha_2)^2}. \tag{7}$$

The goodness of the fit is analysed in this paper using the relative root mean square error (RRMSE) as Schaap and van Genuchten (2006), Weynants et al. (2009), Vereecken et al. (2010) – among others – did in a similar context. We compare models fitted on different amounts of observations considering retention data (Richards vs. μCT data). Besides, at least for $K(h)$ measurements, the measurement variability is greater near saturation. Finally, for the comparison between the models and the measurements for $K(h)$, we rely on three observed points (see Sect. 2.2.3).

RRMSE is calculated (following Alkasawneh et al., 2007) as

$$\text{RRMSE} = \sqrt{\frac{1}{n} \sum_{i=1}^{n} \left(\frac{d_i - D_i}{D_i}\right)^2}$$

with n the number of measurements, d_i the estimated value and D_i the observed value.

Table 1. RRMSE for fitting Richards' data set (R) and combined data set (μCT + R) with van Genuchten (VG) or dual porosity (DP) model on retention functions.

Horizon	"R" VG	"R" DP	"μCT + R" VG	"μCT + R" DP	"μCT + R" DP*
CT	0.035	0.016	0.036	0.013	0.030
RT1	0.039	0.044	0.048	0.012	0.021
RT2	0.026	0.020	0.045	0.004	0.003

* The last column is calculated using parameters from "μCT + R" DP but on Richards' data alone.

2.3.2 Hydraulic macroporosity efficiency

Following Watson and Luxmoore (1986, cited by Imhoff et al., 2010), the number of hydraulically effective pores between two tensions is related to the difference between hydraulic conductivity for these two tensions (K_m, $\mathrm{L\,T^{-1}}$) and can be calculated thanks to Poiseuille's law and the laminar flow equation:

$$N_m = \frac{8\eta K_\mathrm{m}}{\pi \rho g r_\mathrm{a}^4} \tag{8}$$

where η is the water dynamic viscosity ($\mathrm{M\,L^{-1}\,T^{-1}}$) and r_a is the minimum pore radius (L). The associated macroporosity is equivalent to

$$\theta_\mathrm{m} = N_\mathrm{m} \pi r_\mathrm{a}^2. \tag{9}$$

The ratio between effective macroporosity and measured porosity is therefore an indicator of the hydraulic performances of the soil (Buczko et al., 2006 cited by Imhoff et al., 2010), and will be tested in our context.

2.3.3 Principal component analysis

Principal component analysis (PCA) has been widely described in the literature, for example in Benzécri and Benzécri (1980) cited by Palm (1994), or in Jackson (1991). PCA is a multivariate descriptive method. It aims at gathering descriptive parameters in a few components. The use of these components allows a 2-D representation of the data and highlights possible relationships between data and parameters. As Jackson (1980) said: "This method is used to simplify the simultaneous interpretation of a number of related variables."

PCA is performed here taking into account seven parameters (see Sect. 2.2.1): Vol, N_c, $\mathrm{Sf_c}$, SS, R, Def and SC.

3 Results and discussion

3.1 Retention functions

For the retention curves, the three horizons were analysed: CT, RT1 (0–10 cm depth) and RT2 (12–25 cm depth). Figure 4 represents both the data sets for each horizon. We used

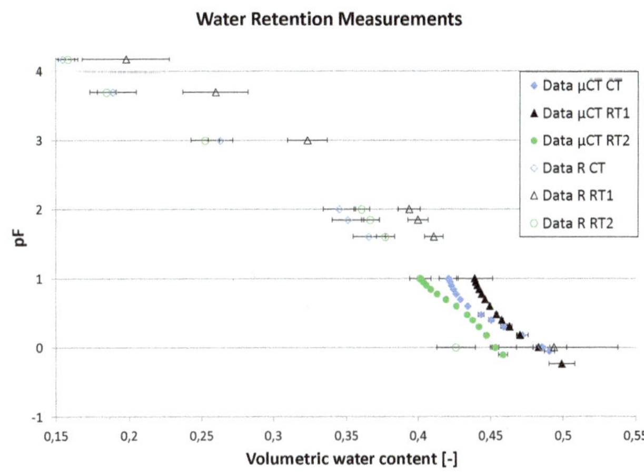

Figure 4. Water retention data for the three objects (CT, RT1, RT2) obtained with the Richards' procedure (empty dots) and with microtomography (full dots) for the entire tension range.

either the van Genuchten (1980) equation (VG) or the dual porosity (DP) model (Durner, 1994) to fit the data points. First, fitting is applied to Richards' measurements alone ("R"). In a second step, fitting is applied to a combination of Richards' (from pF 4.2 to pF 1) and μCT data (from pF 1 to saturation) (called "μCT + R").

With the R fitting (see Fig. 5, dotted lines), RRMSEs are better with the DP model for CT and RT2 but not for RT1 (see Table 1), differences between VG and DP performances being quite important for CT.

Comparing the horizons with DP for R, the only significant difference concerns the total effective porosity: it is significantly greater for CT than for RT2. RT1 is in between, but differences are not significant. We can see that near saturation, curves present different shapes. But the curves for RT are not well fitted for this part, especially for RT1.

The combination of Richards and μCT data is also fitted with the VG and DP models (see Fig. 5, solid lines). First, we can see that RRMSE (see Table 1) are widely better for the DP models than the VG ones. Indeed, the VG model can fit either Richards' data or μCT data, but fails in fitting a coupled data set. Supplementary pore systems have been integrated in retention function by Durner in 1994 and allow in many cases a better fitting of the retention data (see also Durner et al., 1999). In our study and except for RT1,

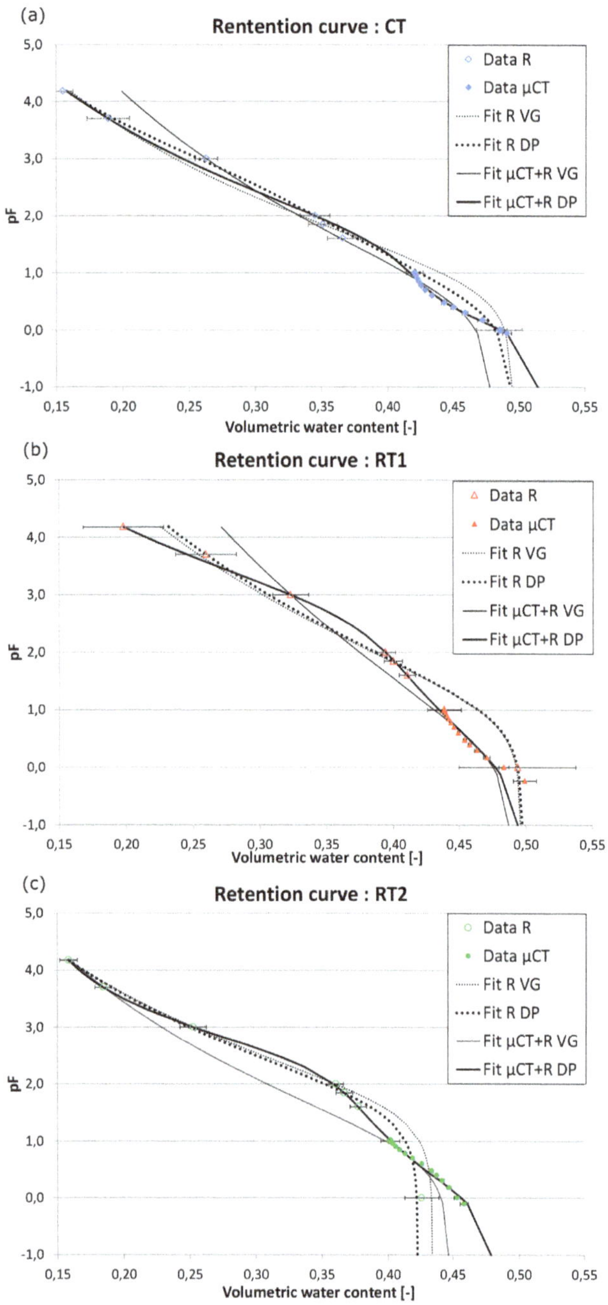

Figure 5. Retention curves. Dotted line: Richards' measurements (R); solid line: combined μCT and Richards' measurements (μCT + R); fine line: van Genuchten adjustment (VG); thick line: dual porosity adjustment (DP).

taking into account this secondary pore system improves (in terms of RRMSE) the match between retention curves and retention measurements even if we consider Richards' measurements alone (see Table 1, columns 1 and 2). Besides, if we use the parameters set coming from the "μCT + R" DP fit to calculate the RRMSE with Richards' measurements alone, the results are better than with VG (see Table 1, column 5).

These elements lead us to think that this secondary pore system is a characteristic of our soil and not an artefact coming from the collation of two data sets.

Looking at the combined DP curves, volumes from pF 1 to saturation are not significantly different for the horizons. But we can see that for RT2, this volume is more important than with Richards' measurements ($p < 0.05$) while for CT and RT1 these volumes match. Considering this match and the elements of the image analysis pre-process (see Sect. 2.2.1), obviously Richards' procedure does not allow a good saturation estimation for RT2. The reason could be linked to the pore distribution.

We can also observe that the DP fit on R and both VG fits have much the same global shape around saturation, but not CT and RT1 μCT + R DP curves. Indeed, comparing DP retention curves designed with μCT + R and designed with R, we observe that for CT and RT1, concavity is inverted from pF 1.2 to saturation. With μCT, the main part of the volume is closer to saturation than with R: For CT curve, the volume increase begins around pF 1 and slows down around pF 0 with R, while with μCT the volume increase begins around pF 0.6 and stops around pF −0.06. As a result, μCT data show more pores with a bigger radius. These conclusions are the same for RT1 and RT2. For RT1, μCT increase is even closer to saturation and consequently we have more and bigger pores than in CT. Furthermore for RT2, it is confirmed that fitting on R data leads to a bad approximation of porosity at saturation.

3.2 Hydraulic conductivity

3.2.1 Hydraulic macroporosity efficiency

If we compare macroporosity estimated from K_{sat} measurements (θ_M) vs. macroporosity estimated from Richards' and μCT measurements, we can use the ratio of these values as an indicator of macroporosity efficiency (ER) – in terms of conductivity ability – and pore network morphology. For both cases, we obtain the following trend: CT > RT2 > RT1 (see Table 2), with the latter being the most distant from Poiseuille's law with respect to structure morphology, i.e. reflecting poor dynamic performances. In fact, RT1 presents the higher near saturation pore volume while its saturated hydraulic conductivity is the lowest.

3.2.2 Hydraulic functions

Using K_{sat} measured with the permeameter and results from fitted parameters with retention data, $K(h)$ curves with the four different adjustments are represented in Fig. 6 for each horizon. TI measurements are indicated as well, and the match between these points and fitting curves are calculated. For RT1 and RT2, the same TI measurements are used. But, because of the lower depth of RT1, measurements are likely to be attributed mostly to RT2.

Table 2. K_s = saturated hydraulic conductivity [$m\,s^{-1}$]; N_M = number of hydraulically effective pores; θ_M = hydraulically effective macroporosity [$cm^3\,cm^{-3}$]; ER = efficiency ratios for μCT (T) and Richards' (R) measurements (m).

Horizon	K_s ($m\,s^{-1}$)	N_M	θ_M	ER T (m)	ER R (m)
CT	1.6×10^{-4}	8.1×10^{-4}	7.0×10^{-1}	1.2×10^{-1}	1.1×10^{-1}
RT1	2.5×10^{-5}	1.3×10^{4}	1.1×10^{-1}	2.3×10^{-2}	2.0×10^{-2}
RT2	4.0×10^{-5}	2.0×10^{4}	1.8×10^{-1}	3.8×10^{-2}	7.2×10^{-2}

Table 3. RRMSE for fitting Richards' data set (R) and combined data set (μCT + R) with van Genuchten (VG) or dual porosity (DP) model on hydraulic functions.

Horizon	"R" VG	"R" DP	"μCT + R" VG	"μCT + R" DP
CT	5.2	0.4	4.2	0.2
RT1	1.9	0.3	0.9	1.0
RT2	25.7	17.2	3.2	0.8

Comparing fitted models, we can see that for CT the unsaturated flow prediction is particularly enhanced when taking into account the presence of a secondary pore system (see Table 3). For RT2, the best prediction is linked with the use of μCT complementary data. But in both cases, it is the combination of μCT data and the DP model that gives the best prediction of the unsaturated flow. These elements seem to validate the adjustment of elliptic cylinders to obtain pore size distribution.

Concerning RT1, results are much the same for "μCT + R" data with VG and DP models, the better prediction being for R DP. The poor match for RT1 is probably caused by two factors. First, the TI measurements are probably more representative of RT2, and this is confirmed by the excellent match for this horizon. This assumption matches as well with our knowledge of the field and the measurements. Secondly, the DP retention curve on combined data does not fit very well between pF 0 and saturation. RT1 measurements present a great variability especially in this pressure range; therefore the mean curve might not be representative of the mean behaviour. Actually, Durner et al. (1999) mention the possible difficulty of averaging dual porosity curves in some cases.

We can see that both fittings on R and μCT + R with the DP model allow a better adjustment for retention curves. Our results show that for R retention curves DP improves the fitting, but not significantly. Actually, without complementary information, it is difficult to choose a model. It is generally accepted that unimodal models – like VG – are adequate and, because of their easier implementation, are therefore often preferred (Durner et al., 1999). But in our case, the DP model proves to better predict the unsaturated conductivity. This is supported by the results of Durner et al. (1999) on silty soils. They show that the more parameters that are involved, the better the fitting on hydraulic functions, i.e. multimodal models. The fact that microtomography demonstrates this, as well as allowing a better prediction of the unsaturated hydraulic conductivity, are evidence for the validation of microtomographic information.

Further, μCT data allow refining retention and hydraulic curves near saturation where Richards' data alone can lead to numerous sets of fitted parameters. However, other methods allow this as well (for example, Hyprop apparatus, UMS®). But microtomographic images processed with an appropriate algorithm may be more powerful. Matching micro and macroscopic measurements allows us to validate μCT information, which is otherwise not so obvious (Baveye et al., 2010). The algorithm we use (Plougonven, 2009; Plougonven and Bernard, 2011) proves to be effective at separating pores individually since both retention and conductivity functions present an enhancement: pore size distribution obtained with this method and the fit of an elliptic cylinder on pores converge to obtain good results. An efficient pore individualization is interesting in soil analyses. Indeed, reproducing accurately the soil behaviour is still a challenge. Notably, characteristics, when calculated on the entire pore space, are influenced by little but numerous pores and are not able to reproduce global behaviour. On the other hand, precise individual characterization could be used to improve our knowledge of the pore space and its modelling. The next step is therefore to make use of potential structural information on individual pores.

3.3 X-ray microtomography: principal component analysis

PCA is performed on our samples, taking into account seven parameters (see Sect. 2.2.1): Vol, N_c, Sf_c, SS, R, Def and SC. The first three components explain about 90 % of the variability between samples; F1 (the first component) alone explains 54 % of the variability. All seven parameters are well correlated at least for one component. We represent in Fig. 7 (a 2-D representation of PCA with the two first components, F1–F2) and Fig. 8 (a 2-D representation of PCA with the first (F1) and third (F3) components) the results for pores with a radius $> 1500\,\mu m$ ($h > -1$ mbar), because of their bigger influence on hydrodynamic behaviour. CT and RT2 are in opposition with RT1 along F1. RT1 and RT2 differ because of a lower surface of connections and a bigger specific surface for RT2, while RT1 and CT are disconnected because of larger volume and radius, and a lower specific surface for

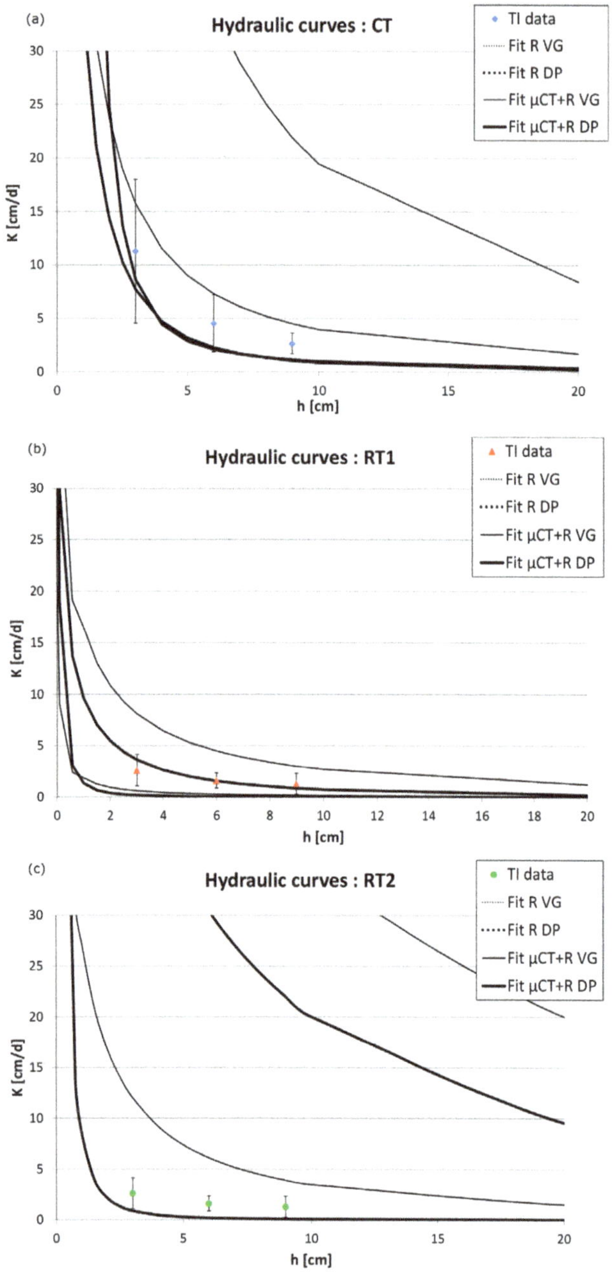

Figure 6. Hydraulic curves. Dotted line: Richards' measurements (R); solid line: combined μCT and Richards' measurements (μCT + R).

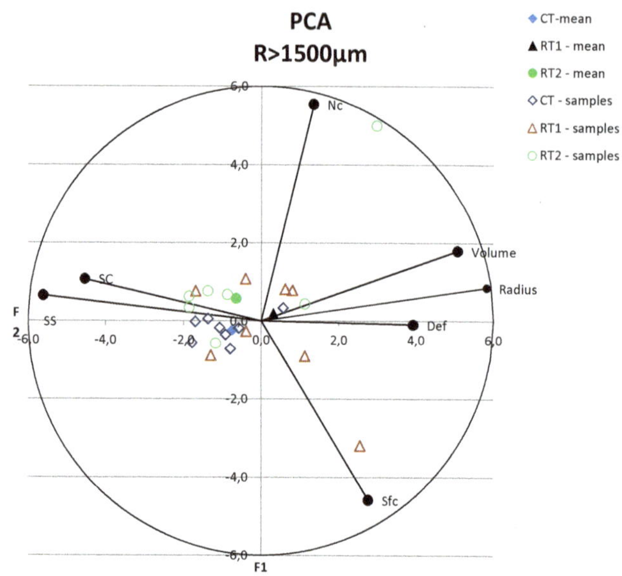

Figure 7. PCA scores for pores with $r > 1500\,\mu$m with the first F1, 54 %) and second (F2, 23 %) components.

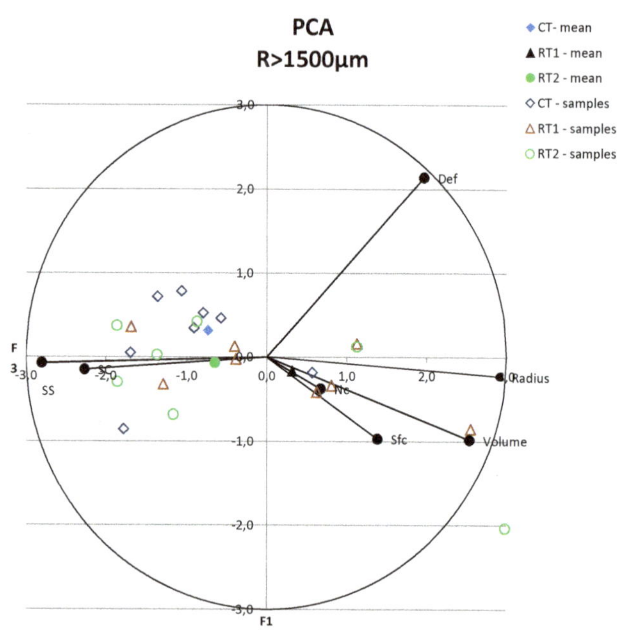

Figure 8. PCA scores for pores with $r > 1500\,\mu$m with the first (F1, 54 %) and third (F3, 11 %) components.

RT1. While this representation of the objects does not underline great differences between them (full dots, Fig. 7), we can still observe along F1 the ranking CT < RT2 < RT1, which is the reverse order as compared to macroporosity conductivity efficiency. Vogel and Roth (1998) demonstrate that realistic pore network representation has to take into account its connectivity and tortuosity. Our PCA is consistent with their results since specific surface and specific connectivity are highly correlated with F1. The importance of these structural parameters in our object differentiation strengthens the idea that macroporosity is not the only parameter to consider for conductivity characterization.

In addition to the conclusions about the mean structural parameters, the position of the samples in PCA representation can be seen as object intra-variability and provides supplementary clues. That is to say, the dispersion coefficient is very low for CT. It is important for RT2, but because of one sample alone, while RT1 is characterized by a dispersed spatial position of the samples (see Table 4). This behaviour seems to be symptomatic for this horizon. If we analyse macroscopic measurements from this point of view (not presented here), we can see that the horizons present the same characteristics as above about variability. CT has a low variation range and its mean and median are often superimposed. RT2 also has a low variation range, but its mean is displaced because of a few outliers. And finally RT1 presents a higher variation range and big differences between mean and median.

We could conclude that the poor dynamic performances – reflected by ratio between macroporosity and saturated hydraulic conductivity – for RT1 can be linked to this great heterogeneity (and therefore agrees with the divergence found in the literature, e.g. Strudley et al., 2008), while for RT2 it is more related to a generally low permeable medium. However, outliers seem to be a rule in RT2 populations, one of the main reasons probably being a greater pedofaunistic activity – which is in accord with results of Cousin et al. (2004). Indeed, pedofaunae produce large pores which can be observed with μCT and permeability measurements – but occur rarely – because of the size of the samples; while Richards' measurements are not able to detect them because of the difficulty to completely saturate the largest macropores.

Table 4. Dispersion coefficient of the samples in the PCA 3-D space for all the samples ("All") and for all the samples except the farthest from the mean ("All – 1").

Horizon	"All"	"All – 1"
CT	0.6	0.2
RT1	3.5	1.6
RT2	4.9	1.2

4 Conclusions

The central point of this paper is to show that X-ray microtomography brings supplementary and complementary information to macroscopic measurements in the context of agricultural practice evolution. It is a bridge between micro- and macroscopic approaches. More precisely, our research shows that: (i) we have an agreement between measurements at the macroscopic and the microscopic scales (ii) behaviours at these different scales can be linked together (iii) microscopic information can enhance macroscopic description and modelling.

In this work, we validate our methodology chain with a relatively quick scan method. Microtomography is still restricted by the numerous steps and choices of image processing. Nevertheless, the good match for both retention and conductivity functions with macroscopic measurements validates globally the μCT information, since the macroscopic measurements match the pore volume distribution through the adjustment of an elliptic cylinder. As a result, we show that a mean macroscopic behaviour can be linked with a mean behaviour at the microscopic scale. Indeed, while most research deals with pore network modelling, we show that X-ray microtomography can be used to enhance plot scale models. More specifically, the pore size distribution derived from microtomography allows us to enhance the near-saturation hydraulic function through parameterization of a double porosity model. Furthermore, we have put in evidence microtomographic parameters which could contribute to explain the mean hydrodynamic behaviour of these soils. Analysis of structural parameters for the biggest pores appears to be significant and offers an additional step in object differentiation. PCA shows that seven structural parameters can explain nearly 90 % of the variability between horizons, and that it is important to differentiate them. The most determining parameters in our context are specific surface, specific connectivity, volume and radius. The next step is the improvement of hydraulic functions via X-ray microtomography structural parameters, for example by including a connectivity and/or a specific surface-based parameter in the function implementation.

Note that with X-ray tomography, the acquired images can constitute a growing database, and that new algorithms can be applied and tested repeatedly. Indeed, images are not damaged by the calculations and more information can thus be obtained as computer resources increase or as algorithms are developed.

Acknowledgements. The authors would like to thank the referees of the paper for their helpful advice. The authors would also like to thank Sébastien Jacqmin for his great work on the illustration of the pore network.

Edited by: A. Shamseldin

References

Alkasawneh, W., Pan, E., and Green, R.: Effect of the Relative Root-Mean-Square error on Pavement Response, Anal. Asphalt Pave. Mater. Syst., 176, 117–132, doi:10.1061/40924(308)10, 2007.

Attix, F. H. R. and Roesch, W. C.: Radiation dosimetry, Academic Press, New York, 1968.

Baveye, P. C., Laba, M., Otten, W., Bouckaert, L., Dello Sterpaio, P., Goswami, R. R., Grinev, D., Houston, A., Hu, Y., Liu, J., Mooney, S., Pajor, R., Sleutel, S., Tarquis, A., Wang, W., Wei, Q., and Sezgin, M.: Observer-dependent variability of the thresholding step in the quantitative analysis of soil images and X-ray microtomography data, Geoderma, 157, 51–63, 2010.

Bayer, A., Vogel, H.-J., and Roth, K.: Direct measurement of the soil water retention curve using X-ray absorption, Hydrol. Earth Syst. Sci., 8, 2–7, doi:10.5194/hess-8-2-2004, 2004.

Beckers, E., Plougonven, E., Roisin, C., Hapca, S., Léonard, A., and Degré, A.: X-ray microtomography: A porosity-based thresholding method to improve soil pore network characterization?, Geoderma, 219–220, 145–154, 2014.

Bhattacharyya, R., Prakash, V., Kundu, S., and Gupta, H. S.: Effect of tillage and crop rotations on pore size distribution and soil hydraulic conductivity in sandy clay loam soil of the Indian Himalayas, Soil Till. Res., 86, 129–140, 2006.

Cousin, I., Vogel, H.-J., and Nicoullaud, B.: Influence de la structure du sol à différentes échelles sur les transferts d'eau : Conséquences d'une réduction du travail du sol, Étude Gestion Sols, 11, 69–81, 2004.

Dal Ferro, N., Delmas, P., Duwig, C., Simonetti, G., and Morari, F.: Coupling X-ray microtomography and mercury intrusion porosimetry to quantify aggregate structures of a cambisol under different fertilisation treatments, Soil Till. Res., 119, 13–21, 2012.

DIN ISO 11274: Soil quality – Determination of the water retention characteristics – Laboratory methods (ISO 11274:1998 + Cor. 1:2009) English translation of DIN ISO 11274:2012-04, Deutsches Institut für Normung, Berlin, Germany, 2012.

Durner, W.: Hydraulic conductivity estimation for soils with heterogeneous pore structure, Water Resour. Res., 30, 211–233, 1994.

Durner, W., Priesack, E., Vogel, H. J., and Zurmühl, T.: Determination of parameters for flexible hydraulic functions by inverse modeling, in: Workshop, Characterization and Measurement of the Hydraulic Properties of Unsaturated Porous Media Proc. Int. , edited by: van Genuchten, M. Th., Leij, F. J., and Wu., L., University of California, Riverside, CA, 817–830, 1999.

Eden, M., Moldrup, P., Schjonning, P., Vogel, H. J., Scow, K. M., and de Jonge, L. W.: Linking soil physical parameters along a density gradient in a loess soil long-term experiment, Soil Sci., 177, 1–11, 2012.

Gerke, H. H.: Macroscopic representation on of the interface between flow domains in structured Soil, Vadose Zone J., 11, doi:10.2136/vzj2011.0125, 2012.

Green, T. R., Ahujaa, L. R., and Benjamin, J. G.: Advances and challenges in predicting agricultural management effects on soil hydraulic properties, Geoderma, 116, 3–27, 2003.

Hapca, S. M., Houston, A. N., Otten, W., and Baveye, P. C.: New objective segmentation method based on minimizing locally the intra-class variance of grayscale images, Vadose Zone J., 12, doi:10.2136/vzj2012.0172, 2013.

Holthusen, D., Jänicke, M., Peth, S., and Horn, R.: Physical properties of a Luvisol for different long-term fertilization treatments: I. Mesoscale capacity and intensity parameters, J. Plant Nutr. Soil Sc., 175, 4–13, 2012.

Houston, A. N., Otten, W., Baveye, P. C., and Hapca S.: Adaptive-window indicator kriging: A thresholding method for computed tomography images of porous media, Comput. Geosci., 54, 239–248, 2013.

Iassonov, P., Gebrenegus, T., and Tuller, M.: Segmentation of X-ray computed tomography images of porous materials: a crucial step for characterization and quantitative analysis of pore structures, Water Resour. Res., 45, W09415, doi:10.1029/2009WR008087, 2009.

Imhoff, S., Ghiberto, P. J., Grioni, A., and Gay, J. P.: Porosity characterization of Argiudolls under different management systems in the Argentine Flat Pampa, Geoderma, 158, 268–274, 2010.

Jackson, J. E.: Principal components and factor analysis: Part I-Principal components, J. Qual. Technol., 12, 201–213, 1980.

Jackson, J. E.: A user's guide to principal components, Wiley, New York, 1991.

Kim, H., Anderson, S. H., Motavalli, P. P., and Gantzer, C. J.: Compaction effects on soil macropore geometry and related parameters for an arable field, Geoderma, 160, 244–251, 2010.

Kumar, S., Anderson, S. H., and Udawatta, R. P.: Agroforestry and grass buffer influences on macropores measured by computed tomography under grazed pasture systems, Soil Sci. Soc. Am. J., 74, 203–212, 2010.

Levenberg, K.: A Method for the Solution of Certain Problems in Least Squares, Quart. Appl. Math., 2, 164–168, 1944.

Marquardt D.: An Algorithm for Least-Squares Estimation of Nonlinear Parameters, SIAM J. Appl. Math., 11, 431–441, 1963.

Mualem, Y.: A new model for predicting the hydraulic conductivity of unsaturated porous media, Water Resour. Res., 12, 513–522, 1976.

Olsen, P. A. and Borresen, T.: Measuring differences in soil properties in soils with different cultivation practices using computer tomography, Soil Till. Res., 44, 1–12, 1997.

Palm, R.: Les méthodes d'analyse factorielle: principes et applications, Biom. Praxim., 34, 35–80, 1994.

Plougonven, E.: Link between the microstructure of porous materials and their permeability, Ph.D. Thesis, Université Bordeaux 1, Bordeaux, 2009.

Plougonven, E. and Bernard, D.: Optimal removal of topological artefacts in microtomographic images of porous materials, Adv. Water Resour., 34, 731–736, 2011.

Quinton, W. L., Elliot, T., Price, J. S., Rezanezhad, F., and Heck, R.: Measuring physical and hydraulic properties of peat from X-ray tomography, Geoderma, 153, 269–277, 2009.

Rachman, A., Anderson, S. H., and Gantzer, C. J.: Computed tomographic measurement of soil macroporosity parameters as affected by stiff-stemmed grass hedges, Soil Sci. Soc. Am. J., 69, 1609–1616, 2005.

Richards, L. A.: Porous plate apparatus for measuring moisture retention and transmission by soils, Soil Sci., 66, 105–110, 1948.

Roisin, C.: A multifractal approach for assessing the structural state of tilled soils, Soil Sci. Soc. Am. J., 71, 15–25, 2007.

Schaap, M. G. and van Genuchten, M. Th.: A Modified Mualem-van Genuchten Formulation for Improved Description of the Hydraulic Conductivity Near Saturation, Vadose Zone J., 5, 27–34, doi:10.2136/vzj2005.0005, 2006.

Schlüter, S., Weller, U., and Vogel, H.-J.: Segmentation of X-ray microtomography images of soil using gradient masks, Comput. Geosci., 36, 1246–1251, 2010.

Schlüter, S., Weller, U., and Vogel, H.-J.: Soil-structure development including seasonal dynamics in a long-term fertilization experiment, J. Plant Nutr. Soil Sc., 174, 395–403, 2011.

Solone, R., Bittelli, M., Tomei, F., and Morari, F.: Errors in water retention curves determined with pressure plates: Effects on the soil water balance, J. Hydrol., 470–471, 65–74, 2012.

Strudley, M. W., Green, T. R., and Ascough, J. C.: Tillage effects on soil hydraulic properties in space and time: State of the science, Soil Till. Res., 99, 4–48, 2008.

Taina, I. A., Heck, R. J., and Elliot, T. R.: Application of X-ray computed tomography to soil science: A literature review, Can. J. Soil Sci., 88, 1–20, 2008.

van Genuchten, M. Th.: A closed-form equation for predicting the hydraulic conductivity of unsaturated soils, Soil Sci. Soc. Am. J., 44, 892–898, 1980.

Vereecken, H., Weynants, M., Javaux, M., Pachepsky, Y., Schaap, M. G., and van Genuchten, M. Th.: Using Pedotransfer Functions to Estimate the van Genuchten-Mualem Soil Hydraulic Properties: A Review, Vadose Zone J., 9, 795–820, doi:10.2136/vzj2010.0045, 2010.

Vogel, H. J. and Roth, K.: A new approach for determining effective soil hydraulic functions, Eur. J. Soil Sci., 49, 547–556, 1998.

Wang, W., Kravchenko, A. N., Smucker, A. J. M., and Rivers, M. L.: Comparison of image segmentation methods in simulated 2D and 3D microtomographic images of soil aggregates, Geoderma, 162, 231–241, 2011.

Weynants, M., Vereecken, H., and Javaux, M.: Revisiting Vereecken pedotransfer functions: Introducing a Closed-Form Hydraulic Model, Vadose Zone J., 8, 86–95, doi:10.2136/vzj2008.0062, 2009.

Wiermann, C., Werner, D., Horn, R., Rostek, J., and Werner, B.: Stress/strain processes in a structured unsaturated silty loam Luvisol under different tillage treatments in Germany, Soil Till. Res., 53, 117–128, 2000.

2

Ensemble modeling of stochastic unsteady open-channel flow in terms of its time–space evolutionary probability distribution – Part 1: theoretical development

Alain Dib and M. Levent Kavvas

Department of Civil and Environmental Engineering, University of California, Davis, 95616, USA

Correspondence: Alain Dib (aedib@ucdavis.edu)

Abstract. The Saint-Venant equations are commonly used as the governing equations to solve for modeling the spatially varied unsteady flow in open channels. The presence of uncertainties in the channel or flow parameters renders these equations stochastic, thus requiring their solution in a stochastic framework in order to quantify the ensemble behavior and the variability of the process. While the Monte Carlo approach can be used for such a solution, its computational expense and its large number of simulations act to its disadvantage. This study proposes, explains, and derives a new methodology for solving the stochastic Saint-Venant equations in only one shot, without the need for a large number of simulations. The proposed methodology is derived by developing the nonlocal Lagrangian–Eulerian Fokker–Planck equation of the characteristic form of the stochastic Saint-Venant equations for an open-channel flow process, with an uncertain roughness coefficient. A numerical method for its solution is subsequently devised. The application and validation of this methodology are provided in a companion paper, in which the statistical results computed by the proposed methodology are compared against the results obtained by the Monte Carlo approach.

1 Introduction

Unsteady open-channel flows are a common occurrence in hydrology and hydraulics problems. They arise as a result of the movement of water waves in natural or artificial channels (Sturm, 2001). Understanding and tracing the movement of such water waves along the channels is of great importance in addressing engineering flow problems, including flood forecasting, flood control, hydrograph generation, and several others (Chow, 1959). The technique used to approximate and trace such water waves is known as flood routing, and the governing equations that are commonly used to solve for the unsteady flows in flood routing problems are known as the Saint-Venant equations (Chanson, 2004).

Various uncertainties may add to the complexity of solving the Saint-Venant equations (Gates and AlZahrani, 1996a; Ercan and Kavvas, 2012a) and these may correspond to several factors. Physical conditions of open channels may be uncertain due to their high degree of variability (Sturm, 2001). One example is Manning's roughness coefficient, which greatly depends on the channel vegetation, bed material, bedforms, and even on the position of the free water surface (Chow, 1959; Sturm, 2001; Ercan and Kavvas, 2012a). With the uncertainties in quantifying or characterizing these factors, the roughness coefficient becomes extremely difficult to estimate (Sturm, 2001), rendering it uncertain. Channel geometric parameters may also be uncertain. This includes the channel bed slope (Ercan and Kavvas, 2012a) and the channel cross section geometry, the latter of which may exhibit significant spatial variability across a river due to its irregular form and due to the changes it may undergo along the direction of flow (Chow, 1959). Other uncertainties may also arise from lateral inflows and initial conditions due to their spatial and/or temporal variability (Liang and Kavvas, 2008), as well as from the upstream boundary conditions due to the temporal variability of the inflows into the channel.

As a result of such uncertainties, the channel and flow parameters may be considered spatially and/or temporally random at the local scale of a river cross section, rendering the system behavior uncertain (Gates and AlZahrani, 1996a). Therefore, deterministically solving the Saint-Venant equations in this case would no longer be providing a representative solution to the flood routing problem being considered. In fact, in this study the governing partial differential equations (PDEs), i.e., the Saint-Venant equations, will be transformed into stochastic PDEs because the channel and flow parameters are now stochastic and can be described as random functions (Liang and Kavvas, 2008). This means that the dependent variables that will be solved for by these equations (e.g., flow velocity and depth) will also be spatiotemporal random functions. Hence, instead of solving for the deterministic values of the dependent variables, the goal will be to solve for their statistical properties (Van Kampen, 1976), which can be obtained at designated discrete time–space positions.

Two popular methods can be used for such a stochastic solution to nonlinear problems: the finite-order analysis and the Monte Carlo (MC) approach. Applying the expectation operator, or any other statistical moment operator, to nonlinear difference equations with stochastic parameters may result in nonlinear expressions that are difficult, or even impossible, to simplify into terms involving the known moments of the model parameters and the unknown moments of the dependent variables (Gates and AlZahrani, 1996a). Finite-order analysis overcomes this by performing a Taylor series expansion of the difference equations about the expected values of the parameters, from which higher-order terms are truncated. For example, truncating the Taylor series after the first-order term is known as the first-order approximation, which is a good approximation when the system nonlinearity is not too high, and when the stochastic parameters have relatively small coefficients of variation (Dettinger and Wilson, 1981). However, with highly nonlinear problems, instead of using higher-order approximations of the finite-order method, it may be required and more efficient to use full-distribution methods such as the MC approach (Dettinger and Wilson, 1981).

The MC approach is well-known for simulating differential equations with stochastic parameters, and is used to determine the distributions of the unknown stochastic dependent variables (Freeze, 1975; Smith and Freeze, 1979; Bellin et al., 1992). This method involves repeatedly solving the governing equations in a deterministic fashion, varying the stochastic parameters for each run, in order to obtain a set of several realizations for each of the dependent variables. When a sufficient number of realizations is obtained, they can be used to determine the required statistical properties, including the mean system behavior and the standard deviation (Gates and AlZahrani, 1996b). Therefore, the MC simulations require two models: one which generates realizations for the stochastic parameters, and another (finite-difference model) which deterministically solves the governing flow equations for each realization (Gates and AlZahrani, 1996b). The MC approach is generally accepted as the most robust approach for uncertainty evaluation, as well as the benchmark for comparing other new methods (Scharffenberg and Kavvas, 2011). The full distribution characteristics may be estimated using the MC approach, which is more intuitive than the finite-order methods (Gates and AlZahrani, 1996a). However, the main drawback of the MC approach is its computational expense due to the usual running of a large number of simulations of the process under study in order to obtain accurate results (Dettinger and Wilson, 1981).

To bypass the need for solving the unsteady open-channel flow governing equations several times, a new methodology is proposed in this study in order to solve for the expected system behavior and variability in only one simulation. This methodology involves upscaling the governing stochastic differential equations from the point scale (at which they are originally valid) to the field scale. Ensemble averaging has been a common approach to upscale hydrologic equations that are linear (Gelhar and Axness, 1983; Kitanidis, 1988; Rubin and Dagan, 1989; Kapoor and Gelhar, 1994; Kavvas and Karakas, 1996; Wood and Kavvas, 1999a, b) or nonlinear (Mantoglou and Gelhar, 1987; Tayfur and Kavvas, 1994; Horne and Kavvas, 1997; Dogrul et al., 1998), in which case these equations are averaged to become deterministic differential equations. These developed deterministic differential equations use statistical descriptions, such as the mean and variance, to represent the values of the stochastic parameters (Liang and Kavvas, 2008). However, most of the studies performing the ensemble averaging technique on nonlinear conservation equations used the regular perturbation method, which includes linearization assumptions and which only works for small fluctuations in the dependent variables (Kavvas, 2003). Other techniques that have been applied, which are not limited by small fluctuations, include the decomposition method (Serrano, 1995), a combination of volume averaging with nonlinear dynamics (Duffy, 1996; Duffy and Cusumano, 1998), as well as the theory of fractals and multifractals (Puente, 1996). Nonetheless, due to some limitations of such methods when used for stochastic nonlinear hydrologic processes, the upscaling method used in this study is chosen to be that of Kavvas (2003).

Kavvas (2003) developed general ensemble average conservation equations (to second order) for nonlinear and linear hydrologic processes in order to determine their probabilistic and mean behavior. The "master key" equations developed may be used on any stochastic hydrologic process after being rewritten as one or more linear/nonlinear stochastic ordinary differential equations (ODEs). This utilization leads to a special Lagrangian–Eulerian form of the Fokker–Planck equation (LEFPE) that models the time–space evolution of the probability density of the dependent variables of any nonlinear/linear stochastic dynamic process (Kavvas, 2003). Such a methodology has been successfully applied to many hydrologic processes, including unsaturated water flow (Kim et al., 2005b), root-water uptake (Kim et al., 2005a), solute transport (Liang and Kavvas, 2008), snow accumulation and melt (Ohara et al., 2008), unconfined groundwater flow (Cayar and Kavvas, 2009a, b), and kinematic open-channel flow (Ercan and Kavvas, 2012a, b).

Noting that the characteristic forms of the Saint-Venant equations are nonlinear ODEs, it is proposed in this study to apply to them their corresponding master key equation from Kavvas (2003). From this operation the corresponding LEFPE of the Saint-Venant open-channel flow equations is obtained, thus providing the ability to model the uncertainties of the channel and flow parameters and to compute their effect on the behavior of the system. Therefore, under the appropriate initial and boundary conditions, the probability density functions (PDFs) of the dependent variables can be computed (to exact second order) through the LEFPE, and the ensemble behavior of the system can be described.

The advantages of using the LEFPE in tackling the flood routing problem greatly echo those of the classical Fokker–Planck equation (FPE). In fact, the LEFPE directly solves for the PDFs of the dependent variables of the system in both time and space, it is linear in the variable being solved for (i.e., in the PDF), and unlike the many simulations usually performed for the MC approach, the LEFPE produces the complete ensemble model results with only one single simulation. As such, the LEFPE provides not only the mean and variance of the process but also a complete description of the evolution of the dependent variables' PDFs in a computationally efficient manner. Note that the LEFPE does not make any linearization assumptions, it works with a wide-ranging parameter space, and the only assumption about the physical process it makes is the finite correlation time for the process (Ercan and Kavvas, 2012a).

Therefore, following from the above discussion, the main objective of this study is to apply the upscaling method based on the LEFPE approach in Kavvas (2003) to the characteristic form of the stochastic Saint-Venant equations in order to derive a new methodology that solves for the probability density of the dependent flow variables, and that quantifies the expected behavior and variability of the system in one shot, instead of running a large number of simulations.

2 Saint-Venant equations for unsteady open-channel flow

The Saint-Venant equations, also known as the spatially varied unsteady flow equations (Sturm, 2001), are the two governing equations used to describe an unsteady open-channel flow problem that will be solved using the hydraulic routing technique (Chow, 1959; Viessman et al., 1977; Sturm, 2001). They consist of the continuity equation and the momentum equation which are used simultaneously in order to solve for the two unknowns (velocity and depth, or discharge and depth). The naming of these equations comes from the French mathematician Adhémar-Jean-Claude Barré de Saint-Venant who published the equations describing one-dimensional unsteady open-channel flow in 1871 (Barré de Saint-Venant, 1871).

Several assumptions are made when deriving these equations (Viessman et al., 1977; Sturm, 2001), including unidirectional flow and uniform cross-sectional velocity, hydrostatic pressure, small channel bed slope, steady state estimation of friction loss, and incompressible flow. Following these assumptions, the Saint-Venant equations for unsteady open-channel flow of an incompressible fluid in a rectangular, prismatic channel (with no lateral inflow/outflow) may be written as follows (Viessman et al., 1977):

$$\text{Continuity}\quad y\frac{\partial V}{\partial x}+V\frac{\partial y}{\partial x}+\frac{\partial y}{\partial t}=0, \tag{1}$$

$$\text{Momentum}\quad \frac{\partial V}{\partial t}+V\frac{\partial V}{\partial x}+g\frac{\partial y}{\partial x}=g\left(S_0-S_f\right), \tag{2}$$

where V is the average flow velocity, y is the flow depth, x is the position, t is the time, S_0 is the slope of the channel bottom, S_f is the friction slope, and g is the acceleration of gravity. When these assumptions are no longer valid, the derived Saint-Venant equations will show some limitations. Such limitations occur, for example, if the flow is not one-dimensional (e.g., in flood plains or large rivers), if the pressure is non-hydrostatic (e.g., presence of sharp geometric variations/bends or hydraulic jumps), if there are sharp discontinuities (e.g., those caused by weirs or gates), or if there are channel irregularities (Litrico and Fromion, 2009). Therefore, keeping those limitations in mind is crucial for the appropriate implementation of the derived Saint-Venant equations.

2.1 Solution methods for the Saint-Venant equations

Since closed-form solutions to the Saint-Venant equations have not been obtained due to the presence of nonlinear terms, it has not been possible to solve these equations analytically except when extreme simplifications are applied (Sturm, 2001; Chaudhry, 2008). As a result, several numerical techniques have been developed in order to solve the Saint-Venant equations deterministically in their full form, without major simplifications. The most frequently used of these techniques are finite-difference methods (Abbott and Ionescu, 1967; Fread, 1973; Beam and Warming, 1976; Fennema and Chaudhry, 1986; Garcia and Kahawita, 1986; Venutelli, 2002), which solve the governing equations explicitly or implicitly along a fixed or adaptive x–t grid (Szymkiewicz, 2010). Finite-element methods are also available for solving such unsteady flow equations (Cooley and Moin, 1976; Szymkiewicz, 1991, 1995; Hicks and Steffler, 1995), though they are usually considered to be more effective for two- and three-dimensional flow problems (Szymkiewicz, 1991). While the standard finite-element method may not be the most suitable or satisfactory method for solving unsteady flow problems (Szymkiewicz, 2010), the modified finite-element method (Szymkiewicz, 1995) seems to be as effective and robust in solving such problems as the high-order finite-difference methods.

Furthermore, another approach to solve the Saint-Venant equations may be utilized after realizing that these equations are hyperbolic PDEs (Chaudhry, 2008). This approach is known as the method of characteristics (MOC; Abbott, 1966), which is one of the earliest and most exact methods for solving hyperbolic PDEs (Tannehill et al., 1997), and which was used early on for solving the Saint-Venant equations (Amein, 1966; Woolhiser and Liggett, 1967; Lai, 1988). The MOC may be used to transform a hyperbolic PDE into a system of ODEs, which may be simpler to solve (Sturm, 2001). These ODEs are usually divided into two equations: the characteristic equation (i.e., the ODE describing the characteristic path), and the compatibility equation (i.e., the ODE that describes the process behavior along that characteristic path) (Hoffman, 2001). After this transformation by the MOC, the finite-difference approximations of the derivatives can then be applied to the characteristic form of the hyperbolic PDE, instead of applying them to its original form. Note that in time and one-space dimensions, the characteristic equations represent curves in the x–t plane along which information propagates through the solution domain (Hoffman, 2001), and along which discontinuities in the derivatives of the dependent variables propagate (Sturm, 2001).

From the several techniques available to solve for the Saint-Venant equations, the MOC is chosen for this study. This is because, as was mentioned in Sect. 1, the upscaling technique based on the LEFPE approach in Kavvas (2003) can be applied to hydrologic processes that are written as one or more ODEs. As such, it is imperative for the progression of this study to transform the Saint-Venant equations into their characteristic form in order to write them as a system of ODEs. With two characteristic directions, the Saint-Venant equations are transformed by the MOC into a system of four ODEs: two characteristic equations and two corresponding compatibility equations. When finite-difference approximations are applied to the characteristic form of the Saint-Venant equations, the results can be numerically computed along an irregular x–t grid formed by the intersection points of the characteristic curves (Gates and AlZahrani, 1996a).

2.2 Characteristic form of the Saint-Venant equations

Through a linear combination of the continuity and momentum equations (Eqs. 1 and 2), the characteristic equations for unsteady open-channel flow of an incompressible fluid in a rectangular, prismatic channel with no lateral inflow can be written as follows (Sturm, 2001):

Positive characteristic curve (C_1)

$$\frac{\mathrm{d}x_1}{\mathrm{d}t} = V + c. \tag{3}$$

Flow process condition to be satisfied along C_1

$$\left(\frac{\mathrm{d}(V+2c)}{\mathrm{d}t}\right)_1 = g\left(S_{0,1} - S_{\mathrm{f},1}\right). \tag{4}$$

Negative characteristic curve (C_2)

$$\frac{\mathrm{d}x_2}{\mathrm{d}t} = V - c. \tag{5}$$

Flow process condition to be satisfied along C_2

$$\left(\frac{\mathrm{d}(V-2c)}{\mathrm{d}t}\right)_2 = g\left(S_{0,2} - S_{\mathrm{f},2}\right), \tag{6}$$

where c is the wave celerity which is equal to $\sqrt{gy}$ for a rectangular channel, and $S_{0,2}$ is equal to $S_0(x_2, t)$ (similarly for the other S variables). The remaining variables are defined as in Eqs. (1) and (2).

Equations (3) and (5) represent two different velocity expressions defining the two characteristic directions of the Saint-Venant equations: the former defining the positive characteristic curve (C_1) with speed $V + c$, and the latter defining the negative characteristic curve (C_2) with speed $V - c$. Equations (4) and (6) represent the compatibility equations for Eqs. (3) and (5), respectively. Each compatibility equation for the flow process behavior should be satisfied along its corresponding characteristic curve. The subscripts in Eqs. (3) to (6) are used to differentiate between the two total derivative operators which correspond to the two different speeds along C_1 and C_2. As such, Eqs. (3) to (6) are seen to describe the change of two functions along two different paths: a function $V + 2c$ that varies along C_1, and another function $V - 2c$ that varies along C_2. These functions are known as the Riemann invariants (Chaudhry, 2008).

Therefore, the MOC transforms the two governing PDEs into a system of four ODEs that are differentiated with respect to time only, and no longer with respect to space. This transformation provides the ability to use the upscaling technique based on the LEFPE approach in Kavvas (2003) on the Saint-Venant equations in order to derive this study's proposed methodology involving the ensemble-averaged equations of stochastic unsteady open-channel flow.

3 Ensemble-averaged equations for the stochastic unsteady open-channel flow

In this section, a new methodology for solving the stochastic Saint-Venant equations will be introduced and derived, and a numerical discretization scheme will be devised for it as well. The proposed methodology aims at obtaining the statistical properties of the dependent variables of the unsteady open-channel flow system in only one simulation, as opposed to the large number of simulations usually involved in the MC approach. The following derivation involves assuming the Manning's roughness coefficient as an uncertain parameter, but similar steps can be followed even when the uncertainty is assumed to arise from other parameters.

3.1 Development of the Fokker–Planck solution methodology for the Saint-Venant equations

Following Kavvas (2003), a system of point-scale conservation equations can be written for a dynamical system as follows:

$$\frac{\partial \boldsymbol{H}(\boldsymbol{x},t)}{\partial t} = \boldsymbol{\eta}(\boldsymbol{H},\mathbf{A},\boldsymbol{f};\boldsymbol{x},t), \tag{7}$$

where $\boldsymbol{H}(\boldsymbol{x}, t)$ is the vector of all state variables of the hydrologic system of equations; $\mathbf{A}(\boldsymbol{x}, t)$ is the tensor of parameters; $\boldsymbol{f}(\boldsymbol{x}, t)$ is the vector of forcing functions; $\boldsymbol{\eta}$ is a function of $\boldsymbol{H}$, $\mathbf{A}$, and $\boldsymbol{f}$; $\boldsymbol{x}$ is the vector of spatial locations, and t is the time. The initial condition for the above system is given as

$$\boldsymbol{H}(\boldsymbol{x},0) = \boldsymbol{H}_0. \tag{8}$$

Under a second-order cumulant expansion, the general Lagrangian–Eulerian form of the Fokker–Planck equation (LEFPE) for the previously defined dynamical system was developed in Kavvas (2003, Eq. 25) to exact second order (i.e., to the order of the covariance time of η) as shown below:

$$\begin{aligned}\frac{\partial P(\boldsymbol{H}(\boldsymbol{x}_t,t);t)}{\partial t} &= -\frac{\partial}{\partial H_j}\Big\{P(\boldsymbol{H}(\boldsymbol{x}_t,t);t)\\ &\Big[\langle \eta_j(\boldsymbol{H}(\boldsymbol{x}_t,t),\mathbf{A}(\boldsymbol{x}_t,t),\boldsymbol{f}(\boldsymbol{x}_t,t))\rangle\\ &+\int_0^t \mathrm{d}s\,\mathrm{Cov}_\mathrm{o}\Big[\frac{\partial \eta_j(\boldsymbol{H}(\boldsymbol{x}_t,t),\mathbf{A}(\boldsymbol{x}_t,t),\boldsymbol{f}(\boldsymbol{x}_t,t))}{\partial H_i};\\ &\eta_i(\boldsymbol{H}(\boldsymbol{x}_{t-s},t-s),\mathbf{A}(\boldsymbol{x}_{t-s},t-s),\boldsymbol{f}(\boldsymbol{x}_{t-s},t-s))\Big]\Big]\Big\}\\ &+\frac{1}{2}\frac{\partial^2}{\partial H_j \partial H_i}\Big\{2P(\boldsymbol{H}(\boldsymbol{x}_t,t);t)\int_0^t \mathrm{d}s\,\mathrm{Cov}_\mathrm{o}\big[\eta_j(\boldsymbol{H}(\boldsymbol{x}_t,t),\\ &\mathbf{A}(\boldsymbol{x}_t,t),\boldsymbol{f}(\boldsymbol{x}_t,t));\eta_i(\boldsymbol{H}(\boldsymbol{x}_{t-s},t-s),\mathbf{A}(\boldsymbol{x}_{t-s},t-s),\\ &\boldsymbol{f}(\boldsymbol{x}_{t-s},t-s))\big]\Big\}\end{aligned} \tag{9}$$

where $P(\boldsymbol{H}(\boldsymbol{x}_t, t); t)$ is the probability density function of the vector of state variables ($\boldsymbol{H}$) at location $\boldsymbol{x}_t$ and at time t, the operator $\langle \bullet \rangle$ is the ensemble average operator, s is a time displacement, and $\mathrm{Cov_o}[\bullet]$ is the time-ordered covariance function as shown in the below equation (Van Kampen, 1974):

$$\begin{aligned}\mathrm{Cov}_\mathrm{o}[\eta_j(\boldsymbol{x},t_1);\eta_i(\boldsymbol{x},t_2)] &= \langle \eta_j(\boldsymbol{x},t_1)\eta_i(\boldsymbol{x},t_2)\rangle\\ &-\langle \eta_j(\boldsymbol{x},t_1)\rangle\langle \eta_i(\boldsymbol{x},t_2)\rangle.\end{aligned} \tag{10}$$

Note that in Eq. (9), the real space location $\boldsymbol{x}_t$ is known, whereas the Lagrangian location $\boldsymbol{x}_{t-s}$ is unknown. This Lagrangian location can be determined from the known location $\boldsymbol{x}_t$ by using a Lie operator as defined in Kavvas and Karakas (1996):

$$\boldsymbol{x}_{t-s} = \overleftarrow{\exp}\left[-\int_{t-s}^{t} \mathrm{d}\tau\, \langle v_l(\boldsymbol{x}_\tau,\tau)\rangle \frac{\partial}{\partial x_l}\right]\boldsymbol{x}_t, \tag{11}$$

where $\overleftarrow{\exp}$ is the time-ordered exponential, and v_l is determined from the characteristic curve equation corresponding to a particular hydrologic/hydraulic conservation equation. In the three-dimensional flow case, l takes on the values 1–3. A first-order approximation of Eq. (11) can be written as follows (Kavvas and Karakas, 1996):

$$\boldsymbol{x}_{t-s} = \boldsymbol{x}_t - \int_{t-s}^{t} \mathrm{d}\tau\,\langle \boldsymbol{v}(\boldsymbol{x}_\tau,\tau)\rangle, \tag{12}$$

where $\boldsymbol{v} = [v_1, v_2, v_3]$ in the general three-dimensional flow case.

Solving the LEFPE, Eq. (9), under the appropriate initial and boundary conditions provides the spatiotemporal evolution of the PDF of the vector of state variables ($\boldsymbol{H}$) for any hydrologic system expressed in terms of Eqs. (7) and (8), thus providing the ensemble behavior and variability of the process with only one simulation. In addition, it is important to note that the LEFPE, a parabolic PDE, is a deterministic equation that is linear in its unknown variable $P(\boldsymbol{H}(\boldsymbol{x}_t, t), t)$, unlike the original hydrologic system which would usually be stochastic and nonlinear. As such, the LEFPE provides great advantages in simplifying the stochastic solution of the hydrologic system being considered.

Since the LEFPE was developed for a system of ODEs (Eq. 7), and since the characteristic form of the Saint-Venant equations is a system of four nonlinear ODEs (Eqs. 3 to 6), it is proposed to apply to these equations the corresponding LEFPE after making some substitutions and adjustments. First, the friction slope (S_f) is computed using Manning's formula (Sturm, 2001). Then, the Riemann invariants are defined as follows:

$$V + 2c = \alpha, \tag{13}$$

$$V - 2c = \beta. \tag{14}$$

As such, Eqs. (3) to (6) can be written as a system of four ODEs in terms of four state variables (x_1, x_2, α, β):

$$\frac{\mathrm{d}x_1}{\mathrm{d}t} = \frac{3}{4}\alpha(x_1,t) + \frac{1}{4}\beta(x_1,t) \equiv \eta_{1,t}(x_1,t) = \eta_{1,t} \tag{15}$$

$$\begin{aligned}\left(\frac{\mathrm{d}\alpha}{\mathrm{d}t}\right)_1 &= gS_0(x_1,t) - \frac{gn^2(x_1,t)}{k^2}\cdot\frac{\frac{1}{4}[\alpha(x_1,t)+\beta(x_1,t)]^2}{R^{4/3}(\alpha(x_1,t),\beta(x_1,t),b;x_1,t)}\\ &\equiv \eta_{\alpha,t}(x_1,t) = \eta_{\alpha,t}\end{aligned} \tag{16}$$

$$\frac{\mathrm{d}x_2}{\mathrm{d}t} = \frac{1}{4}\alpha(x_2,t) + \frac{3}{4}\beta(x_2,t) \equiv \eta_{2,t}(x_2,t) = \eta_{2,t} \tag{17}$$

$$\begin{aligned}\left(\frac{\mathrm{d}\beta}{\mathrm{d}t}\right)_2 &= gS_0(x_2,t) - \frac{gn^2(x_2,t)}{k^2}\cdot\frac{\frac{1}{4}[\alpha(x_2,t)+\beta(x_2,t)]^2}{R^{4/3}(\alpha(x_2,t),\beta(x_2,t),b;x_2,t)}\\ &\equiv \eta_{\beta,t}(x_2,t) = \eta_{\beta,t}\end{aligned} \tag{18}$$

where R denotes the hydraulic radius, n denotes Manning's roughness coefficient, and k denotes the conversion factor between SI and US units for Manning's formula. Note that the width of the channel (b) is not a function of x or t since the equations are derived for a prismatic channel. Moreover, g and k are also not functions of x or t since they are constants. Equations (15) to (18) are now in the form of Eq. (7), with an η function defined for each of the four ODEs, as shown on their right-hand side. As such, it is clear that for a vector of state variables $\boldsymbol{H} = [x_1, x_2, \alpha, \beta]$, there is a vector of functions $\boldsymbol{\eta} = [\eta_{1,t}, \eta_{2,t}, \eta_{\alpha,t}, \eta_{\beta,t}]$. The subscripts for the η functions shown in Eqs. (15) to (18) represent the following: subscript 1 represents the positive characteristic direction (C_1), subscript 2 represents the negative characteristic direction (C_2), subscript α represents the compatibility equation along C_1, and subscript β represents the compatibility equation along C_2.

Note that in equations to follow, the η functions will be interchangeably represented by one of their two forms shown on the right-hand side of Eqs. (15) to (18), whereas $P(x_1, x_2, \alpha, \beta, t)$ may be substituted by P for simplicity. Such typographical simplifications will be used in order to reduce the used space and to increase the readability and simplicity of large equations. Therefore, considering some of the above typographical simplifications, and applying the general LEFPE in Eq. (9) to Eqs. (15) to (18), the LEFPE for the Saint-Venant equations that would solve for the multivariate PDF of the hydrologic state vector can be written to exact second order as shown in Eq. (19) below.

$$
\begin{aligned}
\frac{\partial P(x_1, x_2, \alpha, \beta; t)}{\partial t} = \\
& -\frac{\partial}{\partial x_1}\left\{P(x_1, x_2, \alpha, \beta; t)\left[\begin{array}{l}
\langle \eta_{1,t}(x_1,t)\rangle \\
+\int_0^t ds\,\mathrm{Cov}_0\left[\frac{\partial \eta_{1,t}(x_1,t)}{\partial x_1}; \eta_{1,t-s}(x_1, t-s)\right] \\
+\int_0^t ds\,\mathrm{Cov}_0\left[\frac{\partial \eta_{1,t}(x_1,t)}{\partial x_2}; \eta_{2,t-s}(x_2, t-s)\right] \\
+\int_0^t ds\,\mathrm{Cov}_0\left[\frac{\partial \eta_{1,t}(x_1,t)}{\partial \alpha}; \eta_{\alpha,t-s}(x_1, t-s)\right] \\
+\int_0^t ds\,\mathrm{Cov}_0\left[\frac{\partial \eta_{1,t}(x_1,t)}{\partial \beta}; \eta_{\beta,t-s}(x_2, t-s)\right]
\end{array}\right]\right\} \\
& -\frac{\partial}{\partial x_2}\left\{P(x_1, x_2, \alpha, \beta; t)\left[\begin{array}{l}
\langle \eta_{2,t}(x_2,t)\rangle \\
+\int_0^t ds\,\mathrm{Cov}_0\left[\frac{\partial \eta_{2,t}(x_2,t)}{\partial x_2}; \eta_{2,t-s}(x_2, t-s)\right] \\
+\int_0^t ds\,\mathrm{Cov}_0\left[\frac{\partial \eta_{2,t}(x_2,t)}{\partial x_1}; \eta_{1,t-s}(x_1, t-s)\right] \\
+\int_0^t ds\,\mathrm{Cov}_0\left[\frac{\partial \eta_{2,t}(x_2,t)}{\partial \alpha}; \eta_{\alpha,t-s}(x_1, t-s)\right] \\
+\int_0^t ds\,\mathrm{Cov}_0\left[\frac{\partial \eta_{2,t}(x_2,t)}{\partial \beta}; \eta_{\beta,t-s}(x_2, t-s)\right]
\end{array}\right]\right\} \\
& -\frac{\partial}{\partial \alpha}\left\{P(x_1, x_2, \alpha, \beta; t)\left[\begin{array}{l}
\langle \eta_{\alpha,t}(x_1,t)\rangle \\
+\int_0^t ds\,\mathrm{Cov}_0\left[\frac{\partial \eta_{\alpha,t}(x_1,t)}{\partial \alpha}; \eta_{\alpha,t-s}(x_1, t-s)\right] \\
+\int_0^t ds\,\mathrm{Cov}_0\left[\frac{\partial \eta_{\alpha,t}(x_1,t)}{\partial x_1}; \eta_{1,t-s}(x_1, t-s)\right] \\
+\int_0^t ds\,\mathrm{Cov}_0\left[\frac{\partial \eta_{\alpha,t}(x_1,t)}{\partial x_2}; \eta_{2,t-s}(x_2, t-s)\right] \\
+\int_0^t ds\,\mathrm{Cov}_0\left[\frac{\partial \eta_{\alpha,t}(x_1,t)}{\partial \beta}; \eta_{\beta,t-s}(x_2, t-s)\right]
\end{array}\right]\right\} \\
& -\frac{\partial}{\partial \beta}\left\{P(x_1, x_2, \alpha, \beta; t)\left[\begin{array}{l}
\langle \eta_{\beta,t}(x_2,t)\rangle \\
+\int_0^t ds\,\mathrm{Cov}_0\left[\frac{\partial \eta_{\beta,t}(x_2,t)}{\partial \beta}; \eta_{\beta,t-s}(x_2, t-s)\right] \\
+\int_0^t ds\,\mathrm{Cov}_0\left[\frac{\partial \eta_{\beta,t}(x_2,t)}{\partial x_1}; \eta_{1,t-s}(x_{1,t-s}, t-s)\right] \\
+\int_0^t ds\,\mathrm{Cov}_0\left[\frac{\partial \eta_{\beta,t}(x_2,t)}{\partial x_2}; \eta_{2,t-s}(x_2, t-s)\right] \\
+\int_0^t ds\,\mathrm{Cov}_0\left[\frac{\partial \eta_{\beta,t}(x_2,t)}{\partial \alpha}; \eta_{\alpha,t-s}(x_{1,t-s}, t-s)\right]
\end{array}\right]\right\} \\
& +\frac{1}{2}\frac{\partial^2}{\partial x_1^2}\left\{2P(x_1, x_2, \alpha, \beta; t)\int_0^t ds\,\mathrm{Cov}_0\left[\eta_{1,t}(x_1,t); \eta_{1,t-s}(x_1, t-s)\right]\right\} \\
& +\frac{1}{2}\frac{\partial^2}{\partial x_2^2}\left\{2P(x_1, x_2, \alpha, \beta; t)\int_0^t ds\,\mathrm{Cov}_0\left[\eta_{2,t}(x_2,t); \eta_{2,t-s}(x_2, t-s)\right]\right\} \\
& +\frac{1}{2}\frac{\partial^2}{\partial \alpha^2}\left\{2P(x_1, x_2, \alpha, \beta; t)\int_0^t ds\,\mathrm{Cov}_0\left[\eta_{\alpha,t}(x_1,t); \eta_{\alpha,t-s}(x_1, t-s)\right]\right\} \\
& +\frac{1}{2}\frac{\partial^2}{\partial \beta^2}\left\{2P(x_1, x_2, \alpha, \beta; t)\int_0^t ds\,\mathrm{Cov}_0\left[\eta_{\beta,t}(x_2,t); \eta_{\beta,t-s}(x_2, t-s)\right]\right\} \\
& + \text{other cross-covariance dispersion terms.} \qquad (19)
\end{aligned}
$$

Note that the LEFPE has the form of an advection–diffusion equation. In Eq. (19), the first four terms on the right-hand side represent the advection terms, while the remaining terms represent the diffusion terms. Within the advection terms, the expected values of the η functions are the mean advection coefficients, while the integrals of the ordered covariance functions added to them are the advection correction terms. However, it was shown in a study by Kavvas and Wu (2002), which used a similar approach but was applied to solute transport, that the advection correction terms are negligible when compared to the mean advection term. As a result, the magnitudes of the expectations of the η functions are much larger than those of the integral terms in the advection portion of the LEFPE, thus allowing the removal of these integral terms. As an example, this simplification can be mathematically represented for the first advection term as shown in Eq. (20) below, and is similarly applied to the other advection terms.

$$
\begin{aligned}
&\langle \eta_{1,t}\rangle \gg \mathrm{Cov}_0\left[\frac{\partial \eta_{1,t}}{\partial x_1}; \eta_{1,t-s}\right], \\
&\langle \eta_{1,t}\rangle \gg \mathrm{Cov}_0\left[\frac{\partial \eta_{1,t}}{\partial \alpha}; \eta_{\alpha,t-s}\right], \\
&\langle \eta_{1,t}\rangle \gg \mathrm{Cov}_0\left[\frac{\partial \eta_{1,t}}{\partial x_2}; \eta_{2,t-s}\right], \\
&\langle \eta_{1,t}\rangle \gg \mathrm{Cov}_0\left[\frac{\partial \eta_{1,t}}{\partial \beta}; \eta_{\beta,t-s}\right]. \qquad (20)
\end{aligned}
$$

Moreover, note that the cross-covariance dispersion terms have not been explicitly written in Eq. (19). These terms involve the ordered covariance between two different η functions, η_i and η_j, where $i \neq j$. Examples of those include:

$$\mathrm{Cov}_0\left[\eta_{1,t}(x_1,t); \eta_{2,t-s}(x_2, t-s)\right],$$

$$\mathrm{Cov_o}\left[\eta_{1,t}\left(x_1,t\right);\eta_{\beta,t-s}\left(x_2,t-s\right)\right]. \tag{21}$$

However, all of the four η functions ($\eta_{1,t}$; $\eta_{2,t}$; $\eta_{\alpha,t}$; $\eta_{\beta,t}$) are functions of the state stochastic variables (x_1; x_2; α; β). It has been shown in Liang and Kavvas (2008) that the covariance between any two of the different η functions is substantially smaller in magnitude when compared to the autocovariance of the η function of any one state variable. This leads to another simplification in which all cross-covariance terms, similar to those in Eq. (21), are neglected and removed from the main equation. Such an approximation may break down when the functions are similar, in the sense of their behavior or even the existence of periodicity with close frequencies. In such cases, the cross-covariance terms of the two η functions may become closer in magnitude to the autocovariance values, possibly invalidating their neglection. As a result of these simplifications, Eq. (19) can be written as shown in Eq. (22).

$$\begin{aligned}\frac{\partial P\left(x_1,x_2,\alpha,\beta;t\right)}{\partial t}&=-\frac{\partial}{\partial x_1}\left\{P\left\langle\eta_{1,t}\left(x_1,t\right)\right\rangle\right\}\\&+\frac{1}{2}\frac{\partial^2}{\partial x_1^2}\left\{2P\int_0^t\mathrm{d}s\mathrm{Cov_o}\left[\eta_{1,t}\left(x_1,t\right);\eta_{1,t-s}\left(x_1,t-s\right)\right]\right\}\\&-\frac{\partial}{\partial x_2}\left\{P\left\langle\eta_{2,t}\left(x_2,t\right)\right\rangle\right\}\\&+\frac{1}{2}\frac{\partial^2}{\partial x_2^2}\left\{2P\int_0^t\mathrm{d}s\mathrm{Cov_o}\left[\eta_{2,t}\left(x_2,t\right);\eta_{2,t-s}\left(x_2,t-s\right)\right]\right\}\\&-\frac{\partial}{\partial \alpha}\left\{P\left\langle\eta_{\alpha,t}\left(x_1,t\right)\right\rangle\right\}\\&+\frac{1}{2}\frac{\partial^2}{\partial \alpha^2}\left\{2P\int_0^t\mathrm{d}s\mathrm{Cov_o}\left[\eta_{\alpha,t}\left(x_1,t\right);\eta_{\alpha,t-s}\left(x_1,t-s\right)\right]\right\}\\&-\frac{\partial}{\partial \beta}\left\{P\left\langle\eta_{\beta,t}\left(x_2,t\right)\right\rangle\right\}\\&+\frac{1}{2}\frac{\partial^2}{\partial \beta^2}\left\{2P\int_0^t\mathrm{d}s\mathrm{Cov_o}\left[\eta_{\beta,t}\left(x_2,t\right);\eta_{\beta,t-s}\left(x_2,t-s\right)\right]\right\}\end{aligned} \tag{22}$$

Until this point, there has been no approximation regarding the covariance expressions in the diffusion coefficients of Eq. (22). However, note that the η functions describe motions occurring in opposite directions, in a similar manner to how α may describe a forward propagation motion and β may describe a backward propagation motion, based on their equations. Such backward and forward propagation directions would be expected to have a relatively weak correlation. As a result of this, and as a result of the order of magnitude analysis performed in Liang and Kavvas (2008), one final simplification can be applied to the LEFPE. This simplification involves the approximation that the η_i random functions have short memory with respect to t, and thus may be approximated as delta-correlated. As a result, the covariance integral term for any of the four functions can be written as follows (where $\delta(s)$ is the Dirac delta function):

$$\begin{aligned}\int_0^t \mathrm{d}s\mathrm{Cov_o}\left[\eta_{i,t};\eta_{i,t-s}\right]&=\int_0^t\mathrm{Cov_o}\left[\eta_{i,t};\eta_{i,t-s}\right]\delta(s)\mathrm{d}s\\&=\mathrm{Var}\left[\eta_{i,t}\right].\end{aligned} \tag{23}$$

Note that under the approximation of a delta-correlated covariance, the nonlocal LEFPE reduces to the classical FPE (as it will be called henceforth), which is simpler to apply. Including all of the simplifications discussed above, Eq. (19) can be written as shown in Eq. (24).

$$\begin{aligned}\frac{\partial P\left(x_1,x_2,\alpha,\beta;t\right)}{\partial t}&=\\&-\frac{\partial}{\partial x_1}\left\{P\left\langle\eta_{1,t}\left(x_1,t\right)\right\rangle\right\}+\frac{1}{2}\frac{\partial^2}{\partial x_1^2}\left\{2P\mathrm{Var}\left[\eta_{1,t}\left(x_1,t\right)\right]\right\}\\&-\frac{\partial}{\partial x_2}\left\{P\left\langle\eta_{2,t}\left(x_2,t\right)\right\rangle\right\}+\frac{1}{2}\frac{\partial^2}{\partial x_2^2}\left\{2P\mathrm{Var}\left[\eta_{2,t}\left(x_2,t\right)\right]\right\}\\&-\frac{\partial}{\partial \alpha}\left\{P\left\langle\eta_{\alpha,t}\left(x_1,t\right)\right\rangle\right\}+\frac{1}{2}\frac{\partial^2}{\partial \alpha^2}\left\{2P\mathrm{Var}\left[\eta_{\alpha,t}\left(x_1,t\right)\right]\right\}\\&-\frac{\partial}{\partial \beta}\left\{P\left\langle\eta_{\beta,t}\left(x_2,t\right)\right\rangle\right\}+\frac{1}{2}\frac{\partial^2}{\partial \beta^2}\left\{2P\mathrm{Var}\left[\eta_{\beta,t}\left(x_2,t\right)\right]\right\}\end{aligned} \tag{24}$$

The validity of the preceding approximations will be checked when the results of the proposed FPE methodology are compared against the corresponding results obtained from the MC approach. With the final version of the FPE being ready, the last step is to determine the detailed expressions of the expectations and variances of the η functions in Eq. (24). After expanding these expressions based on their equivalence as denoted in Eqs. (15) to (18), and after some manipulation, Eq. (24) may be written as follows:

$$\begin{aligned}\frac{\partial P\left(x_1,x_2,\alpha,\beta;t\right)}{\partial t}&=-\frac{\partial}{\partial x_1}\left\{P\left[\frac{3}{4}\left\langle\alpha\left(x_1,t\right)\right\rangle+\frac{1}{4}\left\langle\beta\left(x_1,t\right)\right\rangle\right]\right\}\\&-\frac{\partial}{\partial x_2}\left\{P\left[\frac{1}{4}\left\langle\alpha\left(x_2,t\right)\right\rangle+\frac{3}{4}\left\langle\beta\left(x_2,t\right)\right\rangle\right]\right\}\\&-\frac{\partial}{\partial\alpha}\left\{P\left[gS_0-\frac{g}{4k^2}\left(\frac{2}{b}\right)^{4/3}\left\langle n^2\left(x_1,t\right)\cdot\left[\alpha\left(x_1,t\right)+\beta\left(x_1,t\right)\right]^2\right.\right.\right.\\&\left.\left.\left.\cdot\left\{\frac{8gb}{\left[\alpha\left(x_1,t\right)-\beta\left(x_1,t\right)\right]^2}+1\right\}^{4/3}\right\rangle\right]\right\}\\&-\frac{\partial}{\partial\beta}\left\{P\left[gS_0-\frac{g}{4k^2}\left(\frac{2}{b}\right)^{4/3}\left\langle n^2\left(x_2,t\right)\cdot\left[\alpha\left(x_2,t\right)+\beta\left(x_2,t\right)\right]^2\right.\right.\right.\\&\left.\left.\left.\cdot\left\{\frac{8gb}{\left[\alpha\left(x_2,t\right)-\beta\left(x_2,t\right)\right]^2}+1\right\}^{4/3}\right\rangle\right]\right\}\\&+\frac{\partial^2}{\partial x_1^2}\left\{P\left[\left(\frac{9}{16}\right)\mathrm{Var}\left[\alpha\left(x_1,t\right)\right]+\left(\frac{1}{16}\right)\mathrm{Var}\left[\beta\left(x_1,t\right)\right]\right.\right.\\&\left.\left.+\left(\frac{3}{8}\right)\mathrm{Cov}\left[\alpha\left(x_1,t\right),\beta\left(x_1,t\right)\right]\right]\right\}\end{aligned}$$

$$+\frac{\partial^2}{\partial x_2^2}\left\{P\left[\left(\frac{1}{16}\right)\mathrm{Var}[\alpha(x_2,t)]+\left(\frac{9}{16}\right)\mathrm{Var}[\beta(x_2,t)]\right.\right.$$
$$\left.\left.+\left(\frac{3}{8}\right)\mathrm{Cov}[\alpha(x_2,t)\beta(x_2,t)]\right]\right\}$$
$$+\frac{\partial^2}{\partial\alpha^2}\left\{P\left[\frac{g^2}{16k^4}\left(\frac{2}{b}\right)^{8/3}\mathrm{Var}\left[n^2(x_1,t)\right.\right.\right.$$
$$\left.\left.\left.\cdot[\alpha(x_1,t)+\beta(x_1,t)]^2\cdot\left\{\frac{8gb}{[\alpha(x_1,t)-\beta(x_1,t)]^2}+1\right\}^{4/3}\right]\right]\right\}$$
$$+\frac{\partial^2}{\partial\beta^2}\left\{P\left[\frac{g^2}{16k^4}\left(\frac{2}{b}\right)^{8/3}\mathrm{Var}\left[n^2(x_2,t)\right.\right.\right.$$
$$\left.\left.\left.\cdot[\alpha(x_2,t)+\beta(x_2,t)]^2\cdot\left\{\frac{8gb}{[\alpha(x_2,t)-\beta(x_2,t)]^2}+1\right\}^{4/3}\right]\right]\right\}. \tag{25}$$

Denoting the advection terms with F and the diffusion terms with D, Eq. (25) can be written in a simplified form as follows:

$$\frac{\partial P(x_1,x_2,\alpha,\beta;t)}{\partial t}=-\frac{\partial}{\partial x_1}F_1P-\frac{\partial}{\partial x_2}F_2P-\frac{\partial}{\partial\alpha}F_\alpha P-\frac{\partial}{\partial\beta}F_\beta P$$
$$+\frac{\partial^2}{\partial x_1^2}D_1P+\frac{\partial^2}{\partial x_2^2}D_2P+\frac{\partial^2}{\partial\alpha^2}D_\alpha P+\frac{\partial^2}{\partial\beta^2}D_\beta P. \tag{26}$$

Equation (26) is the final analytical form of the FPE methodology proposed in this study for the probabilistic solution of the stochastic Saint-Venant equations in one simulation. The advection–diffusion form of Eq. (26) is clear, in which the F terms are the advection coefficients, and the D terms are their corresponding diffusion coefficients. With the mathematical equations for the FPE methodology being derived, the next step is to find a numerical scheme with which Eq. (26) may be computed.

3.2 Numerical solution for the proposed Fokker–Planck equation methodology

In order to apply the derived FPE methodology, the FPE represented in Eq. (26) must be solved using an appropriate numerical scheme. In general, finite-difference and finite-element methods, among others, have been widely used to solve FPEs numerically. Many studies have compared several such methods to determine how they perform against each other in solving different FPEs. In their study, Pichler et al. (2011) looked at the central finite-difference method, the alternating directions implicit (ADI) method, as well as finite-element methods. They mentioned that finite-difference methods are computationally more economical than finite-element methods as the number of dimensions increases. Park and Petrosian (1996) performed a comparison between the Chang and Cooper (1970) scheme, the schemes presented in Larsen et al. (1985), and some implicit schemes (including a fully implicit mid-point difference method). They also studied the semi-implicit forms of these schemes (e.g., the Crank–Nicholson scheme). They concluded that, among these schemes, the best finite-difference method for solving their FPEs was the Chang–Cooper scheme, as it was the most robust and most stable over a wide range of parameters among the other methods tested in their study.

In fact, the Chang–Cooper scheme has been cited as one of the most widely known schemes for solving the classical FPE numerically. In their paper, Chang and Cooper (1970) developed a practical numerical differencing scheme for the solution of the one-dimensional classical FPE. This scheme uses a centered difference for the diffusion term, a weighted difference for the advection term, and requires that the quasi-equilibrium solution to the FPE be satisfied exactly at any given time along the mesh nodes. The Chang-Cooper scheme, which has first-order convergence in space and time, is a widely used scheme that ensures the non-negativity of the solution, the conservation of the probability mass (in the absence of any external sources or sinks), and the exact representation of the analytical solution upon equilibration. As a result, the Chang–Cooper scheme is highly accurate with a relatively small number of required mesh nodes. While Chang and Cooper (1970) developed and applied that scheme to a one-dimensional FPE, Kim et al. (2005a) generalized and applied the Chang–Cooper scheme to a two-dimensional FPE. In a similar manner, for the case of this study, an attempt is made to generalize the Chang–Cooper scheme to the four-dimensional FPE shown in Eq. (26).

First, Eq. (26) is rewritten as follows:

$$\frac{\partial P(x_1,x_2,\alpha,\beta;t)}{\partial t}=-\frac{\partial}{\partial x_1}\left[F_1P-D_1\frac{\partial}{\partial x_1}P\right]$$
$$-\frac{\partial}{\partial x_2}\left[F_2P-D_2\frac{\partial}{\partial x_2}P\right]-\frac{\partial}{\partial\alpha}\left[F_\alpha P-D_\alpha\frac{\partial}{\partial\alpha}P\right]$$
$$-\frac{\partial}{\partial\beta}\left[F_\beta P-D_\beta\frac{\partial}{\partial\beta}P\right]$$
$$=-\frac{\partial}{\partial x_1}J_1-\frac{\partial}{\partial x_2}J_2-\frac{\partial}{\partial\alpha}J_\alpha-\frac{\partial}{\partial\beta}J_\beta. \tag{27}$$

Equation (27) is in the form of the continuity equation, in which the J parameters may be interpreted as the probability flux or probability current, whereas P (i.e., the probability density function) is considered as the state variable. This equation can then be discretized in the following implicit manner:

$$\frac{P^{n+1}_{i,j,k,l}-P^{n}_{i,j,k,l}}{\Delta t}=-\frac{J^{n+1}_{1;i+\frac{1}{2},j,k,l}-J^{n+1}_{1;i-\frac{1}{2},j,k,l}}{\Delta x_1}$$
$$-\frac{J^{n+1}_{2;i,j+\frac{1}{2},k,l}-J^{n+1}_{2;i,j-\frac{1}{2},k,l}}{\Delta x_2}$$
$$-\frac{J^{n+1}_{\alpha;i,j,k+\frac{1}{2},l}-J^{n+1}_{\alpha;i,j,k-\frac{1}{2},l}}{\Delta\alpha}$$
$$-\frac{J^{n+1}_{\beta;i,j,k,l+\frac{1}{2}}-J^{n+1}_{\beta;i,j,k,l-\frac{1}{2}}}{\Delta\beta}, \tag{28}$$

where i: 0, 1, 2, …, N_I denotes the domain of x_1 in the direction of the C_1 curve; j: 0, 1, 2, …, N_J denotes the domain of x_2 in the direction of the C_2 curve; k: 0, 1, 2, …, N_K denotes the domain of α; l: 0, 1, 2, …, N_L denotes the domain of β; and n: 0, 1, 2, … denotes the domain of time t. Following the Chang–Cooper scheme, the expression for computing variables between two nodes (e.g., at $i+1/2$) is defined by the following expression, which is an analogue for the other dimensions:

$$P^{n+1}_{i+\frac{1}{2},j,k,l} = \left(1-\lambda^{n+1}_{1;i}\right) P^{n+1}_{i+1,j,k,l} + \lambda^{n+1}_{1;i} P^{n+1}_{i,j,k,l}, \tag{29}$$

where λ is a weighting factor. In the one-dimensional case, Chang and Cooper (1970) developed the expression of the weighting factor in a manner that would ensure the nonnegativity of the PDF solution and that would give proper equilibration. In a similar manner, the same steps are followed for this study in order to derive the expression for λ. This expression, corresponding to Eq. (29), is shown below and is an analogue for the other dimensions:

$$\lambda^{n+1}_{1;i} = \frac{D^n_{1;i+\frac{1}{2},j,k,l} - \left(D^n_{1;i+\frac{1}{2},j,k,l} - \Delta x_1 F^n_{1;i+\frac{1}{2},j,k,l}\right)\exp\left[\Delta x_1 \frac{F^n_{1;i+\frac{1}{2},j,k,l}}{D^n_{1;i+\frac{1}{2},j,k,l}}\right]}{\Delta x_1 F^n_{1;i+\frac{1}{2},j,k,l}\left\{\exp\left[\Delta x_1 \frac{F^n_{1;i+\frac{1}{2},j,k,l}}{D^n_{1;i+\frac{1}{2},j,k,l}}\right] - 1\right\}}. \tag{30}$$

Moreover, following the Chang–Cooper scheme, the expression for the J parameter in the x_1 direction may be derived to be represented as follows:

$$J^{n+1}_{1;i+\frac{1}{2},j,k,l} = \left[F^n_{1;i+\frac{1}{2},j,k,l}\left(1-\lambda^{n+1}_{1;i}\right) - \frac{D^n_{1;i+\frac{1}{2},j,k,l}}{\Delta x_1}\right] P^{n+1}_{i+1,j,k,l} + \left[F^n_{1;i+\frac{1}{2},j,k,l}\lambda^{n+1}_{1;i} + \frac{D^n_{1;i+\frac{1}{2},j,k,l}}{\Delta x_1}\right] P^{n+1}_{i,j,k,l}. \tag{31}$$

Equation (31), and its analogous equations for the other dimensions, can then be substituted into the discretized FPE, Eq. (28), in order to provide the implicit finite-difference form of the FPE methodology, shown in Eq. (32) below, which can be numerically solved. Note that in the expression of Eq. (32), each subscript (i, j, k, l) that is not followed by a $+1/2$ or $-1/2$ is dropped from the expressions of F and D for readability purposes; e.g., $F^n_{2;i,j+\frac{1}{2},k,l}$ is simplified and written as $F^n_{2;j+\frac{1}{2}}$.

$$\begin{aligned}
P^n_{i,j,k,l} = {} & \left\{\begin{array}{l}
1 + \frac{\Delta t}{\Delta x_1}\lambda^{n+1}_{1;i} F^n_{1;i+\frac{1}{2}} + \frac{\Delta t}{(\Delta x_1)^2} D^n_{1;i+\frac{1}{2}} \\
- \frac{\Delta t}{\Delta x_1}\left(1-\lambda^{n+1}_{1;i-1}\right) F^n_{1;i-\frac{1}{2}} + \frac{\Delta t}{(\Delta x_1)^2} D^n_{1;i-\frac{1}{2}} \\
+ \frac{\Delta t}{\Delta x_2}\lambda^{n+1}_{2;j} F^n_{2;j+\frac{1}{2}} + \frac{\Delta t}{(\Delta x_2)^2} D^n_{2;j+\frac{1}{2}} \\
- \frac{\Delta t}{\Delta x_2}\left(1-\lambda^{n+1}_{2;j-1}\right) F^n_{2;j-\frac{1}{2}} + \frac{\Delta t}{(\Delta x_2)^2} D^n_{2;j-\frac{1}{2}} \\
+ \frac{\Delta t}{\Delta \alpha}\lambda^{n+1}_{\alpha;k} F^n_{\alpha;k+\frac{1}{2}} + \frac{\Delta t}{(\Delta \alpha)^2} D^n_{\alpha;k+\frac{1}{2}} \\
- \frac{\Delta t}{\Delta \alpha}\left(1-\lambda^{n+1}_{\alpha;k-1}\right) F^n_{\alpha;k-\frac{1}{2}} + \frac{\Delta t}{(\Delta \alpha)^2} D^n_{\alpha;k-\frac{1}{2}} \\
+ \frac{\Delta t}{\Delta \beta}\lambda^{n+1}_{\beta;l} F^n_{\beta;l+\frac{1}{2}} + \frac{\Delta t}{(\Delta \beta)^2} D^n_{\beta;l+\frac{1}{2}} \\
- \frac{\Delta t}{\Delta \beta}\left(1-\lambda^{n+1}_{\beta;l-1}\right) F^n_{\beta;l-\frac{1}{2}} + \frac{\Delta t}{(\Delta \beta)^2} D^n_{\beta;l-\frac{1}{2}}
\end{array}\right\} P^{n+1}_{i,j,k,l} \\
& + \left[\frac{\Delta t}{\Delta x_1}\left(1-\lambda^{n+1}_{1;i}\right) F^n_{1;i+\frac{1}{2}} - \frac{\Delta t}{(\Delta x_1)^2} D^n_{1;i+\frac{1}{2}}\right] P^{n+1}_{i+1,j,k,l} \\
& + \left[\frac{\Delta t}{\Delta x_2}\left(1-\lambda^{n+1}_{2;j}\right) F^n_{2;j+\frac{1}{2}} - \frac{\Delta t}{(\Delta x_2)^2} D^n_{2;j+\frac{1}{2}}\right] P^{n+1}_{i,j+1,k,l} \\
& + \left[\frac{\Delta t}{\Delta \alpha}\left(1-\lambda^{n+1}_{\alpha;k}\right) F^n_{\alpha;k+\frac{1}{2}} - \frac{\Delta t}{(\Delta \alpha)^2} D^n_{\alpha;k+\frac{1}{2}}\right] P^{n+1}_{i,j,k+1,l} \\
& + \left[\frac{\Delta t}{\Delta \beta}\left(1-\lambda^{n+1}_{\beta;l}\right) F^n_{\beta;l+\frac{1}{2}} - \frac{\Delta t}{(\Delta \beta)^2} D^n_{\beta;l+\frac{1}{2}}\right] P^{n+1}_{i,j,k,l+1} \\
& + \left[-\frac{\Delta t}{\Delta x_1}\lambda^{n+1}_{1;i-1} F^n_{1;i-\frac{1}{2}} - \frac{\Delta t}{(\Delta x_1)^2} D^n_{1;i-\frac{1}{2}}\right] P^{n+1}_{i-1,j,k,l} \\
& + \left[-\frac{\Delta t}{\Delta x_2}\lambda^{n+1}_{2;j-1} F^n_{2;j-\frac{1}{2}} - \frac{\Delta t}{(\Delta x_2)^2} D^n_{2;j-\frac{1}{2}}\right] P^{n+1}_{i,j-1,k,l} \\
& + \left[-\frac{\Delta t}{\Delta \alpha}\lambda^{n+1}_{\alpha;k-1} F^n_{\alpha;k-\frac{1}{2}} - \frac{\Delta t}{(\Delta \alpha)^2} D^n_{\alpha;k-\frac{1}{2}}\right] P^{n+1}_{i,j,k-1,l} \\
& + \left[-\frac{\Delta t}{\Delta \beta}\lambda^{n+1}_{\beta;l-1} F^n_{\beta;l-\frac{1}{2}} - \frac{\Delta t}{(\Delta \beta)^2} D^n_{\beta;l-\frac{1}{2}}\right] P^{n+1}_{i,j,k,l-1}
\end{aligned} \tag{32}$$

Note that the derived FPE discretized in Eq. (32) was originally described and represented using the characteristic method. As a result, the computed values of the state variables at a new time step would be solved at the intersection of the characteristic curves C_1 and C_2 (see Eqs. 3 and 5). In a similar manner, the values of P to be solved for in Eq. (32) should be those corresponding to the positions of intersection between C_1 and C_2, i.e., when $x_1 = x_2$. Hence, additional simplifications can be applied to Eq. (32), including $x_1 = x_2 = x$ and $\Delta x_1 = \Delta x_2 = \Delta x$. As such, since the variables x_1 and x_2 are now represented by one variable x, which is the intersection position, their corresponding i and j subscript representations can be merged into a single representation, h, thus reducing the equation from four to three dimensions (x, α, β). Therefore, the PDF $P_{i,j,k,l}$ can now be represented as $P_{h,k,l}$, where h represents the domain of the intersection position x. However, note that this does not affect the computations of the F, D, and λ parameters for x_1 and x_2 since each one has its own different expression for its calculation. With the above changes, Eq. (32) can finally be rewritten as shown in Eq. (33).

$$
\begin{aligned}
P^{n}_{h,k,l} = & \left\{ \begin{array}{l}
1+\frac{\Delta t}{\Delta x}\lambda^{n+1}_{1;h}F^{n}_{1;h+\frac{1}{2}}+\frac{\Delta t}{(\Delta x)^2}D^{n}_{1;h+\frac{1}{2}} \\
-\frac{\Delta t}{\Delta x}\left(1-\lambda^{n+1}_{1;h-1}\right)F^{n}_{1;h-\frac{1}{2}}+\frac{\Delta t}{(\Delta x)^2}D^{n}_{1;h-\frac{1}{2}} \\
+\frac{\Delta t}{\Delta x}\lambda^{n+1}_{2;h}F^{n}_{2;h+\frac{1}{2}}+\frac{\Delta t}{(\Delta x)^2}D^{n}_{2;h+\frac{1}{2}} \\
-\frac{\Delta t}{\Delta x}\left(1-\lambda^{n+1}_{2;h-1}\right)F^{n}_{2;h-\frac{1}{2}}+\frac{\Delta t}{(\Delta x)^2}D^{n}_{2;h-\frac{1}{2}} \\
+\frac{\Delta t}{\Delta \alpha}\lambda^{n+1}_{\alpha;k}F^{n}_{\alpha;k+\frac{1}{2}}+\frac{\Delta t}{(\Delta \alpha)^2}D^{n}_{\alpha;k+\frac{1}{2}} \\
-\frac{\Delta t}{\Delta \alpha}\left(1-\lambda^{n+1}_{\alpha;k-1}\right)F^{n}_{\alpha;k-\frac{1}{2}}+\frac{\Delta t}{(\Delta \alpha)^2}D^{n}_{\alpha;k-\frac{1}{2}} \\
+\frac{\Delta t}{\Delta \beta}\lambda^{n+1}_{\beta;l}F^{n}_{\beta;l+\frac{1}{2}}+\frac{\Delta t}{(\Delta \beta)^2}D^{n}_{\beta;l+\frac{1}{2}} \\
-\frac{\Delta t}{\Delta \beta}\left(1-\lambda^{n+1}_{\beta;l-1}\right)F^{n}_{\beta;l-\frac{1}{2}}+\frac{\Delta t}{(\Delta \beta)^2}D^{n}_{\beta;l-\frac{1}{2}}
\end{array} \right\} P^{n+1}_{h,k,l} \\
& + \left\{ \begin{array}{l}
\frac{\Delta t}{\Delta x}\left(1-\lambda^{n+1}_{1;h}\right)F^{n}_{1;h+\frac{1}{2}}-\frac{\Delta t}{(\Delta x)^2}D^{n}_{1;h+\frac{1}{2}} \\
+\frac{\Delta t}{\Delta x}\left(1-\lambda^{n+1}_{2;h}\right)F^{n}_{2;h+\frac{1}{2}}-\frac{\Delta t}{(\Delta x)^2}D^{n}_{2;h+\frac{1}{2}}
\end{array} \right\} P^{n+1}_{h+1,k,l} \\
& + \left[\frac{\Delta t}{\Delta \alpha}\left(1-\lambda^{n+1}_{\alpha;k}\right)F^{n}_{\alpha;k+\frac{1}{2}}-\frac{\Delta t}{(\Delta \alpha)^2}D^{n}_{\alpha;k+\frac{1}{2}}\right] P^{n+1}_{h,k+1,l} \\
& + \left[\frac{\Delta t}{\Delta \beta}\left(1-\lambda^{n+1}_{\beta;l}\right)F^{n}_{\beta;l+\frac{1}{2}}-\frac{\Delta t}{(\Delta \beta)^2}D^{n}_{\beta;l+\frac{1}{2}}\right] P^{n+1}_{h,k,l+1} \\
& + \left\{ \begin{array}{l}
-\frac{\Delta t}{\Delta x}\lambda^{n+1}_{1;h-1}F^{n}_{1;h-\frac{1}{2}}-\frac{\Delta t}{(\Delta x)^2}D^{n}_{1;h-\frac{1}{2}} \\
-\frac{\Delta t}{\Delta x}\lambda^{n+1}_{2;h-1}F^{n}_{2;h-\frac{1}{2}}-\frac{\Delta t}{(\Delta x)^2}D^{n}_{2;h-\frac{1}{2}}
\end{array} \right\} P^{n+1}_{h-1,k,l} \\
& + \left[-\frac{\Delta t}{\Delta \alpha}\lambda^{n+1}_{\alpha;k-1}F^{n}_{\alpha;k-\frac{1}{2}}-\frac{\Delta t}{(\Delta \alpha)^2}D^{n}_{\alpha;k-\frac{1}{2}}\right] P^{n+1}_{h,k-1,l} \\
& + \left[-\frac{\Delta t}{\Delta \beta}\lambda^{n+1}_{\beta;l-1}F^{n}_{\beta;l-\frac{1}{2}}-\frac{\Delta t}{(\Delta \beta)^2}D^{n}_{\beta;l-\frac{1}{2}}\right] P^{n+1}_{h,k,l-1} \qquad (33)
\end{aligned}
$$

Equation (33) is the discretized version of the FPE that represents the proposed methodology of this study. A comparison of Eqs. (25) and (26) provides the expressions for the F and D parameters, while Eq. (30) and its analogues for the other dimensions provide the expressions for the λ parameters, all of which would then complete the solution of Eq. (33). This equation has to be solved implicitly in order to compute the ensemble behavior and variability of a hydrologic system defined by the stochastic Saint-Venant equations, and it does that by solving for the joint PDF of the state variables within the x–α–β domain. Equation (33) provides an effective approach to solve for the ensemble behavior and variability of the stochastic unsteady open channel flow in a rectangular, prismatic channel under uncertain roughness coefficient, by running only one simulation. This proposed FPE methodology can also be expanded to problems with uncertainties in other channel or flow parameters. The performance of the proposed FPE methodology is evaluated in a companion paper by Dib and Kavvas (2018) which compares its results to those obtained by the MC approach.

4 Summary and conclusions

This study proposed a new methodology to model the expected behavior and variability of a system described by the stochastic open-channel flow equations. The governing equations that were used to represent the flood routing problem in this study are the continuity and momentum equations, otherwise known as the Saint-Venant equations. Many uncertainties can add to the complexity of solving the Saint-Venant equations in engineering routing problems. These uncertainties may include uncertainties in the channel's physical and geometric properties, as well as uncertainties in the lateral inflows and upstream boundary conditions, all of which render the Saint-Venant equations stochastic. As such, the dependent variables that will be solved for by these equations will also become stochastic, thus requiring that their statistical properties be solved for at specific time–space locations. Therefore, with uncertain parameters, the Saint-Venant equations have to be solved within a stochastic framework in order to quantify the ensemble behavior and variability of the system being considered. While the Mote Carlo method is a viable approach for the solution of such a stochastic unsteady open-channel flow problem, its computational expense and its large number of simulations act to its disadvantage. Hence, a new methodology was proposed in this study by which the statistical properties of the dependent variables of the considered hydrologic problem may be obtained in only one single simulation.

The proposed FPE methodology derived in this study involved upscaling the governing stochastic differential equations by developing their corresponding Lagrangian–Eulerian Fokker–Planck Equation (LEFPE), thus transforming the original stochastic equations into the framework of a deterministic differential equation. The deterministic LEFPE that describes the time–space evolution of the probability density function of the unsteady open-channel flow state variables, was developed following the method in Kavvas (2003) after the governing Saint Venant equations were transformed into their characteristic form by using the method of characteristics. Through simplifications, this LEFPE was reduced to a classical FPE that could be solved deterministically for the evolution of the probability density of the state variables of the system. The obtained linear FPE was, then, discretized in an implicit manner following Chang and Cooper (1970). This provided the equations that may be solved numerically, through only one simulation, in order to determine the ensemble behavior and variability of a system described by the stochastic open-channel flow equations. The application and validation of this methodology, applied to an open-channel flow problem with an uncertain roughness coefficient, is provided in a companion paper by Dib and Kavvas (2018), in which the statistical results of the proposed FPE methodology are compared against the results obtained by the MC approach.

The open-channel flow problem considered in this study was for a rectangular, prismatic channel under an uncertain roughness coefficient. However, the proposed methodology can be expanded to problems which assume uncertainties that arise from other flow or channel parameters. For instance, when parameters such as the channel bed slope or the channel width are assumed to be uncertain, their corresponding representations in the equations will simply have to be included in the expectation and variance expressions of the advection and diffusion coefficients for the α and β directions. When lateral inflows/outflows exist and are uncertain, additional terms corresponding to the lateral inflows/outflows will have to be added to the Saint-Venant equations, and will subsequently appear in the derived FPE. These additional terms will also have to be included inside the expectation and variance expressions. Hence, future research could entail investigating the uncertainties due to the channel slope, channel cross section, lateral inflows, and initial and boundary conditions. Moreover, any number of such uncertainties may be incorporated into this methodology at the same time by simultaneously applying their corresponding changes to the necessary equations. Therefore, applying the proposed FPE methodology to systems which include more than one source of uncertainty could be a further extension of the methodology in its attempt to effectively describe such highly nonlinear and stochastic systems.

Competing interests. The authors declare that they have no conflict of interest.

Edited by: Sabine Attinger

References

Abbott, M. B.: An introduction to the method of characteristics, American Elsevier, New York, 243 pp., 1966.

Abbott, M. B. and Ionescu, F.: On the numerical computation of nearly horizontal flows, J. Hydraul. Res., 5, 97–117, https://doi.org/10.1080/00221686709500195, 1967.

Amein, M.: Streamflow routing on computer by characteristics, Water Resour. Res., 2, 123–130, https://doi.org/10.1029/WR002i001p00123, 1966.

Barré de Saint-Venant, A.-J.-C.: Théorie du mouvement non permanent des eaux, avec application aux crues des rivières et à l'introduction des marées dans leur lit, C. R. Acad. Sci., 73, 147–154, 1871.

Beam, R. M. and Warming, R. F.: Implicit finite-difference algorithm for hyperbolic systems in conservation-law form, J. Comput. Phys., 22, 87–110, https://doi.org/10.1016/0021-9991(76)90110-8, 1976.

Bellin, A., Salandin, P., and Rinaldo, A.: Simulation of dispersion in heterogeneous porous formations – statistics, 1st-order theories, convergence of computations, Water Resour. Res., 28, 2211–2227, https://doi.org/10.1029/92wr00578, 1992.

Cayar, M. and Kavvas, M. L.: Symmetry in nonlinear hydrologic dynamics under uncertainty: ensemble modeling of 2d Boussinesq equation for unsteady flow in heterogeneous aquifers, J. Hydrol. Eng., 14, 1173–1184, https://doi.org/10.1061/(Asce)He.1943-5584.0000112, 2009a.

Cayar, M. and Kavvas, M. L.: Ensemble average and ensemble variance behavior of unsteady, one-dimensional groundwater flow in unconfined, heterogeneous aquifers: an exact second-order model, Stoch. Env. Res. Risk. A., 23, 947–956, https://doi.org/10.1007/s00477-008-0263-1, 2009b.

Chang, J. S. and Cooper, G.: A practical difference scheme for Fokker–Planck equations, J. Comput. Phys., 6, 1–16, https://doi.org/10.1016/0021-9991(70)90001-X, 1970.

Chanson, H.: The hydraulics of open channel flow: an introduction, in: 2nd Edn., Butterworth-Heinemann, Oxford, 585 pp., 2004.

Chaudhry, M. H.: Open-channel flow, in: 2nd Edn., Springer US, New York, 523 pp., 2008.

Chow, V. T.: Open-channel hydraulics, McGraw-Hill civil engineering series, McGraw-Hill, New York, 680 pp., 1959.

Cooley, R. L. and Moin, S. A.: Finite-element solution of Saint-Venant equations, J. Hydrol. Eng. Div.-Asce, 102, 759–775, 1976.

Dettinger, M. D. and Wilson, J. L.: 1st-order analysis of uncertainty in numerical-models of groundwater-flow .1. mathematical development, Water Resour. Res., 17, 149–161, https://doi.org/10.1029/WR017i001p00149, 1981.

Dib, A. and Kavvas, M. L.: Ensemble modeling of stochastic unsteady open-channel flow in terms of its time–space evolutionary probability distribution – Part 2: numerical application, Hydrol. Earth Syst. Sci., 22, 2007–2021, https://doi.org/10.5194/hess-22-2007-2018, 2018.

Dogrul, E. C., Kavvas, M. L., and Chen, Z. Q.: Prediction of subsurface stormflow in heterogeneous sloping aquifers, J. Hydrol. Eng., 3, 258–267, https://doi.org/10.1061/(Asce)1084-0699(1998)3:4(258), 1998.

Duffy, C. J.: A two-state integral-balance model for soil moisture and groundwater dynamics in complex terrain, Water Resour. Res., 32, 2421–2434, https://doi.org/10.1029/96wr01049, 1996.

Duffy, C. J. and Cusumano, J.: A low-dimensional model for concentration-discharge dynamics in groundwater stream systems, Water Resour. Res., 34, 2235–2247, https://doi.org/10.1029/98wr01705, 1998.

Ercan, A. and Kavvas, M. L.: Ensemble modeling of hydrologic and hydraulic processes at one shot: application to kinematic open-channel flow under uncertain channel properties by the stochastic method of characteristics, J. Hydrol. Eng., 17, 168–181, https://doi.org/10.1061/(Asce)He.1943-5584.0000425, 2012a.

Ercan, A. and Kavvas, M. L.: Ensemble modeling of hydrologic and hydraulic processes at one shot: application to kinematic open-channel flow under uncertain channel properties and uncertain lateral flow conditions by the stochastic method of characteristics, J. Hydrol. Eng., 17, 414–423, https://doi.org/10.1061/(Asce)He.1943-5584.0000434, 2012b.

Fennema, R. J. and Chaudhry, M. H.: Explicit numerical schemes for unsteady free-surface flows with shocks, Water Resour. Res., 22, 1923–1930, https://doi.org/10.1029/WR022i013p01923, 1986.

Fread, D. L.: Technique for implicit dynamic routing in rivers with tributaries, Water Resour. Res., 9, 918–926, https://doi.org/10.1029/WR009i004p00918, 1973.

Freeze, R. A.: Stochastic-conceptual analysis of one-dimensional groundwater flow in nonuniform homogeneous media, Water Resour. Res., 11, 725–741, https://doi.org/10.1029/WR011i005p00725, 1975.

Garcia, R. and Kahawita, R. A.: Numerical-solution of the St-Venant equations with the Maccormack finite-difference scheme, Int. J. Numer. Meth. Fluids, 6, 259–274, https://doi.org/10.1002/fld.1650060502, 1986.

Gates, T. K. and AlZahrani, M. A.: Spatiotemporal stochastic open-channel flow. 1. model and its parameter data, J. Hydraul. Eng., 122, 641–651, https://doi.org/10.1061/(Asce)0733-9429(1996)122:11(641), 1996a.

Gates, T. K. and AlZahrani, M. A.: Spatiotemporal stochastic open-channel flow. 2. simulation experiments, J. Hydraul. Eng., 122, 652–661, https://doi.org/10.1061/(Asce)0733-9429(1996)122:11(652), 1996b.

Gelhar, L. W. and Axness, C. L.: 3-dimensional stochastic-analysis of macrodispersion in aquifers, Water Resour. Res., 19, 161–180, https://doi.org/10.1029/WR019i001p00161, 1983.

Hicks, F. E. and Steffler, P. M.: Comparison of finite-element methods for the St Venant equations, Int. J. Numer. Meth. Fluids, 20, 99–113, https://doi.org/10.1002/fld.1650200202, 1995.

Hoffman, J. D.: Numerical methods for engineers and scientists, in: 2nd Edn., Marcel Dekker, New York, 823 pp., 2001.

Horne, F. E. and Kavvas, M. L.: Physics of the spatially averaged snowmelt process, J. Hydrol., 191, 179–207, https://doi.org/10.1016/S0022-1694(96)03063-6, 1997.

Kapoor, V. and Gelhar, L. W.: Transport in 3-dimensionally heterogeneous aquifers. 1. dynamics of concentration fluctuations, Water Resour. Res., 30, 1775–1788, https://doi.org/10.1029/94wr00076, 1994.

Kavvas, M. L.: Nonlinear hydrologic processes: conservation equations for determining their means and probability distributions, J. Hydrol. Eng., 8, 44–53, https://doi.org/10.1061/(Asce)1084-0699(2003)8:2(44), 2003.

Kavvas, M. L. and Karakas, A.: On the stochastic theory of solute transport by unsteady and steady groundwater flow in heterogeneous aquifers, J. Hydrol., 179, 321–351, 1996.

Kavvas, M. L. and Wu, J.: Conservation equations for solute transport by unsteady and steady flows in heterogeneous aquifers: the cumulant expansion/Lie operator methodology, in: Stochastic Methods in Subsurface Contaminant Hydrology, edited by: Govindaraju, R. S., ASCE Press, New York, 281–306, 2002.

Kim, S., Kavvas, M. L., and Chen, Z. Q.: Root-water uptake model at heterogeneous soil fields, J. Hydrol. Eng., 10, 160–167, https://doi.org/10.1061/(Asce)1084-0699(2005)10:2(160), 2005a.

Kim, S., Kavvas, M. L., and Yoon, J.: Upscaling of vertical unsaturated flow model under infiltration condition, J. Hydrol. Eng., 10, 151–159, https://doi.org/10.1061/(Asce)1084-0699(2005)10:2(151), 2005b.

Kitanidis, P. K.: Prediction by the method of moments of transport in a heterogeneous formation, J. Hydrol., 102, 453–473, https://doi.org/10.1016/0022-1694(88)90111-4, 1988.

Lai, C.: Comprehensive method of characteristics models for flow simulation, J. Hydraul. Eng., 114, 1074–1097, 1988.

Larsen, E. W., Levermore, C. D., Pomraning, G. C., and Sanderson, J. G.: Discretization Methods for One-Dimensional Fokker-Planck Operators, J. Comput. Phys., 61, 359–390, https://doi.org/10.1016/0021-9991(85)90070-1, 1985.

Liang, L. and Kavvas, M. L.: Modeling of solute transport and macrodispersion by unsteady stream flow under uncertain conditions, J. Hydrol. Eng., 13, 510–520, https://doi.org/10.1061/(Asce)1084-0699(2008)13:6(510), 2008.

Litrico, X. and Fromion, V.: Modeling and control of hydrosystems, Springer, Dordrecht, New York, 409 pp., 2009.

Mantoglou, A. and Gelhar, L. W.: Capillary tension head variance, mean soil-moisture content, and effective specific soil-moisture capacity of transient unsaturated flow in stratified soils, Water Resour. Res., 23, 47–56, https://doi.org/10.1029/WR023i001p00047, 1987.

Ohara, N., Kavvas, M. L., and Chen, Z. Q.: Stochastic upscaling for snow accumulation and melt processes with PDF approach, J. Hydrol. Eng., 13, 1103–1118, https://doi.org/10.1061/(Asce)1084-0699(2008)13:12(1103), 2008.

Park, B. T. and Petrosian, V.: Fokker-Planck equations of stochastic acceleration: A study of numerical methods, Astrophys. J. Suppl. S., 103, 255–267, https://doi.org/10.1086/192278, 1996.

Pichler, L., Masud, A., and Bergman, L. A.: Numerical Solution of the Fokker–Planck Equation by Finite Difference and Finite Element Methods – A Comparative Study, in: Computational Methods in Stochastic Dynamics, edited by: Papadrakakis, M., Stefanou, G., and Papdopoulos, V., Springer, Dordrecht, 2011.

Puente, C. E.: A new approach to hydrologic modeling: derived distributions revisited, J. Hydrol., 187, 65–80, https://doi.org/10.1016/S0022-1694(97)87978-4, 1996.

Rubin, Y. and Dagan, G.: Stochastic-analysis of boundaries effects on head spatial variability in heterogeneous aquifers. 2. impervious boundary, Water Resour. Res., 25, 707–712, https://doi.org/10.1029/WR025i004p00707, 1989.

Scharffenberg, W. A. and Kavvas, M. L.: Uncertainty in flood wave routing in a lateral-inflow-dominated stream, J. Hydrol. Eng., 16, 165–175, https://doi.org/10.1061/(Asce)He.1943-5584.0000298, 2011.

Serrano, S. E.: Analytical solutions of the nonlinear groundwater-flow equation in unconfined aquifers and the effect of heterogeneity, Water Resour. Res., 31, 2733–2742, https://doi.org/10.1029/95wr02038, 1995.

Smith, L. and Freeze, R. A.: Stochastic-analysis of steady-state groundwater-flow in a bounded domain. 2. two-dimensional simulation, Water Resour. Res., 15, 1543–1559, https://doi.org/10.1029/WR015i006p01543, 1979.

Sturm, T. W.: Open channel hydraulics, McGraw-Hill series in water resources and environmental engineering, McGraw-Hill, Boston, 493 pp., 2001.

Szymkiewicz, R.: Finite-element method for the solution of the Saint Venant equations in an open channel network, J. Hydrol., 122, 275–287, https://doi.org/10.1016/0022-1694(91)90182-H, 1991.

Szymkiewicz, R.: Method to solve 1d unsteady transport and flow equations, J. Hydraul. Eng., 121, 396–403, https://doi.org/10.1061/(Asce)0733-9429(1995)121:5(396), 1995.

Szymkiewicz, R.: Numerical modeling in open channel hydraulics, in: Water science and technology library, 83, Springer, Dordrecht, the Netherlands, 419 pp., 2010.

Tannehill, J. C., Anderson, D. A., and Pletcher, R. H.: Computational fluid mechanics and heat transfer, in: 2nd Edn., Series in computational and physical processes in mechanics and thermal sciences, Taylor & Francis, Washington, D.C., 792 pp., 1997.

Tayfur, G. and Kavvas, M. L.: Spatially averaged conservation equations for interacting rill-interrill area overland flows, J. Hydraul. Eng., 120, 1426–1448, https://doi.org/10.1061/(Asce)0733-9429(1994)120:12(1426), 1994.

Van Kampen, N. G.: A cumulant expansion for stochastic linear differential equations. II, Physica, 74, 239–247, https://doi.org/10.1016/0031-8914(74)90122-0, 1974.

Van Kampen, N. G.: Stochastic differential equations, Phys. Rep., 24, 171–228, https://doi.org/10.1016/0370-1573(76)90029-6, 1976.

Venutelli, M.: Stability and accuracy of weighted four-point implicit finite difference schemes for open channel flow, J. Hydraul. Eng., 128, 281–288, https://doi.org/10.1061/(Asce)0733-9429(2002)128:3(281), 2002.

Viessman, W., Knapp, J. W., Lewis, G. L., and Harbaugh, T. E.: Introduction to hydrology, in: 2nd Edn., Series in civil engineering, IEP-Dun-Donnelley, Harper & Row, New York, 704 pp., 1977.

Wood, B. D. and Kavvas, M. L.: Ensemble-averaged equations for reactive transport in porous media under unsteady flow conditions, Water Resour. Res., 35, 2053–2068, https://doi.org/10.1029/1999wr900113, 1999a.

Wood, B. D. and Kavvas, M. L.: Stochastic solute transport under unsteady flow conditions: Comparison of theory, Monte Carlo simulations, and field data, Water Resour. Res., 35, 2069–2084, https://doi.org/10.1029/1999wr900112, 1999b.

Woolhiser, D. A. and Liggett, J. A.: Unsteady 1-dimensional flow over a plane – rising hydrograph, Water Resour. Res., 3, 753–771, https://doi.org/10.1029/WR003i003p00753, 1967.

3

Ensemble modeling of stochastic unsteady open-channel flow in terms of its time–space evolutionary probability distribution – Part 2: numerical application

Alain Dib and M. Levent Kavvas

Department of Civil and Environmental Engineering, University of California, Davis, 95616, USA

Correspondence: Alain Dib (aedib@ucdavis.edu)

Abstract. The characteristic form of the Saint-Venant equations is solved in a stochastic setting by using a newly proposed Fokker–Planck Equation (FPE) methodology. This methodology computes the ensemble behavior and variability of the unsteady flow in open channels by directly solving for the flow variables' time–space evolutionary probability distribution. The new methodology is tested on a stochastic unsteady open-channel flow problem, with an uncertainty arising from the channel's roughness coefficient. The computed statistical descriptions of the flow variables are compared to the results obtained through Monte Carlo (MC) simulations in order to evaluate the performance of the FPE methodology. The comparisons show that the proposed methodology can adequately predict the results of the considered stochastic flow problem, including the ensemble averages, variances, and probability density functions in time and space. Unlike the large number of simulations performed by the MC approach, only one simulation is required by the FPE methodology. Moreover, the total computational time of the FPE methodology is smaller than that of the MC approach, which could prove to be a particularly crucial advantage in systems with a large number of uncertain parameters. As such, the results obtained in this study indicate that the proposed FPE methodology is a powerful and time-efficient approach for predicting the ensemble average and variance behavior, in both space and time, for an open-channel flow process under an uncertain roughness coefficient.

1 Introduction

One of the most important types of unsteady open-channel flow problems is that of flood routing (Scharffenberg and Kavvas, 2011). This problem considers an initially uniform flow rate through an open channel, after which a flood wave enters upstream of the channel reach, and is translated downstream. The routing process, which predicts the spatial shape and temporal development of this flood wave as it traverses downstream (Viessman et al., 1977), involves solving for two dependent flow variables as a function of time and space through the river reach: velocity and depth, or discharge and depth. Solving for these two unknowns by means of the hydraulic routing technique involves using two governing equations (Chow, 1959), the continuity equation and the momentum equation, which are jointly known as the Saint-Venant equations (Sturm, 2001).

While it may be possible to assume that the flow and channel parameters are deterministic so as to obtain a deterministic solution to the Saint-Venant equations, such parameters usually exhibit high uncertainty in the real world (Gates and AlZahrani, 1996; Ercan and Kavvas, 2012a). In fact, uncertainties in these parameters may originate from several factors, including a channel's physical conditions and geometric parameters, its upstream boundary and initial conditions, and any lateral inflows (Chow, 1959; Sturm, 2001; Liang and Kavvas, 2008; Ercan and Kavvas, 2012a). As such, the parameters may be formulated as random functions, thus becoming spatially and/or temporally random, in which case the system can no longer be assumed deterministic. This necessitates the solution of the governing equations within a stochastic framework, from which a quantitative description of the ensemble behavior and variability of the process is obtained. In this manner, the two dependent flow variables are solved for their statistical properties, not for their deterministic values, at designated time–space positions.

Among the available approaches that can be used to solve the Saint-Venant equations within a stochastic framework, the Monte Carlo (MC) approach is one of the most well-known due to its abundant use in simulating differential equations with stochastic parameters (Freeze, 1975; Smith and Freeze, 1979; Bellin et al., 1992). It is also generally accepted as the most robust approach for uncertainty evaluation, as well as the benchmark for comparing other new methods (Scharffenberg and Kavvas, 2011). However, one of the main disadvantages of the MC approach is its computational expense, which results from the large number of simulations that it usually involves.

In order to circumvent having to solve the Saint-Venant equations repeatedly for a large number of times, this study uses a new methodology that solves for the time–space evolutionary probability distribution of the unsteady open-channel flow process in only one simulation. From this probabilistic solution, one can then obtain the ensemble mean and ensemble variance of the process as they evolve in time and space. This new methodology, which is proposed, explained, and derived in the companion paper by Dib and Kavvas (2018), makes use of the ensemble averaging technique developed in Kavvas (2003) to obtain a Fokker–Planck Equation (FPE) that specifically describes an unsteady open-channel flow process. Some other hydrologic processes have been successfully simulated by following a similar procedure, which involved applying the ensemble averaging technique of Kavvas (2003) to their corresponding governing equations and obtaining the FPEs specific to their case. These include unsaturated water flow (Kim et al., 2005a), root-water uptake (Kim et al., 2005b), solute transport (Liang and Kavvas, 2008), snow accumulation and melt (Ohara et al., 2008), unconfined groundwater flow (Cayar and Kavvas, 2009a, b), and kinematic open-channel flow (Ercan and Kavvas, 2012a, b).

Note that in addition to producing the statistical properties in a computationally efficient manner through one simulation, the FPE methodology developed for the unsteady open-channel flow process directly solves for, and is linear in, the probability density of the dependent variables. Moreover, while this methodology assumes a finite correlation time for the considered process, it does not make any linearization assumptions and it does not have limitations on the working range of the parameter space.

Therefore, in the wake of the preceding discussions, the first objective of this study is to use the FPE methodology derived in the companion paper for the unsteady open-channel flow process (Dib and Kavvas, 2018) and to apply it to a representative stochastic unsteady open-channel flow problem in order to solve for the probability density of the state variables of the flow process and to provide a quantitative description of the expected behavior and variability of the system in one single simulation. The second objective is to evaluate the performance of the proposed methodology and to validate its results by comparing the statistical properties of the flow variables computed by the FPE methodology against those calculated by the MC approach.

2 Saint-Venant equations: characteristic form and ensemble-averaged form

The Saint-Venant equations solved in this study are written for the unsteady open-channel flow of an incompressible fluid in a rectangular, prismatic channel with no lateral inflow. The method of characteristics is used to transform these equations from two partial differential equations into a system of four ordinary differential equations (ODEs). These four ODEs include two characteristic equations describing the two characteristic paths, and two compatibility equations which must be satisfied along their corresponding characteristic path. The characteristic form of the Saint-Venant equations is shown below (Sturm, 2001):

Positive characteristic curve (C_1)

$$\frac{\mathrm{d}x_1}{\mathrm{d}t} = V + c. \tag{1}$$

Flow process condition to be satisfied along C_1

$$\left(\frac{\mathrm{d}(V+2c)}{\mathrm{d}t}\right)_1 = g\left(S_{0,1} - S_{\mathrm{f},1}\right). \tag{2}$$

Negative characteristic curve (C_2)

$$\frac{\mathrm{d}x_2}{\mathrm{d}t} = V - c. \tag{3}$$

Flow process condition to be satisfied along C_2

$$\left(\frac{\mathrm{d}(V-2c)}{\mathrm{d}t}\right)_2 = g\left(S_{0,2} - S_{\mathrm{f},2}\right), \tag{4}$$

where V is the average flow velocity, c is the wave celerity which is equal to $\sqrt{gy}$ for a rectangular channel where y is the flow depth, x is the position, t is the time, S_0 is the slope of the channel bottom, S_f is the friction slope, and g is the acceleration of gravity. Note that $S_{0,i}$ denotes $S_0(x_i, t)$; the same applies to S_f. The positive and negative characteristic curves are defined as C_1 and C_2, respectively, and the variables or derivatives corresponding to C_1 and C_2 are denoted by the subscripts 1 and 2, respectively.

The Lagrangian–Eulerian form of the Fokker–Planck equation (LEFPE), corresponding to the Saint-Venant equations, that can solve for the multivariate probability density function (PDF) of the hydrologic state variables of an unsteady open-channel flow problem was obtained through the following steps: (i) using the two substitutions $V+2c=\alpha$ and $V-2c=\beta$ on Eqs. (1) to (4), (ii) applying the technique in Kavvas (2003) while assuming the uncertainty arising from the Manning's roughness coefficient, and (iii) performing several simplifying assumptions. The result is shown in Eq. (5), while its detailed derivation can be found in the companion paper by Dib and Kavvas (2018).

$$
\begin{aligned}
&\frac{\partial P(x_1,x_2,\alpha,\beta;t)}{\partial t} = \\
&-\frac{\partial}{\partial x_1}\left\{P\left[\frac{3}{4}\langle\alpha(x_1,t)\rangle+\frac{1}{4}\langle\beta(x_1,t)\rangle\right]\right\} \\
&-\frac{\partial}{\partial x_2}\left\{P\left[\frac{1}{4}\langle\alpha(x_2,t)\rangle+\frac{3}{4}\langle\beta(x_2,t)\rangle\right]\right\} \\
&-\frac{\partial}{\partial \alpha}\left\{P\left[gS_0-\frac{g}{4k^2}\left(\frac{2}{b}\right)^{4/3}\left\langle n^2(x_1,t)\cdot[\alpha(x_1,t)+\beta(x_1,t)]^2\right.\right.\right. \\
&\left.\left.\left.\cdot\left\{\frac{8gb}{[\alpha(x_1,t)-\beta(x_1,t)]^2}+1\right\}^{4/3}\right\rangle\right]\right\} \\
&-\frac{\partial}{\partial \beta}\left\{P\left[gS_0-\frac{g}{4k^2}\left(\frac{2}{b}\right)^{4/3}\left\langle n^2(x_2,t)\cdot[\alpha(x_2,t)+\beta(x_2,t)]^2\right.\right.\right. \\
&\left.\left.\left.\cdot\left\{\frac{8gb}{[\alpha(x_2,t)-\beta(x_2,t)]^2}+1\right\}^{4/3}\right\rangle\right]\right\} \\
&+\frac{\partial^2}{\partial x_1^2}\left\{P\left[\left(\frac{9}{16}\right)\mathrm{Var}[\alpha(x_1,t)]+\left(\frac{1}{16}\right)\mathrm{Var}[\beta(x_1,t)]\right.\right. \\
&\left.\left.+\left(\frac{3}{8}\right)\mathrm{Cov}[\alpha(x_1,t),\beta(x_1,t)]\right]\right\} \\
&+\frac{\partial^2}{\partial x_2^2}\left\{P\left[\left(\frac{1}{16}\right)\mathrm{Var}[\alpha(x_2,t)]+\left(\frac{9}{16}\right)\mathrm{Var}[\beta(x_2,t)]\right.\right. \\
&\left.\left.+\left(\frac{3}{8}\right)\mathrm{Cov}[\alpha(x_2,t),\beta(x_2,t)]\right]\right\} \\
&+\frac{\partial^2}{\partial \alpha^2}\left\{P\left[\frac{g^2}{16k^4}\left(\frac{2}{b}\right)^{8/3}\mathrm{Var}\left[n^2(x_1,t)\cdot[\alpha(x_1,t)+\beta(x_1,t)]^2\right.\right.\right. \\
&\left.\left.\left.\cdot\left\{\frac{8gb}{[\alpha(x_1,t)-\beta(x_1,t)]^2}+1\right\}^{4/3}\right]\right]\right\} \\
&+\frac{\partial^2}{\partial \beta^2}\left\{P\left[\frac{g^2}{16k^4}\left(\frac{2}{b}\right)^{8/3}\mathrm{Var}\left[n^2(x_2,t)\cdot[\alpha(x_2,t)+\beta(x_2,t)]^2\right.\right.\right. \\
&\left.\left.\left.\cdot\left\{\frac{8gb}{[\alpha(x_2,t)-\beta(x_2,t)]^2}+1\right\}^{4/3}\right]\right]\right\} \qquad (5)
\end{aligned}
$$

In Eq. (5), b denotes the width of the channel, n denotes Manning's roughness coefficient, and k denotes the conversion factor between SI and US units for Manning's formula. The remaining variables are as defined previously. Note that in Eq. (5), and other later equations, $P(x_1, x_2, \alpha, \beta, t)$ is sometimes substituted by P for simplicity. Moreover, note that Eq. (5) resembles an advection–diffusion equation, in which the first four bracketed terms multiplied by P on their left-hand sides resemble the advection coefficients, and the last four bracketed terms resemble the diffusion coefficients. Hence, after denoting the advection coefficients by F and the diffusion coefficients by D, Eq. (5) was simplified into Eq. (6) below, which is the final analytical form of the proposed FPE methodology.

$$
\begin{aligned}
\frac{\partial P(x_1,x_2,\alpha,\beta;t)}{\partial t} &= -\frac{\partial}{\partial x_1}F_1P-\frac{\partial}{\partial x_2}F_2P-\frac{\partial}{\partial \alpha}F_\alpha P-\frac{\partial}{\partial \beta}F_\beta P \\
&+\frac{\partial^2}{\partial x_1^2}D_1P+\frac{\partial^2}{\partial x_2^2}D_2P+\frac{\partial^2}{\partial \alpha^2}D_\alpha P+\frac{\partial^2}{\partial \beta^2}D_\beta P. \qquad (6)
\end{aligned}
$$

An appropriate numerical scheme to solve Eq. (6) was derived following Chang and Cooper (1970), as discussed in detail in the companion paper. This scheme involves implicitly discretizing the equation, while noting that the dependent variables are to be solved at the intersection of the characteristic curves C_1 and C_2, which renders $x_1 = x_2 = x$. As such, the discretized version of Eq. (6) was derived in the companion paper to be as follows:

$$
\begin{aligned}
P^n_{h,k,l} = &\left\{\begin{array}{l}
1+\frac{\Delta t}{\Delta x}\lambda^{n+1}_{1;h}F^n_{1;h+\frac{1}{2}}+\frac{\Delta t}{(\Delta x)^2}D^n_{1;h+\frac{1}{2}} \\
-\frac{\Delta t}{\Delta x}\left(1-\lambda^{n+1}_{1;h-1}\right)F^n_{1;h-\frac{1}{2}}+\frac{\Delta t}{(\Delta x)^2}D^n_{1;h-\frac{1}{2}} \\
+\frac{\Delta t}{\Delta x}\lambda^{n+1}_{2;h}F^n_{2;h+\frac{1}{2}}+\frac{\Delta t}{(\Delta x)^2}D^n_{2;h+\frac{1}{2}} \\
-\frac{\Delta t}{\Delta x}\left(1-\lambda^{n+1}_{2;h-1}\right)F^n_{2;h-\frac{1}{2}}+\frac{\Delta t}{(\Delta x)^2}D^n_{2;h-\frac{1}{2}} \\
+\frac{\Delta t}{\Delta \alpha}\lambda^{n+1}_{\alpha;k}F^n_{\alpha;k+\frac{1}{2}}+\frac{\Delta t}{(\Delta \alpha)^2}D^n_{\alpha;k+\frac{1}{2}} \\
-\frac{\Delta t}{\Delta \alpha}\left(1-\lambda^{n+1}_{\alpha;k-1}\right)F^n_{\alpha;k-\frac{1}{2}}+\frac{\Delta t}{(\Delta \alpha)^2}D^n_{\alpha;k-\frac{1}{2}} \\
+\frac{\Delta t}{\Delta \beta}\lambda^{n+1}_{\beta;l}F^n_{\beta;l+\frac{1}{2}}+\frac{\Delta t}{(\Delta \beta)^2}D^n_{\beta;l+\frac{1}{2}} \\
-\frac{\Delta t}{\Delta \beta}\left(1-\lambda^{n+1}_{\beta;l-1}\right)F^n_{\beta;l-\frac{1}{2}}+\frac{\Delta t}{(\Delta \beta)^2}D^n_{\beta;l-\frac{1}{2}}
\end{array}\right\}P^{n+1}_{h,k,l} \\
&+\left\{\begin{array}{l}
\frac{\Delta t}{\Delta x}\left(1-\lambda^{n+1}_{1;h}\right)F^n_{1;h+\frac{1}{2}}-\frac{\Delta t}{(\Delta x)^2}D^n_{1;h+\frac{1}{2}} \\
+\frac{\Delta t}{\Delta x}\left(1-\lambda^{n+1}_{2;h}\right)F^n_{2;h+\frac{1}{2}}-\frac{\Delta t}{(\Delta x)^2}D^n_{2;h+\frac{1}{2}}
\end{array}\right\}P^{n+1}_{h+1,k,l} \\
&+\left[\frac{\Delta t}{\Delta \alpha}\left(1-\lambda^{n+1}_{\alpha;k}\right)F^n_{\alpha;k+\frac{1}{2}}-\frac{\Delta t}{(\Delta \alpha)^2}D^n_{\alpha;k+\frac{1}{2}}\right]P^{n+1}_{h,k+1,l} \\
&+\left[\frac{\Delta t}{\Delta \beta}\left(1-\lambda^{n+1}_{\beta;l}\right)F^n_{\beta;l+\frac{1}{2}}-\frac{\Delta t}{(\Delta \beta)^2}D^n_{\beta;l+\frac{1}{2}}\right]P^{n+1}_{h,k,l+1} \\
&+\left\{\begin{array}{l}
-\frac{\Delta t}{\Delta x}\lambda^{n+1}_{1;h-1}F^n_{1;h-\frac{1}{2}}-\frac{\Delta t}{(\Delta x)^2}D^n_{1;h-\frac{1}{2}} \\
-\frac{\Delta t}{\Delta x}\lambda^{n+1}_{2;h-1}F^n_{2;h-\frac{1}{2}}-\frac{\Delta t}{(\Delta x)^2}D^n_{2;h-\frac{1}{2}}
\end{array}\right\}P^{n+1}_{h-1,k,l} \\
&+\left[-\frac{\Delta t}{\Delta \alpha}\lambda^{n+1}_{\alpha;k-1}F^n_{\alpha;k-\frac{1}{2}}-\frac{\Delta t}{(\Delta \alpha)^2}D^n_{\alpha;k-\frac{1}{2}}\right]P^{n+1}_{h,k-1,l} \\
&+\left[-\frac{\Delta t}{\Delta \beta}\lambda^{n+1}_{\beta;l-1}F^n_{\beta;l-\frac{1}{2}}-\frac{\Delta t}{(\Delta \beta)^2}D^n_{\beta;l-\frac{1}{2}}\right]P^{n+1}_{h,k,l-1}, \qquad (7)
\end{aligned}
$$

where h: 0, 1, 2, …, N_H denotes the domain of x; k: 0, 1, 2, …, N_K denotes the domain of α; l: 0, 1, 2, …, N_L denotes the domain of β; and n: 0, 1, 2, … denotes the domain of time t. The expression for the λ_α parameter is given in Eq. (8) below, while the expressions for the other λ parameters can be written in a similar manner.

$$\lambda_{\alpha;k}^{n+1} = \frac{D^n_{\alpha;i,j,k+\frac{1}{2},l} - \left(D^n_{\alpha;i,j,k+\frac{1}{2},l} - \Delta\alpha F^n_{\alpha;i,j,k+\frac{1}{2},l}\right)\exp\left[\Delta\alpha \frac{F^n_{\alpha;i,j,k+\frac{1}{2},l}}{D^n_{\alpha;i,j,k+\frac{1}{2},l}}\right]}{\Delta\alpha F^n_{\alpha;i,j,k+\frac{1}{2},l}\left\{\exp\left[\Delta\alpha \frac{F^n_{\alpha;i,j,k+\frac{1}{2},l}}{D^n_{\alpha;i,j,k+\frac{1}{2},l}}\right] - 1\right\}} \tag{8}$$

where i: 0, 1, 2 ..., N_I denotes the domain of x_1 in the direction of the C_1 curve; j: 0, 1, 2 ..., N_J denotes the domain of x_2 in the direction of the C_2 curve. Equation (7) can be solved implicitly to compute the joint PDF of the state variables within the x–α–β domain. From this, a quantitative probabilistic description of the ensemble behavior and variability can be determined for the hydrologic system represented by the stochastic Saint-Venant equations with uncertain parameters.

3 Application of the Monte Carlo approach and the new Fokker–Planck equation methodology to a hydraulic routing problem

A hypothetical hydraulic routing problem will be the illustrative example which the new proposed methodology will be tested on. The hypothetical problem involves a river reach that is 2.70 km long, sloping at 0.0015 throughout the whole reach, having a rectangular cross section with a constant width of 6.1 m, and having no lateral inflow. Initially (at $t = 0$), the river is assumed to have a steady uniform flow of 15.5 $m^3\,s^{-1}$ throughout the reach. At $t > 0$, the upstream flow increases linearly to reach 56 $m^3\,s^{-1}$ at $t = 20$ min, then decreases linearly back to reach the initial flow value at $t = 60$ min, after which it remains at the same constant flow value for $t > 60$ min. For the purpose of this study, the problem is made stochastic through the uncertainty in the Manning's roughness coefficient, whose statistical characteristics must be estimated. All simulations for the MC approach and the FPE methodology were run on a computer having 16 GB of RAM and an Intel i7 processor with four cores, each core having a base frequency of 2.40 GHz and a maximum frequency of 3.40 GHz.

3.1 Estimating the statistical characteristics of Manning's roughness coefficient

Manning's roughness coefficient (n) represents the resistance of the bed of a channel to the flow of water in it (Chow, 1959). Generally, a higher n value defines a greater resistance to the flow. The value of this coefficient depends on several factors, including surface roughness, vegetation, obstructions, and channel irregularity and alignment (Chow, 1959), all of which may exhibit considerable variability along the length of an open channel. It is, therefore, crucial to account for such variability in order to better represent the behavior of the unsteady open-channel flow system being solved.

The published and technical studies with sizeable datasets to address the variability of n in such open channels are few. However, some studies have provided median and range values, while others have attempted to fit different probability distributions to the data (Gates and AlZahrani, 1996). Using such information, Manning's n was assumed for this study to be a random variable with a normal distribution having a mean of 0.035 and a standard deviation of 0.005. The mean and standard deviation were chosen in a way that provides realistic values and distributions of n that fall within the ranges and statistics provided by Gates and AlZahrani (1996) and that conform to the table of typical values in Chow (1959). Moreover, the selected mean and standard deviation allowed the generation of n values which never fell below 0.01, thus complying with the fact that the roughness coefficients for flows in natural streams and excavated channels are always greater than 0.01 (Chow, 1959). As such, no truncation or discarding of any generated n values was required. The chosen PDF for the roughness coefficient was used by the MC approach as well as by the new FPE methodology when solving for the ensemble behavior and variability of the hypothetical flow problem.

3.2 Application of the Monte Carlo approach

For the hypothetical routing problem of this study, the MC approach requires repeatedly solving the Saint-Venant equations in a deterministic manner for a large number of different roughness coefficient (n) realizations. To deterministically solve the Saint-Venant equations in their full form, several numerical techniques have been developed because their analytical solution has not been possible due to the presence of nonlinear terms (Sturm, 2001; Chaudhry, 2008). For this study, the characteristic form of the Saint-Venant equations was discretized in an explicit manner by substituting the time derivatives with their first-order finite-difference forms, as detailed in several references, e.g., Viessman et al. (1977) and Sturm (2001). The values of the dependent variables at the new time steps were computed at the points of intersection of the positive and negative characteristic curves, which rendered the final solution on an irregular x–t grid. The solution was then interpolated onto a rectangular grid, with a Δx of 75 m and a Δt of 3 min. The simulations were performed using a parallelized process which optimized the computational time by running the simulations over the total number of available cores (with no hyper-threading).

The stability of the numerical method being used was ensured by checking the Courant condition (Sturm, 2001). Furthermore, two boundary conditions, one at each end of the reach, and two initial conditions were defined in this study, since the problem deals with the subcritical unsteady nonuniform flow case (Sturm, 2001). As initial conditions, the discharge at every location along the river was provided (taken as 15.5 $m^3\,s^{-1}$, as explained in the problem description), and the flow was assumed to be initially uniform and steady. As for the boundary conditions, the flow hydrograph at the channel entrance was given, while at the downstream end, the channel was assumed to be hydraulically long so that the flow can be taken as normal flow, thus satisfying Manning's equation. As such, the downstream boundary condition was chosen as the depth–discharge relationship represented by Manning's equation. This equation, along with the equations corresponding to the positive characteristic (i.e., Eqs. 1 and 2), was used to compute the flow variables at the downstream boundary.

Following the preceding discussion, the Saint-Venant equations were deterministically solved for a total of 1000 times, each time using a different realization of n that was generated based on the PDF chosen in Sect. 3.1. While a lower number of realizations may have been sufficient for accurate computations of the first moment of the flow variables, it would not have been sufficient for the accurate computations of the second moment. In fact, the standard deviation of the flow discharge computed using 50, 100, 200, and 500 realizations showed absolute relative differences that reached 65, 35, 20, and 15 %, respectively, when compared to the results of the 1000-realization run. Therefore, the number of realizations in this study was selected to be large enough to numerically approximate, with sufficient accuracy, both the first and the second moments of the stochastic quantities of the problem at hand.

3.3 Application of the proposed Fokker–Planck equation methodology

Unlike the MC approach, the FPE methodology aims at solving for the ensemble behavior and variability of the stochastic unsteady open-channel flow system in one shot by computing the PDF of its dependent variables over time and space. Solving the hypothetical routing problem following the FPE methodology involved solving Eq. (7). Since this equation is the result of implicit discretization, it is unconditionally stable and requires no constraint on the size of the time step for its stability. However, to obtain sufficiently accurate solutions, the time step must still be limited with the Courant condition (Zheng and Bennett, 2002). Therefore, at every time position, the multidimensional Courant condition was checked in order to determine the appropriate size of the next time step.

Furthermore, to correctly solve the hydraulic routing problem, both the initial and boundary conditions of the problem in the physical space must be accurately represented in the probability space for use in the FPE methodology. Note that when the initial condition in the physical space is taken as deterministic, the Dirac delta function ($\delta(s)$) can be used to represent it in the probability space. Assuming that $\boldsymbol{H}$ is a vector of m state variables (m dimensional), and $\boldsymbol{H}_0$ is the corresponding vector of initial conditions, the initial condition PDF of $\boldsymbol{H}$ can be written as

$$P(\boldsymbol{H},t) = P(\boldsymbol{H},0) = \delta(\boldsymbol{H} - \boldsymbol{H}_0). \quad (9)$$

For the purpose of this study, $\boldsymbol{H}$ corresponds to a point in the x–α–β domain (3 dimensional), and $\boldsymbol{H}_0$ represents the initial condition that defines the deterministic flow at time $t = 0$ in the x–α–β domain. However, since it is not possible to numerically represent the Dirac delta function as a function with infinite value at only one position, it was estimated in the probability domain as a function with a very high value over a small bottom width, while preserving an area of unity.

Concerning the boundary conditions, they are usually divided for FPEs into two categories: accessible and inaccessible. Inaccessible boundaries are defined as those boundaries that could never be reached if the process starts from any interior point of the domain. On the other hand, for accessible boundaries, there is a positive probability that these boundaries will be reached from the interior of the domain within a finite amount of time (Feller, 1954). Accessible boundaries can be further subdivided into absorbing and reflecting, or no-flux, boundaries.

Note that since the FPE can be considered as the conservation equation for probability mass, a probability mass of unity needs to be conserved in the probability domain of the system. Therefore, using a reflecting (no-flux) boundary condition would be the most suitable choice for this study in order to ensure the completeness of the probability domain and to prevent any probability mass from leaving the domain (Gardiner, 1985). Such a condition was used to describe the boundaries of both the α and β dimensions, noting that the minimum and maximum boundaries of both α and β were chosen to be far enough so that they would encompass all possibilities that could occur for the considered routing problem. As for the upstream boundary, recall that the discharge hydrograph was assumed to be known upstream, in which case the probability densities at the upstream boundary of the x dimension were known for all $t > 0$. Finally, the downstream boundary in the x dimension was formulated to replicate that of the MC model, and it was extended downstream much further than the required length of the reach so as to eliminate any of its effect on the numerical solution.

4 Numerical results and discussion

Both the MC approach and the proposed FPE methodology were used to solve the stochastic Saint-Venant equations of the hypothetical hydraulic routing problem presented in Sect. 3, in order to determine the ensemble behavior of the system and the statistical distributions of the flow variables. While the state variables directly solved for were the velocity and depth (or celerity), the discharge was easily computed from these two variables.

A plot of the ensemble average discharge over time and space computed by the FPE methodology can be seen in

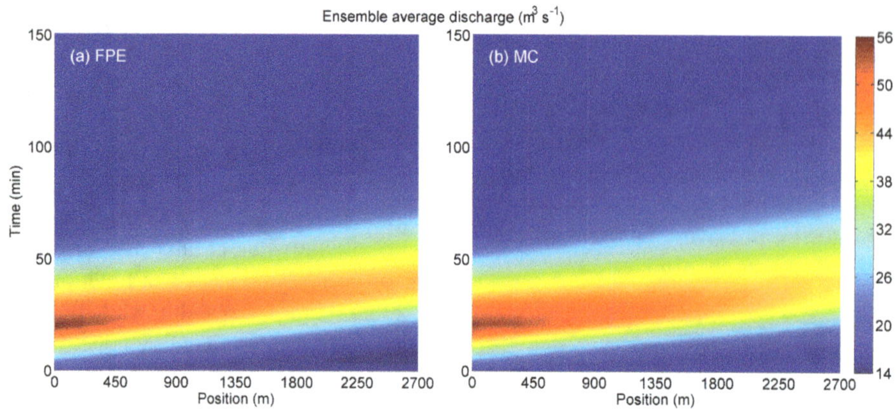

Figure 1. Ensemble average discharge over channel position and time obtained by **(a)** the FPE methodology and **(b)** the MC approach.

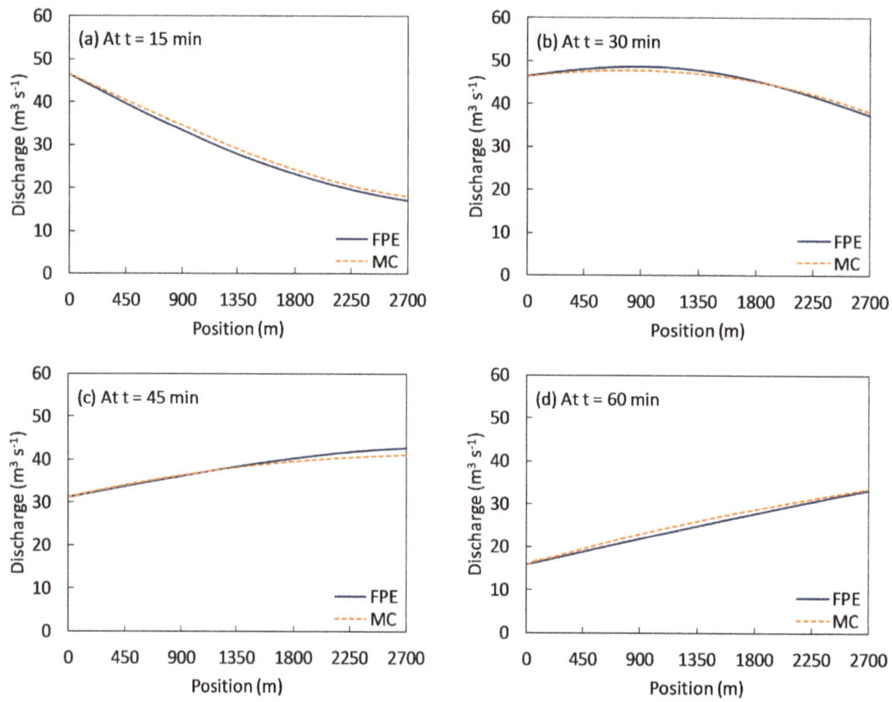

Figure 2. Comparison of the ensemble average discharge obtained by the FPE methodology and the MC approach as a function of channel location (Position), at different times.

Fig. 1 alongside a plot of the same results that were obtained by the MC simulations. From this figure, it is clear that the ensemble average discharge computed by the FPE methodology resembles the one obtained from the MC simulations quite well, while showing the same behavior and evolution of the mean discharge in both time and space as a result of the applied upstream wave. From the average behavior of the system, both plots show that the wave that was initiated upstream is observed to be transmitted downstream through the reach. In this case, the discharge at every location is seen to increase from the initial value of $15.5\,m^3\,s^{-1}$ to some peak value, after which it decreases back again to the initial flow value. However, it is noticeable in both plots that the average peak discharge becomes lower at locations further downstream. This decrease comes as a result of the dissipation of energy through viscous effects as the water flows along the reach (Jeppson, 2011). In fact, the shearing stresses due to the vertical velocity gradients within the water lead to a loss of the fluid energy (kinetic and potential) as non-recoverable energy (e.g., increase in temperature), causing the peak velocity and discharge to decrease along the reach.

For a clearer comparison between the FPE and MC mean discharge results, cross sections of both plots from Fig. 1 at specific times (Fig. 2) and at specific channel locations (Fig. 3) were compared individually. From Fig. 2, it is clear that the change in the ensemble average discharge as a func-

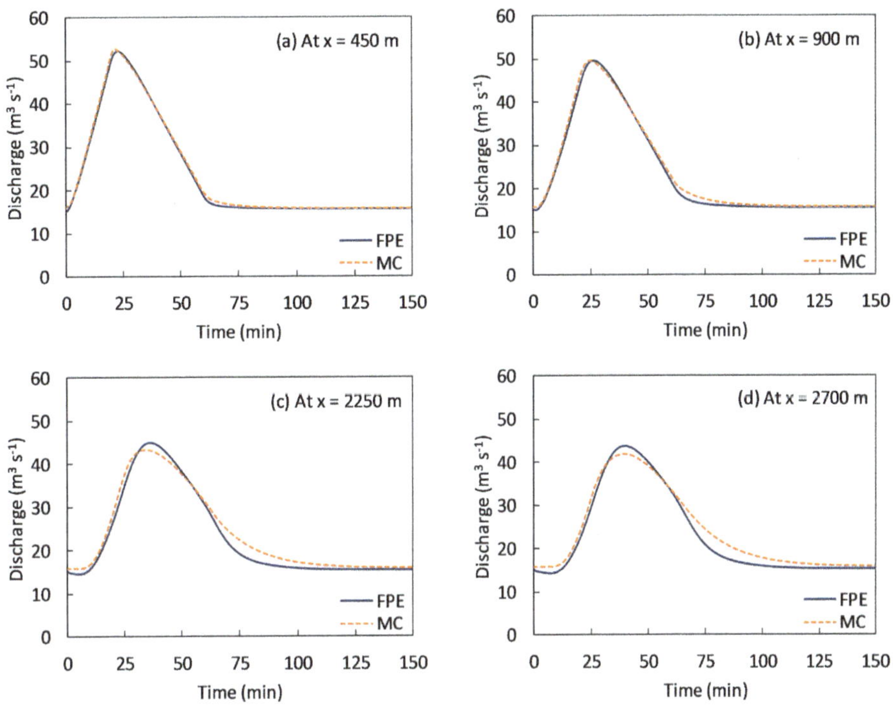

Figure 3. Comparison of the ensemble average discharge obtained by the FPE methodology and the MC approach as a function of time, at different channel locations.

tion of position (location within the channel reach along the flow direction) as computed by the FPE methodology almost perfectly replicates the corresponding results obtained by the MC simulations. In both methods, the effect of the wave is clear as it causes an increase in the discharge first upstream, and then throughout the reach over time. Comparing the mean discharge at specific positions over time (Fig. 3) reveals that the FPE methodology also predicts the temporal change in the mean discharge very well at different channel locations. While a slight decrease in the discharge can be seen at early times of Fig. 3c and d, which may be attributed to the numerical discretization, the timing and positions of the peak discharges predicted by the FPE methodology are very similar to those of the MC approach in all plots of Fig. 3, with a maximum relative difference of only around 6 %.

Similar results were also plotted for the ensemble average flow depth. Figure 4 shows the mean river channel flow profile at different times as computed by the FPE and the MC methods. This figure shows how the applied upstream wave affects the depth profile of the river at different times, which is similarly predicted in both the FPE and MC simulations. Compared to the MC results, the ensemble average depth is predicted quite well by the FPE methodology, which only slightly overestimates the mean depth values at times between $t = 30$ min and $t = 45$ min. A comparison of the changes in depth as a function of time (Fig. 5) reveals that the FPE methodology provides a good match to the MC results while showing the same peaking pattern as the wave reaches a specific channel location. The timings of the peaks of the FPE methodology greatly match those of the MC simulations. Similarly to Fig. 4, a slight overestimation can be noticed from the FPE methodology, especially around the peak depths, but with a maximum relative difference of only around 7.5 %.

As for the velocity, Figs. 6 and 7 show the same two kinds of plots as before by comparing the mean ensemble average velocity results of the FPE methodology to those of the MC approach. Figure 6 shows that the mean velocity computed by the FPE methodology provides a good match to the MC results, with a slight underestimation. The pattern of change in the mean velocity as a function of position that is presented by the FPE methodology is very similar to that shown by the MC simulations, which demonstrates the effects of the applied upstream wave as it travels downstream with time. Figure 7 also reveals a good match between the two modeling approaches, with the FPE methodology providing the same pattern of velocity change over time as computed by the MC simulations. Similarly, a slight underestimation of the FPE mean velocity is also seen in Fig. 7, but the maximum absolute relative difference between both methods was only around 11 %. Therefore, Figs. 1 to 7 show the capability of the FPE methodology in predicting the ensemble averages of the flow variables quite well over space and time, which in turn provides support and validation to the simplifications and assumptions applied to the FPE methodology.

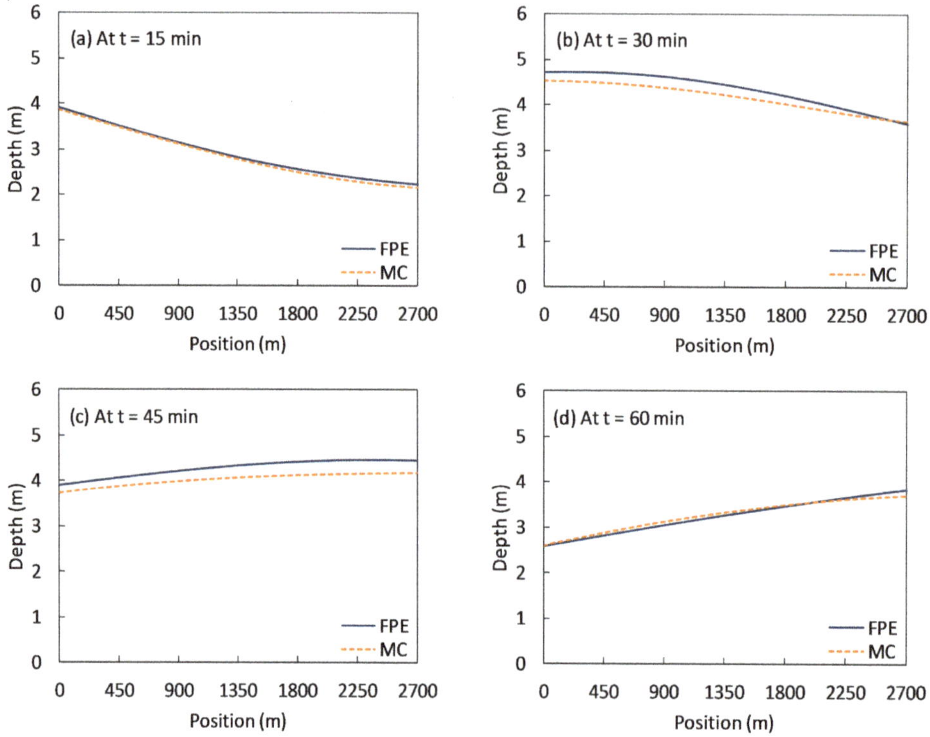

Figure 4. Comparison of the ensemble average flow depth obtained by the FPE methodology and the MC approach as a function of channel location, at different times.

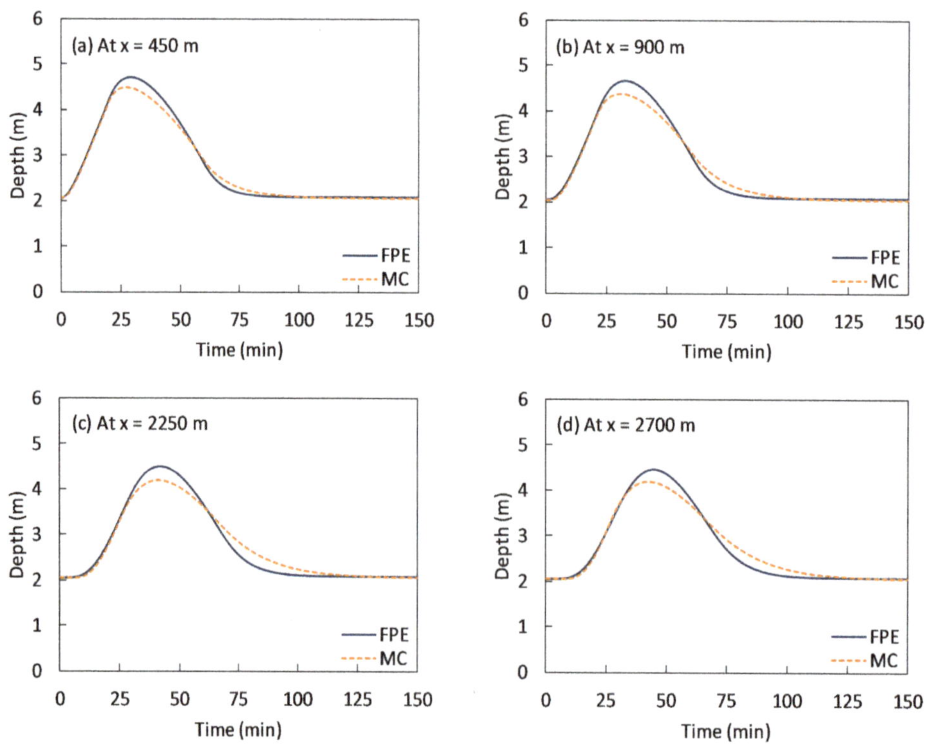

Figure 5. Comparison of the ensemble average flow depth obtained by the FPE methodology and the MC approach as a function of time, at different channel locations.

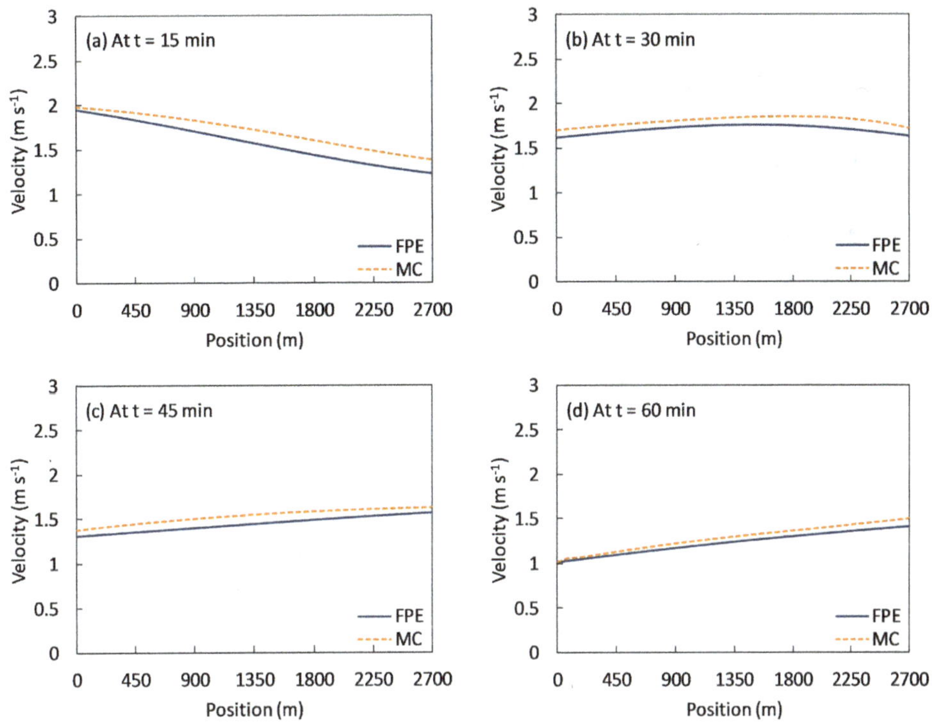

Figure 6. Comparison of the ensemble average velocity obtained by the FPE methodology and the MC approach as a function of channel location, at different times.

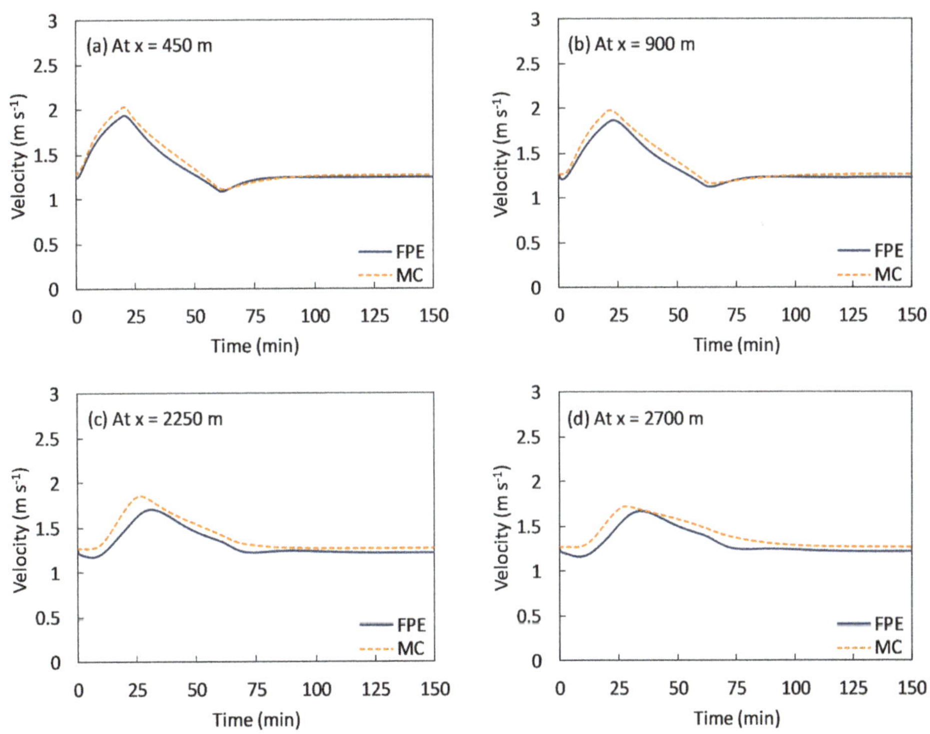

Figure 7. Comparison of the ensemble average velocity obtained by the FPE methodology and the MC approach as a function of time, at different channel locations.

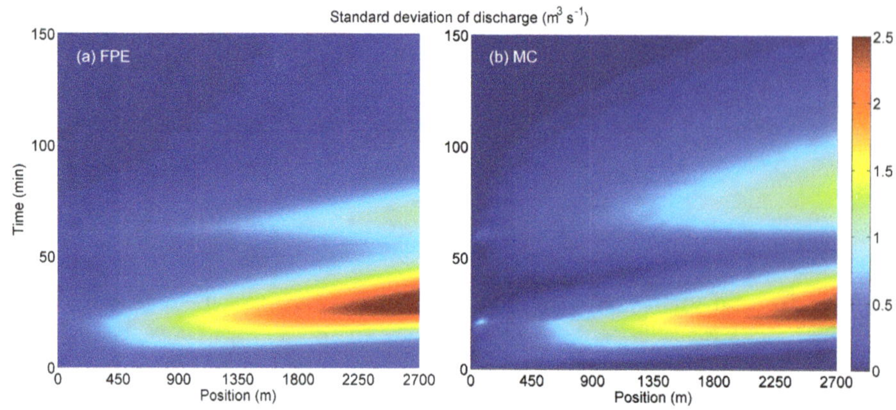

Figure 8. Standard deviation of flow discharge over channel position and time obtained by **(a)** the FPE methodology and **(b)** the MC approach.

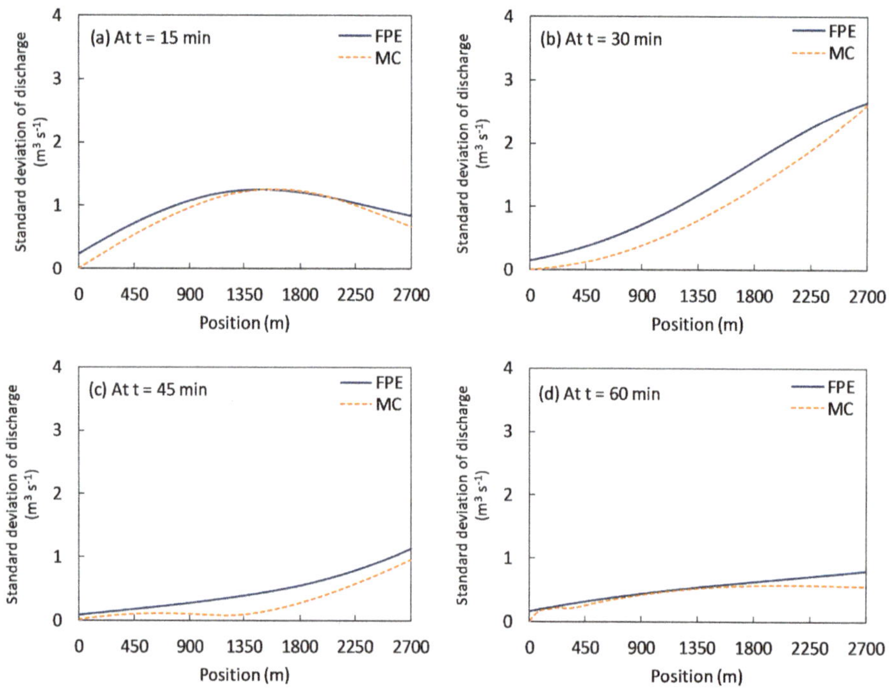

Figure 9. Comparison of the standard deviation of the flow discharge obtained by the FPE methodology and the MC approach as a function of channel location, at different times.

In a similar manner to the ensemble averages, the relative performance of the FPE methodology in predicting the system's variability was examined, this time by checking the standard deviations. Figure 8 shows a comparison of the standard deviation of the flow discharge over space and time as computed by the FPE methodology and by the MC simulations. Both plots of this figure reveal that the standard deviation experiences two triangular areas of high values, the earlier in time being generally higher than the later, and both areas showing a stronger intensity further downstream. While the general resemblance of the FPE plot to the MC plot is good, the second area of elevated standard deviations is more compact in the FPE results and shows slightly lower values than those of the MC results. In an attempt to study such differences more closely, cross sections of both plots from Fig. 8 at specific times and at specific channel locations were compared individually (Figs. 9 and 10), in a manner similar to the ensemble average plots discussed previously.

Looking at the change in the standard deviation of the discharge over position, Fig. 9 reveals how the movement of the wave in the downstream direction causes an increase in the standard deviation of the discharge, thus increasing the variability in the discharge. After the whole channel is affected by the moving wave, the variability is seen to be highest at the downstream end of the channel (e.g., Fig. 9b–d). As the wave subsides, the standard deviation starts to decrease

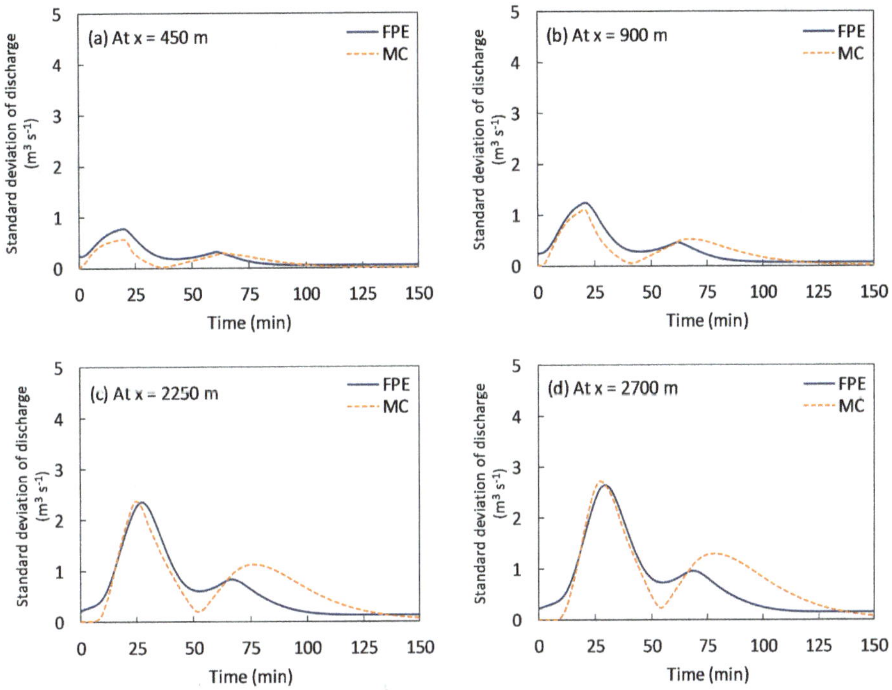

Figure 10. Comparison of the standard deviation of the flow discharge obtained by the FPE methodology and the MC approach as a function of time, at different channel locations.

(Fig. 9c and d). Comparing the results of the FPE methodology to those of the MC approach shows that the FPE results correctly predict the patterns of change in the standard deviation as a function of position at each of the different times (Fig. 9). While the values of the standard deviations are slightly overestimated in some of the FPE results, especially at $t = 30$ min, the FPE results are still quite close to the MC results.

Concerning the changes in the standard deviation of the discharge over time, Fig. 10 shows that the standard deviation forms an M-shaped pattern at all the plotted locations, which can also be deduced from Fig. 8. For the MC results, the magnitudes of the standard deviation forming this M-shaped pattern are observed to be generally increasing as a function of the distance away from the upstream boundary. It is noticeable that the results of the FPE methodology also predict an M-shaped pattern and an increase in the magnitude of the standard deviation of the discharge as a function of location. However, the standard deviation of the FPE methodology shows an offset at time $t = 0$ when compared to the MC simulations, which can also be noticed at the upstream positions of Fig. 9. Recall that the initial and upstream flow discharge is assumed to be known. Nonetheless, a single known value of the flow discharge, when joined with a spread of roughness coefficients due to the uncertainty involved, leads to an unavoidable spread in the velocity and depth values. Since α and β are functions of the velocity and celerity (and in turn, depth), this spread is translated onto the α–β plane of the FPE methodology. As a result, with an uncertain roughness coefficient, the only way to numerically represent a deterministic discharge in the α–β plane is to have a spread of probability mass over the values involved. The existence of this spread on a non-continuous, discretized α–β plane may have had the most contribution to the offset of the standard deviation at the initial times and positions of Figs. 9 and 10.

Nonetheless, when looking at the standard deviation after the initial times, one can see that the timings, shapes, and magnitudes of the first peak for the MC results are generally matched well by the FPE results as shown by the plots of Fig. 10. Moreover, the locations and magnitudes of the second peak are well represented by the FPE results in Fig. 10a and b. However, differences arise in the position and magnitude of the second peak for locations further downstream (Fig. 10c and d). In fact, while moving further downstream, the second peak is observed to be underestimated by the FPE results and slightly shifted back in time when compared to the MC simulations. Moreover, the standard deviation following the second peak shows a faster decline in time for the FPE methodology than for the MC approach, while the local minimum between the two peaks is seen to be overestimated by the FPE results.

Several factors may have caused the discrepancies in the standard deviation of the discharge as computed by the FPE methodology. Among those factors may be the approximations and assumptions that were applied in order to simplify the FPE methodology to an easily solvable degree. These approximations have simplified the computations of the drift and diffusion coefficients of the equation, thus causing some discrepancies in the movement and the spread of the probability mass within the domain. Other factors may include the numerical method selected for computing the FPE methodology, as well as the associated spatial and temporal discretizations that were used. All of these may have had an effect on the representation of the probability flow within the domain, thus leading to some discrepancies in the variability results. Such discrepancies may be faced when using the FPE methodology in engineering flow applications, such as flood forecasting and flood control. The variability of the flow in flood forecasting applications, for example, may be underestimated at the downstream end of the reach, specifically during the later time periods. This would impact the range of flows that are forecasted to occur at the downstream end.

Nonetheless, it may still be inferred that the FPE methodology performs satisfactorily in predicting the variability of the discharge in both space and time. In fact, the problem being solved is highly nonlinear, and involves a large uncertainty in one of its parameters. As such, estimations in the mean and variance are even more difficult to accurately quantify. The FPE methodology was capable of not only correctly matching the general patterns of the spatial and temporal changes in the discharge standard deviation but also correctly providing the standard deviation values within a range that is very similar to the range computed by the MC simulations.

Concerning the standard deviations of the velocity and depth, it is important to note that their behavior over position is somewhat different from that of the flow discharge. In fact, the standard deviations of the velocity at the same four time positions of Fig. 9 seem to be relatively constant at each time position, having a value between 0.015 and $0.02\,\mathrm{m\,s^{-1}}$. On the other hand, the standard deviation of the depth showed a greater range of values at each time position, as a function of location, with values ranging between 0.15 and 0.5 m. When looking at the standard deviations as a function of time at the same four locations of Fig. 10, both standard deviations seem to show that their values increase to reach a maximum and then decrease to levels similar to original levels, not unlike their corresponding ensemble average plots over time. Again, the range of change in the standard deviation of the velocity is much smaller (0.015 to $0.02\,\mathrm{m\,s^{-1}}$) than that of the depth (0.15 to 0.5 m). Note that the relative differences in the FPE results when compared to the MC results reach up to 23 and 29 % for velocity and depth, respectively.

A final comparison between the FPE methodology and the MC approach is to compare the probability density functions (PDFs) of the flow discharge at different times and locations. The large number of simulations (1000) done using the MC approach provided an equal number of flow discharge results at each specific x–t position in the space–time plane. These values were then used in order to estimate the PDF of the discharge at that specific x–t position. The FPE methodology, on the other hand, directly solved for the evolution of the PDFs of the state variables through time and space. However, recall that the FPE was solved in the x–α–β domain. With the discharge being a function of both α and β, the PDF of the discharge at each specific x–t position was deduced from the x–α–β PDF provided by the FPE methodology.

For comparison, the PDFs of the flow discharge obtained from the FPE and MC methods were plotted at three different channel locations ($x = 900$, 2250, and 2700 m) and at four different times for each location ($t = 15$, 30, 45, and 60 min), as shown in Fig. 11. For most of the plots, the peak values of the PDFs predicted by the FPE methodology are similar to those computed by the MC approach. However, taking Fig. 11g as an example, while the peak of the MC PDF has a value of around 0.35, that of the FPE methodology has a value of around 0.065. This is because the PDF of the FPE methodology has a larger spread, which causes the reduction in its PDF values to preserve an area of unity. It may be noted that this plot corresponds to a point close to the local minimum of the plot in Fig. 10b, in which the FPE methodology is seen to predict a larger standard deviation than the MC approach. Therefore, while both of the PDFs, corresponding to the FPE and MC approaches, seem to be located within the same range and thus providing similar expected values of the discharge, the greater variability of the FPE PDF causes its lower peak values.

Many different factors may have affected the evolution and calculation of the PDFs for the flow discharge in space and time. Such factors may even include the step sizes in the α and β directions whose values directly affect the computation of the discharge PDFs. However, the PDFs predicted by the FPE methodology are generally seen to be following similar trends to those computed by the MC approach (Fig. 11) while also satisfactorily predicting the ranges and locations of the PDFs (along the x-axis). Thus, with all the variability encompassing the routing problem considered, and with all the assumptions and approximations used during the application of the FPE methodology, the probability densities predicted by the proposed FPE methodology are considered to be rather encouraging.

It should be noted that these FPE results required significantly less time for computation as opposed to the MC results. Recall that the 1000 MC simulations were parallelized and run over all four cores (with no hyper-threading), thus noticeably reducing the computational time as compared to an un-parallelized run. With such parallelization, the MC simulations ran for over 2 days. On the other hand, the results of the FPE methodology, which was not parallelized, were obtained in about 7 h.

If we observe the computational times of the implicit numerical solution of the FPE methodology, the portion of the simulation requiring the greatest time is filling out the coefficient matrix, especially for small α and β discretizations. Parallelizing this portion over the four cores would allow one to considerably reduce the time to fill out the coefficient matrix, thus reducing the total computational time of this

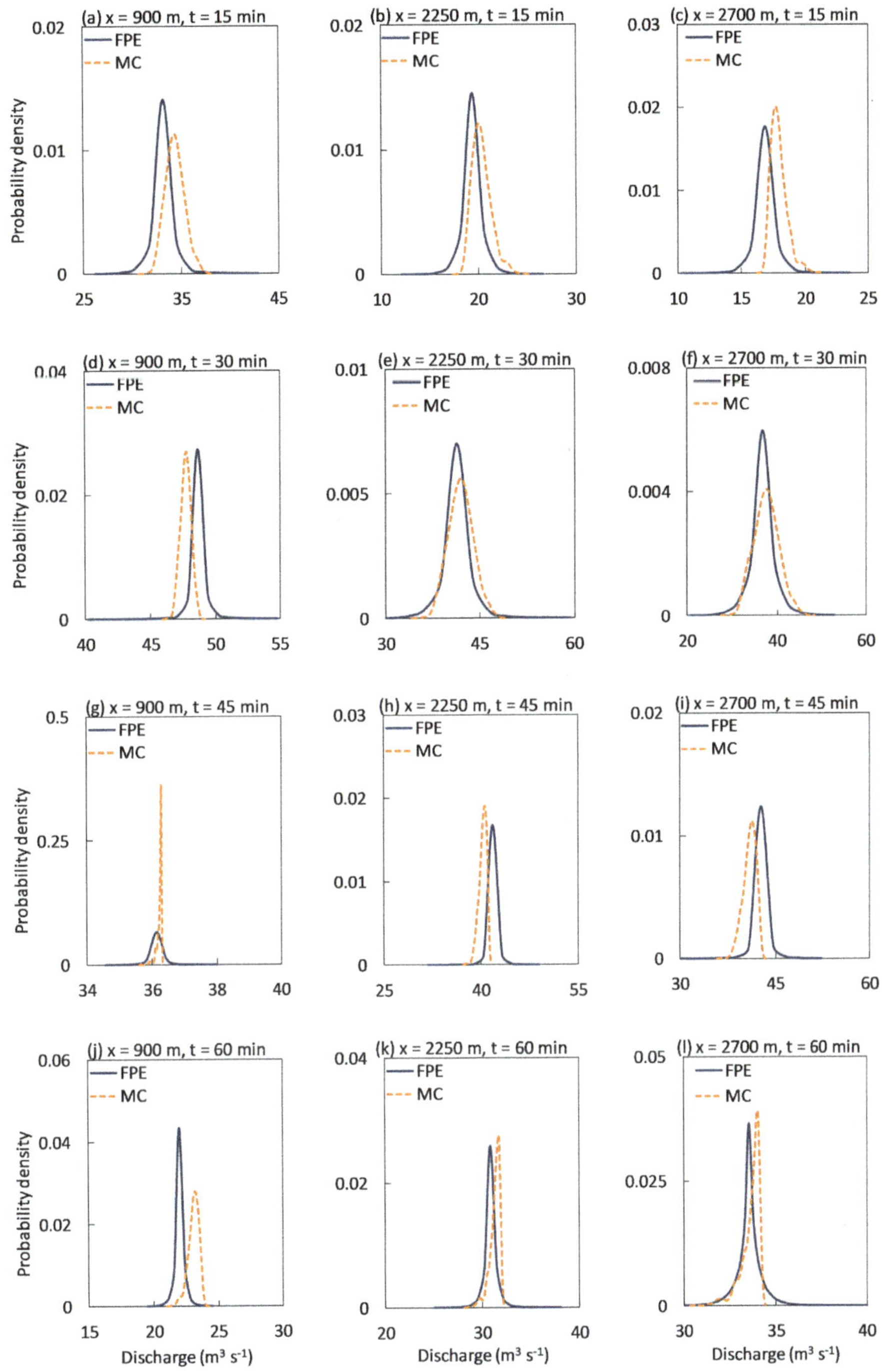

Figure 11. Comparison of the probability density functions of the flow discharge obtained by the FPE methodology and the MC approach, plotted at different times and channel locations.

method. Without the parallelization of the FPE methodology, its one simulation may still not seem to provide an immense advantage when only one uncertain parameter is involved, especially with the possibility of parallelizing the MC simulations among a much larger number of cores. Nonetheless, when the problem being solved involves a greater number of uncertain parameters and boundary conditions, or even a larger system, such an advantage may prove to be crucial. In fact, the computational expense of the MC simulations for such a case would be expected to increase exponentially due to the higher number of simulations needed to maintain the desired accuracy in the results, thus significantly increasing the computational time regardless of parallelization. On the other hand, such additional uncertainties can be easily implemented into the FPE methodology by making simple changes and additions that will be reflected in Eq. (5), after which the FPE would be solved following similar steps as discussed for this study, with minimal implications on the computational expense.

5 Summary and conclusions

This study applied the proposed FPE methodology derived in the companion paper by Dib and Kavvas (2018) to a stochastic unsteady open-channel flow problem, with an uncertain roughness coefficient. The equations used to describe the open-channel flow problem were the Saint-Venant equations, transformed into their characteristic form by the method of characteristics. The proposed FPE methodology was applied in order to solve for the probability density of the flow state variables (velocity and depth/celerity, as well as discharge) and to provide a quantitative description of the expected behavior and variability of the stochastic system in one single simulation, as opposed to the large number of simulations usually performed by the MC approach. The unsteady open-channel flow problem was also solved using the MC approach in order to use its results for evaluating the performance of the proposed FPE methodology.

Comparisons of the FPE results to those of the MC simulations revealed the effectiveness of the proposed FPE methodology in describing the ensemble behavior and variability of the stochastic flow problem. In fact, the FPE methodology was found to replicate the ensemble average discharge of the MC simulations quite well in both space and time. In addition, it was also capable of effectively representing well the temporal and spatial change in the ensemble average depth and the ensemble average velocity. Furthermore, this method provided a good representation of the patterns and ranges of the standard deviation of the discharge over time and space, and showed a satisfactory prediction of the spatial and temporal trends and ranges of the flow discharge PDFs. These encouraging results were obtained despite simplifications applied to the FPE methodology. Furthermore, with the FPE approach, the computational time was significantly less than the time taken by the MC approach, and the FPE methodology results were obtained by running only one simulation, as opposed to the large number of simulations performed by the MC approach. Such an advantage becomes even more prominent with a greater number of uncertain parameters and boundary conditions, in which case the computational expense of the MC simulations that is needed to preserve the desired accuracy would exponentially increase. On the other hand, only simple adjustments would be required for the FPE, which could then be solved as was done in this study, with minor implications on its computational expense. Therefore, the results obtained in this study indicate that the proposed FPE methodology may be a powerful and time-efficient approach for predicting the ensemble average and variance behavior, in both space and time, for the unsteady open-channel flow process under an uncertain roughness coefficient, hence being an approach that would be essential for engineering flow problems.

While the FPE methodology satisfactorily described the ensemble average and variability of the open-channel flow system in this study, this methodology is open to improvements especially with regard to reducing any discrepancies in its numerical results. Running a more comprehensive version of the FPE methodology, by including only some of the simplifying assumptions used in this study, may be one option. Another option may involve using a higher-order and more accurate numerical scheme for the discretization of the multidimensional FPE. As such, numerous opportunities present themselves for future research within this topic, all of which would be of great benefit in the further improvement of the proposed methodology.

Competing interests. The authors declare that they have no conflict of interest.

Edited by: Sabine Attinger

References

Bellin, A., Salandin, P., and Rinaldo, A.: Simulation of dispersion in heterogeneous porous formations - statistics, 1st-order theories, convergence of computations, Water Resour. Res., 28, 2211–2227, https://doi.org/10.1029/92wr00578, 1992.

Cayar, M. and Kavvas, M. L.: Symmetry in nonlinear hydrologic dynamics under uncertainty: ensemble modeling of 2d Boussinesq equation for unsteady flow in heterogeneous aquifers, J. Hydrol. Eng., 14, 1173–1184, https://doi.org/10.1061/(Asce)He.1943-5584.0000112, 2009a.

Cayar, M. and Kavvas, M. L.: Ensemble average and ensemble variance behavior of unsteady, one-dimensional groundwater flow in unconfined, heterogeneous aquifers: an exact second-order model, Stoch. Environ. Res. Risk. A., 23, 947–956, https://doi.org/10.1007/s00477-008-0263-1, 2009b.

Chang, J. S. and Cooper, G.: A practical difference scheme for Fokker–Planck equations, J. Comput. Phys., 6, 1–16, https://doi.org/10.1016/0021-9991(70)90001-X, 1970.

Chaudhry, M. H.: Open-channel flow, in: 2nd Edn., Springer US, New York, 523 pp., 2008.

Chow, V. T.: Open-channel hydraulics, McGraw-Hill civil engineering series, McGraw-Hill, New York, 680 pp., 1959.

Dib, A. and Kavvas, M. L.: Ensemble modeling of stochastic unsteady open-channel flow in terms of its time–space evolutionary probability distribution – Part 1: theoretical development, Hydrol. Earth Syst. Sci., 22, 1993–2005, https://doi.org/10.5194/hess-22-1993-2018, 2018.

Ercan, A. and Kavvas, M. L.: Ensemble modeling of hydrologic and hydraulic processes at one shot: application to kinematic open-channel flow under uncertain channel properties by the stochastic method of characteristics, J. Hydrol. Eng., 17, 168–181, https://doi.org/10.1061/(Asce)He.1943-5584.0000425, 2012a.

Ercan, A. and Kavvas, M. L.: Ensemble modeling of hydrologic and hydraulic processes at one shot: application to kinematic open-channel flow under uncertain channel properties and uncertain lateral flow conditions by the stochastic method of characteristics, J. Hydrol. Eng., 17, 414–423, https://doi.org/10.1061/(Asce)He.1943-5584.0000434, 2012b.

Feller, W.: Diffusion processes in one dimension, T. Am. Math. Soc., 77, 1–31, https://doi.org/10.2307/1990677, 1954.

Freeze, R. A.: Stochastic-conceptual analysis of one-dimensional groundwater flow in nonuniform homogeneous media, Water Resour. Res., 11, 725–741, https://doi.org/10.1029/WR011i005p00725, 1975.

Gardiner, C. W.: Handbook of stochastic methods for physics, chemistry, and the natural sciences, in: 2nd Edn., Springer series in synergetics 13, Springer-Verlag, Berlin, New York, 442 pp., 1985.

Gates, T. K. and AlZahrani, M. A.: Spatiotemporal stochastic open-channel flow. 1. model and its parameter data, J. Hydraul. Eng., 122, 641–651, https://doi.org/10.1061/(Asce)0733-9429(1996)122:11(641), 1996.

Jeppson, R.: Open channel flow: numerical methods and computer applications, CRC Press, Boca Raton, FL, 1258 pp., 2011.

Kavvas, M. L.: Nonlinear hydrologic processes: conservation equations for determining their means and probability distributions, J. Hydrol. Eng., 8, 44–53, https://doi.org/10.1061/(Asce)1084-0699(2003)8:2(44), 2003.

Kim, S., Kavvas, M. L., and Yoon, J.: Upscaling of vertical unsaturated flow model under infiltration condition, J. Hydrol. Eng., 10, 151–159, https://doi.org/10.1061/(Asce)1084-0699(2005)10:2(151), 2005a.

Kim, S., Kavvas, M. L., and Chen, Z. Q.: Root-water uptake model at heterogeneous soil fields, J. Hydrol. Eng., 10, 160–167, https://doi.org/10.1061/(Asce)1084-0699(2005)10:2(160), 2005b.

Liang, L. and Kavvas, M. L.: Modeling of solute transport and macrodispersion by unsteady stream flow under uncertain conditions, J. Hydrol. Eng., 13, 510–520, https://doi.org/10.1061/(Asce)1084-0699(2008)13:6(510), 2008.

Ohara, N., Kavvas, M. L., and Chen, Z. Q.: Stochastic upscaling for snow accumulation and melt processes with PDF approach, J. Hydrol. Eng., 13, 1103–1118, https://doi.org/10.1061/(Asce)1084-0699(2008)13:12(1103), 2008.

Scharffenberg, W. A. and Kavvas, M. L.: Uncertainty in flood wave routing in a lateral-inflow-dominated stream, J. Hydrol. Eng., 16, 165–175, https://doi.org/10.1061/(Asce)He.1943-5584.0000298, 2011.

Smith, L. and Freeze, R. A.: Stochastic-analysis of steady-state groundwater-flow in a bounded domain. 2. two-dimensional simulation, Water Resour. Res., 15, 1543–1559, https://doi.org/10.1029/WR015i006p01543, 1979.

Sturm, T. W.: Open channel hydraulics, McGraw-Hill series in water resources and environmental engineering, McGraw-Hill, Boston, 493 pp., 2001.

Viessman, W., Knapp, J. W., Lewis, G. L., and Harbaugh, T. E.: Introduction to hydrology, in: 2nd Edn., Series in civil engineering, IEP-Dun-Donnelley, Harper & Row, New York, 704 pp., 1977.

Zheng, C. and Bennett, G. D.: Applied contaminant transport modeling, in: 2nd Edn., Wiley-Interscience, New York, 621 pp., 2002.

4

Advantages of analytically computing the ground heat flux in land surface models

Valentijn R. N. Pauwels and Edoardo Daly

Monash University, Department of Civil Engineering, Clayton, Victoria, Australia

Correspondence to: Valentijn R. N. Pauwels (valentijn.pauwels@monash.edu)

Abstract. It is generally accepted that the ground heat flux accounts for a significant fraction of the surface energy balance. In land surface models, the ground heat flux is typically estimated through a numerical solution of the heat conduction equation. Recent research has shown that this approach introduces errors in the estimation of the energy balance. In this paper, we calibrate a land surface model using a numerical solution of the heat conduction equation with four different vertical spatial resolutions. It is found that the thermal conductivity is the most sensitive parameter to the spatial resolution. More importantly, the thermal conductivity values are directly related to the spatial resolution, thus rendering any physical interpretation of this value irrelevant. The numerical solution is then replaced by an analytical solution. The results of the numerical and analytical solutions are identical when fine spatial and temporal resolutions are used. However, when using resolutions that are typical of land surface models, significant differences are found. When using the analytical solution, the ground heat flux is directly calculated without calculating the soil temperature profile. The calculation of the temperature at each node in the soil profile is thus no longer required, unless the model contains parameters that depend on the soil temperature, which in this study is not the case. The calibration is repeated, and thermal conductivity values independent of the vertical spatial resolution are obtained. The main conclusion of this study is that care must be taken when interpreting land surface model results that have been obtained using numerical ground heat flux estimates. The use of exact analytical solutions, when available, is recommended.

1 Introduction

An accurate estimate of the surface energy balance is very important for climate modeling and numerical weather prediction (Orth and Seneviratne, 2014). Of the three components of the net radiation (the latent, sensible and ground heat fluxes), the latent, and sensible heat fluxes provide a direct coupling of the surface energy balance to the atmosphere. For this reason, and also because typically the amplitude of the ground heat flux is smaller than the turbulent fluxes, it can be argued that climate and land surface modelers have paid more attention to an accurate estimation of these fluxes than to the ground heat flux.

However, Gentine et al. (2011) showed that the ground heat flux acts as a high-pass filter because of the strong contrast in the soil- and air-heat capacities and thermal conductivities. Because numerical solutions of the heat conduction equation can miss high-frequency fluctuations, errors in the estimation of the surface energy balance may arise. Gentine et al. (2012) showed that both models and measurements indeed miss these high-frequency fluctuations, and suggested a correction method. Wang and Bou-Zeid (2012) also noted the difficulty of accurately measuring ground heat fluxes, and used an analytical solution of the heat diffusion equation to correct ground heat flux measurements. The problem of ground heat flux estimation errors when using spatially discrete data was also shown by Lunati et al. (2012), who derived an analytical expression for the energy residual, using a cosine boundary condition at the surface. The impact of the ground heat flux parameterization on the energy balance was also studied by Russell et al. (2015). They concluded that the methods that allowed for the most variation in inputs between time steps outperformed the use of diurnal or constant input values. Other studies that intercompared ground heat flux parameterization schemes can be found in Liebethal and Foken (2007) and Venegas et al. (2013).

The problem with the numerical estimation of the ground heat flux in land surface models is that their vertical spatial resolution is too coarse to accurately estimate the soil temperature gradient. This gradient can be very steep near the soil surface, and errors in its estimation are compensated for by adopting fictitious values for the soil thermal parameters (the thermal conductivity and heat capacity). The use of analytical solutions of the heat conduction equation can be expected to partially solve this problem.

A few attempts have been undertaken to derive analytical solutions of the heat conduction equation that can easily be implemented in land surface models. A number of solutions can be found in Carslaw and Jaeger (1959). Shao et al. (1998) solved the equation analytically and compared the solution to temperature observations. Gao et al. (2003) derived an analytical solution as well, in order to determine thermal conductivity values. Cichota et al. (2004) compared analytical and numerical solutions for specific conditions. In an application for large-scale modeling, Bennett et al. (2008) used an analytical solution to estimate the global ground heat flux. Another example can be found in Nunez et al. (2010), where an analytical solution to model the ground heat flux was used. Wang et al. (2012) derived an analytical solution using a sine wave as a boundary condition, and Hu et al. (2016) used the Fourier transform to derive an analytical solution. What these studies have in common is that the solutions have been derived for specific initial and boundary conditions. These solutions also assume vertical homogeneity in the soil thermal parameters, which is very rarely the case. This makes it very difficult to apply them in land surface models, where an analytical expression for the boundary conditions is impossible to determine.

This paper focuses on the estimation of ground heat fluxes and soil thermal properties using a land surface model. It is first examined whether or not calibrated soil thermal properties are independent of the vertical spatial resolution of the model, if the heat conduction equation is solved numerically. An analytical solution of the heat conduction equation is then derived, with temporally varying boundary conditions, which can be applied in the model. Using this analytical solution instead of the numerical approximation, the dependence of the obtained soil thermal properties on the model spatial resolution is then investigated.

2 Site and data description

The data used in this study have been acquired in the framework of the AgriSAR (AGRIcultural bio/geophysical retrieval from frequent repeat pass SAR (Synthetic Aperture Radar) and optical imaging) 2006 campaign, for which the test site was located in Mecklenburg-Vorpommern in northeastern Germany, approximately 150 km North of Berlin. More specifically, time-domain reflectometry (TDR)-based soil moisture observations and Bowen ratio-energy balance (BREB)-based observations of the energy balance components in a large winter wheat field were available from 20 April to 5 July 2006, with the Bowen ratio data containing a number of gaps. The soil moisture was measured at a depth of 5, 9, 15, and 25 cm. Meteorologic data from the weather station at Görmin were available as well and can be used as model forcing from 2005 onwards. All observations were converted to an hourly time step by averaging the 10 min observations. For this study, all model simulations were performed from 1 April 2006 to 5 July 2006, with an hourly time step, unless differently stated. A detailed description of this data set is given in Pauwels et al. (2008).

Table 1. The model parameters that need to be estimated.

Parameter	Description	Units
λ	Pore size distribution index	–
ψ_c	Air entry pressure head	m
K_s	Saturated hydraulic conductivity	$\mathrm{mm\,h^{-1}}$
f	TOPMODEL exponential decay	–
α	Surface albedo	–
κ	Thermal conductivity	$\mathrm{W\,m^{-1}\,K^{-1}}$
C	Volumetric heat capacity	$\mathrm{J\,m^{-3}\,K^{-1}}$
r_c	Surface resistance	$\mathrm{s\,m^{-1}}$
f_d	Zero-plane displacement height fraction	–
f_h	Roughness length for heat transfer fraction	–
f_v	Roughness length for vapor transfer fraction	–

3 Model description

For the purpose of this study, the water and energy balance model developed in Scheerlinck et al. (2009) and applied in Pauwels and De Lannoy (2011) was used. Only a short description will be provided in this section, and for a full model description we refer to Scheerlinck et al. (2009).

The model couples three physical equations. The movement of soil water in the unsaturated zone is modeled using a numerical solution of the Richards equation (Richards, 1931), which results in the profile of the pressure head (ψ; m). This equation requires the evapotranspiration as a boundary condition, which is calculated through an iteration of the surface energy balance, resulting in the surface skin temperature (Shuttleworth, 1992). This skin temperature is then used as a boundary condition for a numerical solution to the heat conduction equation, which results in the soil temperature profile (T; K).

Table 1 lists the eleven parameters that need to be estimated. All parameters are constant in time. The soil parameters are also assumed to be homogeneous throughout the profile. We acknowledge the fact that the model represents a very strong simplification of the physical reality. However, Scheerlinck et al. (2009) and Pauwels and De Lannoy (2011) obtained excellent results for the test site with this model. For this reason, the model is deemed sufficiently realistic.

The model is applied with four different uniform vertical spatial resolutions, namely, 0.01, 0.025, 0.05, and 0.1 m.

4 The parameter estimation algorithm: particle swarm optimization

4.1 General description

The parameter estimation algorithm used in this paper, particle swarm optimization (PSO), is based on the collective behavior of individuals in decentralized, self-organizing systems. These systems are created through a population of individuals that interact locally with each other and with the community. These interactions lead to global behavior, which can result in the achievement of certain objectives. Examples of such systems in nature are abundant: ant colonies, swarms of birds, and schools of fish (Kennedy and Eberhart, 1995). For a complete description of the algorithm we refer to Scheerlinck et al. (2009).

4.2 Application in this study

In order to estimate the model parameters, observations of the net radiation (R_n), latent heat flux (LE), sensible heat flux (H), ground heat flux (G), and soil moisture at 5 (θ_1), 9 (θ_2), 15 (θ_3), and 25 cm (θ_4) are used. The energy balance variables are in W m^{-2}, and the soil moisture values are dimensionless. A global objective function is calculated:

$$\mathrm{OF} = \frac{\mathrm{RMSE}_{R_n}}{\sigma_{R_n}} + \frac{\mathrm{RMSE}_{\mathrm{LE}}}{\sigma_{\mathrm{LE}}} + \frac{\mathrm{RMSE}_{\mathrm{H}}}{\sigma_{\mathrm{H}}} + \frac{\mathrm{RMSE}_{\mathrm{G}}}{\sigma_{\mathrm{G}}} + \frac{\mathrm{RMSE}_{\theta_1}}{\sigma_{\theta_1}} + \frac{\mathrm{RMSE}_{\theta_2}}{\sigma_{\theta_2}} + \frac{\mathrm{RMSE}_{\theta_3}}{\sigma_{\theta_3}} + \frac{\mathrm{RMSE}_{\theta_4}}{\sigma_{\theta_4}}. \quad (1)$$

The RMSE (root mean square error) values for each variable are calculated as

$$\mathrm{RMSE}_x = \sqrt{\frac{1}{N_x}\sum_{i=1}^{N_x}(x_o(i) - x_s(i))^2}, \quad (2)$$

where x_o and x_s are the observed and simulated values, respectively, and σ_x is the standard deviation of the variable. The global objective function is then minimized through the application of PSO; 36 iterations are performed, ensuring convergence of the algorithm, and the method is repeated 24 times. In order to ensure an analysis of the most optimal parameter values, the parameter sets corresponding to the eight lowest objective function values are retained for further analysis. More specifically, the average parameter and objective function values over these eight repetitions are examined.

5 Model calibration using the numerically calculated ground heat flux

The model simulations resulting in the lowest objective function values for the different spatial resolutions will be analyzed in this section. Figure 1 shows the comparison between the modeled energy balance components and the observations, for the spatial resolution of 0.01 m. Table 2 shows the statistics of the linear regressions between the observed and simulated energy balance components for the four spatial resolutions. Figure 2 shows the comparison between the modeled and the observed soil moisture values for the four spatial resolutions. From these figures and tables, it can be concluded that the model adequately simulated the water and energy balance processes, and that the results are very similar for the four different resolutions.

Table 3 shows the parameter values obtained from the model calibration with a spatial resolution of 0.01, 0.025, 0.05, and 0.1 m. In order to determine which parameters are significantly effected by the model spatial resolution, we applied a t test to the slopes from the linear regressions between the spatial resolution (x axis) and the parameter values (y axis), at the 95 % confidence level. In other words, it is tested whether or not there is a significant linear trend between Δz (the model resolution) and the parameter values. The objective function value has been found to change significantly with the resolution, as well as λ, ψ_c, K_s, f, and κ. All the other parameters are not significantly dependent on the spatial resolution. Of all the parameters affected by the resolution, the parameter that shows the largest variation in values is the thermal conductivity κ, with the value at $\Delta z = 0.1$ being more than 4 times the value at $\Delta z = 0.01$ m. No other parameter shows such a dramatic variation.

This result can be explained by the independence of the albedo (α) and the parameters determining the roughness lengths and zero-plane displacement height (f_v, f_h, and f_d) on the model spatial resolution. Since these parameters are similar for the different resolutions, and the model is calibrated using R_n and the three heat fluxes, the resulting skin temperatures are similar for the different resolutions. A similar skin temperature leads to similar sensible and ground heat fluxes, as is proven in Table 2. The ground heat flux is defined as the gradient in the soil temperature multiplied by the thermal conductivity. Since the skin temperatures and the ground heat fluxes (the latter used in the calibration) are equal, but the model spatial resolution is different, the same ground heat flux can only be obtained with a different thermal conductivity. Because the finest spatial resolution will result in the steepest gradient of the soil temperature, it can be expected that this will result in the lowest thermal conductivity value. Table 3 shows that this is indeed the case. This renders the physical interpretation of the thermal conductivity values impossible.

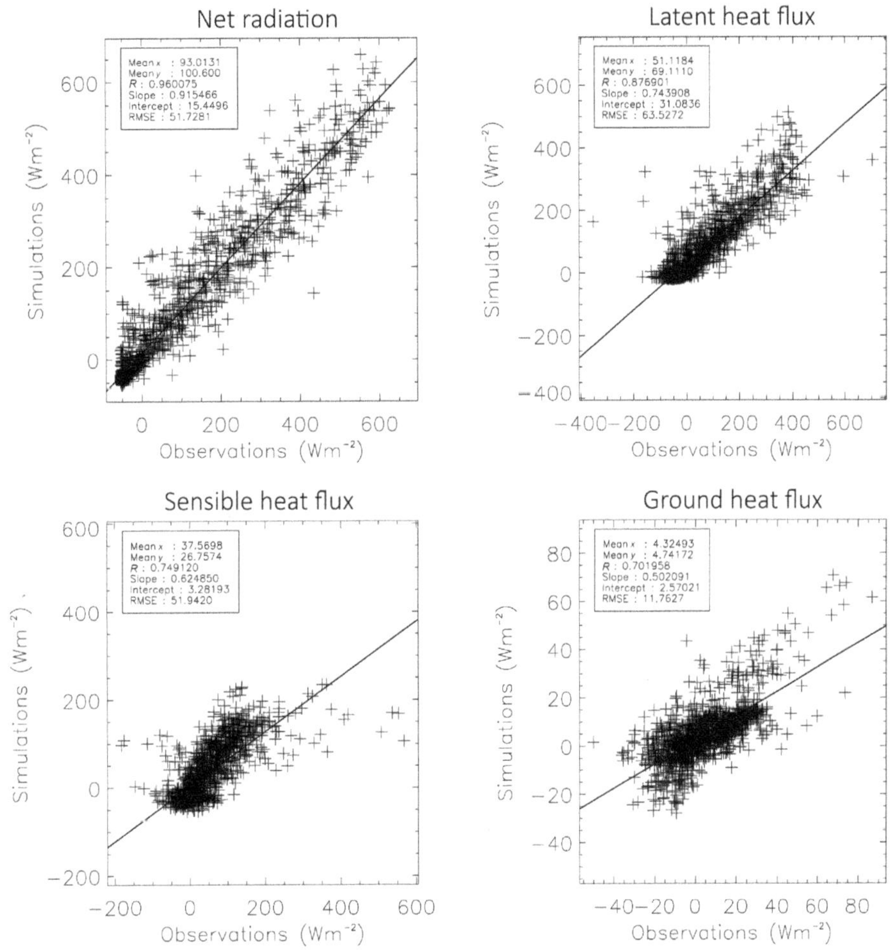

Figure 1. Comparison between the modeled and the observed energy balance terms for the simulation with $\Delta z = 0.01$ m and a numerical solution of the heat conduction equation.

6 Analytical solution of the heat conduction equation

6.1 Derivation of the solution

In order to solve the issue related to the dependence on the grid resolution in the use of a numerical solution of the heat conduction equation, we propose the use of an analytical solution. First, the steady-state temperature profile for a constant temperature at the bottom ($T_{\mathrm{b},0}$) and top ($T_{\mathrm{u},0}$) of the profile is calculated. The depths of the top and bottom of the profile are denoted as z_{u} and z_{b}, respectively. This solution is then used as the initial condition for the same equation, now with different bottom ($T_{\mathrm{b},1}$) and top ($T_{\mathrm{u},1}$) temperatures as boundary conditions. It should be clarified that the time t starts at zero for this new solution. The temperature profile at time Δt is then calculated, and used as the initial condition for the same equation, again with different temperatures ($T_{\mathrm{b},2}$ and $T_{\mathrm{u},2}$) as boundary conditions, and time starting at zero. The solution at time Δt is then again used as the initial condition for the same equation with different boundary conditions ($T_{\mathrm{b},3}$ and $T_{\mathrm{u},3}$) and time starting at zero, and so on.

Using this recursive methodology, the temperature profile for the Mth time step can be written as

$$\begin{aligned}
T(z,t) &= \left(T_{\mathrm{b,M}}-T_{\mathrm{b,0}}\right)\frac{z-z_{\mathrm{u}}}{z_{\mathrm{b}}-z_{\mathrm{u}}}+\left(T_{\mathrm{u,M}}-T_{\mathrm{b,0}}\right)\frac{z-z_{\mathrm{b}}}{z_{\mathrm{u}}-z_{\mathrm{b}}} \\
&+T_{\mathrm{b,0}}+2\left(T_{\mathrm{b,M}}-T_{\mathrm{b,M-1}}\right)\sum_{n=1}^{\infty}\frac{(-1)^n}{x_n}\sin\left(\frac{z-z_{\mathrm{u}}}{z_{\mathrm{b}}-z_{\mathrm{u}}}x_n\right) \\
&e^{y_n t}+2\left(T_{\mathrm{u,M}}-T_{\mathrm{u,M-1}}\right)\sum_{n=1}^{\infty}\frac{(-1)^n}{x_n}\sin\left(\frac{z-z_{\mathrm{b}}}{z_{\mathrm{u}}-z_{\mathrm{b}}}x_n\right) \\
&e^{y_n t}+2\sum_{m=1}^{M-1}\left(T_{\mathrm{b,m}}-T_{\mathrm{b,m-1}}\right)\sum_{n=1}^{\infty}\frac{(-1)^n}{x_n}\sin\left(\frac{z-z_{\mathrm{u}}}{z_{\mathrm{b}}-z_{\mathrm{u}}}x_n\right) \\
&e^{y_n[t+(M-m-1)\Delta t]}+2\sum_{m=1}^{M-1}\left(T_{\mathrm{u,m}}-T_{\mathrm{u,m-1}}\right)\sum_{n=1}^{\infty}\frac{(-1)^n}{x_n} \\
&\sin\left(\frac{z-z_{\mathrm{b}}}{z_{\mathrm{u}}-z_{\mathrm{b}}}x_n\right)e^{y_n[t+(M-m-1)\Delta t]}, \qquad (3)
\end{aligned}$$

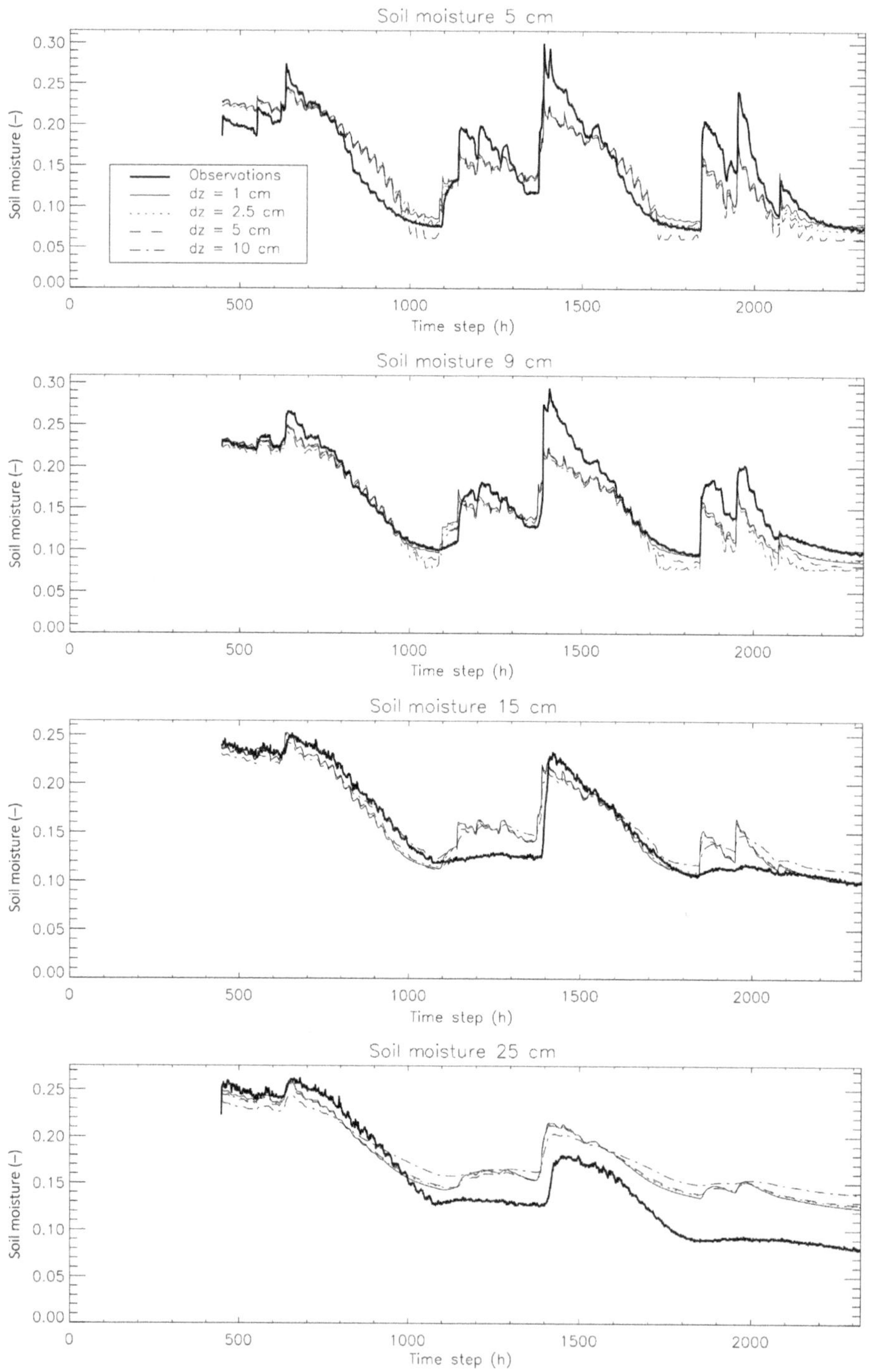

Figure 2. Comparison between the modeled and the observed soil moisture values for the four different spatial resolutions, and a numerical solution of the heat conduction equation. The model results for all four spatial resolutions are very similar and therefore difficult to distinguish.

Table 2. Results of the linear regressions between the energy balance observations (x axis) and the simulations (y axis) for the simulations with a numerical solution of the heat conduction equation. Units are W m^{-2}.

Δz (m)	$\overline{x}$	$\overline{y}$	Slope	Intercept	R	RMSE
			R_n			
0.01	93.0131	100.600	0.915466	15.4496	0.960075	51.7281
0.025		98.3913	0.914448	13.3357	0.960305	51.3149
0.05		95.7262	0.885549	13.3586	0.959990	51.7374
0.1		98.1781	0.902939	14.1929	0.959981	51.5954
			LE			
0.01	51.1184	69.1110	0.743908	31.0836	0.876901	63.5272
0.025		68.2937	0.713683	31.8114	0.867113	65.5609
0.05		65.2262	0.703889	29.2445	0.875136	63.4643
0.1		68.7922	0.739403	30.9951	0.875059	63.8546
			H			
0.01	37.5698	26.7574	0.624850	3.28193	0.749120	51.9420
0.025		25.4533	0.680222	−0.102505	0.757655	52.2046
0.05		26.0201	0.627508	2.44475	0.749720	52.0724
0.1		25.0475	0.617920	1.83234	0.752516	51.9020
			G			
0.01	4.32493	4.74172	0.502091	2.57021	0.701958	11.7627
0.025		4.64757	0.505230	2.46248	0.709196	11.6390
0.05		4.48276	0.483105	2.39337	0.703112	11.7385
0.1		4.34807	0.471090	2.31064	0.702533	11.7558

where $T_{\mathrm{u,M}}$ and $T_{\mathrm{b,M}}$ are the top and bottom temperatures (boundary conditions) for the Mth time step, with $M > 0$, and

$$\begin{cases} x_n = n\pi \\ y_n = -\dfrac{x_n^2 \kappa}{C(z_\mathrm{u} - z_\mathrm{b})^2} \end{cases}. \tag{4}$$

The surface ground heat flux then becomes

$$\begin{aligned} G(t) = {} & \kappa \left.\frac{\partial T}{\partial z}\right|_{z=z_\mathrm{u}} = \kappa \frac{T_{\mathrm{u,M}} - T_{\mathrm{b,M}}}{z_\mathrm{u} - z_\mathrm{b}} \\ & + \frac{2\kappa}{z_\mathrm{u} - z_\mathrm{b}} \left(T_{\mathrm{b,M}} - T_{\mathrm{b,M-1}}\right) \sum_{n=1}^{\infty} (-1)^n e^{y_n t} \\ & + \frac{2\kappa}{z_\mathrm{u} - z_\mathrm{b}} \left(T_{\mathrm{u,M}} - T_{\mathrm{u,M-1}}\right) \sum_{n=1}^{\infty} e^{y_n t} \\ & + \frac{2\kappa}{z_\mathrm{u} - z_\mathrm{b}} \sum_{m=1}^{M-1} \left(T_{\mathrm{b,m}} - T_{\mathrm{b,m-1}}\right) \\ & \sum_{n=1}^{\infty} (-1)^n e^{y_n [t + (M-m-1)\Delta t]} + \frac{2\kappa}{z_\mathrm{u} - z_\mathrm{b}} \\ & \sum_{m=1}^{M-1} \left(T_{\mathrm{u,m}} - T_{\mathrm{u,m-1}}\right) \sum_{n=1}^{\infty} e^{y_n [t + (M-m-1)\Delta t]}. \end{aligned} \tag{5}$$

Appendix A describes the details of the derivation of this solution as well as a methodology to apply these equations in a computationally efficient manner.

It should be noted that with this analytical solution it is no longer necessary to calculate the soil temperature profile in order to calculate the ground heat flux. In the original model formulation, the heat conduction equation needed to be solved numerically, using the surface skin temperature as a boundary condition, so the temperature of the first soil layer could be calculated, and the ground heat flux could be computed. However, Eq. (5) shows that this first layer temperature is no longer a variable in the calculation of the ground heat flux. More complex land surface models may contain parameters that depend on the soil temperature profile, and thus would require the application of Eq. (3). However, in this model, this is not required.

6.2 Comparison to the numerical solution

A synthetic test case is used to intercompare the analytical and the numerical solutions. Equations (3) and (5) are applied to a soil column of 1 m depth, with a thermal conductivity of 0.5 W m^{-1} K^{-1}, and a heat capacity of 2.5×10^6 J m^{-3} K^{-1}. In a first application, the spatial resolution is 0.01 m, and the temporal resolution is 60 s. The numerical solution thus needs to be applied 60 times per 1 h time step in this case.

In a second application, a spatial resolution of 0.05 and a temporal resolution of 1 h (3600 s) are used. The air temperature values from the AgriSAR data set are used as boundary conditions at the top of the profile, and the bottom boundary conditions are assumed to linearly increase from 3 to 13° throughout the simulation.

Figure 3 shows the comparison of the temperature profiles obtained from the numerical and analytical solutions, for the test case with the fine spatial and temporal resolutions. An excellent agreement between both methods can be seen. Figure 4 shows the same comparison for the coarse resolutions. In many of these profiles, especially when sharp changes of temperature occur, a strong deviation of the numerical solution from the analytical solution can be observed. Since these coarse resolutions correspond to values typically used in land surface models, this leads to the conclusion that care must be taken when interpreting ground heat flux simulations from these models. This is demonstrated in Fig. 5, in which the ground heat fluxes from both solutions are compared. For the fine resolutions, a very good agreement is obtained, but relatively strong differences between both methods are found when coarse spatial and temporal resolutions are used.

7 Model calibration using the analytically calculated ground heat flux

Figure 6 shows the comparison between the modeled energy balance terms and the observations, for the simulations with a spatial resolution of 0.01 m, and for the parameter set corresponding to the lowest objective function value. A similar model performance as obtained with the numerical solution of the heat conduction equation is achieved. Table 4 shows the results of the linear regressions between the modeled energy balance terms and the observations, again obtained using the parameter values corresponding to the lowest objective function value, for the four different spatial resolutions. Comparing Tables 4 to 2 leads to the conclusion that the energy balance terms are simulated practically identically when the model uses the numerical or the analytical solution of the heat conduction equation.

Figure 7 shows the comparison between the simulated soil moisture values and the observations for the same parameter sets. The comparison of Figs. 7 and 2 shows that the analytical and numerical solutions of the heat conduction equation lead to very similar simulations of the soil water balance as well.

Table 3 shows the average parameter values from the eight retained PSO results. A slope t test for the linear regressions between the spatial resolutions (x axis) and the parameter values (y axis), at the 95 % confidence level, showed that the objective function value changes significantly with the resolution, as well as λ, ψ_c, K_s, and f. The only difference with the results obtained with the numerical solution of the heat conduction equation is that κ is no longer dependent on

Table 3. Averages of the eight lowest objective function values and the corresponding parameter values.

Parameter	Units	Numerical				Analytical			
		$\Delta z = 0.01$ m	$\Delta z = 0.025$ m	$\Delta z = 0.05$ m	$\Delta z = 0.1$ m	$\Delta z = 0.01$ m	$\Delta z = 0.025$ m	$\Delta z = 0.05$ m	$\Delta z = 0.1$ m
OF	–	3.908	3.894	3.959	4.127	3.981	3.971	4.041	4.210
λ	–	1.068	0.950	0.747	0.499	1.081	0.926	0.740	0.541
ψ_c	m	−0.970	−0.893	−0.717	−0.420	−0.964	−0.854	−0.677	−0.453
K_s	$\mathrm{mm\,h^{-1}}$	190.164	168.582	167.567	141.790	173.642	169.079	150.140	157.642
f	–	17.692	16.738	15.082	9.733	17.353	15.858	13.632	10.302
α	–	0.074	0.074	0.105	0.085	0.077	0.066	0.104	0.108
κ	$\mathrm{W\,m^{-1}\,K^{-1}}$	0.089	0.183	0.269	0.469	0.198	0.189	0.210	0.210
C	$\mathrm{J\,m^{-3}\,K^{-1}}$	2 640 742.750	1 660 719.125	2 967 731.000	2 315 448.750	5 529 063.000	5 318 158.000	5 728 717.000	5 079 298.500
r_c	$\mathrm{s\,m^{-1}}$	130.684	121.139	110.999	107.160	110.621	114.010	101.966	110.756
f_d	–	0.594	0.563	0.528	0.568	0.507	0.610	0.641	0.603
f_h	–	0.152	0.154	0.183	0.175	0.153	0.136	0.184	0.210
f_v	–	0.368	0.316	0.359	0.313	0.241	0.207	0.257	0.347

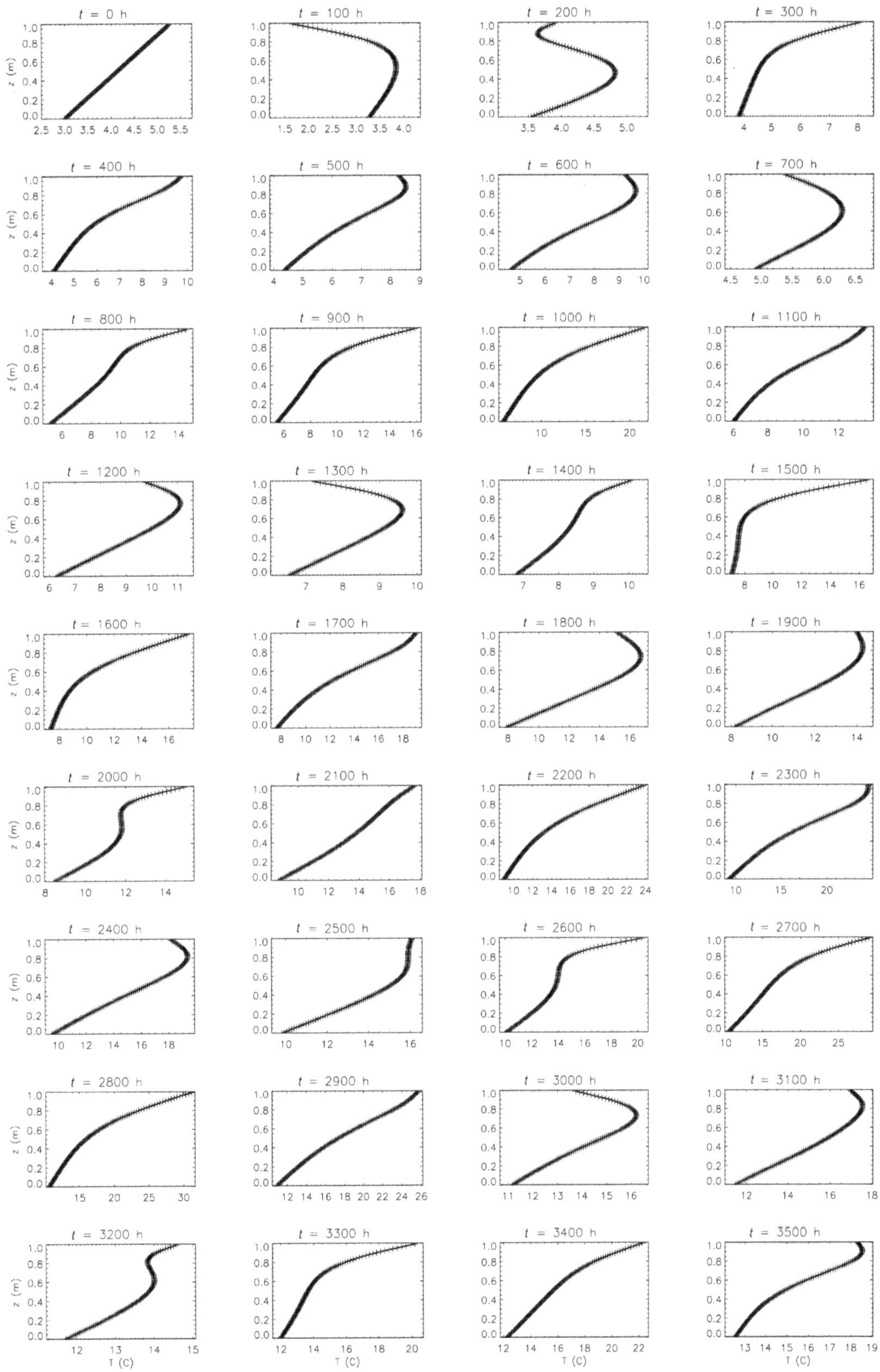

Figure 3. Comparison between soil temperature profiles obtained using the numerical (+) and analytical (solid line) solutions for fine spatial and temporal resolutions. Both solutions are very similar and therefore difficult to distinguish.

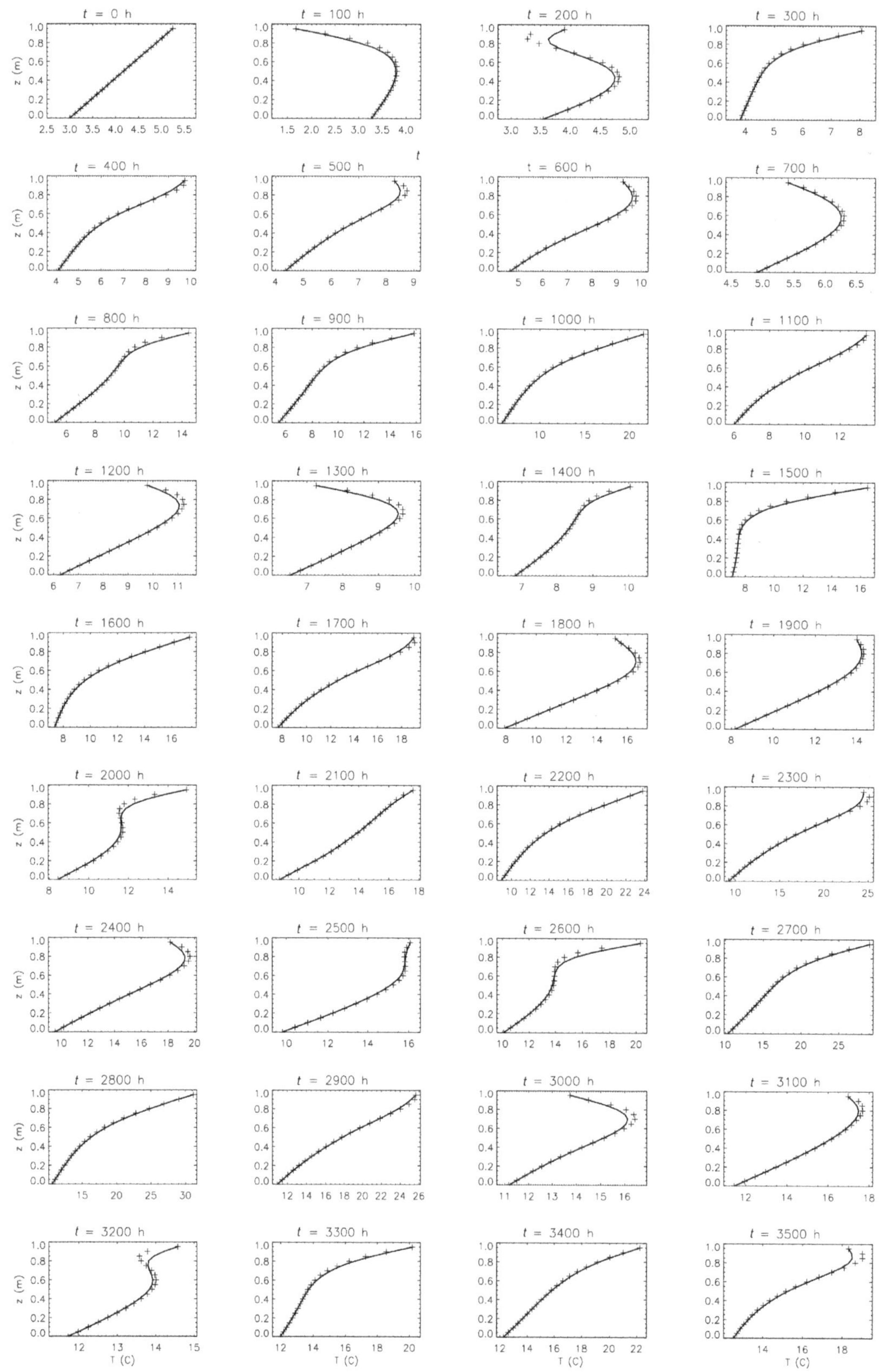

Figure 4. Comparison between soil temperature profiles obtained using the numerical (+) and analytical (solid line) solutions for coarse spatial and temporal resolutions.

Table 4. Results of the linear regressions between the energy balance observations (x axis) and the simulations (y axis) for the simulations with an analytical solution of the heat conduction equation. Units are W m^{-2}.

Δz (m)	$\overline{x}$	$\overline{y}$	Slope	Intercept	R	RMSE
			R_n			
0.01	93.0131	96.8658	0.907984	12.4113	0.960480	51.1205
0.025		97.1449	0.898791	13.5455	0.960282	51.3837
0.05		94.7747	0.889366	12.0521	0.960323	51.4117
0.1		100.133	0.922799	14.3001	0.960243	51.5437
			LE			
0.01	51.1184	69.4063	0.714640	32.8750	0.866276	66.0089
0.025		68.8670	0.731788	31.4592	0.875547	63.8527
0.05		66.1448	0.694242	30.6563	0.870716	64.6836
0.1		74.1509	0.776766	34.4438	0.873384	65.8182
			H			
0.01	37.5698	27.1087	0.672448	1.84497	0.755206	51.9761
0.025		27.8972	0.619521	4.62193	0.748911	51.6733
0.05		28.3065	0.654199	3.72834	0.751075	51.8782
0.1		25.7319	0.611265	2.76678	0.753079	51.6224
			G			
0.01	4.32493	0.341580	0.425440	−1.49842	0.680172	12.7680
0.025		0.357694	0.427544	−1.49141	0.674257	12.8336
0.05		0.338068	0.453744	−1.62434	0.673366	12.8353
0.1		0.323368	0.433978	−1.55356	0.671378	12.8747

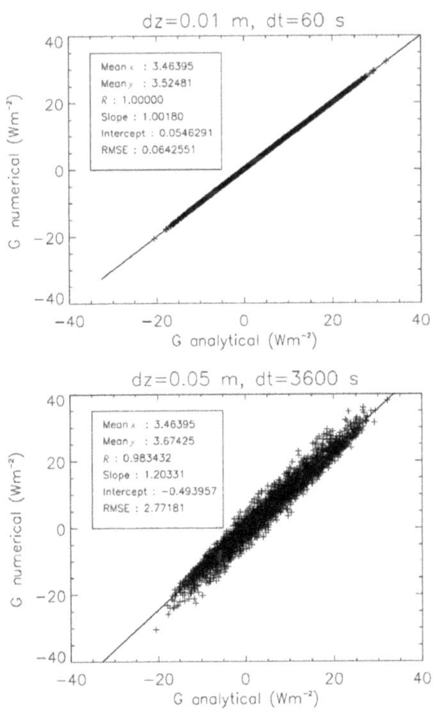

Figure 5. Comparison of the resulting ground heat fluxes from the fine and coarse spatial and temporal resolutions to the analytical solution.

the spatial resolution. This could be expected, because the expression for the ground heat flux is not dependent on the temperature of the first soil layer, and thus on the spatial resolution. Furthermore, the value for the heat capacity (C) is now less variable than when the ground heat flux was calculated numerically. More specifically, the standard deviation has been reduced from 557 943 to 278 984 J m^{-3} K^{-1}.

A pooled variance t test with 95 % confidence was applied to the parameter values obtained with the analytical and numerical solutions, to investigate which parameters are significantly different. This test showed that the only parameters that are significantly different are the objective function value and the heat capacity for all spatial resolutions, and the thermal conductivity for a spatial resolution of 0.01 and 0.1 m. Interestingly, the objective function values obtained with the analytical solution are slightly higher than with the numerical solution. However, since this objective function is composed of eight rescaled RMSE values, the resulting model performance is very similar, as is shown in Table 4 and Fig. 7. Since the thermal conductivity values no longer depend on the spatial resolution, on which they depended when the numerical solution was used, it can be expected that the obtained thermal conductivity values are different. Because of the different manner of solving the heat conduction equation, it can also be expected that the obtained soil-heat capacity values are different as well.

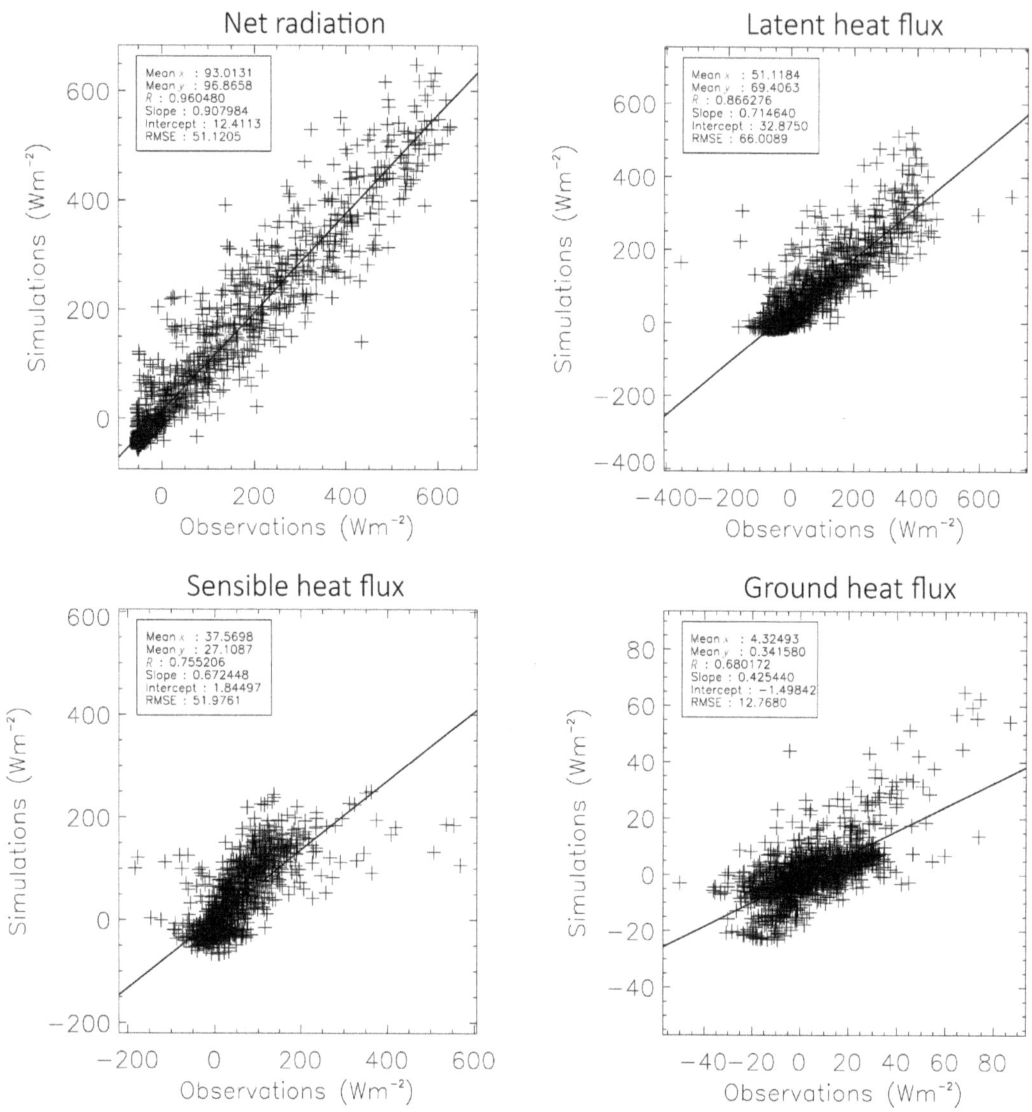

Figure 6. Comparison between the modeled and the observed energy balance terms for the simulation where $\Delta z = 0.01$ m and an analytical solution of the heat conduction equation (Eq. A31).

The key conclusion from these simulations is that the overall model performance is independent of the type of calculation of the ground heat flux (analytically or numerically), but that the results of the model calibration are more robust (i.e., independent of the spatial resolution) if an analytical solution of the ground heat flux equation is used.

8 Conclusions

A water and energy balance model, using a numerical solution of the heat conduction equation, has been calibrated against energy balance and soil moisture observations, for four different vertical spatial resolutions (0.01, 0.025, 0.05, and 0.1 m). It has been found that a number of parameters are dependent on this resolution, with the soil thermal conductivity values showing the largest dependence. An analytical solution of the heat conduction equation has then been derived, allowing for the bottom and top boundary conditions (i.e., the bottom and surface skin temperatures) to vary over time. Using this analytical solution has the advantage that the soil temperature profile no longer needs to be computed. For fine spatial and temporal resolutions the analytical and numerical solutions cannot be distinguished, but different results are obtained for resolutions typically used in land surface models. When the ground heat flux is calculated using this analytical solution, and the model is calibrated, the obtained soil thermal conductivity is no longer dependent on the model spatial resolution. Furthermore, the variability in the obtained soil-heat capacity is also strongly reduced.

The results in this paper indicate that a similar model performance is obtained when the ground heat flux is calculated analytically or numerically. However, the calibration is

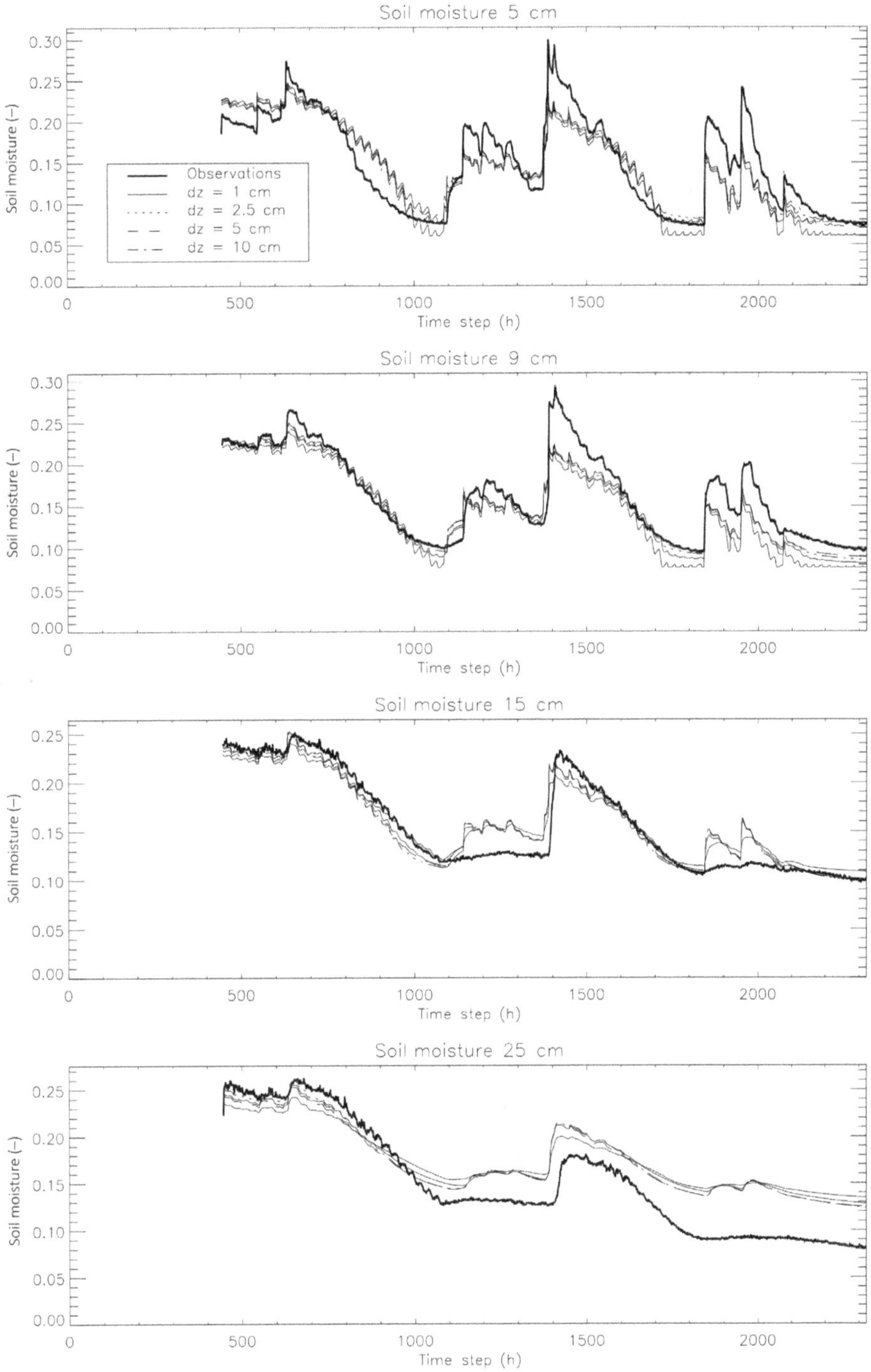

Figure 7. Comparison between the modeled and the observed soil moisture values for the four different spatial resolutions, and an analytical solution of the heat conduction equation.

more robust, and the parameter values more physically interpretable, if the analytical solution is used. One must thus be careful when using numerical solutions of the heat conduction equation in land surface models, and preference should be given to the use of analytical solutions. The solution derived in this paper does not allow for temporally varying soil thermal properties, and ongoing research is focusing on the derivation of an analytical solution that is straightforward to apply in land surface models in these conditions.

9 Data availability

The data can be accessed by contacting the AgriSAR team: http://www.dlr.de/hr/desktopdefault.aspx/tabid-2511/3772_read-5673/.

Appendix A: Analytical solution

A1 The governing equation

A solution of the heat conduction–convection equation is derived first, since this equation is analytically more straightforward to solve than the heat conduction equation because of the easier inversion from the Laplace domain. Furthermore, this general solution can be used for purposes outside the scope of this paper. The limit case with zero convection is then calculated. The governing equation is

$$\frac{\partial CT}{\partial t}=\frac{\partial}{\partial z}\left[\kappa\frac{\partial T}{\partial z}\right]-vC\frac{\partial T}{\partial z}, \tag{A1}$$

where t is time (s), z is the depth positive upwards (m), T is the soil temperature (K), C is the volumetric heat capacity of the soil ($\mathrm{J\,m^{-3}\,K^{-1}}$), κ is the soil thermal conductivity ($\mathrm{W\,m^{-1}\,K^{-1}}$), and v is the water velocity ($\mathrm{m\,s^{-1}}$; positive upwards). We assume the parameters are uniform throughout the profile, so the equation becomes

$$C\frac{\partial T}{\partial t}=\kappa\frac{\partial^2 T}{\partial z^2}-vC\frac{\partial T}{\partial z}. \tag{A2}$$

A2 Steady-state solution

In order to obtain a realistic initial condition, we will calculate the steady-state solution. For example, the profile at the end of a very long, hot day. The equation becomes

$$\kappa\frac{\mathrm{d}^2 T}{\mathrm{d}z^2}-vC\frac{\partial T}{\partial z}=0, \tag{A3}$$

with the boundary conditions

$$\begin{cases} T = T_{\mathrm{b},0}, & z=z_{\mathrm{b}} \\ T = T_{\mathrm{u},0}, & z=z_{\mathrm{u}} \end{cases}. \tag{A4}$$

The solution of this equation is

$$T=T_{\mathrm{b},0}+\left(T_{\mathrm{b},0}-T_{\mathrm{u},0}\right)\frac{e^{\frac{vC}{\kappa}z}-e^{\frac{vC}{\kappa}z_{\mathrm{b}}}}{e^{\frac{vC}{\kappa}z_{\mathrm{b}}}-e^{\frac{vC}{\kappa}z_{\mathrm{u}}}}. \tag{A5}$$

A3 Solution for first new boundary conditions

We will use a constant time steps Δt. For the first new boundary conditions, we will use the steady-state profile as the initial condition. The boundary conditions are

$$\begin{cases} T = T_{\mathrm{b},1}, & z=z_{\mathrm{b}} \\ T = T_{\mathrm{u},1}, & z=z_{\mathrm{u}} \end{cases}. \tag{A6}$$

We will solve this equation through a Laplace transform. We will denote the transform of $T(z,t)$ as $F(z,y)$, with y the Laplace variable. The differential equation becomes

$$CyF-C\left(T_{\mathrm{b},0}+\left(T_{\mathrm{u},0}-T_{\mathrm{b},0}\right)\frac{e^{\frac{vC}{\kappa}z}-e^{\frac{vC}{\kappa}z_{\mathrm{b}}}}{e^{\frac{vC}{\kappa}z_{\mathrm{u}}}-e^{\frac{vC}{\kappa}z_{\mathrm{b}}}}\right)$$

$$=\kappa\frac{\mathrm{d}^2F}{\mathrm{d}z^2}-vC\frac{\mathrm{d}F}{\mathrm{d}z}. \tag{A7}$$

In the Laplace domain, the boundary conditions are

$$\begin{cases} F = \frac{T_{\mathrm{b},1}}{y}, & z=z_{\mathrm{b}} \\ F = \frac{T_{\mathrm{u},1}}{y}, & z=z_{\mathrm{u}} \end{cases}. \tag{A8}$$

The solution of this differential equation is

$$\begin{aligned} F=&\left(T_{\mathrm{b},1}-T_{\mathrm{b},0}\right)\frac{e^{b(z-z_{\mathrm{b}})}}{y}\frac{\sinh\left(\sqrt{b^2+\frac{Cy}{\kappa}}\,(z-z_{\mathrm{u}})\right)}{\sinh\left(\sqrt{b^2+\frac{Cy}{\kappa}}\,(z_{\mathrm{b}}-z_{\mathrm{u}})\right)} \\ &+\left(T_{\mathrm{u},1}-T_{\mathrm{u},0}\right)\frac{e^{b(z-z_{\mathrm{u}})}}{y}\frac{\sinh\left(\sqrt{b^2+\frac{Cy}{\kappa}}\,(z-z_{\mathrm{b}})\right)}{\sinh\left(\sqrt{b^2+\frac{Cy}{\kappa}}\,(z_{\mathrm{u}}-z_{\mathrm{b}})\right)} \\ &+\frac{1}{y}\left(T_{\mathrm{u},0}-T_{\mathrm{b},0}\right)\frac{e^{\frac{vC}{\kappa}z}-e^{\frac{vC}{\kappa}z_{\mathrm{b}}}}{e^{\frac{vC}{\kappa}z_{\mathrm{u}}}-e^{\frac{vC}{\kappa}z_{\mathrm{b}}}}+\frac{T_{\mathrm{b},0}}{y}, \end{aligned} \tag{A9}$$

where b is defined as

$$b=\frac{Cv}{2\kappa}. \tag{A10}$$

We will calculate the poles of this equation (the values for y for which the denominator is zero), and calculate the residuals of each pole. The analytical solution is then simply the sum of the residuals (Brutsaert, 1994). Following this theorem, writing $F(y)$ as $P(y)/T(y)$, we can write

$$R_i=\lim_{y\to y_i}\left[(y-y_i)\frac{P(y)e^{yt}}{T(y)}\right]. \tag{A11}$$

If the pole cannot be factored out, this becomes

$$R_i=\frac{P(y_i)\,e^{y_it}}{\left.\frac{\partial T(y)}{\partial y}\right|_{y=y_i}}. \tag{A12}$$

Equation (A9) has poles for y equal to zero and for the hyperbolic sine equal to zero. For y equal to zero this simply becomes

$$\begin{aligned} T_1(z)=&\left(T_{\mathrm{b},1}-T_{\mathrm{b},0}\right)e^{b(z-z_{\mathrm{b}})}\frac{\sinh\left(b\,(z-z_{\mathrm{u}})\right)}{\sinh\left(b\,(z_{\mathrm{b}}-z_{\mathrm{u}})\right)} \\ &+\left(T_{\mathrm{u},1}-T_{\mathrm{u},0}\right)e^{b(z-z_{\mathrm{u}})}\frac{\sinh\left(b\,(z-z_{\mathrm{b}})\right)}{\sinh\left(b\,(z_{\mathrm{u}}-z_{\mathrm{b}})\right)} \\ &+\left(T_{\mathrm{u},0}-T_{\mathrm{b},0}\right)\frac{e^{\frac{vC}{\kappa}z}-e^{\frac{vC}{\kappa}z_{\mathrm{b}}}}{e^{\frac{vC}{\kappa}z_{\mathrm{u}}}-e^{\frac{vC}{\kappa}z_{\mathrm{b}}}}+T_{\mathrm{b},0}. \end{aligned} \tag{A13}$$

For the hyperbolic sine, we define

$$\sqrt{b^2+\frac{Cy}{\kappa}}\,(z_{\mathrm{u}}-z_{\mathrm{b}})=jx, \tag{A14}$$

where j is the complex variable ($\sqrt{-1}$). We can now write

$$\sinh\left(\sqrt{b^2+\frac{Cy}{\kappa}}\,(z_u-z_b)\right)=j\sin(x). \tag{A15}$$

This is equal to zero for $x=n\pi$, with $n=0$, 1, 2, etc. This means that y is equal to

$$\left(b^2+\frac{Cy_n}{\kappa}\right)(z_u-z_b)^2=-x_n^2\Rightarrow y_n = -\frac{x_n^2\kappa}{C(z_u-z_b)^2}-\frac{\kappa b^2}{C}, \tag{A16}$$

where $x_n=n\pi$. We can thus write the solution as

$$T_2(z,t)=-\left(T_{b,1}-T_{b,0}\right)\frac{2\kappa}{C(z_u-z_b)^2}e^{b(z-z_b)}\sum_{n=1}^{\infty}(-1)^n \frac{x_n}{y_n}\sin\left(\frac{z-z_u}{z_b-z_u}x_n\right)e^{y_nt}-\left(T_{u,1}-T_{u,0}\right)\frac{2\kappa}{C(z_u-z_b)^2}e^{b(z-z_u)}\sum_{n=1}^{\infty}(-1)^n\frac{x_n}{y_n}\sin\left(\frac{z-z_b}{z_u-z_b}x_n\right)e^{y_nt}. \tag{A17}$$

The analytical solution is then simply

$$T(z,t)=T_1(z)+T_2(z,t). \tag{A18}$$

A4 Solution for second new inputs

We will calculate the temperature profile at $t=\Delta t$, and use this as the initial condition for the same equation. The boundary conditions are

$$\left\{\begin{array}{lll} T & = & T_{b,2}, \quad z=z_b \\ T & = & T_{u,2}, \quad z=z_u. \end{array}\right. \tag{A19}$$

In the Laplace domain, the differential equation becomes

$$CyF-C\left(T_{b,1}-T_{b,0}\right)e^{b(z-z_b)}\frac{\sinh(b(z-z_u))}{\sinh(b(z_b-z_u))}-C\left(T_{u,1}-T_{u,0}\right)e^{b(z-z_u)}\frac{\sinh(b(z-z_b))}{\sinh(b(z_u-z_b))}-C\left(T_{u,0}-T_{b,0}\right)\frac{e^{\frac{vC}{\kappa}z}-e^{\frac{vC}{\kappa}z_b}}{e^{\frac{vC}{\kappa}z_u}-e^{\frac{vC}{\kappa}z_b}}-CT_{b,0}+C\left(T_{b,1}-T_{b,0}\right)\frac{2\kappa}{C(z_u-z_b)^2}e^{b(z-z_b)}\sum_{n=1}^{\infty}(-1)^n\frac{x_n}{y_n}\sin\left(\frac{z-z_u}{z_b-z_u}x_n\right)e^{y_n\Delta t}+C\left(T_{u,1}-T_{u,0}\right)\frac{2\kappa}{C(z_u-z_b)^2}e^{b(z-z_u)}\sum_{n=1}^{\infty}(-1)^n\frac{x_n}{y_n}\sin\left(\frac{z-z_b}{z_u-z_b}x_n\right)e^{y_n\Delta t}=\kappa\frac{\mathrm{d}^2F}{\mathrm{d}z^2}-vC\frac{\mathrm{d}F}{\mathrm{d}z}. \tag{A20}$$

The boundary conditions are, in the Laplace domain

$$\left\{\begin{array}{lll} F & = & \frac{T_{b,2}}{y}, \quad z=z_b \\ F & = & \frac{T_{u,2}}{y}, \quad z=z_u \end{array}\right. . \tag{A21}$$

F can then be written as

$$F=\left(T_{b,2}-T_{b,1}\right)\frac{e^{b(z-z_b)}}{y}\frac{\sinh\left(\sqrt{b^2+\frac{Cy}{\kappa}}(z-z_u)\right)}{\sinh\left(\sqrt{b^2+\frac{Cy}{\kappa}}(z_b-z_u)\right)}+\left(T_{u,2}-T_{u,1}\right)\frac{e^{b(z-z_u)}}{y}\frac{\sinh\left(\sqrt{b^2+\frac{Cy}{\kappa}}(z-z_b)\right)}{\sinh\left(\sqrt{b^2+\frac{Cy}{\kappa}}(z_u-z_b)\right)}+\frac{1}{y}\left(T_{b,1}-T_{b,0}\right)e^{b(z-z_b)}\frac{\sinh(b(z-z_u))}{\sinh(b(z_b-z_u))}+\frac{1}{y}\left(T_{u,1}-T_{u,0}\right)e^{b(z-z_u)}\frac{\sinh(b(z-z_b))}{\sinh(b(z_u-z_b))}+\frac{1}{y}\left(T_{u,0}-T_{b,0}\right)\frac{e^{\frac{vC}{\kappa}z}-e^{\frac{vC}{\kappa}z_b}}{e^{\frac{vC}{\kappa}z_u}-e^{\frac{vC}{\kappa}z_b}}+\frac{T_{b,0}}{y}-\left(T_{b,1}-T_{b,0}\right)\frac{2\kappa}{C(z_u-z_b)^2}e^{b(z-z_b)}\sum_{n=1}^{\infty}(-1)^n\frac{x_n}{y_n(y-y_n)}\sin\left(\frac{z-z_u}{z_b-z_u}x_n\right)e^{y_n\Delta t}-\left(T_{u,1}-T_{u,0}\right)\frac{2\kappa}{C(z_u-z_b)^2}e^{b(z-z_u)}\sum_{n=1}^{\infty}(-1)^n\frac{x_n}{y_n(y-y_n)}\sin\left(\frac{z-z_b}{z_u-z_b}x_n\right)e^{y_n\Delta t}. \tag{A22}$$

Through calculating the poles and the residuals, the inverse transform is

$$T(z,t)=\left(T_{b,2}-T_{b,0}\right)e^{b(z-z_b)}\frac{\sinh(b(z-z_u))}{\sinh(b(z_b-z_u))}+\left(T_{u,2}-T_{u,0}\right)e^{b(z-z_u)}\frac{\sinh(b(z-z_b))}{\sinh(b(z_u-z_b))}+\left(T_{u,0}-T_{b,0}\right)\frac{e^{\frac{vC}{\kappa}z}-e^{\frac{vC}{\kappa}z_b}}{e^{\frac{vC}{\kappa}z_u}-e^{\frac{vC}{\kappa}z_b}}+T_{b,0}-\left(T_{b,2}-T_{b,1}\right)\frac{2\kappa}{C(z_u-z_b)^2}e^{b(z-z_b)}\sum_{n=1}^{\infty}(-1)^n\frac{x_n}{y_n}\sin\left(\frac{z-z_u}{z_b-z_u}x_n\right)e^{y_nt}-\left(T_{u,2}-T_{u,1}\right)\frac{2\kappa}{C(z_u-z_b)^2}e^{b(z-z_u)}\sum_{n=1}^{\infty}(-1)^n$$

$$
\begin{aligned}
&\frac{x_n}{y_n}\sin\left(\frac{z-z_\mathrm{b}}{z_\mathrm{u}-z_\mathrm{b}}x_n\right)e^{y_n t} \\
&-\left(T_{\mathrm{b},1}-T_{\mathrm{b},0}\right)\frac{2\kappa}{C\left(z_\mathrm{u}-z_\mathrm{b}\right)^2}e^{b(z-z_\mathrm{b})}\sum_{n=1}^{\infty}(-1)^n \\
&\frac{x_n}{y_n}\sin\left(\frac{z-z_\mathrm{u}}{z_\mathrm{b}-z_\mathrm{u}}x_n\right)e^{y_n(t+\Delta t)} \\
&-\left(T_{\mathrm{u},1}-T_{\mathrm{u},0}\right)\frac{2\kappa}{C\left(z_\mathrm{u}-z_\mathrm{b}\right)^2}e^{b(z-z_\mathrm{u})}\sum_{n=1}^{\infty}(-1)^n \\
&\frac{x_n}{y_n}\sin\left(\frac{z-z_\mathrm{b}}{z_\mathrm{u}-z_\mathrm{b}}x_n\right)e^{y_n(t+\Delta t)}. \qquad \text{(A23)}
\end{aligned}
$$

A5 Solution for M sequential boundary conditions

Through again calculating the temperature profile at time t equal to Δt, using this as input for the governing equation with new boundary conditions, and applying this methodology recursively, the temperature profile for boundary conditions $T_{\mathrm{u,M}}$ and $T_{\mathrm{b,M}}$ as

$$
\begin{aligned}
T(z,t) = &\left(T_{\mathrm{b,M}}-T_{\mathrm{b},0}\right)e^{b(z-z_\mathrm{b})}\frac{\sinh\left(b\left(z-z_\mathrm{u}\right)\right)}{\sinh\left(b\left(z_\mathrm{b}-z_\mathrm{u}\right)\right)} \\
&+\left(T_{\mathrm{u,M}}-T_{\mathrm{u},0}\right)e^{b(z-z_\mathrm{u})}\frac{\sinh\left(b\left(z-z_\mathrm{b}\right)\right)}{\sinh\left(b\left(z_\mathrm{u}-z_\mathrm{b}\right)\right)} \\
&+\left(T_{\mathrm{u},0}-T_{\mathrm{b},0}\right)\frac{e^{\frac{vC}{\kappa}z}-e^{\frac{vC}{\kappa}z_\mathrm{b}}}{e^{\frac{vC}{\kappa}z_\mathrm{u}}-e^{\frac{vC}{\kappa}z_\mathrm{b}}}+T_{\mathrm{b},0} \\
&-\left(T_{\mathrm{b,M}}-T_{b,M-1}\right)\frac{2\kappa}{C\left(z_\mathrm{u}-z_\mathrm{b}\right)^2}e^{b(z-z_\mathrm{b})}\sum_{n=1}^{\infty}(-1)^n \\
&\frac{x_n}{y_n}\sin\left(\frac{z-z_\mathrm{u}}{z_\mathrm{b}-z_\mathrm{u}}x_n\right)e^{y_n t} \\
&-\left(T_{\mathrm{u,M}}-T_{\mathrm{u,M-1}}\right)\frac{2\kappa}{C\left(z_\mathrm{u}-z_\mathrm{b}\right)^2}e^{b(z-z_\mathrm{u})}\sum_{n=1}^{\infty}(-1)^n \\
&\frac{x_n}{y_n}\sin\left(\frac{z-z_\mathrm{b}}{z_\mathrm{u}-z_\mathrm{b}}x_n\right)e^{y_n t} \\
&-\frac{2\kappa}{C\left(z_\mathrm{u}-z_\mathrm{b}\right)^2}e^{b(z-z_\mathrm{b})}\sum_{m=1}^{M-1}\left(T_{\mathrm{b,m}}-T_{\mathrm{b,m-1}}\right) \\
&\sum_{n=1}^{\infty}(-1)^n\frac{x_n}{y_n}\sin\left(\frac{z-z_\mathrm{u}}{z_\mathrm{b}-z_\mathrm{u}}x_n\right)e^{y_n[t+(M-m-1)\Delta t]} \\
&-\frac{2\kappa}{C\left(z_\mathrm{u}-z_\mathrm{b}\right)^2}e^{b(z-z_\mathrm{u})}\sum_{m=1}^{M-1}\left(T_{\mathrm{u,m}}-T_{\mathrm{u,m-1}}\right) \\
&\sum_{n=1}^{\infty}(-1)^n\frac{x_n}{y_n}\sin\left(\frac{z-z_\mathrm{b}}{z_\mathrm{u}-z_\mathrm{b}}x_n\right)e^{y_n[t+(M-m-1)\Delta t]}. \qquad \text{(A24)}
\end{aligned}
$$

Calculating the first derivative with respect to z at z_u and multiplying by the thermal conductivity leads to the ground heat flux:

$$
\begin{aligned}
G(t) = &\,\kappa\left(T_{\mathrm{u,M}}-T_{\mathrm{b,M}}-T_{\mathrm{u},0}+T_{\mathrm{b},0}\right)\frac{be^{b(z_\mathrm{u}-z_\mathrm{b})}}{\sinh\left(b\left(z_\mathrm{u}-z_\mathrm{b}\right)\right)} \\
&+vC\left(T_{\mathrm{u},0}-T_{\mathrm{b},0}\right)\frac{e^{\frac{vC}{\kappa}z_\mathrm{u}}}{e^{\frac{vC}{\kappa}z_\mathrm{u}}-e^{\frac{vC}{\kappa}z_\mathrm{b}}} \\
&\frac{2\kappa^2}{C\left(z_\mathrm{u}-z_\mathrm{b}\right)^3}e^{b(z_\mathrm{u}-z_\mathrm{b})}\sum_{n=1}^{\infty}(-1)^n-\left(T_{\mathrm{b,M}}-T_{\mathrm{b,M-1}}\right) \\
&\frac{x_n^2}{y_n}e^{y_n t}-\left(T_{\mathrm{u,M}}-T_{\mathrm{u,M-1}}\right)\frac{2\kappa^2}{C\left(z_\mathrm{u}-z_\mathrm{b}\right)^3}\sum_{n=1}^{\infty}\frac{x_n^2}{y_n}e^{y_n t} \\
&-\frac{2\kappa^2}{C\left(z_\mathrm{u}-z_\mathrm{b}\right)^3}e^{b(z_\mathrm{u}-z_\mathrm{b})}\sum_{m=1}^{M-1}\left(T_{\mathrm{b,m}}-T_{\mathrm{b,m-1}}\right)\sum_{n=1}^{\infty}(-1)^n \\
&\frac{x_n^2}{y_n}e^{y_n[t+(M-m-1)\Delta t]}-\frac{2\kappa^2}{C\left(z_\mathrm{u}-z_\mathrm{b}\right)^3}\sum_{m=1}^{M-1} \\
&\left(T_{\mathrm{u,m}}-T_{\mathrm{u,m-1}}\right)\sum_{n=1}^{\infty}\frac{x_n^2}{y_n}e^{y_n[t+(M-m-1)\Delta t]}. \qquad \text{(A25)}
\end{aligned}
$$

A6 Computationally efficient formulation

For the temperature profile, we define two variables $\tau_\mathrm{t}(z, n)$ and $\tau_\mathrm{b}(z, n)$ for each value of z, which are initially zero. In general, we write the solution for temperature input M as

$$
\begin{aligned}
T(z,t) = &\left(T_{\mathrm{b,M}}-T_{\mathrm{b},0}\right)e^{b(z-z_\mathrm{b})}\frac{\sinh\left(b\left(z-z_\mathrm{u}\right)\right)}{\sinh\left(b\left(z_\mathrm{b}-z_\mathrm{u}\right)\right)} \\
&+\left(T_{\mathrm{u,M}}-T_{\mathrm{u},0}\right)e^{b(z-z_\mathrm{u})}\frac{\sinh\left(b\left(z-z_\mathrm{b}\right)\right)}{\sinh\left(b\left(z_\mathrm{u}-z_\mathrm{b}\right)\right)} \\
&+\left(T_{\mathrm{u},0}-T_{\mathrm{b},0}\right)\frac{e^{\frac{vC}{\kappa}z}-e^{\frac{vC}{\kappa}z_\mathrm{b}}}{e^{\frac{vC}{\kappa}z_\mathrm{u}}-e^{\frac{vC}{\kappa}z_\mathrm{b}}}+T_{\mathrm{b},0} \\
&-\left(T_{\mathrm{b,M}}-T_{\mathrm{b,M-1}}\right)\frac{2\kappa}{C\left(z_\mathrm{u}-z_\mathrm{b}\right)^2}e^{b(z-z_\mathrm{b})}\sum_{n=1}^{\infty}(-1)^n \\
&\frac{x_n}{y_n}\sin\left(\frac{z-z_\mathrm{u}}{z_\mathrm{b}-z_\mathrm{u}}x_n\right)e^{y_n t}-\left(T_{\mathrm{u,M}}-T_{\mathrm{u,M-1}}\right) \\
&\frac{2\kappa}{C\left(z_\mathrm{u}-z_\mathrm{b}\right)^2}e^{b(z-z_\mathrm{u})}\sum_{n=1}^{\infty}(-1)^n\frac{x_n}{y_n}\sin\left(\frac{z-z_\mathrm{b}}{z_\mathrm{u}-z_\mathrm{b}}x_n\right) \\
&e^{y_n t}-\frac{2\kappa}{C\left(z_\mathrm{u}-z_\mathrm{b}\right)^2}e^{b(z-z_\mathrm{b})}\sum_{n=1}^{\infty}\tau_{\mathrm{b},n}e^{y_n t} \\
&-\frac{2\kappa}{C\left(z_\mathrm{u}-z_\mathrm{b}\right)^2}e^{b(z-z_\mathrm{u})}\sum_{n=1}^{\infty}\tau_{\mathrm{t},n}e^{y_n t}. \qquad \text{(A26)}
\end{aligned}
$$

At the end of the time step we update the variables:

$$
\begin{cases}
\tau_\mathrm{b}(z,n)\rightarrow\tau_\mathrm{b}(z,n)e^{y_n\Delta t}+\left(T_{\mathrm{b,M}}-T_{\mathrm{b,M-1}}\right)(-1)^n\frac{x_n}{y_n}\sin\left(\frac{z-z_\mathrm{u}}{z_\mathrm{b}-z_\mathrm{u}}x_n\right)e^{y_n\Delta t} \\
\tau_\mathrm{t}(z,n)\rightarrow\tau_\mathrm{t}(z,n)e^{y_n\Delta t}+\left(T_{\mathrm{u,M}}-T_{\mathrm{u,M-1}}\right)(-1)^n\frac{x_n}{y_n}\sin\left(\frac{z-z_\mathrm{b}}{z_\mathrm{u}-z_\mathrm{b}}x_n\right)e^{y_n\Delta t}
\end{cases}. \qquad \text{(A27)}
$$

For the ground heat flux, we define two variables $\psi_\mathrm{t}(n)$ and $\psi_\mathrm{b}(n)$, again initially zero. We then write the ground heat flux as

$$
\begin{aligned}
G(t) = &\,\kappa\left(T_{\mathrm{u,M}}-T_{\mathrm{b,M}}-T_{\mathrm{u},0}+T_{\mathrm{b},0}\right)\frac{be^{b(z_\mathrm{u}-z_\mathrm{b})}}{\sinh\left(b\left(z_\mathrm{u}-z_\mathrm{b}\right)\right)} \\
&+vC\left(T_{\mathrm{u},0}-T_{\mathrm{b},0}\right)\frac{e^{\frac{vC}{\kappa}z_\mathrm{u}}}{e^{\frac{vC}{\kappa}z_\mathrm{u}}-e^{\frac{vC}{\kappa}z_\mathrm{b}}}
\end{aligned}
$$

$$
\begin{aligned}
&-\left(T_{\mathrm{b,M}}-T_{\mathrm{b,M-1}}\right)\frac{2\kappa^2}{C\left(z_{\mathrm{u}}-z_{\mathrm{b}}\right)^3}e^{b(z_{\mathrm{u}}-z_{\mathrm{b}})}\sum_{n=1}^{\infty}(-1)^n\\
&\frac{x_n^2}{y_n}e^{y_nt}-\left(T_{\mathrm{u,M}}-T_{\mathrm{u,M-1}}\right)\frac{2\kappa^2}{C\left(z_{\mathrm{u}}-z_{\mathrm{b}}\right)^3}\sum_{n=1}^{\infty}\frac{x_n^2}{y_n}e^{y_nt}\\
&-\frac{2\kappa^2}{C\left(z_{\mathrm{u}}-z_{\mathrm{b}}\right)^3}e^{b(z_{\mathrm{u}}-z_{\mathrm{b}})}\sum_{n=1}^{\infty}\psi_{\mathrm{b}}(n)e^{y_nt}\\
&-\frac{2\kappa^2}{C\left(z_{\mathrm{u}}-z_{\mathrm{b}}\right)^3}\sum_{n=1}^{\infty}\psi_{\mathrm{t}}(n)e^{y_nt}.
\end{aligned}
\tag{A28}
$$

At the end of each time step we then update

$$
\left\{
\begin{array}{l}
\psi_{\mathrm{b}}(n)\rightarrow\psi_{\mathrm{b}}(n)e^{y_n\Delta t}+\left(T_{\mathrm{b,M}}-T_{\mathrm{b,M-1}}\right)(-1)^n\frac{x_n^2}{y_n}e^{y_n\Delta t}\\
\psi_{\mathrm{t}}(n)\rightarrow\psi_{\mathrm{t}}(n)e^{y_n\Delta t}+\left(T_{\mathrm{u,M}}-T_{\mathrm{u,M-1}}\right)\frac{x_n^2}{y_n}e^{y_n\Delta t}
\end{array}
\right. .
\tag{A29}
$$

A7 Limit case for zero convection

In this case b becomes zero. We can rewrite Eq. (A24) as (using de l'Hôpitals's rule)

$$
\begin{aligned}
T(z,t)=&\left(T_{\mathrm{b,M}}-T_{\mathrm{b,0}}\right)\frac{z-z_{\mathrm{u}}}{z_{\mathrm{b}}-z_{\mathrm{u}}}+\left(T_{\mathrm{u,M}}-T_{\mathrm{b,0}}\right)\frac{z-z_{\mathrm{b}}}{z_{\mathrm{u}}-z_{\mathrm{b}}}\\
&+T_{\mathrm{b,0}}+2\left(T_{\mathrm{b,M}}-T_{\mathrm{b,M-1}}\right)\sum_{n=1}^{\infty}\frac{(-1)^n}{x_n}\sin\left(\frac{z-z_{\mathrm{u}}}{z_{\mathrm{b}}-z_{\mathrm{u}}}x_n\right)\\
&+e^{y_nt}2\left(T_{\mathrm{u,M}}-T_{\mathrm{u,M-1}}\right)\sum_{n=1}^{\infty}\frac{(-1)^n}{x_n}\sin\left(\frac{z-z_{\mathrm{b}}}{z_{\mathrm{u}}-z_{\mathrm{b}}}x_n\right)\\
&e^{y_nt}+2\sum_{m=1}^{M-1}\left(T_{\mathrm{b,m}}-T_{\mathrm{b,m-1}}\right)\sum_{n=1}^{\infty}\frac{(-1)^n}{x_n}\\
&\sin\left(\frac{z-z_{\mathrm{u}}}{z_{\mathrm{b}}-z_{\mathrm{u}}}x_n\right)e^{y_n[t+(M-m)\Delta t]}+2\sum_{m=1}^{M-1}\\
&\left(T_{\mathrm{u,m}}-T_{\mathrm{u,m-1}}\right)\sum_{n=1}^{\infty}\frac{(-1)^n}{x_n}\sin\left(\frac{z-z_{\mathrm{b}}}{z_{\mathrm{u}}-z_{\mathrm{b}}}x_n\right)\\
&e^{y_n[t+(M-m)\Delta t]}.
\end{aligned}
\tag{A30}
$$

The ground heat flux then becomes

$$
\begin{aligned}
G(t)=&\kappa\frac{T_{\mathrm{u,M}}-T_{\mathrm{b,M}}}{z_{\mathrm{u}}-z_{\mathrm{b}}}+\frac{2\kappa}{z_{\mathrm{u}}-z_{\mathrm{b}}}\left(T_{\mathrm{b,M}}-T_{\mathrm{b,M-1}}\right)\sum_{n=1}^{\infty}(-1)^n\\
&e^{y_nt}+\frac{2\kappa}{z_{\mathrm{u}}-z_{\mathrm{b}}}\left(T_{\mathrm{u,M}}-T_{\mathrm{u,M-1}}\right)\sum_{n=1}^{\infty}e^{y_nt}+\frac{2\kappa}{z_{\mathrm{u}}-z_{\mathrm{b}}}\\
&\sum_{m=1}^{M-1}\left(T_{\mathrm{b,m}}-T_{\mathrm{b,m-1}}\right)\sum_{n=1}^{\infty}(-1)^ne^{y_n[t+(M-m)\Delta t]}\\
&+\frac{2\kappa}{z_{\mathrm{u}}-z_{\mathrm{b}}}\sum_{m=1}^{M-1}\left(T_{\mathrm{u,m}}-T_{\mathrm{u,m-1}}\right)\sum_{n=1}^{\infty}e^{y_n[t+(M-m)\Delta t]}.
\end{aligned}
\tag{A31}
$$

The computationally efficient algorithm can be applied to these limit cases as well.

Author contributions. Edoardo Daly and Valentijn R. N. Pauwels developed the idea to use an analytical solution to model the ground heat flux in land surface models, instead of numerical solutions. Valentijn R. N. Pauwels solved the equation, prepared the manuscript, and performed the model simulations. The results were extensively discussed and analyzed by both authors.

Acknowledgements. Valentijn R. N. Pauwels is funded through Australian Research Council Future Fellowship grant number FT130100545.

Edited by: P. Gentine

References

Bennett, W. B., Wang, J. F., and Bras, R. L.: Estimation of global ground heat flux, J. Hydrometeorol., 9, 744–759, 2008.

Brutsaert, W.: The unit response of groundwater outflow from a hillslope, Water Resour. Res., 30, 2759–2763, 1994.

Carslaw, H. S. and Jaeger, J. C.: Conduction of heat in solids, Clarendon Press, Oxford, 1959.

Cichota, R., Elias, E. A., and van Lier, Q. D.: Testing a finite-difference model for soil heat transfer by comparing numerical and analytical solutions, Environ. Model. Softw., 19, 495–506, 2004.

Gao, Z. Q., Fan, X. G., and Bian, L. G.: An analytical solution to one-dimensional thermal conduction-convection in soil, Soil Sci., 168, 99–107, doi:10.1097/00010694-200302000-00004, 2003.

Gentine, P., Polcher, J., and Entekhabi, D.: Harmonic propagation of variability in surface energy balance within a coupled soil–vegetation–atmosphere system, Water Resour. Res., 48, W05525, doi:10.1029/2010WR009268, 2011.

Gentine, P., Entekhabi, D., and Heusinkveld, B.: Systematic errors in ground heat flux estimation and their correction, Water Resour. Res., 48, W09541, doi:10.1029/2010WR010203, 2012.

Hu, G., Zhao, L., Wu, X., Li, R., Wu, T., Xie, C., Qiao, Y., Shi, J., Li, W., and Cheng, G.: New Fourier-series-based analytical solution to the conduction-convection equation to calculate soil temperature, determine soil thermal properties, or estimate water flux, Int. J. Heat Mass Transf., 95, 815–823, doi:10.1016/j.ijheatmasstransfer.2015.11.078, 2016.

Kennedy, J. and Eberhart, R. C.: Particle Swarm Optimization, in: Proc. IEEE In. Conf. Neur. Netw., Piscataway, NJ, 1942–1948, 1995.

Liebethal, C. and Foken, T.: Evaluation of six parameterization approaches for the ground heat flux, Theor. Appl. Climatol., 88, 43–56, doi:10.1007/s00704-005-0234-0, 2007.

Lunati, I., Ciocca, F., and Parlange, M. B.: On the use of spatially discrete data to compute energy and mass balance, Water Resour. Res., 48, W05542, doi:10.1029/2012WR012061, 2012.

Nunez, C. M., Varas, E. A., and Meza, F. J.: Modelling soil heat flux, Theor. Appl. Climatol., 100, 251–260, doi:10.1007/s00704-009-0185-y, 2010.

Orth, R. and Seneviratne, S.: Using soil moisture forecasts for sub-seasonal summer temperature predictions in Europe, Clim. Dynam., 43, 3403–3418, doi:10.1007/s00382-014-2112-x, 2014.

Pauwels, V. R. N. and De Lannoy, G. J. M.: Multivariate calibration of a water and energy balance model in the spectral domain,

Water Resour. Res., 47, W07523, doi:10.1029/2010WR010292, 2011.

Pauwels, V. R. N., Timmermans, W., and Loew, A.: Comparison of the estimated water and energy budgets of a large winter wheat field during AgriSAR 2006 by multiple sensors and models, J. Hydrol., 349, 425–440, 2008.

Richards, L.: Capillary conduction of liquids through porous media, Physics, 1, 318–333, 1931.

Russell, E. S., Liu, H. P., Gao, Z. M., Finn, D., and Lamb, B.: Impacts of soil heat flux calculation methods on the surface energy balance closure, Agr. Forest Meteorl., 214, 189–200, doi:10.1016/j.agrformet.2015.08.255, 2015.

Scheerlinck, K., Pauwels, V. R. N., Vernieuwe, H., and De Baets, B.: Calibration of a water and energy balance model: Recursive Parameter Estimation versus Particle Swarm Optimization, Water Resour. Res., 10, W10422, doi:10.1029/2009WR008051, 2009.

Shao, M., Horton, R., and Jaynes, D. B.: Analytical Solution for One-Dimensional Heat Conduction-Convection Equation, Soil Sci. Am. J., 62, 123–128, 1998.

Shuttleworth, W. J.: Evaporation, McGraw-Hill, New York, NY, 4.1–4.53, 1992.

Venegas, P., Grandon, A., Jara, J., and Paredes, J.: Hourly estimation of soil heat flux density at the soil surface with three models and two field methods, Theor. Appl. Climatol., 112, 45–59, doi:10.1007/s00704-012-0705-z, 2013.

Wang, L., Gao, Z., Horton, R., Lenschow, D. H., Meng, K., and Jaynes, D. B.: An Analytical Solution to the One-Dimensional Heat Conduction-Convection Equation in Soil, Soil Sci. Am. J., 76, 1978–1986, doi:10.2136/sssaj2012.0023N, 2012.

Wang, Z. H. and Bou-Zeid, E.: A novel approach for the estimation of soil ground heat flux, Agr. Forest Meteorol., 154, 214–221, doi:10.1016/j.agrformet.2011.12.001, 2012.

5

A two-parameter design storm for Mediterranean convective rainfall

Rafael García-Bartual and Ignacio Andrés-Doménech

Universitat Politècnica de València, Instituto Universitario de Investigación de Ingeniería del Agua y Medio Ambiente (IIAMA), Camí de Vera s/n, 46022 Valencia, Spain

Correspondence to: Ignacio Andrés-Doménech (igando@hma.upv.es)

Abstract. The following research explores the feasibility of building effective design storms for extreme hydrological regimes, such as the one which characterizes the rainfall regime of the east and south-east of the Iberian Peninsula, without employing intensity–duration–frequency (IDF) curves as a starting point. Nowadays, after decades of functioning hydrological automatic networks, there is an abundance of high-resolution rainfall data with a reasonable statistic representation, which enable the direct research of temporal patterns and inner structures of rainfall events at a given geographic location, with the aim of establishing a statistical synthesis directly based on those observed patterns. The authors propose a temporal design storm defined in analytical terms, through a two-parameter gamma-type function. The two parameters are directly estimated from 73 independent storms identified from rainfall records of high temporal resolution in Valencia (Spain). All the relevant analytical properties derived from that function are developed in order to use this storm in real applications. In particular, in order to assign a probability to the design storm (return period), an auxiliary variable combining maximum intensity and total cumulated rainfall is introduced. As a result, for a given return period, a set of three storms with different duration, depth and peak intensity are defined. The consistency of the results is verified by means of comparison with the classic method of alternating blocks based on an IDF curve, for the above mentioned study case.

1 Introduction

Design storms are of paramount importance for hydrologic engineering and remain mainstream practice as they provide a simple and apparently appropriate tool for the design of hydraulic infrastructure. Design storms have been used for more than a century if we consider the block rainfall as input of the rational method (Watt and Marsalek, 2013). They experienced an important development during the 1970s and 1980s with more realistic approaches being implemented (Pilgrim and Cordery, 1975; Walesh et al., 1979; Hogg, 1980, 1982; Pilgrim, 1987).

The need for design storms in hydrologic engineering must be analysed according to the spatial scale of the problem, which might range from typical urban drainage designs to small and intermediate catchment basins. As reported by Watt and Marsalek (2013), one of the earliest applications of design storms to urban drainage took place in Rochester, New York (Kuichling, 1889). It followed the rational method which is still widely used today. In the urban context, the City of Los Angeles method (Hicks, 1944) and the Chicago Hydrograph Method (Keifer and Chu, 1957) represented an important step towards the development of hydrograph methods. On the watershed scale, design storms are needed to obtain design floods when streamflow data are scarce or do not exist (Watt and Marsalek, 2013) for the design of culverts, bridges and small dams, drainage systems, drainage planning, and flood management.

Design storms usually fall into two different categories. The first one considers models based on intensity–duration–frequency (IDF) relations. The second one corresponds to synthetic events where the temporal distribution is derived from observed storms.

Within the first category, the most widely used synthetic storms are probably the National Resource Conservation Service (NRCS, formerly SCS) dimensionless storms and the so-called alternating-block method storms. Standard rainfall patterns for 24 h storms are available for four different geographic regions of the United States (Froehlich, 2009). The NRCS design storms are appropriate for catchments smaller than 250 km^2, and they are considered to be applicable to storms of any average return period. Temporal distributions within this method are based on depth–duration–frequency relations available for the US territory, divided into four different climatic regions (McCuen, 1989).

The alternating-block method (Chow et al., 1988) is solely based on an IDF curve. These design storms display a maximum intensity block in the centre of the event and a total rainfall depth at any time that coincides with the total depth given by the IDF relation. The method is simple but has also been widely criticized, because it does not represent any observed rainfall internal structure. Another noticeable weak point of the method, already pointed out by McPherson (1978), is the arbitrary selection of the storm duration, which causes total rainfall depth to also be arbitrarily selected. The Chicago design storm (Keifer and Chu, 1957) is a special case of an alternating-block storm. In Spain, the use of this method is still today concretized through local or regional IDF curves such as those proposed by Témez for all the Iberian Peninsula (Témez, 1978). Recent publications demonstrate that, generally, peak-flow calculations using these design storms tend to overestimate the results (Alfieri et al., 2008).

The second category of design storms corresponds to temporal patterns derived from observed records. One of the first temporal distributions using this approach was developed by Huff (1967) in Illinois (US). The method determines in which time quartile the maximum intensity occurs. This work eventually became the Illinois State Water Survey Design Storm (Huff and Angel, 1989), extensively used by state and local agencies in the US Midwest. Following the same methodology, Hogg (1980) presented his findings on temporal patterns depending on the storm duration for different regions in Canada. Results led to the AES design storm (Hogg, 1982), widely used in urban drainage design. The former design storm reproduces the maximum intensity, the time of this maximum and the rainfall depth that occurs before the peak on the basis of observed records. Other works into this category are those developed in Australia (Pilgrim, 1987; French and Jones, 2012) or the UK (Packman and Kidd, 1980). In Spain, García-Bartual and Marco (1990) studied hyetographs of extreme convective precipitation where the intensity resulting from the activity of each rainfall cell was represented by a gamma-type function with maximum intensity and volume as random variables.

Some authors point out that the design storm concept itself is fraught with conceptual error when used to simplify engineer analysis with unrealistic assumptions (Adams and Howard, 1986). Indeed, many of the concerns about classic design storms arise from the storm duration selection, the IDF concept limitations, the temporal distribution and the difficulties of relating the synthetic storm event to a specific return period.

The design storm duration is not a determining factor if the purpose is to determine a peak flow to design conveyance infrastructures. Consequently, it is common practice to fix it around the concentration time of the catchment basin. Nevertheless, when storage elements are to be analysed, the influence of storm duration and temporal pattern becomes critical (Ball, 1994).

As has been shown in the past (Watt and Marsalek, 2013), uncertainties arising from existing IDF relations have strong consequences. First, record series used to fit IDF expressions are usually short for low-frequency occurrences. Second, IDF curves are considered to represent worst maxima regardless of the physical nature of the storm. García-Bartual and Schneider (2001) exposed the inherent uncertainty in the process, which significantly affects the definition of the IDF curves' shape in the interval 0–10 min. Finally, there is enough reason to deem data acquisition insufficiently accurate in providing robust data for IDF analysis, especially in urban areas (Hoppe, 2008). Moreover, as is the case in Spain, outdated IDF curves are still regularly used, as they are still found in guidance and regulations. The above mentioned uncertainties in IDF curves' estimation can significantly affect the reliability of derived design storms, especially in the definition of its peak rainfall intensities, with undesirable consequences when used for hydrologic design purposes.

For the simplest applications (i.e. rational method), a temporal pattern is not required for the design storm. However, for most hydrologic engineering applications, a design hyetograph is necessary. Selecting this temporal trend is one of the most uncertain steps of the design storm definition, since the physical nature of the process cannot be disregarded.

A storm event presents many characteristics, so it cannot be fully described by the statistics of only one of them. For a return period definition, a common practice is to assign a given frequency to a specific event feature (i.e. its maximum intensity). But, given that a design storm is composed of many variables (depth, duration, temporal pattern, antecedent conditions), assigning a single return period may not be appropriate.

The objective herein is to formulate an analytical approach in order to describe rainfall intensities in time, as an alternative for practical design storm definition in Mediterranean areas. Another aim is to develop all required analytical properties to ensure their applicability under usual criteria and requirements of design storm approaches for hydrological design. These include a methodology for return period assignment based on both total depth and peak intensity of the storm. Also, a practical methodology to build the storm, applied to a given case study to validate it. For illustrative purposes, a comparison with most extended design storms in Mediterranean areas will be developed and discussed.

2 Design storm

The temporal pattern of rainfall intensities representing the design storm is expressed in terms of a continuous analytical function of the form given as follows:

$$i(t) = i_0 f(t), \quad (1)$$

where t ≥ 0 (min) is the time elapsed from the start of the rainfall episode ($t = 0$), $i(t)$ (mm h^{-1}) represents the rainfall intensity at instant t, i_0 (mm h^{-1}) is the instantaneous peak intensity of the storm and $f(t)$ is a convenient non-dimensional, continuous and differentiable analytical function, which will be defined below.

The adopted function $f(t)$ must reproduce the activity life cycle of a convective cell, i.e. an initial development until the maturity stage is reached, during which maximum intensities are attained, followed by a stage of dissipation in time, typified by a progressive attenuation of rainfall.

Several recent studies characterize the physical dynamics of convective cells from radar-provided data. More precisely, these data correspond to relevant characteristics such as duration, spatial extension or the importance of the above-mentioned stages, (Capsoni et al., 2009; Rigo and Llasat, 2005). On the basis of high-resolution rainfall data, some authors report statistical evidence of the predominance of temporal patterns where the attenuation or temporal dissipation stages tend to last longer than the initial growing and development stage (Brummer, 1984). This characteristic supports the use of relationships like the gamma function, successfully employed in previous mathematical models of rainfall (García-Bartual and Marco, 1990; Salsón and Garcia-Bartual, 2003) since it represents more accurately the patterns observed in the temporal registers of convective rainfall events in the east and south-east of the Iberian Peninsula. Nonetheless, there are other mathematical models where an analytic function $f(t)$ is postulated, and where the maximum value is located precisely at half the total duration of the event produced by the convective cell (Northrop and Stone, 2005).

In terms of the proposed design storm, the adopted temporal pattern shows an evolution described in a parametrical way with a function $f(t)$: a non-dimensional gamma-type function with a single parameter which describes a fast initial growing stage of intensities until reaching the maximum value, followed by a slower diminishing stage, asymptotic in time and tending towards a null value when time is growing towards infinity.

$$f(t) = \varphi t e^{1-\varphi t}, \quad (2)$$

where φ (min^{-1}) is a parameter.

This model proved to be an acceptable and consistent representation of the rainfall intensities from convective Mediterranean storms (Andrés-Doménech et al., 2016)

Table 1. Parameters η_1 and η_2 for different truncation criteria.

Truncation criterion as a % of the intensity peak value	η_1	η_2
1 %	0.01	7.6386
5 %	0.05	5.7439
10 %	0.10	4.8897

2.1 Analytical properties

Some interesting analytical properties of the $f(t)$ function are revised, which will prove useful in subsequent development. The following can be deduced from Eq. (2):

$$f(0) = 0, \quad (3)$$

$$\lim_{t\to\infty} f(t) = 0. \quad (4)$$

In addition, as

$$f'(t) = \varphi(1 - \varphi t) e^{1-\varphi t}, \quad (5)$$

function $f(t)$ displays a relative maximum at point $t = t_0 = \varphi^{-1}$. The corresponding value of this maximum is as follows:

$$f(t_0) = 1. \quad (6)$$

Given that the duration, t_C, of the cell is finite, and in order to establish a finite duration of the process, a simple truncating criteria is adopted for the asymptote of this function. To do so, a final or residual value is established as a fraction η_1 of the maximum so that

$$f(t_C) = \eta_1, \quad (7)$$

where t_C (min) represents the total storm duration, with $t_C > t_0$ and $0 < \eta_1 < 1$. Convenient η_1 values are shown in Table 1. Introducing conditions given in Eq. (7) into Eq. (2), we obtain the following:

$$f(t_C) = \varphi t_C e^{1-\varphi t_C} = \eta_1. \quad (8)$$

Equation (8) admits the following solution:

$$t_C = \frac{\eta_2}{\varphi}, \quad (9)$$

and thus verifies the condition

$$\eta_2 e^{1-\eta_2} = \eta_1. \quad (10)$$

Table 1 shows some of the solution values for this equation, for chosen values of the parameter η_1.

In other words, once the truncating criteria is defined, for example 5 %, the duration of the rainfall event is automatically defined as a function of parameter φ through Eq. (9) with $\eta_2 = 5.7439$.

2.2 Properties of the aggregated process

The suggested analytical function can be integrated, yielding to the following result:

$$F_{[t_1;t_2]} = \int_{t_1}^{t_2} f(t)\,\mathrm{d}t = \int_{t_1}^{t_2} \varphi t e^{1-\varphi t}\,\mathrm{d}t = \left(t_1 + \frac{1}{\varphi}\right) e^{1-\varphi t_1} - \left(t_2 + \frac{1}{\varphi}\right) e^{1-\varphi t_2}, \tag{11}$$

where $0 \le t_1 < t_2 \le t_C$. In this way, the integrated value of $F_{[t_1;t_2]}$ is expressed in minutes. By applying Eqs. (9) and (11), the following particular results are easily obtained:

$$F_{[0;t_C]} = \frac{e}{\varphi} - \left(t_C + \frac{1}{\varphi}\right) e^{1-\varphi t_C} = \frac{e}{\varphi}\left[1 - (1+\eta_2)e^{-\eta_2}\right], \tag{12}$$

$$F_{[0;\infty]} = \frac{e}{\varphi}, \tag{13}$$

$$\frac{F_{[0;t_C]}}{F_{[0;\infty]}} = 1 - (1+\eta_2)e^{-\eta_2}. \tag{14}$$

It must be noted that the result of Eq. (14) is independent of parameter φ. For instance, if a truncating value of 5 % is adopted ($\eta_1 = 0.05$), it automatically leads to $\eta_2 = 5.7439$ as shown in Table 1, and therefore

$$\frac{F_{[0;t_C]}}{F_{[0;\infty]}} = 0.98. \tag{15}$$

That is, the truncating criteria of 5 % for $f(t)$ is equivalent to establishing the total duration of the cell when 98 % of the cumulative rainfall has already taken place with respect to the hypothetical 100 % linked to a cell whose intensities are asymptotic to 0 and have infinite duration, according to the known analytical properties of the tail of $f(t)$.

From Eqs. (1) and (11), the total cumulative rainfall (mm) can be obtained, for a given time interval, $[t_1; t_2]$, as follows:

$$P_{[t_1;t_2]} = \int_{t_1}^{t_2} i(t)\,\mathrm{d}t = \frac{i_0}{60}\int_{t_1}^{t_2} f(t)\,\mathrm{d}t = \frac{i_0}{60}\left[\left(t_1 + \frac{1}{\varphi}\right) e^{1-\varphi t_1} - \left(t_2 + \frac{1}{\varphi}\right) e^{1-\varphi t_2}\right]. \tag{16}$$

The average rainfall intensity ($\mathrm{mm\,h^{-1}}$) during such a given time interval can be calculated as follows:

$$i_{[t_1;t_2]} = \frac{i_0}{t_2 - t_1}\left[\left(t_1 + \frac{1}{\varphi}\right) e^{1-\varphi t_1} - \left(t_2 + \frac{1}{\varphi}\right) e^{1-\varphi t_2}\right]. \tag{17}$$

In the same manner, the total cumulative rainfall for the time interval $[0; t]$ results in the following:

$$P_{[0;t]} = \frac{i_0}{60}\left[\left(\frac{e}{\varphi}\right) - \left(t + \frac{1}{\varphi}\right) e^{1-\varphi t}\right]. \tag{18}$$

Replacing $t = t_C$ in Eq. (18) and substituting Eq. (9), we obtain the total rainfall for the theoretical storm, given by the following expression:

$$P_{[0;t_C]} = \frac{i_0}{60}\left[\left(\frac{e}{\varphi}\right) - \left(\frac{\eta_2}{\varphi} + \frac{1}{\varphi}\right) e^{1-\eta_2}\right]. \tag{19}$$

If we assume a truncating criteria of 5 % ($\eta_1 = 0.05$) a straightforward expression is obtained for the total cumulative rainfall associated with the analytical storm:

$$P_{[0;t_C]} = 0.0443\frac{i_0}{\varphi}. \tag{20}$$

2.3 Maximum intensity for a given Δt

For practical applications, a given time interval of aggregation Δt is used, conveniently chosen depending on the type of hydrological application, the rainfall–runoff model to be used, and the characteristics of the urban hydrology application to be carried out.

Once a given Δt (in minutes) is selected, it is convenient to locate the most intense rainfall interval along the time axes, so that

$$I_{\Delta t} = \frac{i_0}{60}\max\left\{F_{[t;t+\Delta t]}\right\}, \tag{21}$$

where $t < t_0 < t + \Delta t$ and $I_{\Delta t}$ is the maximum rainfall intensity ($\mathrm{mm\,h^{-1}}$), for the most intense interval of the storm, as shown in Fig. 1.

If the above-mentioned central interval is

$$[t_L; t_U] = \left[\frac{1}{\varphi} - \xi\Delta t;\ \frac{1}{\varphi} + (1-\xi)\Delta t\right], \tag{22}$$

as indicated in Fig. 1, the optimization problem has a solution in terms of the auxiliary variable ξ, being $0 < \xi < 1$. Such a solution is given by the following:

$$\xi = \frac{1}{\varphi\Delta t} - \frac{e^{-\varphi\Delta t}}{1 - e^{-\varphi\Delta t}}. \tag{23}$$

Consequently, according to Eq. (17), the maximum intensity of the storm, once it has been discretized in time intervals of Δt minutes, can be calculated as follows:

$$I_{\Delta t} = \frac{i_0}{\Delta t}\left[\left(t_L + \frac{1}{\varphi}\right) e^{1-\varphi t_L} - \left(t_U + \frac{1}{\varphi}\right) e^{1-\varphi t_U}\right]. \tag{24}$$

In summary, the main derived properties of the chosen analytical shape of the storm are total duration of the storm given a truncation criterion (Eq. 9), total cumulative rainfall (Eq. 20) and maximum intensity for a given time level of aggregation Δt (Eq. 24). All these relations are uniquely expressed as functions of the two parameters of the storm, i_0 and φ.

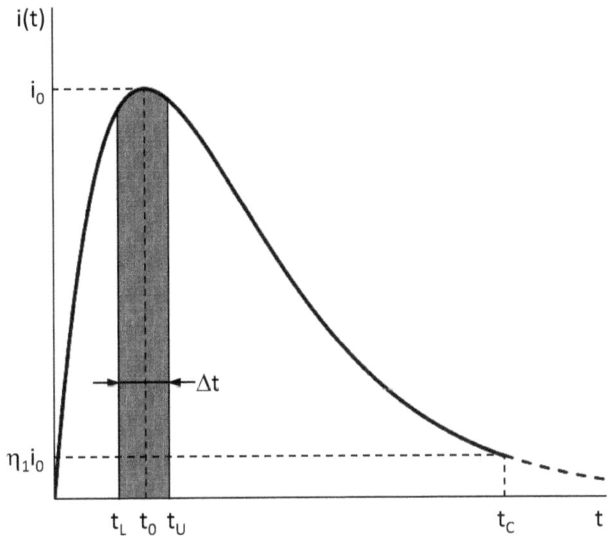

Figure 1. Most intense interval of the storm defined by $[t_L; t_U]$ for a Δt time interval of aggregation.

3 Rainfall data processing

Valencia is a Mediterranean city, located on the eastern coast of the Iberian Peninsula. It presents a typical temperate Mediterranean climate (Csa, according to Köppen climate classification). This type of climate is characterized by mild temperatures (annual average of 17 °C), without marked extremes and with a rainfall of about 450 mm yr^{-1}. Rainfall is very unevenly distributed throughout the year, with very marked minima during the months of June, July and August and maxima happening during the months of September and October, these two months concentrating almost a third of the annual rainfall.

Another important characteristic of the rainfall regime is its irregularity, alternating dry and more humid intervals. These dry or humid periods tend to last several years due to the Mediterranean climatic inertia. The torrential character of storms is also a main feature of the rainfall regime of the region, with frequent convective rainfall mesoscale episodes, most widely known as cut-offs, characterized by very localized high-intensity storms.

The rainfall series used in this study were recorded by the Júcar River Basin Authority during the period 1990–2012. The rainfall gauge is installed in the city centre and the data time step is 5 min. Previous studies demonstrated the validity of this data set for similar purposes (Andrés-Doménech et al., 2010). The continuous rainfall series are processed to identify and extract convective storms. First, statistically independent rainfall events are identified. Then, amongst them, only convective events are extracted. Finally, convective storms are identified from convective events and finally selected to estimate model parameters.

3.1 Convective storms set

3.1.1 Identification of statistically independent rainfall episodes

Before undertaking the storm analysis, a preliminary step is required in order to separate the original continuous series of rainfall records in statistically independent rainfall events. There is no universal method for identifying the minimum inter-event time of a rainfall regime and, thus, independent storms. Dunkerley (2008) presents an interesting review of the range of approaches used in the recognition of main events. Early works by Restrepo-Posada and Eagleson (1982) are still in force, and according to them the identification of independent events is based on considering events such as statistically independent events, so that the minimum inter-event time must be an outcome of a Poisson process. Bonta and Rao (1988) bore out this theory, studying some other aspects in depth. Andrés-Doménech et al. (2010) completed the original methodology based on the coefficient of variation analysis and established for Valencia a minimum inter-event time equal to 22 h. The latter implies that if two rainfall pulses are separated by more than 22 h, then, they belong to different events. Under this premise, 987 statistically independent events are identified for the period 1990–2012.

3.1.2 Identification of convective episodes

The required rainfall episodes must have a certain convective character. Therefore, only storms that verify the following conditions can be taken into account: maximum intensity over 35 mm h^{-1} and convectivity index $\beta^* > 0.3$. The convectivity index introduced by Llasat (2001) reflects in an objective way the greater or lesser convectivity degree of a rainfall episode, on the sole basis of the registered 5 min data, with no additional meteorological information being required. The value β^* depends on a convectivity threshold which depends itself on the record time step. This convectivity threshold was estimated for the Spanish Mediterranean coastline by Llasat (2001). For a 5 min resolution data series, the threshold was set to 35 mm h^{-1}. Consequently, this index represents the proportion of total rainfall fallen with an intensity higher than 35 mm h^{-1}. Events with $\beta^* > 0.3$ represent convective storms at this location. Thus, according to this additional criterion, only 64 convective events from the complete set are selected.

3.1.3 Selection of convective storms

Some of the independent convective events selected above can correspond to long or very long episodes with important dry intra-periods (always less than 22 h). Concatenation of some convective cells can lead to this situation, resulting in long episodes on some days.

Often, these rainfall cells (storms) can be linked by very slight background intensity (around 2 mm h^{-1}). Usually, these convective cells only correspond to a small duration within the whole episode. Nevertheless, they can represent more than 80 % of the total rainfall amount. According to this fact, the convective events set is classified as follows.

a. Type I events: these storms consist of a single convective cell. They are characterized by a moderate duration and a considerable average intensity. They can present low-intensity intervals before and/or after the larger part of rainfall.

b. Type II events: long-lasting rainfall events consisting of two or more storms separated in time.

Following this classification, 58 events are type I and 6 events are type II. These 6 type II events are carefully examined and analysed to extract storms within them. The following criteria to select individual storms are adopted:

a. Identify the event peak intensity, always over 35 mm h^{-1}, and its near range.

b. The first storm time interval corresponds to the prior interval to 9.6 mm h^{-1} intensity (3 times the rain gauge sensitivity).

c. The last storm time interval is defined by a shift in the sign of the hyetograph derivative, always around intensities lower than 9.6 mm h^{-1}.

Finally, and according to this methodology, 73 storms are defined for the period 1990–2012. Table 2 shows a basic report of the empirical statistics of this sample. Andrés-Doménech et al. (2016) also pointed out a strong correlation between the storm volume and duration (0.839) and also an evident correlation between storm volume and its maximum intensity (0.639).

3.2 Relations between cumulative rainfall and maximum intensity of the storm

Three different sets of events were identified, according to their duration. As shown in Fig. 2, each of them can be characterized in terms of a representative value of the following ratio.

$$\alpha_i = \frac{P}{I_{10}} \tag{25}$$

Figure 2 shows the three different ratios empirically found: $\alpha_1 = 0.1993$ h, $\alpha_2 = 0.2919$ h and $\alpha_3 = 0.5299$ h.

Such distinction allows for the identification of three different families, depending on α_i. Each of them is characterized by its corresponding storm pattern. In accordance with this, a given return period T should yield to three storms, one per family, all of them with equivalent magnitude, but with different time patterns. Low α values typically correspond with storms with peak intensity occurring shortly after the initiation of the storm, while higher α values are found for longer events and usually higher cumulative rainfall depths.

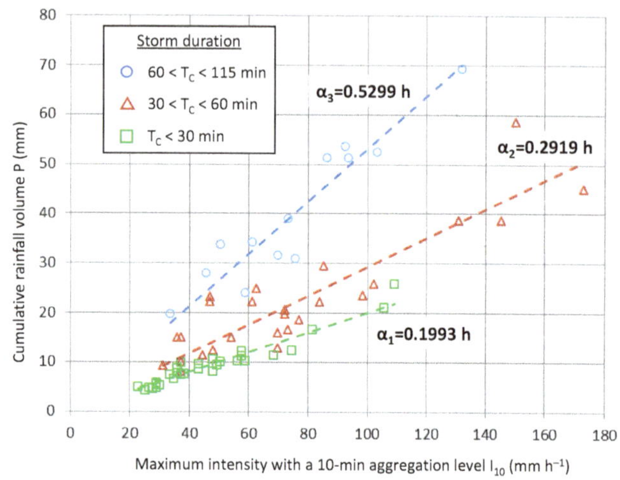

Figure 2. Relations between cumulative rainfall and maximum intensity of the storm depending on the storm duration.

3.3 Storm magnitude

The question of determining the magnitude of a given storm is undertaken through a principal-component analysis (PCA), over the observed sample (I_{10}; P). This strategy is based on the fact that both maximum intensity and cumulative rainfall are directly related to the magnitude of the event, and are thus relevant to it, while the preliminary statistical analysis showed a significant correlation among them as stated before (Andrés-Doménech et al., 2016).

Table 3 shows the results of the principal-component analysis, resulting in the two new variables, X_1 and X_2.

It can be noted that the first main component, X_1, explains 92.1 % of the variance observed in the sample. This main component is defined as

$$X_1 = \beta_P P + \beta_I I_{10} = 0.3704P + 0.9289 I_{10}. \tag{26}$$

According to the relationships between the cumulative rainfall depth and the storm maximum intensity, both variables are used together to define a new combined variable that is able to represent the storm magnitude in terms of volume and maximum intensity. X_1 can be considered a measurement of the magnitude of the rainfall event, as both initial variables, P and I_{10}, contribute to it. This new variable after the PCA analysis, in statistical terms, contains more information by itself than either P or I_{10}, and thus represents an adequate variable in order to establish a return period, T, linked to a given design storm.

3.4 Return period

The process of assigning a return period T to a given design storm should be based on a previous statistical analysis of the selected variable, X_1. To do so, an appropriate extreme value distribution function is sought. For the given set of rainfall episodes, several distribution functions were

Table 2. Storm univariate statistics (adapted from Andrés-Doménech et al., 2016).

	Rainfall volume P (mm)	Maximum intensity I_{10} (mm h^{-1})	Storm duration T_C (min)
Mean	20.0	76.4	38.0
Maximum	69.2	206.4	115.0
Minimum	4.2	36.0	10.0
Median	15.0	64.8	30.0
Standard deviation	15.9	37.3	21.9
Bias	1.39	1.46	1.21
Kurtosis	1.36	2.09	1.18

Table 3. Principal-component eigenvectors resulting from the PCA analysis.

Original variable	Principal component X_1	Principal component X_2
P	0.3704	0.9289
I_{10}	0.9289	−0.3704

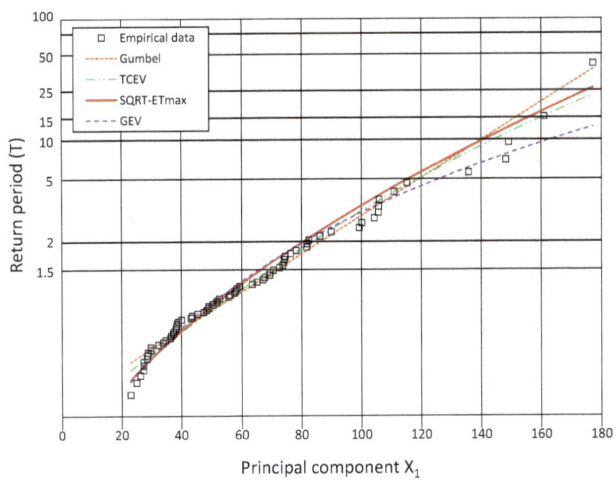

Figure 3. Extreme value distribution analysis for principal component X_1.

tested, including Gumbel, two-component extreme value (TCEV), squared root exponential type distribution of maximum (SQRT-ETmax) and general extreme value (GEV). In all cases, maximum likelihood was used to estimate the corresponding parameters. Figure 3 shows the results of this extreme value function analysis. Best fit was obtained with the SQRT-ETmax distribution, with the advantage of being more parsimonious than TCEV and GEV functions. This result is in accordance with what usually occurs on the eastern coastline of Spain.

4 Construction of the design storm

If $X_1(T)$ is the quantile of the extreme value distribution corresponding to a given return period T, the two variables P and I_{10}, which define the design storm for that given return period, are obtained by solving Eqs. (25) and (26) for each family $i = 1$, 2 and 3. That is,

$$\begin{cases} I_{10,i}(T) = \dfrac{X_1(T)}{\beta_I + \beta_P \alpha_i} \\ P_i(T) = \dfrac{\alpha_i X_1(T)}{\beta_I + \beta_P \alpha_i} \end{cases} \tag{27}$$

In order to define, in practice, the design storm associated with $I_{10,i}$ and P_i values, and once chosen a convenient time level of aggregation (i.e. $\Delta t = 10$ min), it is necessary to obtain the two parameters, i_0 and φ, which analytically define the design storm. To do so, Eqs. (20) and (24) are used, and they result as follows, for each $i = 1$, 2 and 3:

$$P_i(T) = 0.0443 \frac{i_{0,i}}{\varphi_i}, \tag{28}$$

$$I_{10,i}(T) = \frac{i_0}{\Delta t}\left[\left(t_L + \frac{1}{\varphi_i}\right) e^{1-\varphi_i t_L} - \left(t_U + \frac{1}{\varphi_i}\right) e^{1-\varphi_i t_U}\right], \tag{29}$$

where t_L and t_U are calculated according to Eqs. (22) and (23).

5 Comparison with the alternating-block design storm

After formulating the practical steps to build a synthetic storm, a comparison of the former with the most widely used storm (built with alternating blocks obtained from an IDF curve), is performed. In order to carry out this comparison, storms corresponding to a return period of 25 years are built. The choice of 25 years corresponds to the requirements set by the Municipality of Valencia regulations for the design of urban drainage hydraulic infrastructures.

Before obtaining the alternating-block design storm, an ID (intensity–duration) curve for 25 years must be determined, from the very same sample of storms previously used for the development of the Gamma storm and described in Sect. 3.

Table 4. Parameters for the three synthetic storms.

Storm parameters	Storm 1 (short)	Storm 2 (intermediate)	Storm 3 (long)
X_1	175.5	175.5	175.5
α (h)	0.1993	0.2919	0.5299
P_i (mm)	34.9	49.4	82.7
$I_{i,10}$ (mm h^{-1})	175.0	169.2	156.0
φ (min^{-1})	0.3047	0.1699	0.0862
i_0 (mm h^{-1})	239.8	189.3	160.8
t_c (min)	18.85	33.81	66.61
ξ	0.2783	0.3648	0.4290
Number of blocks	2	4	6

To do this, the usual procedure for obtaining IDF curves is followed, adjusting the empirical sample to the following IDF relation:

$$i(t) = \frac{a}{(b+t)^c}, \tag{30}$$

where i (mm h^{-1}) is the maximum intensity corresponding to a rainfall duration t (min), while a, b and c are the parameters of the IDF curve. Vaskova (2001) demonstrated the fitness of this expression for adjusting local IDF curves in Valencia. With the data employed in the present paper, the following coefficients result for the 25-year return period ID curve: $a = 8198$ mm h^{-1}, $b = 29.8$ min and $c = 1.06$. Then, for each case, the alternating-block design storm is built from the ID curve defined by Eq. (31), following the usual methodology (Chow et al., 1988). To allow for a proper comparison with the Gamma storm, the same number of blocks is kept for every case.

To perform the comparison, first, the three synthetic storms corresponding to each of the families defined by α_1 (short storms), α_2 (medium duration storm) and α_3 (long storms) are built. In order to do this, once the truncating level has been set η_1 (0.05 in the present paper), the method summarized in Sect. 4 is followed. For a return period of 25 years, it results in a storm magnitude $X_1 = 175.5$ (Fig. 3). A continuous storm for each of the three families is obtained and, after being discretized in blocks of $\Delta t = 10$ min, generates, for each family, a storm of 2, 4 and 6 blocks respectively, as once the truncation criterion is selected, the storm duration is established (Eq. 9), so that, for a given time level of aggregation (ΔT), the number of blocks can be derived. Table 4 summarizes the essential parameters of each of the three storms.

Figure 4 represents, for each family, both the continuous and the aggregated Gamma storms, along with the alternating-block storm obtained from the ID curve.

Both methods lead to consistent and relatively similar results, those being particularly alike for the longer storms. However, for short- and medium-duration storms, it becomes clear that the classic method offers significantly more pessimistic results. In other words, the common method displays higher intensities. This result is coherent with the well-known process of defining the storm. Indeed, given that the alternating-block method assumes the simultaneous occurrence of maximum intensities for different durations, even when those values had not been encountered historically in the same rainfall event, overestimated intensities seem to be an unsurprising outcome. On the contrary, the Gamma storm is built directly from the temporal pattern observed in real episodes. That is, as demonstrated by Andrés-Doménech et al. (2016), the Gamma storm is coherent with the temporal structure of the rain process and that is why the proposed synthetic storm reproduces the observed rainfall more accurately. Table 5 gathers the quantitative differences found for each of the three storms.

As expected, the higher the duration of the storm, the lesser the difference between the maximum instant intensity of the continuous storm and the one of the maximum block. Furthermore, differences between the maximum block intensities between the aggregated Gamma storm and the alternating block storm are also reduced as the duration of the storm increases.

Nonetheless, the most remarkable differences lie on rainfall volumes. Given a return period, the alternating-block method combines in a single theoretical storm the most adverse statistics for several durations, which originally derive from different historical rainfall events. Conceptually, this is a worst-case-scenario storm ignoring actual rainfall patterns found in the rainfall registers, yielding to a volume overestimation (Di Baldassarre et al., 2006). For the aggregated Gamma storm, differences with regard to the continuous model are more limited, in all cases, which supports the conclusion of having generated a synthetic storm that not only reproduces peak intensities properly but also respects the observed temporal patterns and, consequently, reproduces better storm volumes. Concerning variable X_1, results are very similar for both methods, as shown in Table 5.

6 Conclusions

The use of design storms has been a common worldwide practice for many years, employed to solve a range of hydrologic engineering problems in a direct way. These synthetic storms represent an appropriate statistical synthesis of historical rainfall records and therefore, are of maximal utility in their application to problems of urban drainage infrastructure design. In many European and North and South American countries, they are directly obtained from IDF curves, which are usually pre-established for a given area. This simplifies notably the setting of the design storm, making this a straightforward and fast process. Moreover, it presents the huge advantage of being applicable to places where is little or no rainfall information, inasmuch as it is possible to assume as a starting point certain IDF curves, deemed to be suffi-

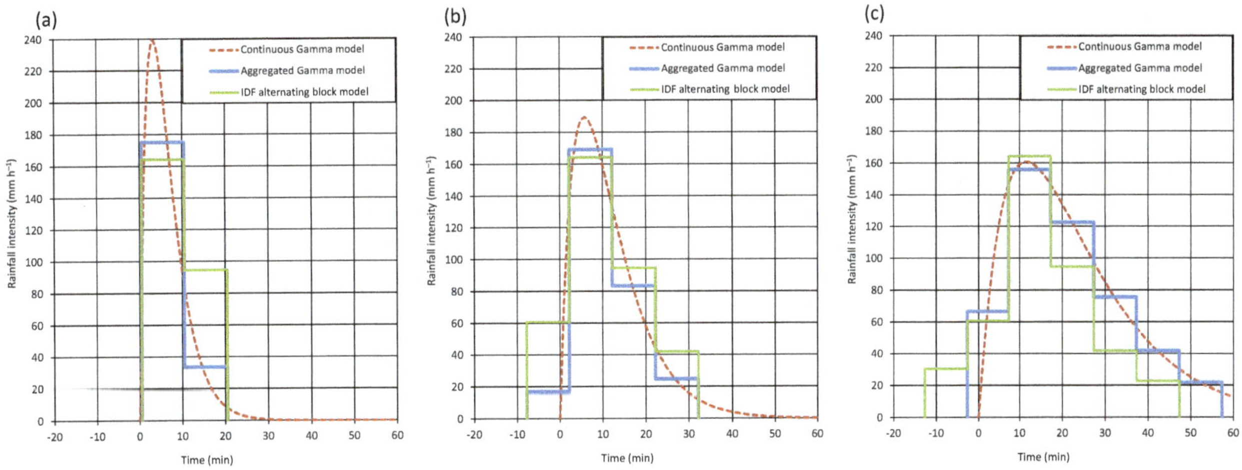

Figure 4. Comparison of the continuous and aggregated Gamma model with the IDF alternating-block model for the three families $\alpha_1 = 0.1993$ **(a)**, $\alpha_2 = 0.2919$ **(b)** and $\alpha_3 = 0.5299$ **(c)** for $T = 25$ years.

Table 5. Comparison of volume, peak intensity and magnitude of the Gamma-aggregated and IDF alternating-block design storms.

		Duration (min)	Maximum intensity ($\mathrm{mm\,h^{-1}}$)	Volume (mm)	Magnitude X_1
Storm α_1	Gamma aggregated	20	175.0	34.8	175.45
	IDF alternating block	20	164.4	43.2	168.71
Storm α_2	Gamma aggregated	40	169.2	45.0	173.84
	IDF alternating block	40	164.4	60.3	175.05
Storm α_3	Gamma aggregated	60	156.0	80.9	174.87
	IDF alternating block	60	164.4	69.3	178.38

ciently reliable or representative of the maximum rainfall of that location.

One of the downsides of this process is the fact that it ignores, in its approach, aspects relative to the actual duration and structure – or inner pattern – of intensities of rain, visible in high-resolution rainfall registers. In some countries, automatic pluviometer networks have been working for decades and thus, detailed information is now available, allowing engineers to undertake such matters with statistical representativity (De Luca, 2014).

However, the diversity of hydraulic elements of current drainage systems (e.g. storm tanks, sustainable drainage systems) means that most conditioning storm parameters for the design are not only rain intensities but also duration, total cumulated rainfall and temporal structure of the storm. This makes the exploration of new strategies for building design storms particularly interesting, starting directly from the observed patterns in the high-resolution registers, instead of using IDF curves. This research explores the possibilities in this sense, for the case of convective-type Mediterranean storms and proposes a case study from the automatic pluviometer register of the city of Valencia.

The design storm is defined in an analytical way through a two-parameter function (i_0 and φ), already substantiated by previous studies for the Mediterranean area. The former parameters are estimated directly from independent rainfall events, identified in the original temporal series. The assignment of a return period is done through an auxiliary variable which describes the magnitude of the event, and incorporates simultaneously both the total cumulated rainfall and the maximum intensity. In practice, this criterion leads to three different design storms for each return period, of a similar magnitude but with different temporal patterns and durations. Those storms, exclusively defined in terms of the two pointed parameters, are easily discretized in time intervals Δt, in view of their application to practical cases.

For illustrative purposes, the construction of these storms for Valencia is developed and then compared with the classical alternating-block storm, obtained by the usual methods from the same records. This enables the verification of the consistency of the proposed method, resulting in three storms for every return period, with temporal patterns derived from the observation and direct analysis of high-resolution rainfall series. Besides, they are exclusively defined through the value of their only two parameters in each case. While it is true that the process is clearly more laborious than the alternating-block method, the feasibility of the process in a real case is verified, starting from the principle of direct determination of the storm without using IDF curves. Naturally, it has the important limitation of being only applicable in geographic locations where there is high-resolution rainfall information of sufficient quality and appropriate length of historical record series. In the future, for a higher statistical representativity, it will become necessary to count with a longer register.

The proposed method herein, as well as other simple design storm approaches, present some inherent limitations for certain hydrological engineering applications, as they are not suitable for case studies where a more detailed or comprehensive description of the rainfall process is required. Some examples are continuous-time hydrological-system evaluation, hydrological applications in large catchments, or applications where ensembles or stochastic generation of events are needed to account for a number of possible scenarios (Frances et al., 2012).

Despite that, the proposed analytical definition defines a feasible work framework to provide the design storm with the spatio-temporal dimension of the event, through the addition of a component that considers the decline of intensities from the centre of the cell. By following the practical strategy contained in the present paper, the characterization and estimation of parameters of such a component must be founded on the direct observation of radar data for the most significant storms, with the goal of parametrizing the most characteristic spatial patterns (Barnolas et al., 2010).

Competing interests. The authors declare that they have no conflict of interest.

Acknowledgements. This work was supported by the Regional Government of Valencia (Generalitat Valenciana, Conselleria d'Educació, Investigació, Cultura i Esport) through the project "Formulación de un hietograma sintético con reproducción de las relaciones de dependencia entre variables de evento y de la estructura interna espacio-temporal" (reference GV/2015/064).

Edited by: M. Mikos

References

Adams, B. J. and Howard, C. D. D.: Design Storm Pathology, Can. Water Resour. J., 11, 49–55, doi:10.4296/cwrj1103049, 1986.

Alfieri, L., Laio, F., and Claps, P.: A simulation experiment for optimal design hyetograph selection, Hydrol. Process., 22, 813–820, doi:10.1002/hyp.6646, 2008.

Andrés-Doménech, I., Montanari, A., and Marco, J. B.: Stochastic rainfall analysis for storm tank performance evaluation, Hydrol. Earth Syst. Sci., 14, 1221–1232, doi:10.5194/hess-14-1221-2010, 2010.

Andrés-Doménech, I., García-Bartual, R., Rico Cortés, M., and Albentosa Hernández, E.: A Gaussian design-storm for Mediterranean convective events. Sustainable Hydraulics in the Era of Global Change, edited by: Erpicum, S., Dewals, B., Archambeau, P., and Pirotton, M., Taylor & Francis, London, ISBN 978-1-138-02977-4, 2016.

Ball, J. E.: The influence of storm temporal patterns on catchment response, J. Hydrol., 158, 285–303, 1994.

Barnolas, M., Rigo, T., and Llasat, M. C.: Characteristics of 2-D convective structures in Catalonia (NE Spain): an analysis using radar data and GIS, Hydrol. Earth Syst. Sci., 14, 129–139, doi:10.5194/hess-14-129-2010, 2010.

Bonta, J. V. and Rao, R.: Factors affecting the identification of independent storm events, J. Hydrol., 98, 275–293, 1988.

Brummer, J.: Rainfall events as paths of a stochastic process: Problems of design storm analysis, Water Sci. Technol., 16, 131–138, 1984.

Capsoni, C., Luini, L., Paraboni, A., Riva, C., and Martellucci A.: A new prediction model of rain attenuation that separately accounts for stratiform and convective rain, IEEE T. Antenn. Propag., 57, 196–204, 2009.

Chow, V. T., Maidment, D. R., and Mays, L. W.: Applied hydrology, Mc Graw-Hill, New York, 1988.

De Luca, D. L.: Analysis and modelling of rainfall fields at different resolutions in southern Italy, Hydrolog. Sci. J., 59, 1536–1558, doi:10.1080/02626667.2014.926013, 2014.

Di Baldassarre, G., Brath, A., and Montanari, A.: Reliability of different depth-duration-frequency equations for estimating short-duration design storms, Water Resour. Res., 42, W12501, doi:10.1029/2006WR004911, 2006.

Dunkerley, D.: Identifying individual rain events from pluviography records: a review with analysis of data from an Australian dryland site, Hydrol. Process., 22, 5024–5036, 2008.

Frances, F., García-Bartual, R., and Bussi, G.: High return period annual maximum reservoir water level quantiles estimation using synthetic generated flood events, in: "Risk Analysis, Dam Safety, Dam Security and Critical Infrastructure Management", Taylor and Francis, ISBN 978-0-415-62078-9, 185–190, 2012.

French, R. and Jones, M.: Design rainfall temporal patterns in Australian Rainfall and Runoff: Durations exceeding one hour, Australian Journal of Water Resources, 16, 21–27, 2012.

Froehlich, D. C.: Mathematical formulations of NRCS 24-hour design storms, J. Irrig. Drain E.-ASCE, 135, 241–247, doi:10.1061/(ASCE)0733-9437(2009)135:2(241), 2009.

García-Bartual, R. and Marco, J.: A stochastic model of the internal structure of convective precipitation in time at a raingauge site, J. Hydrol., 118, 129–142, doi:10.1016/0022-1694(90)90254-U, 1990.

García-Bartual, R. and Schneider, M.: Estimating maximum expected short-duration rainfall intensities from extreme convective storms, Phys. Chem. Earth Pt. B, 26, 675–681, doi:10.1016/S1464-1909(01)00068-5, 2001.

Hicks, W. I.: A method of computing urban runoff, T. Am. Soc. Civ. Eng., 109, 1217–1253, 1944.

Hogg, W. D.: Time distribution of short duration rainfall in Canada, in: Proceedings Canadian Hydrology Symposium, 80, Ottawa, Ontario, 53–63, 1980.

Hogg, W. D.: Distribution of design rainfall with time: design considerations. American Geophysical Union Chapman on Rainfall Rates, Urbana, Illinois, 27–29 April 1982.

Hoppe, H.: Impact of input data uncertainties on urban drainage models: climate change – a crucial issue? In Proceedings of the

11th International Conference on Urban Drainage, Edinburgh, UK, 31 August–5 September, 10 pp., 2008.

Huff, F. A.: Time distribution of rainfall in heavy storms, Water Resour. Res., 3, 1007–1019, doi:10.1029/WR003i004p01007, 1967.

Huff, F. A. and Angel, J. R.: Rainfall Distributions and Hydroclimatic Characteristics of Heavy Rainstorms in Illinois (Bulletin 70), Illinois State Water Survey, 1989.

Keifer, C. J. and Chu, H. H.: Synthetic storm pattern for drainage design, J. Hydraul. Eng-ASCE, 83, 1–25, 1957.

Kuichling, E.: The relation between rainfall and the discharge in sewers in populous districts, T. Am. Soc. Civ. Eng., 20, 37–40, 1889.

Llasat, M. C.: . An objective classification of rainfall events on the basis of their convective features: application to rainfall intensity in the northeast of Spain, Int. J. Climatol., 21, 1385–1400, 2001.

McCuen, R. H.: Hydrologic analysis and design, Prentice-Hall, Englewood Cliffs, N. J., 1989.

McPherson, M. B.: Urban runoff control planning, EPA-600/9-78-035, Environmental Protection Agency, Washington D.C., 1978.

Northrop, P. J. and Stone, T. M.: A point process model for rainfall with truncated gaussian rain cells. Research Report No. 251, Department of Statistical Science, University College London, 2005.

Packman, J. C. and Kidd, C. H. R.: A logical approach to the design storm concept, Water Resour. Res., 16, 994–1000, doi:10.1029/WR016i006p00994, 1980.

Pilgrim, D. H.: Australian rainfall and runoff, a guide to flood estimation. The Institution of Engineers, ACT, Australia, 1987.

Pilgrim, D. H. and Cordery, I.: Rainfall temporal patterns for design floods, J. Hydr. Eng. Div.-ASCE, 101, 81–95, 1975.

Restrepo-Posada, P. J. and Eagleson, P. S.: Identification of independent rainstorms, J. Hydrol., 55, 303–319, 1982.

Rigo, T. and Llasat, M. C.: Radar analysis of the life cycle of Mesoscale Convective Systems during the 10 June 2000 event, Nat. Hazards Earth Syst. Sci., 5, 959–970, doi:10.5194/nhess-5-959-2005, 2005.

Salsón, S. and Garcia-Bartual, R.: A space-time rainfall generator for highly convective Mediterranean rainstorms, Nat. Hazards Earth Syst. Sci., 3, 103–114, doi:10.5194/nhess-3-103-2003, 2003.

Témez, J.: Cálculo Hidrometeorológico de caudales máximos en pequeñas cuencas naturales, Dirección General de Carreteras, Madrid, España, 1978.

Vaskova, I.: Cálculo de las curvas IDF mediante la incorporación de las propiedades de escala y de dependencia temporales, PhD Thesis, Universitat Politècnica de València, 2001 (in Spanish).

Walesh, S. G., Lau, D. H., and Liebman, M. D.: Statistically based use of event models. Proceedings of the International Symposium on Urban Storm Runoff, University of Kentucky, Lexington, 75–81, 1979.

Watt, E. and Marsalek, J.: Critical review of the evolution of design storm event concept, Can. J. Civil. Eng., 40, 105–113, doi:10.1139/cjce-2011-0594, 2013.

6

From engineering hydrology to Earth system science: milestones in the transformation of hydrologic science

Murugesu Sivapalan[1,*]

[1]Department of Civil and Environmental Engineering, Department of Geography and Geographic Information Science, University of Illinois at Urbana-Champaign, Urbana, Illinois 61801, USA
[*] *Invited contribution by Murugesu Sivapalan, recipient of the EGU Alfred Wegener Medal & Honorary Membership 2017.*

Correspondence: Murugesu Sivapalan (sivapala@illinois.edu)

Abstract. Hydrology has undergone almost transformative changes over the past 50 years. Huge strides have been made in the transition from early empirical approaches to rigorous approaches based on the fluid mechanics of water movement on and below the land surface. However, progress has been hampered by problems posed by the presence of heterogeneity, including subsurface heterogeneity present at all scales. The inability to measure or map the heterogeneity everywhere prevented the development of balance equations and associated closure relations at the scales of interest, and has led to the virtual impasse we are presently in, in terms of development of physically based models needed for hydrologic predictions. An alternative to the mapping of heterogeneity everywhere is a new Earth system science view, which sees the heterogeneity as the end result of co-evolutionary hydrological, geomorphological, ecological, and pedological processes, each operating at a different rate, which help to shape the landscapes that we find in nature, including the heterogeneity that we do not readily see. The expectation is that instead of specifying exact details of the heterogeneity in our models, we can replace it (without loss of information) with the ecosystem function that they perform. Guided by this new Earth system science perspective, development of hydrologic science is now addressing new questions using novel holistic co-evolutionary approaches as opposed to the physical, fluid mechanics based reductionist approaches that we inherited from the recent past. In the emergent Anthropocene, the co-evolutionary view has expanded further to involve interactions and feedbacks with human-social processes as well. In this paper, I present my own perspective of key milestones in the transformation of hydrologic science from engineering hydrology to Earth system science, drawn from the work of several students and colleagues of mine, and discuss their implication for hydrologic observations, theory development, and predictions.

> எப்பொருள் யார்யார்வாய்க் கேட்பினும்
> அப்பொருள் மெய்ப்பொருள் காண்ப தறிவு
> In whatever matter and from whomever heard,
> Wisdom will witness its true meaning.
> Thirukkural, Verse 423 (Thiruvalluvar, c. 132 BC).
> Translated by GU Pope, 1886

1 Introduction

Hydrology has undergone an almost complete transformation over the past century, from its empirical origins in the early 20th century, to become a fully fledged and key component of Earth system science by the early 21st century (Chow, 1964; Sivapalan and Blöschl, 2017). In this paper, I use the precious opportunity given to me by the European Geosciences Union, as part of my 2017 Alfred Wegener Lecture, to reflect on the evolution of the field, as both interested observer and participant, as well as a student of hydrology for the past 50 years.

I hope that the issues I raise, and the trends and milestones I recognize in the evolution of the science that I recount here, resonate with the experiences of many readers, even if they might not fully agree with my interpretations of the events and/or their antecedents. This is one person's view of the vast landscape that hydrologists have collectively traversed, so perspectives abound. This is also not the first commentary of its kind: indeed, my thinking during my formative years was guided by the writings of Vit Klemeš and James Dooge, including their classic papers (Klemeš, 1983, 1986; Dooge, 1986). Their writings have provided unique perspectives on the field of hydrological science and its evolution over the past several decades. I hope that by re-telling this story through the admittedly narrow prism of my own observations and experience as a catchment hydrologist, I might be able to provide fresh impetus to early career hydrologists and new entrants to the field, and encourage them to reflect on and chart the course of their own future education and research.

2 Engineering hydrology: legacies of the past to new beginnings

My earliest introduction to hydrology and later work experience are probably similar to those of many hydrologists of my generation, i.e., 1970s vintage, especially those coming out of engineering schools. Going back in time, 50 years ago hydrology was mostly non-existent as a subject of study in many engineering schools (certainly not in Sri Lanka where I grew up). Most water-related education was then centered on fluid mechanics, e.g., open channel or river hydraulics, with a strong focus on applications to irrigation or hydraulic engineering. In 1972 when I walked into a hydrology class at the then University of Ceylon (later Sri Lanka), it was taught by a recent convert from fluid mechanics, who had subsequently received specialist training in hydrology.

Hydrology education in those days covered three key topics: infiltration and runoff generation (e.g., SCS curve number method), the unit hydrograph method to runoff routing, and the flood frequency curve approach to flood estimation (Linsley et al., 1958). The focus of the training was on estimation, especially flood estimation (event scale, recipe style) and the approach was lumped and empirical. Compared to now, these involved relatively much simpler problems, framed directly at the catchment scale, and formed the *bread and butter* of engineering hydrology, not just in Sri Lanka, but all over the world. The three building blocks of engineering hydrology were indeed legacies of three different previous eras in the growth of the field up until the 1970s (Sivapalan and Blöschl, 2017): empirical era (flood frequency analysis), rationalization era (SCS curve number method) and systems era (unit hydrograph method). These estimation methods remain relevant to this day, very much the core of engineering hydrology, judging by the popularity of advanced textbooks such as Chow et al. (1988), which is still widely used for the training of hydrologists in engineering schools.

For some of us who were then being trained in mechanics – solid mechanics, fluid mechanics, soil mechanics, etc. – the hydrology being taught and practiced using traditional estimation methods might have appeared strange and even mysterious. In the early days, I remember being frustrated that I could not make much physical sense of the cookbook treatment of the hydrologic estimation procedures, being unable to connect it to fundamental fluid mechanics principles. However, there was also wide appreciation for these otherwise simpler methods that efficiently solved practical problems of that day and age. In fact, even today's standard engineering practice would still be impossible without these simpler approaches.

There was also a profound fascination. How is it possible that something so inherently complex, as hydrology is, still comes out so simple (apparently) in practice? Over the years, there was a grudging realization that perhaps hydrology is more than about fluid mechanics (Yevjevich, 1968). Indeed, the search for simplicity in amongst the enormous mechanistic complexity has been a constant theme in hydrologic research over the last 50 years (Rodriguez-Iturbe and Valdes, 1979; Sivapalan, 2003; Savenije, 2017). The answer to this question, we now know in hindsight, is tied up with the issue of scale (both space and time) and the objects of our study, catchments, being not simple physical (or mechanical) objects but complex ecosystems (Gaál et al., 2012), consisting of component parts that have co-evolved together over time and are thus co-dependent (Dooge, 1986; Davies, 1992). I return to this theme throughout this paper.

3 Spatial heterogeneity and scale: promise of Newtonian mechanics

While the simple black-box methods of engineering hydrology did solve practical problems efficiently, they had limitations when extrapolating to circumstances beyond those from which they were developed. For example, what is the effect of antecedent wetness on runoff generation (Mein and Larson, 1973)? How does one account for the nonlinearity of catchment response to the size of rainfall events (Minshall, 1960)? How does one deal with the climatic and geologic controls when extrapolating flood frequencies to ungauged catchments (IACWD, 1982)? Also, new and more complex problems were emerging, such as the effects of land use changes on streamflows and flooding, and increased concerns about river water quality, which demanded application of more process understanding (Woolhiser, 1973). There was increasing realization that only a recourse to fluid mechanics and process physics would help address such questions, bring about needed improvements to the methods and make them applicable more generally.

Hydrologists thus began to open up the black-box to explicitly capture space, and to characterize water movement in catchments through application of physical (fluid mechanics) principles. The physically based models that were starting to be developed at the time were based on Newtonian mechanics valid at the laboratory or hydrodynamic scale (e.g., Richards equation, St.-Venant equations; Chow et al., 1988), and not at the scale of a catchment. For example, Darcy's law embedded in the Richards equation is based on a local equilibrium assumption to warrant a well-defined potential, implying that internal mixing in the pore spaces is faster than external disturbance, but this is only possible at small scales (Or et al., 2015). This also brought with it the enormous challenge of specifying the (highly heterogeneous) parameter values to go with the equations to represent the landscape heterogeneity (e.g., subsurface soil formations, surface features such as micro-topography, macropores, rills and the stream channel network, and vegetation cover), and the complexity of the resulting flow processes (Freeze, 1974). These problems are at the heart of the scale issue, which remains a major challenge in hydrology in spite of considerable progress made to date, of which I will speak more later.

A key development in this line of reasoning was Peter Eagleson's 1970 book titled *Dynamic Hydrology* (Eagleson, 1970), which provided a bold theoretical framework for a new hydrology based on the consistent and rigorous application of fluid mechanics principles, i.e., Newtonian mechanics, along with an appreciation of an organized structure behind soil, land surface, and vegetation heterogeneity. Eagleson's treatment of hydrology was a major departure, a paradigm shift, away from the lumped, black-box treatment practiced until then.

Another major milestone was the computer implementation of the *coupled* governing equations by Allan Freeze (Freeze and Harlan, 1969) that framed the catchment hydrologic modeling problem as a boundary value problem. Progress in this area was advanced by the increasing availability of digital terrain information on topography, soils, vegetation, etc., and computational tools that allowed them to be processed and visualized (Band and Wood, 1988). These data allowed the natural heterogeneity of landscapes to be captured in a more realistic way. Also, it was exciting that new runoff generation mechanisms such as saturation excess overland flow and shallow subsurface stormflow were being discovered in the field (Hewlett and Hibbert, 1967; Dunne and Black, 1970) and could now be faithfully replicated in models using the newly available spatial information, e.g., digital elevation models (DEMs) (Beven and Kirkby, 1979; Beven, 1981; Freeze, 1980; Band and Wood, 1988).

3.1 Heterogeneity: to resolve or to parameterize?

The Freeze and Harlan modeling paradigm has remained in force to this day, and several formal models have been developed based on the paradigm, such as the SHE model and variants of the same (Abbott et al., 1986). Due to data limitations and limits on computing power, early modeling efforts were restricted to small catchments, and their goal was limited to generating process understanding, and not so much to making predictions in real catchments (Stephenson and Freeze, 1974; Freeze, 1974). Their adoption and use in real-world applications did not take off for a long time because of their enormous data needs and insufficient computing power and also concerns about the appropriateness of the process physics that could be included in the models at the model element scale (Beven, 1989; Beven and Germann, 1982; Grayson et al., 1992).

As computing power increased many-fold and the landscapes began to be mapped at increased resolution for soil, vegetation and topographic characteristics in many parts of the world, there has been a revival of sorts in recent times, with several models being developed and beginning to be applied to *larger* (e.g., meso-scale) catchments using more realistic data. Some (of the many) examples include ParFlow (Kollet and Maxwell, 2008), FIHM (Kumar et al., 2009), PAWS (Shen and Phanikumar, 2010), and HydroGeoSphere (Brunner and Simmons, 2012). These modeling efforts are expanding and I will not be surprised if, before long, such models are used not just to reproduce field observations, but also, as in the case of Freeze (1980), to discover and explain previously unobserved phenomena at all scales.

In recent times, as part of climate change studies, modeling has been extended to continental and global scales to serve as the land-surface hydrology components of regional or global climate models. Because of the large scales, spatial resolution of such models in the past tended to be very coarse, on the order of hundreds of kilometers (e.g., bucket model, Manabe, 1969). However, progress is continually being made to improve the resolution from hundreds of kilometers to hundreds of meters, helped along by increased computational power and the availability of terrain information at finer resolutions (e.g., Community Land Model, Oleson et al., 2013). There is now a new thrust to improve the resolution of these models to 30 m for the whole globe (note: 30 m is the resolution of terrain information globally available), under the theme of "hyper-resolution modeling" (Wood et al., 2011; Bierkens, 2015; Wing et al., 2017). Concerns and debates about the appropriateness of the physics used in these models, the appropriateness of constitutive relations being adopted (Loritz et al., 2017), and the resulting uncertainty in hydrologic predictions, have not gone away, however (Clement, 2011; Beven and Cloke, 2011; Wood et al., 2012).

An ostensible reason for going for finer resolution, is of course, since now we can do it, i.e., with data availability and computing power, why not, and also since the benefits of hyper-resolution modeling for both science and practice are increasingly becoming evident. For example, hyper-resolution models over regional or continental scale domains can track large-scale storm movements and the resulting impacts at scales relevant to human wellbeing, as well as generate insights about large-scale teleconnections between or within regions (Senatore et al., 2015; Fang and Shen, 2017). A more immediate reason, however, is that as one improves the resolution, the (fervent) hope is that all heterogeneity will disappear and the governing equations used will match the spatial scales of the processes one is trying to model (e.g., Clark et al., 2015, 2017; Peters-Lidard et al., 2017). In reality, of course, field evidence has strongly indicated that the heterogeneity does not disappear even if one goes to finer scales, and there is a limit to how far we can split the landscape to achieve any desired level of homogeneity and still maintain a continuum (Beven, 1989; Blöschl and Sivapalan, 1995). Besides, the presence of macro-pores and other preferred pathways (Beven and Germann, 1982), and explicit treatment of runoff phenomena relating to the "old water, new water" concept (McDonnell, 1990), present major challenges to traditional continuum representations of water movement on, and particularly below, the land surface based on Newtonian mechanics alone (Beven, 2006). One therefore reaches the alternative conclusion that, whatever grid scale one chooses, there is no alternative but to parameterize the effects of any remaining heterogeneity at the sub-grid scale.

The end result for spatially distributed, physically based models of catchment hydrology is that model development has been caught between two temptations (Hrachowitz and Clark, 2017):

1. split the landscape more and more down to the continuum (hydrodynamic) scale so that heterogeneity disappears as an issue (i.e., heterogeneity is completely resolved);

2. aggregate to some scale at which the effects of heterogeneity are taken care of through simplified treatments with process parameterizations that can account for the effects of all sub-grid heterogeneity.

Advancing technological capability and increased process knowledge at small scales are the drivers towards the former option (Wood et al., 2011). Appreciation of scale effects, including expectation of simplicity through averaging (Sivapalan, 2003), and concerns about predictive uncertainty (Beven and Cloke, 2011) and about a dominant focus on techniques and other quick fixes (e.g., model–data assimilation) over the important search for hydrological understanding (Klemeš, 1986), are the key drivers towards the latter option. The problem of heterogeneity is not unique to hydrology, and applies to all environmental sciences, and approaches to mediate the alternative perspectives presented above may draw inspiration from a quote from ecologist Simon Levin (1992):

> To scale from the leaf to the ecosystem to the landscape and beyond... we must understand how information is transferred from fine scales, and vice versa. We must learn how to aggregate and simplify, retaining essential information without getting bogged down in unnecessary detail. The essence of modeling is, in fact, to facilitate the acquisition of this understanding, by abstracting and incorporating just enough detail to produce observed patterns... the objective of a model should be to ask how much detail can be ignored without producing results that contradict specific sets of observations, on particular scales of interest.

Levin's (1992) is an argument for "coarse graining", i.e., formulating models with only the necessary degree of complexity, and against attempts to fully resolve the enormous heterogeneity and complexity of hydrologic processes at progressively smaller timescales and space scales, a perspective which is echoed by hydrologist Thomas Dunne as well (Dunne, 1998).

3.2 Organized heterogeneity and preferred space scales: REA

Inspired by the organization that nature exhibits around a hierarchy of spatial scales, i.e., hillslope, catchment, region, etc. (Blöschl and Sivapalan, 1995), there have been concerted efforts to identify whether a characteristic space scale, which reflects the spatial organization and can thus serve as the building block of distributed models, exists in catchment hydrology. This idea is similar to the continuum or REV concept long used in groundwater hydrology with a considerable degree of success (Bear, 1972; Hassanizadeh and Gray, 1979). The argument has been that aggregating the governing equations or process descriptions to this building block scale might lead to simplified (effective) lumped or continuum treatments, obviating the need to split the catchment into smaller elements to capture the effects of heterogeneity. This way of thinking reflects the long-standing conviction, supported by observations, that in spite of the enormous complexity of hydrologic processes in landscapes, catchment-scale hydrologic responses can often be described by simpler models with only a few parameters (Jakeman and Hornberger, 1993; Sivapalan, 2003). Dooge (1986) has argued that catchments are complex systems with some level of organization, and indeed "simplicity out of complexity" is a useful property of such complex systems (Davies, 1992). However, when one opts for the parameterization approach inspired by this reasoning, one still needs to know key features of the underlying heterogeneity, e.g., their statistical distributions or organizational structure (and not necessarily the actual observed patterns). Furthermore, we also need to have information about cross-scale process interactions that might lead to the simplicity we desire (Dunne and Black, 1970; Hassanizadeh and Gray, 1979), and must utilize efficient approaches to incorporate them into models through appropriate model structures and parameterizations (Beven and Kirkby, 1979; Zehe et al., 2014).

Motivated by this reasoning, Wood et al. (1988) pursued this scale question in the context of runoff generation responses at the catchment scale. On the basis of spatial averaging of numerical simulation outcomes from a distributed

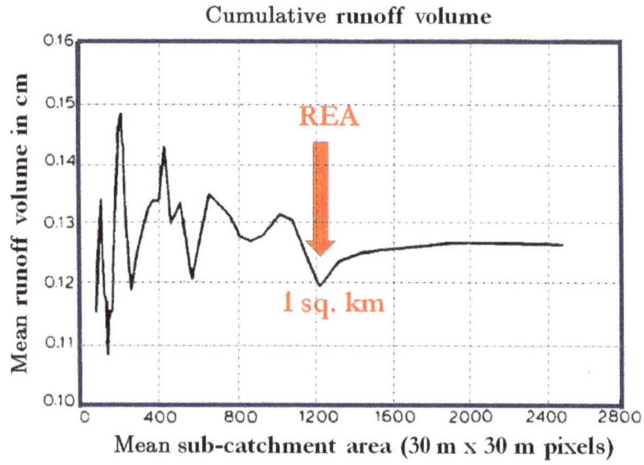

Figure 1. Mean event runoff volume as a function of mean sub-catchment area (for the case of variable topography but uniform soils and precipitation). Adapted from Wood et al. (1988).

rainfall–runoff model of hypothetical catchments, Wood et al. (1988) postulated that such a spatial scale could exist, which they called the representative elementary area (REA), and estimated its size to be about $1\,\text{km}^2$ (Fig. 1). Subsequent field observations and simulations in New Zealand and Germany supported the idea with some reservations and further refinements (Woods et al., 1995; Didszun and Uhlenbrook, 2008). The claims about both the existence and size of the REA have been questioned by others, based on perceived limitations of the averaging approach used to resolve these questions (Fan and Bras, 1995; Blöschl et al., 1995). A further limitation of the REA of Wood et al. (1988) is that it focused on surface runoff volume, and not its timing. Robinson et al. (1995) explored the change in dominant controls on runoff timing (e.g., hydrograph dispersion) with increased catchment area, and showed that the scale at which the dispersion is maximally reproduced can be affected by whether (fast) surface or (slower) subsurface runoff pathways dominate. The significance of the existence of the REA, whatever its size, is that it is small enough to still account for relevant spatial hydrological variations, and yet large enough to avoid the overwhelming data and parameter needs of small-scale (distributed) models. As a result, despite the reservations, the notion of using a representative watershed scale as the building block towards the development of distributed models (now at a sub-catchment scale) has remained as a working hypothesis for organizing our thought processes in respect of modeling.

Inspired by the possibility that adoption of such a representative scale might lead to simplified yet physically based hydrologic prediction models, Reggiani (1999) developed a new theory of hydrology around this building block, now named the representative elementary watershed or REW, to account for some of the limitations of the earlier REA concept. Reggiani's theory was expressed in the form of balance equations for mass, momentum, energy and entropy written down directly at the scale of the REW (Reggiani et al., 1998), and an accompanying constitutive theory (Reggiani et al., 1999) to ensure the theory gives rise to a determinate set of governing equations. Subsequently, Zehe et al. (2006) derived soil moisture characteristic (i.e., constitutive) relationships at the REW scale through the upscaling of corresponding point-scale observations. Likewise, Lee et al. (2007) developed a set of closure relations for the many boundary fluxes for the catchment system – between different REWs and between different sub-regions within each REW – again, to account for the effects of sub-grid heterogeneity and process complexity, thus helping to complete the specification of the governing equations representing the catchment response.

The net advantage of the REW approach is that it ends up with the solution of a system of coupled ordinary differential equations (as opposed to a system of partial differential equations, which one obtains if they were to split a catchment into rectangular or triangular elements). Both Tian et al. (2006) and Lee et al. (2007) developed numerical schemes to solve these governing equations, thus forming a new generation of models based on the REW approach that parallel distributed models based on rectangular or triangular finite elements following the Freeze and Harlan paradigm. In this way Tian et al. (2006) and Lee et al. (2007) have contributed to a new class of models of intermediate complexity, resolving processes occurring at scales larger than the size of the REW, and parameterizing those happening at smaller scales (Beven, 2012). Their potential as a new modeling framework has also been highlighted through several applications. For example, models based on the REW approach have been applied to real catchments, generating space–time predictions of the rainfall–runoff response for catchments in Australia (near Darwin, Lee et al., 2007) and in Oklahoma in the United States (Li et al., 2012; Tian et al., 2012). In recent years, these models have been extended to make distributed predictions of sediment and nutrient transport and export at the catchment scale and across the corresponding stream networks (Patil et al., 2012; Ye et al., 2012). It should be noted that in spite of this success, the REW approach has still not taken off, nor has it been widely adopted as would be expected, due to the lack of progress in the development of constitutive and closure relations to account for the effects of sub-REW heterogeneity and process complexity in catchments everywhere; this is discussed next.

4 Catchments as ecosystems: limits of Newtonian mechanics

In spite of its promise, and in spite of the progress made so far, the REW approach shares the as yet unsolved problem confronted by all distributed models, which is how to derive or estimate parameters a priori at the scale of the chosen grid (in this case, the REW) without first fully resolving sub-

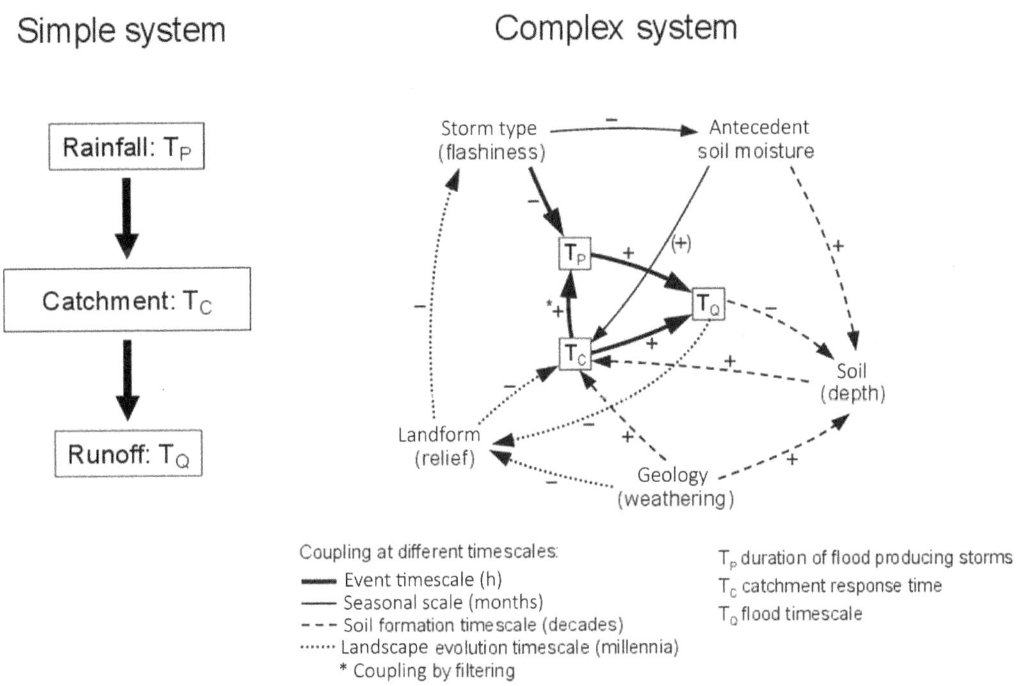

Figure 2. Simple and complex system representations of timescales contributing to floods. Interactions of multiple timescales make the response an emergent outcome of a complex system. Taken from Gaál et al. (2012).

grid *heterogeneity* and process *complexity*, which was the key problem in the first place. The spatial heterogeneity and the resulting cross-scale process interactions can introduce enormous complexity and richness to catchment hydrological responses. However, processes can become even more complex when we extend the analysis to include multiple timescales. When we attempt to make predictions over longer timescales, i.e., longer than events (e.g., months or years), the media within which hydrologic processes operate (i.e., soils, topography, vegetation) may themselves change dynamically or transform permanently (e.g., time-variant macroporosity due to the life cycle of earthworms, Zehe and Sivapalan, 2009; soil swelling and cracking, Savenije and Hrachowitz, 2017). In other words these are not inert objects fixed for all time, as is commonly assumed in deterministic models. Furthermore, if the hydrologic prediction problem is deemed a boundary value problem, as it often is, it also means that the location of the boundary and its condition itself are not fixed, but evolve dynamically as part of the water dynamics, the very dynamics we are trying to predict (see Fig. 2). The end result is that the hydrologic response can no longer be described by Newtonian mechanics alone, and is now the emergent outcome of two dynamic, coupled processes: water flow itself and the changes to the media, both occurring over multiple timescales (Gaál et al., 2012). Note that all of the complex spatial processes we talked about operate in the time domain as well.

The most obvious example of a catchment subsystem that evolves dynamically is vegetation, which is directly involved in the crucial process of evapotranspiration. Vegetation adapts itself dynamically over multiple timescales in response to changing climate and hydrology, even as it has an impact on the hydrology. Also, over much longer timescales it can modify the environment around it (e.g., soil structure, macro-pores, and topography, to name a few). In this way the temporal dynamics of vegetation can also indirectly impact runoff generation processes. As well as evolving with the dynamic hydrology, such biologic activity renders the system even more internally dynamic. The adaptations of vegetation and their modifications of landscapes may be governed by biological or ecological processes, and cannot be captured by universal descriptions based on Newtonian mechanics that focus on flow processes alone; presently their effects on flow can only be described by empirical, place-based relationships. As in the case of spatial upscaling problems discussed before, it is always tempting to ask the question whether the *temporal* organization of a catchment's hydrologic response, arising from vegetation adaptations to the multi-scale temporal variability of climate and hydrology, and the resulting landscape modifications, could also be averaged to produce a simplified description of catchment response. This question is explored next by highlighting different examples of vegetation (and even landscape) adaptations to natural climate variability and hydrology.

4.1 Adaptation strategies by vegetation

Understanding vegetation adaptation processes is a rich and multifaceted problem, but here I will use examples from my own modeling experience to only illustrate the point about simplicity in spite of complexity, and how this might arise as a result of vegetation adaptations in the time domain. Many existing catchment models, including the ones mentioned before, approximate evapotranspiration rates by vegetation (e.g., forests, grasses) as if they function like human-engineered agricultural crops; i.e., they transpire at a potential rate (e.g., potential evaporation) when the soil is wet, and reduce their evapotranspiration rate when soil moisture falls below the saturation value. In reality, this is not normally the case under native vegetation, and such approximations can lead to serious discrepancies when evapotranspiration is a large fraction of annual water balance, as is the case in most dry places around the world (Sivapalan et al., 1996; Thompson et al., 2011a). This was highlighted in the Darling Range region of south-western Western Australia during field experiments carried out by Silberstein et al. (2001).

South-western Western Australia experiences a strong Mediterranean climate, with warm dry summers and cold wet winters. Field measurements made by Silberstein et al. (2001) in a forested catchment in this region showed that daily evapotranspiration rate (bare soil evaporation plus transpiration) was roughly 2.5 mm day^{-1} during a 2-week period in the wet winter phase of the experiment. Remarkably, when the experiment was repeated 6 months later during a 2-week period following a very dry summer, the evapotranspiration rate turned out to be very much the same, about 2.5 mm day^{-1} every day. The assumption that trees, like agricultural crops, might reduce their transpiration rates during dry periods (due to their experiencing water stress) was found not to be the case. It turned out that native Eucalyptus trees in this part of Western Australia have adapted to the strong seasonality of climate (hot dry summers and cold wet winters) by growing dynamic deep roots that tap into a groundwater aquifer present that is over 30 m deep. Indeed, it was found that tree roots remained always in contact with the deep water table even when the latter fluctuated between summer and winter and between years. The net result is an apparently constant transpiration response by the vegetation, i.e., 2.5 mm day^{-1} every day.

In a subsequent study at a site near Darwin, in Northern Territory, Australia, Schymanski (2007) reported an altogether different response by native vegetation. This area experiences a monsoon climate, with a wet 4-month period of heavy monsoon rains, followed by 8 months of dry (non-rainy) conditions. Being tropical, potential evaporation is uniformly high during both periods. Water table is present at a depth of about 10 m all year. Schymanski reported that measured total evapotranspiration rate appeared to follow the variation of surface soil wetness, regardless of the energy available, i.e., potential evaporation (Schymanski, 2007). However, the adaptation strategy by the prevailing vegetation was different. In Darwin, the more permanent, deep-rooted trees tap into the groundwater table, just as in Western Australia, and were found to transpire at a rate of 1 mm day^{-1} all year, regardless of surface soil moisture. However, in addition, during the wet season, a dense under-story of grasses develops which, like crops, transpires at rates proportional to the surface soil moisture: a peak of 2.5–3.0 mm day^{-1} during the wet season, decreasing as the soil dries during the subsequent dry season, and dropping to negligible values as the grasses completely senesce and dry out. Once again, a slightly more complex, almost predictable transpiration pattern that arises via complex adaptation strategies adopted by the vegetation to the climate and hydrology.

Two different places, two different adaptation strategies by vegetation that has developed in each setting, yet permitting a simplified transpiration pattern in each case: in both cases, the strategies used by the vegetation are different from what is normally assumed for agricultural crops. This raises two issues: if one wants to develop a predictive model of evapotranspiration for a specific place, then the adaptation strategy adopted by the vegetation that develops locally must be known a priori. Newtonian theory by itself cannot predict the vegetation and its adaptation strategy one is likely to encounter in a given place, both of which may be governed by (the as yet unknown) biological/ecological laws at the scales of interest to us. The adaptation strategy adopted by the vegetation would be relatively easy to determine, given the kind of observational evidence available to Silberstein et al. (2001) or Schymanski (2007). Few places would have that kind of field evidence, however. In a catchment context, many more places will have rainfall–runoff data only, from which the adaptation strategy adopted by vegetation may have to be inferred (with a lot more ambiguity, given the kind of limited data, especially in the absence of evapotranspiration data). The fact that prevailing vegetation in different places may use different adaptation strategies, in the absence of such supporting data, also opens the way for much freedom or pluralism in the development of predictive models, as opposed to a universal model that would be expected from Newtonian mechanics.

4.2 Ecosystem function and data-based inference

A key observation from the last section is that the dynamics of evapotranspiration is intimately connected to that of vegetation adaptation. We can broaden this argument to include other catchment responses as well, such as runoff. We are quite used to treating runoff generation as a physical process, governed by Newtonian mechanics (Larsen et al., 1994). This presumes that we know a priori what mechanism of runoff generation dominates in a given place. However, Newtonian mechanics alone may not be able to determine a priori (i.e., without field evidence) the dominant runoff generation mechanism in any given place. Based on field evidence Dunne (1978) provided a perspective on climate, soil, topography and vegetation controls on dominant runoff generation mechanisms, popularly known as the Dunne diagram. To this day, the Dunne diagram has continued to defy explanations based on Newtonian mechanics alone (Larsen et al., 1994; Li et al., 2014). Vegetation adaptation may again be the chief cause of this phenomenon, e.g., vegetation perhaps adapting itself and adapting the environment around it (i.e., soils, topography) in such a way that it can retain soil moisture longer in arid environments and drain water sooner in wet environments. Also, forest soils are known to have higher surface infiltration capacities, which also impact their stormflow response. Thus, "forest" vegetation type becomes a property linked to infiltration and runoff generation, as well as to evapotranspiration.

Given this field evidence on evapotranspiration and runoff generation, and given the limitations of the more reductionist approach based on Newtonian mechanics, one is tempted to look for a complementary but holistic approach that can accommodate the adaptation of catchment properties (i.e., vegetation, soils) to environmental conditions. What if, instead of treating catchments as physical or mechanical systems, we consider them more broadly as ecosystems. Then the same processes, i.e., evapotranspiration and runoff generation, could be deemed a catchment's ecological responses, as part of its overall ecosystem "function". This was the wisdom behind the "functional" approach proposed by Black (1997), who framed a catchment's hydrological responses to precipitation more broadly as partition, storage, transmission, and release (see also Wagener et al., 2007), as opposed to point-scale processes (e.g., infiltration), in this way aligning them to the catchment's ecosystem function.

Another observation from the last section is that in spite of the known complexity of hydrologic processes at small timescales (rain events) and space (point) scales, which Newtonian mechanics may well be able to capture under some circumstances, catchment organization and ecosystem adaptation impart a level of simplification to hydrological responses at the catchment scale. One example is the dynamics of evapotranspiration seen in forested catchments in Perth and Darwin. Another example is the well-known fact that the discharge of the groundwater aquifer to streams can often be represented by simple linear reservoir theory, characterized by a mean residence time, in spite of the known complexity of flow processes operating at small scales, governed by Darcy's law (Savenije, 2017). This intriguing connection between micro-scale complexity and macro-scale simplicity has been an unsolved problem in catchment hydrology (Dooge, 1986; Sivapalan, 2003). The challenge for predictions is that the mechanisms through which such transformations from complexity to simplicity happen are largely unknown. If the biological or ecological laws or principles that govern them at the catchment scale were known, we might then use them to derive simpler, holistic parameterizations of hydrologic responses. Until such laws become available and lead to the development of universal predictive models, however, the problem lends itself to a plurality of modeling ideas, including data-driven approaches that rely on inferences from available rainfall–runoff data.

Wittenberg and Sivapalan (1999) presented an interesting and insightful implementation of the data-driven inference idea. They performed diagnostic analysis of rainfall–runoff data in several catchments in Western Australia (WA). This first involved analysis of streamflow recessions (i.e., following the end of individual rainfall events) during different times of the year, and in WA's seasonal climate, attributing differences in the slopes of the recession curves to changing evapotranspiration (ET) within the year. This made it possible to back-calculate the unknown ET from the recession curve slopes, and thus piece together a simple groundwater balance model for the catchments. This approach to data-based inference was later extended by Kirchner (2009) to UK catchments, calling the approach "doing hydrology backwards". In both cases, however, the power of the approach was due to the apparent simplicity of the hydrologic response manifesting itself at the catchment scale, which results from the spatial landscape organization and the temporal adaptation of the catchment ecosystem to the prevailing climatic variability.

Following the lead of Wittenberg and Sivapalan, Jothityangkoon et al. (2001) and Atkinson et al. (2002) extended the data-based inference idea further by treating the catchment's observed rainfall–runoff response as a reflection of its ecosystem functions (i.e., partition, storage, transmission, and release) but manifesting differently at different timescales. They assumed that the ecosystem functions are reflected collectively in several runoff signatures (i.e., temporal patterns extracted from streamflow data at different timescales), through which it was assumed that the catchment reveals its internal dynamics or functioning (see Fig. 3). These streamflow signatures, e.g., the regime curve (monthly timescale), flow duration curve (daily), or flood frequency curve (multiple timescales), are thus seen as outward manifestations of internal ecosystem functioning, and can therefore be deemed emergent (temporal) patterns.

The functional insight and the realization that catchment responses can show organization along timescales gave rise to a new approach to the modeling of catchment rainfall–runoff responses, called the top–down or downward approach (Klemeš, 1983; Sivapalan et al., 2003). We might as well call it the functional approach, following the lead of Black (1997). This modeling approach is diametrically opposed but complementary to the bottom–up or upward approach to model development based on Newtonian mechanics (Klemeš, 1983), an example of which is the REW approach of Reggiani et al. (1998, 1999) discussed earlier. The idea behind the downward or functional approach is to develop models step by step, but starting at long timescales (e.g., the annual timescale, as in the case of Jothityangkoon

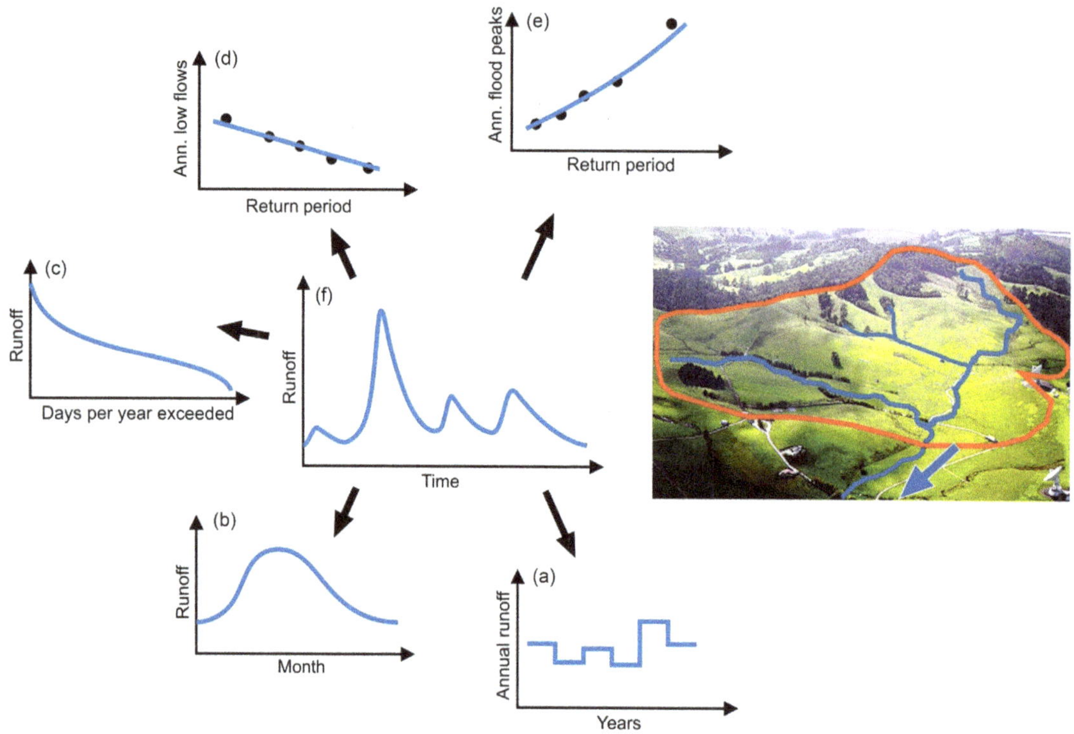

Figure 3. Runoff signatures deemed emergent patterns reflecting the functioning of the catchment ecosystem. **(a–f)**: annual runoff, seasonal runoff, flow duration curves, low flows, floods, and runoff hydrographs. Taken from Blöschl et al. (2013).

et al., 2001). Once a simple (simplest possible) model is developed that reproduces the runoff signature at the annual scale, i.e., inter-annual variability, one then moves down to the monthly timescale, and adds just enough functional complexity to the model to match a new signature, now at the monthly timescale, i.e., the regime curve. If partitioning is the main function that is accommodated at the annual timescale, storage and release could be functions added at the monthly scale, and so on. This process continues until a model is developed that is able to reproduce signatures extracted from streamflow data simultaneously at all timescales (Jothityangkoon et al., 2001; Atkinson et al., 2002). The top–down modeling approach bears some similarity to the data-based mechanistic modeling approach proposed earlier by Peter Young of Lancaster University, and illustrated in Young and Beven (1994) and Young (2003), as discussed in Sivapalan and Young (2005).

Interestingly, the "functional" models, developed in a top–down way through inference from data (e.g., Jothityangkoon et al., 2001; Atkinson et al., 2002), are in many ways similar to what have long been known as conceptual models, which have been developed from the 1960s, beginning with the Stanford Watershed Model (Crawford and Linsley, 1966). These conceptual model types can be lumped (at the catchment scale) or distributed (at the sub-catchment scale), and are also developed without recourse to Newtonian mechanics. Examples (among numerous other models) include the Tank Model (Sugawara, 1967), HBV (Bergström, 1976), LASCAM (Sivapalan et al., 1996), and the FLEX generation of models proposed in recent times (Fenicia et al., 2011; Gharari et al., 2011). Both conceptual and top–down model types take the form of a combination of inter-connected storage reservoirs that mimic the functioning of different parts of a catchment, both attempt to provide some kind of mapping between landscape structure and model structure (Sivapalan, 2005), and both types contain strong subjective elements. The main difference is the thought process that goes into developing these models. Development of top–down models follows a systematic procedure to decipher the model structure from observed rainfall–runoff variability at multiple timescales. In the case of traditional conceptual models, the model structure is chosen, ostensibly but often arbitrarily, to reflect the functioning of parts of the catchment, with the opportunity to further refine it through calibration or confrontation with rainfall–runoff observations. In another sense, both model types are also similar to physically based models of intermediate complexity arising from the REW approach (e.g., Tian et al., 2006; Lee et al., 2007), in that all three model structures are expressed in the form of coupled ordinary differential equations that reflect water balances of individual compartments (or storages).

5 Newtonian mechanics vs. ecosystem function: impasse and reconciliation

In the preceding sections I outlined two complementary approaches that hydrologists have pursued over the past 5 decades to make catchment-scale predictions. Using the terminology of Klemeš (1983), subsequently expanded by Sivapalan et al. (2003), these can be termed the upward (or bottom–up reductionist) and downward (top–down ecosystem or functional) approaches. The upward approach fundamentally involves application of Newtonian mechanics, either through spatially distributed models that attempt to explicitly resolve spatial heterogeneity, or through semi-distributed (e.g., REW) models that attempt to parameterize the effects of sub-grid heterogeneity, while keeping the essential physics. The downward approach to model development, on the other hand, involves making inferences from catchment-scale rainfall–runoff data (and other responses, e.g., evapotranspiration, if available).

The advantage of Newtonian mechanics based models is that the role of spatial gradients in controlling hydrologic processes, at least in principle, can be explicitly captured, and so the fidelity of process descriptions at small scales can be guaranteed (at least in principle). However, they have the disadvantage that they cannot (yet) account for the functioning of the catchment as an ecosystem, which is defined by process interactions in the time domain, the biological/ecological laws behind which are yet to be discovered in the catchment hydrologic context and incorporated into the models. The top–down approach to modeling has the advantage, being derived from rainfall–runoff data, that it can capture the holistic nature of catchment functioning. On the other hand, it cannot (yet) unambiguously account for the physical laws that govern flow processes in landscapes, especially in the spatial domain.

Indeed, it has been said that "bottom–up reductionist-type models are best suited to represent *"known" knowledge*, and are at their weakest to reveal *"unknown" knowledge* (R. Silberstein, personal communication, 2017); vice versa for top–down ecosystem-type models. In this sense, the biggest strength of the top–down approach is also the biggest drawback of the bottom–up approach, and vice versa (Sivapalan et al., 2003). These two perspectives will remain irreconcilable until major breakthroughs are made in our understanding of multi-scale spatial heterogeneity and temporal variability, no doubt governed by Newtonian mechanics at small scales, the resulting cross-scale interactions and adaptations at long timescales governed by ecological laws, and the understanding of how these contribute to whole ecosystem function and manifest themselves in more holistic and simplified parameterizations of catchment responses (Sivapalan, 2005; McDonnell et al., 2007). Presently, in this context, we are at an impasse.

One of the clearest manifestations of the impasse in modeling is in the uncertainty in the resulting hydrologic predictions (Beven and Binley, 1992). Neither approach is complete in terms of its theoretical foundation to generate the predictions we need. In the case of bottom–up models based on Newtonian mechanics, the predictive uncertainty arises from an inability to characterize the heterogeneity of landscape properties accurately, the lack of complete knowledge or understanding of hydrologic processes in the real world, and the inability to capture them in the models. In the case of top–down or functional models that depend on data-based inference, equally there is uncertainty arising from the diversity of catchment adaptations to climate and consequently in the plurality of model structures. In addition, there is further uncertainty due to the inability to unambiguously estimate the parameters through calibration, in the absence of physical guidance or universal laws to constrain the parameter values. In a broad sense, the uncertainty arises from both inadequate understanding of catchment behavior (*epistemic* uncertainty) and an inability to characterize catchment heterogeneity fully (*aleatory* uncertainty) (Beven, 2016). Further progress in modeling will thus depend on our ability to break the impasse between these two different modeling approaches by changing the science questions that have been behind hydrologic research over the past decades (Sivapalan, 2009).

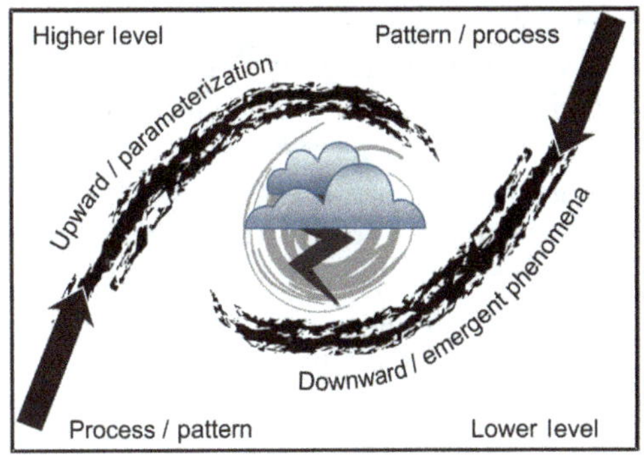

Figure 4. Reconciliation of downward and upward approaches – break from a reliance on parameterization of lower-level features to discovery and explanation of emergent phenomena at the higher level. Taken from Sivapalan (2005).

5.1 Breaking the impasse: change the question

True reconciliation, however, is unlikely to come about if we stay within the existing paradigms, through mere refinements to existing approaches to estimation and prediction. Reconciliation between Newtonian mechanics and ecosystem function based approaches will require a new era of research aimed at explanation and discovery rather than estimation and prediction (see also Burt and McDonnell, 2015). Reconciliation will require, on the one hand, that we understand the functional role of landscape heterogeneity, and on the other hand, the physics (i.e., mechanics or thermodynamics) underpinning ecosystem function. This might require going beyond Newtonian mechanics and searching for other universal laws or organizing principles that explain ecosystem functioning of catchments in particular places. Instead of parameterizing the effects of heterogeneity, the focus should be on recognizing it as an emergent pattern and coming up with explanations to describe how it came about and discover its ecosystem function (Fig. 4, Sivapalan, 2005). The modeling question should no longer be whether we can explicitly account for the effects of heterogeneity mechanistically, but whether its ecosystem function can be reproduced (McDonnell et al., 2007; Schaefli et al., 2011). In other words, to move forward towards both understanding and improved predictions, research goals and associated scientific questions must broaden from just *estimation to explanation*.

5.2 Moving from estimation to explanation: focus on phenomena

As we have seen, the focus on estimation goes back to the beginnings of engineering hydrology. In spite of the enormous progress we have made in our ability to make predictions, these advances have not contributed towards generalized theories that operate universally across catchments and places. General theories are very much needed to bring about the reconciliation of the diversity of modeling approaches we now have and achieve a unification of the field. Hydrologic research will thus have turn to new types of questions (Sivapalan, 2009), which are focused on explanation and discovery, perhaps as a necessary prelude to prediction. Organization of catchment responses and the patterns of ecosystem function that one observes in the real world are no longer objects to be mimicked by our models ("grist to the calibration mill", à la hydrograph fitting), but need to be seen as emergent phenomena. The enormous computational power and data availability that we now have should be utilized to discover and/or explain previously unobserved or unexplained phenomena at all scales and places (Li et al., 2014; Dunne, 1978). This may herald a major culture change in the way we do research.

Phenomena abound in hydrology, differing in complexity and richness, and arising in different contexts, which warrants a separate review article by itself, given their increasing importance in hydrology research. Examples of phenomena in catchment hydrology include the old water-new water concept (McDonnell, 1990), the pan evaporation paradox (Roderick and Farquhar, 2004), the Budyko hypothesis (Budyko, 1974), the proportionality hypothesis that lies behind the success of the SCS-curve number method (Wang et al., 2014), and the linear reservoir approximation to groundwater contributions to streamflow (Savenije, 2017). The goal of research will increasingly revolve around coming up with plausible hypotheses about their causes and testing them out through further observations or targeted modeling. If the explanations hold in several places, then it contributes to accumulation of knowledge and understanding, and eventually to general theories and underlying organizing principles. There is a long history in other branches of hydrology where such a focus on phenomena has led to major advances in hydrologic understanding. One can point as examples the phenomenon of *macrodispersion* in groundwater transport in heterogeneous porous media (Gelhar and Axness, 1983) and that of *hysteresis* in vadose zone hydrology (Hassanizadeh et al., 2002), where sound explanations were found while remaining within the Newtonian and/or thermodynamic frameworks.

Until recently, under the weight of the dominant estimation/prediction (i.e., hydrograph fitting) paradigm, inadequate attention has been given to the study of phenomena in catchment hydrology. Here I will illustrate, through two simple examples, how the physical or functional causes of phenomena can be explored, and how this might contribute to generalized understanding. The two examples I present here involve an apparent power-law relationship with catchment size (area). The first one involves annual maximum flood peak (scaled by catchment area), in this case, taken from several nested catchments in the Appalachian region of the United States (Fig. 5a: Smith, 1992; Robinson and Sivapalan, 1997; Sivapalan, 2005). The second one involves a sediment delivery ratio (the fraction of eroded sediment at an event scale that actually reaches the catchment outlet; Fig. 5b: Lu et al., 2005). Even though these examples are obtained from specific places, the phenomena themselves are universally observed. The question is, why is it a power law? Is there a causal explanation? Answers to such questions test our understanding of the underlying processes and the interpretation of observations of all kinds.

Robinson and Sivapalan (1997) showed that the observed power law can be explained by a simple argument, supported by a simple linear "bucket" model with two different timescales (event duration, and a mean response time that is a function of catchment area, due to the geometric relationship between length of travel and area), and an understanding of rainfall variability. If rainfall intensity is a constant during a single storm event, the magnitude of the flood peak can be analytically derived based on storm duration and mean response time: it does not follow the power law (see the dotted line... in Fig. 5a). Firstly, the introduction of within-storm rainfall variability increases the flood peak for small catchments (due to their fast response), but has no impact in large catchments (denoted by the symbol "***" in the figure). Secondly, the introduction of between-event interactions and seasonality increases the flood peak for large catchments (due to their effect on antecedent conditions), but has no influence for small catchments (denoted by o o o). Their combination leads to a combination of the two effects (denoted by - - - and —): flood peak is increased for small catchments (because of within-storm variability) and for large catchments (because of interaction between events and/or seasonality). So what looks like a power law is really an emergent pattern that falls out as a result of a complex in-

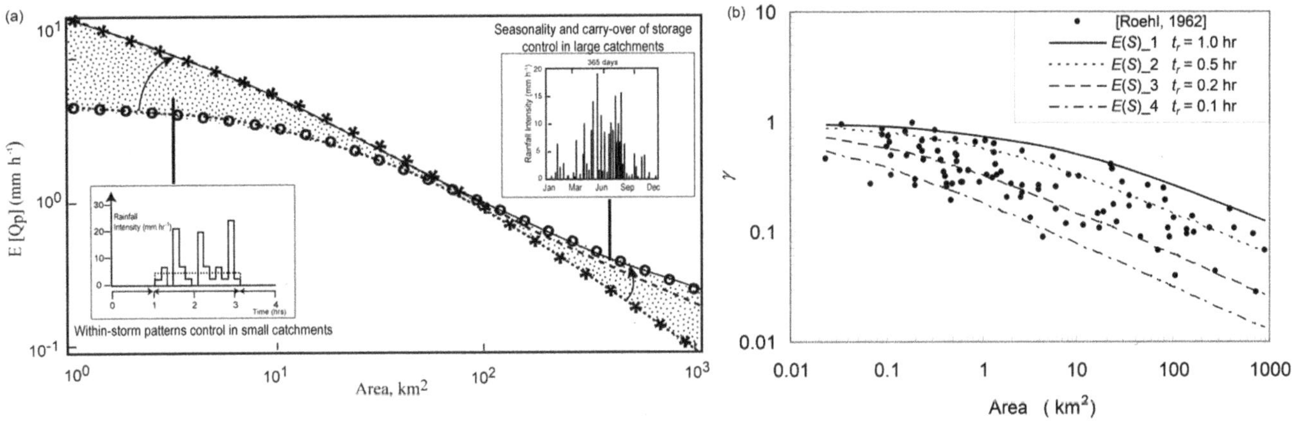

Figure 5. Explaining power-law relationships as emergent phenomena arising from timescale interactions: **(a)** mean annual flood as a function catchment area, A, in the Appalachian region (taken from Sivapalan, 2005); **(b)** sediment delivery ratio (SDR) γ as a function of catchment area for different values of rainfall duration t_r and an assumed power-law relationship of mean channel travel time t_c with A (taken from Lu et al., 2005).

terplay across different timescales, and the geometry of the catchment's organization.

The scaling of the sediment delivery ratio is an even more complex phenomenon to explain (with considerable scatter; see Fig. 5b), because it involves not only runoff processes as discussed above, but also sediment transport processes. Even in this case, however, Lu et al. (2005) proposed an explanation that was similar, indeed one that followed that of Robinson and Sivapalan (1997). Lu et al. too used a simple conceptual model that consisted of two linear stores arranged in series: a hillslope store that addresses transport of sediment to the nearest streams and a channel store that addresses sediment routing in the channel network. They showed analytically that, as with Robinson and Sivapalan (1997), the spatial scaling of the sediment delivery ratio (as an emergent pattern) can be explained, to first order, in terms of the interactions between several timescales (i.e., storm duration, hillslope/channel travel times, and two depositional timescales), and also catchment geometry as before.

I used these simple examples to also highlight the effects of the interactions between multi-scale temporal variability of both climate inputs and catchment response, and how they tend to generate emergent patterns due to the resonance that is generated through these interactions, even in the absence of adaptations of vegetation or catchment properties. In other words, the way that the catchment responds to a sequence of events is different from the way it responds to a single event, indicating that the response of the catchment must be considered holistically across multiple timescales. The importance of timescale interactions has also been highlighted in other contexts to explain several environmental phenomena: power-law-type streamflow recessions resulting from landscape heterogeneity (Harman and Sivapalan, 2009); the effect of rainfall event variability on the pesticide leaching risk of groundwater (McGrath et al., 2010); and climate controls on vadose zone contaminant transport (Harman et al., 2011). This issue will be explored in more detail in later sections using an expanded, co-evolutionary framework.

5.3 Towards explanation and extrapolation: new kinds of models, new kinds of phenomena

As the focus turns from estimation to explanation, the types of models one uses may also need to change. It is not an accident that in the two examples discussed above, I have demonstrated how simple models are able to provide first-order explanations of otherwise complex phenomena. In this age of big data, data mining and hyper-resolution modeling (Wood et al., 2011; Peters-Lidard et al., 2017), one might be surprised that I advocate the use of simple models. I do this deliberately because the focus here is not on prediction of the complete system response in one place with all its gory complexity, but as per the Einstein quote below (see the earlier quote from Simon Levin), on the ability to explain complex, possibly universal phenomena more simply, and thus contribute to accumulation of knowledge and understanding.

> If you can't explain it simply, you don't understand it well enough. – Albert Einstein

Note that the simplicity I advocate here is, in the words of James Dooge, a "search for... *a rational simplicity*, not the simplicity of crude assumptions,... or assuming the difficulty out of the way" (Dooge, 1995). In the spirit of the top–down reasoning, as also argued by Dooge (1995), we will be better off starting with simple models targeted on the phenomenon of interest, not models of everything, and increasing the complexity only as needed. Dooge suggested linearization and similarity analysis as effective ways to simplify models and gain understanding of phenomena, which was also the case in the previously highlighted work of Robinson and Sivapalan (1997) and Lu et al. (2005).

Phenomena of interest to us can be dynamics (temporal patterns) in a single place and on a range of timescales (Burt and McDonnell, 2015). Similarly, in the space domain, phenomena can be spatial patterns in a single place (within a catchment, Blöschl et al., 2016) or spatial patterns between places (catchments) or scales (as in the two examples above), extending all the way to regional or global patterns (Blöschl et al., 2013). Increasingly, as discussed before, the focus turns towards developing a generalized understanding of whole catchment responses and ecosystem functioning, and the discovery of universal laws. Therefore, it is no longer sufficient to study phenomena relating to a single place or a few places. Increasingly, the phenomena of interest to us will come from simultaneous observations in many different places with different histories of the kinds of timescale interactions shown to be critical in the two examples presented above (Hipsey et al., 2015), and the synthesis of data collected from across places, scales and processes (Blöschl et al., 2013).

6 Timescale interactions and catchment co-evolution: the Darwinian approach

Understanding the reasons for differences in required model structure between catchments, including the controls of climate and landscape properties, may be a good starting point in the search for more generalized understanding, by drawing from many different places with different histories. Atkinson et al. (2002) and Farmer et al. (2003) implemented the top–down modeling approach, in a comparative way, to two dozen catchments around Australia and New Zealand. They found systematic variations in model structure, with the differences reflecting differences in climate and in how vegetation and soils may have adapted to the climate. The modeling studies indicated evidence of a hierarchy of required model structures, with changing timescales (i.e., annual, monthly, daily), and with changing aridity. For example, the required model complexity increased with decreasing timescales, and increased with increasing aridity. The argument could thus be made that the required model structure, reflecting the functioning of a catchment, is itself an emergent property in terms of how the catchment has adapted itself to the prevailing climate and geology.

6.1 Catchments "marching to a different drummer": comparative hydrology

In order to shed more light on the required level of model complexity, Jothityangkoon and Sivapalan (2009) compared a dozen catchments from all around Australia and New Zealand, using the same top–down modeling approach described above (i.e., matching streamflow signatures on a range of timescales). They carried out a diagnostic analysis in each catchment that focused on elucidating the climate controls on one particular signature, i.e., inter-annual variability of runoff. In particular, the authors were interested in discovering the aspect of the within-year rainfall variability (i.e., storminess, seasonality) that may have a dominant control on the observed inter-annual runoff variability. The diagnostic analysis adopted involved running calibrated top–down hydrological models in each catchment, but now with artificial rainfall inputs that included storminess and seasonality separately, before being combined. The outputs from the model in each catchment, for each of the artificial climate inputs, were compared to the observed inter-annual variability. The question pursued was: what combination of within-year climate variability (seasonality, storminess) is needed to reproduce the observed inter-annual variability of runoff?

The study found that in catchments in Queensland in north-eastern Australia, inter-annual variability of runoff was most sensitive to storminess (not seasonality), whereas in catchments in Western Australia and South Australia, seasonality was the feature that contributed most to the observed inter-annual variability. This finding is remarkable, in that the climate of Queensland is indeed dominated by storminess (i.e., a small number of large storms), and the climates of South and Western Australia were indeed dominated by strong seasonality. The outcomes from these diagnostic analyses thus indicated that the water balance dynamics of catchments (i.e., model structure and parameters inferred from the data in a top–down way) somehow "resonate" with the dominant within-year variability, i.e., storminess in Queensland and seasonality in Western Australia. In other words, and paraphrasing American naturalist Henry David Thoreau, these results indicated that each catchment is "marching to a different drummer", the drummer being the variability of climate drivers present in a given place.

> If a (wo)man does not keep pace with (her)his companions, perhaps it is because (s)he hears a different drummer. Let (her)him step to the music which (s)he hears, however measured or far away.
> – adapted from Henry David Thoreau: Walden, 1854.

The modeling results also reaffirm the ecosystem or functional view presented earlier, including the argument that there is much to be gained by making inferences from observed rainfall–runoff time series. The results also raise the hope that a more universal understanding of both ecosystem function and catchment response and their relationship with climate and geology may be gained by repeating these modeling and diagnostic studies simultaneously in many more catchments around the world, in a comparative way, across gradients of climate and geology. This leads to the notion of *comparative hydrology*, first introduced by Falkenmark and Chapman (1989), and defined more broadly in the next section by linking it to catchment co-evolution.

6.2 Catchment co-evolution and Earth system science: a Darwinian view

Berghuijs et al. (2014) extended the above comparative hydrology approach to the modeling of seasonal water balances of over 400 catchments across the continental United States. On the basis of these modeling results they came up with a classification of observed seasonal water balances (i.e., regime curve, a key signature of runoff variability), expressed in terms of three driving factors: climate aridity, seasonality (i.e., relative timing of the precipitation within a year), and the fraction of snow as precipitation. On the basis of this similarity analysis, they identified 10 dominant catchment classes among the 400-odd catchments. Not surprisingly, they found a geographic aspect to the locations of the 10 classes, in that they were clustered geographically, influenced by the slow climatic and geologic variations across the continent. In addition they found that similarity of seasonal water balances carried over to have an imprint on between-class differences in several other signatures of runoff variability, such as the flow duration curve or flood frequency curve, and yet there was also considerable within-class variability due to other climatic and landscape factors.

An even more interesting finding from Berghuijs et al.'s (2014) results, however, was that the regional mapping of catchments on the basis of similarity of seasonal water balances was closely aligned with the regional mapping of not only vegetation classes and ecoregions (i.e., the different ways ecologists map vegetation and ecosystems), but also soil orders. This clearly indicated that the seasonal water balance response is not just about the partitioning of incoming precipitation into spatially resolved hydrologic processes, i.e., runoff, evaporation and storage change. In fact, it is both a contributor to, and an outcome of, the co-evolution of physical, biological and pedologic processes governing vegetation establishment and adaptation, and soil formation, as well as water balance, which all together reflect an underlying ecosystem function. Indeed, it is well known in the ecological literature that it is the seasonal water balance that determines vegetation types that become established in a given place, including their functioning and productivity (Stephenson, 1990; Robinson et al., 2012).

One can speculate therefore that the phenomenon of each catchment "marching to a different drummer" is a reflection of vegetation adapting to (and in turn changing) the environment around it (e.g., hydrology, soils), in response to the fluctuating water and energy supply at any place of interest (Sivapalan and Blöschl, 2015). This is tantamount to the soils, vegetation and topography belonging to a catchment co-evolving together in response to the climate above and geology below through both land-forming and life-sustaining processes operating and interacting across multiple timescales (Fig. 6). The focus can no longer be on just hydrological processes, but all land surface processes that operate together and feed back on each other, which thus takes hydrology into the realm of Earth system science. With this broader view, the focus of analysis is therefore no longer on streamflow signatures alone, but also on signatures arising from the co-evolution of Earth system processes, such as vegetation and erosional patterns that one sees in landscapes (Sivapalan, 2005; Brantley et al., 2017). Examples include patterns in measured evapotranspiration at individual (e.g., FLUXNET) sites (Thompson et al., 2011a), regional patterns of evapotranspiration measured by remote sensing (Cheng et al., 2011), and vegetation, soil and micro-topography patterns in landscapes on a range of scales (Thompson et al., 2011b; Saco et al., 2013; Harman et al., 2014). These are the kinds of ecohydrologic and hydrogeomorphic phenomena that will need to be further explored in the future and synthesized to generate the new understanding and new theories required to close the gap between bottom–up (reductionist) and top–down (ecosystem) approaches to modeling.

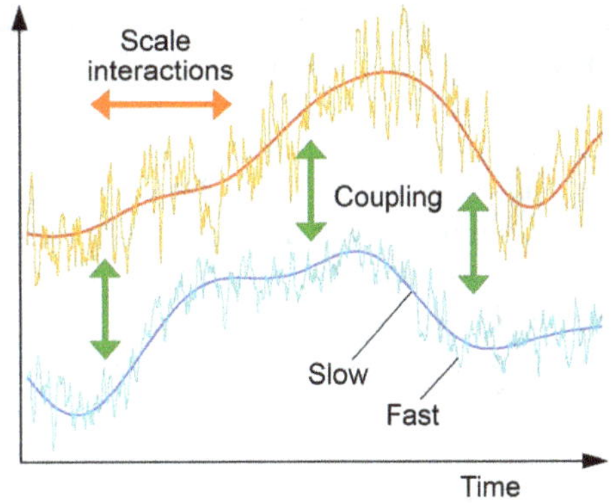

Figure 6. Co-evolution resulting from the interaction of a number of processes (at least two: one represents hydrological processes and the other landscape physical or ecological or social processes), each of which consists of fast and slow components producing emergent dynamics. Redrawn from Sivapalan and Blöschl (2015).

These outcomes of co-evolution, framed as emergent phenomena, may be different in different places, because of differences in the history of co-evolution and initial and boundary conditions of climate and geology. Comparative hydrology, i.e., comparing many catchments in different places along gradients of climate and geology, helps us to shed light on their co-evolution, just as Charles Darwin interpreted differences between animals, birds and plants he found in his travels through the lens of their natural history (Harman and Troch, 2014). The Darwinian approach to comparative hydrology embraces the history of each place, including features that are relics of historical events, as central to understanding both its present and its future (Blöschl et al., 2013). The essence of the Darwinian approach then is to develop generalizations beyond individual catchments through learning from differences between catchments, and interpreting them as legacies of past co-evolution.

6.3 Hydrologic similarity and catchment classification: steps towards generalization

Given the co-evolved nature of catchments, their structure and function, and the resulting hydrologic responses, explanation of the similarity and differences between catchments must take a more Darwinian approach that emphasizes timescale interactions specific to a place, and how the catchment ecosystem may have adapted to these. While there can be considerable diversity and randomness allowed in Darwinian co-evolution of catchments due to contingency effects (Beven, 2015), patterns and connections may yet be discernible which may lead to relationships of more general applicability. One approach to creating order in an otherwise heterogeneous world, widely adopted in chemistry and ecology, is through the process of classification. As in the case of the periodic table in chemistry, classification of catchments based on hydrologic similarity may be used to group catchments, helping to simplify relationships and generalize findings. Hydrologic similarity is the foundation for the transfer of information from one catchment to another, including from gauged to ungauged catchments, and was the basis for the successful outcomes of the Predictions in Ungauged Basins (PUB) initiative (Sivapalan et al., 2003; Blöschl et al., 2013; Hrachowitz et al., 2013). The PUB synthesis study by Blöschl et al. (2013) framed hydrologic similarity in terms of several signatures of runoff variability, such as the flow duration curve and the flood frequency curve. The diversity of catchment responses is such that no two catchments are so similar that all signatures are the same between them, and so for prediction purposes one concentrates only on the signature of interest. However, as in the case of the comparative hydrology study by Berghuijs et al. (2014), the synthesis project by Blöschl et al. (2013) found that similarity of seasonal water balance had a significant influence on the similarity of other signatures too, and may be the strongest indicator used to delineate hydrologically similar regions (Weingartner and Aschwanden, 1992).

The organization of catchments into distinct classes based on hydrologic similarity can open the way to understanding the differences between the classes themselves in terms of the underlying climatic, geologic and landscape controls. The most celebrated example of progress in this area is the empirically derived Budyko curve, famously named after Russian hydrologist Mikhail Budyko. The Budyko curve graphically expresses the long term (or mean annual) water balance, defined by the ratio of long-term average evapotranspiration to precipitation, E/P, as a universal function of climatic aridity, the ratio of long-term average potential evaporation to precipitation, E_p/P (Budyko, 1974). The existence of the Budyko curve is usually explained as resulting from a competition between water available (precipitation) and energy available (potential evaporation). In the case of the MOPEX catchments across the United States, Berghuijs et al. (2014) found that the different catchment seasonal water balance (and also vegetation/soil) classes clustered along different segments of the Budyko curve, with little overlap. On the basis of this and other evidence, one can then argue that the competition between water and energy available is mediated by the vegetation that forms through natural selection, highlighting the key role of ecohydrology in hydrologic theory development (Eagleson, 2002; Rodriguez-Iturbe, 2000).

The fact that most of the MOPEX catchments (and other catchments of the world) fall on or about the Budyko curve (leaving a large part of the Budyko space unoccupied) can be seen as an empirical organizing principle, i.e., only some combinations of climate, soil and vegetation characteristics may exist in reality (i.e., are *behavioral*, Schaefli et al., 2011), indicating a mutual co-dependence through Darwinian co-evolution and selection (Eagleson, 2002). Inspired by these observations, in recent times several hydrologists have embarked on a search for physical explanations for the existence of the Budyko curve (Wang and Tang, 2014; Wang et al., 2015; Westhoff et al., 2016). Wang and Tang (2014) and Wang et al. (2015) studied the conceptual bases for empirical models developed by engineers (and geographers) from the 1930s to the 1950s to describe water balances across three different timescales: event, seasonal and annual. These were the SCS curve number method for runoff estimation at the event scale (Mockus, 1949), the *abcd* model for seasonal (monthly) water balance developed during the Harvard Water Program in the 1960s (Thomas, 1981), and the L'vovich method developed in the former Soviet Union for partitioning of annual precipitation into fast and slow runoff and evaporation (L'vovich, 1979). In the first phase of their work, Wang and Tang (2014) and Wang et al. (2015) showed that all three of them can be described in terms of the so-called *proportionality hypothesis*, which is long understood as providing the conceptual basis for the SCS curve number method. In the second phase of their work, they showed that the proportionality hypothesis, in turn, can be explained thermodynamically, in terms of the principle of maximum entropy production (MEP). One can thus see that the empirical proportionality hypothesis and the thermodynamic MEP principle appear to provide a vehicle to extend the hydrologic similarity arising from co-evolution across multiple timescales. Admittedly, these are only the first tentative steps in the quest for new theories of hydrology, and yet more creative approaches are needed to extend the notions of similarity and optimality to cover not just the Budyko curve but all other signatures, in order that we fully exploit the order or simplicity that may arise out of Darwinian co-evolution.

The MEP principle, or the equivalent maximum power principle proposed by Kleidon and Renner (2013) and used by Westhoff et al. (2016) to derive the Budyko curve, can be viewed as providing a constraint to the diversity exhibited during Darwinian co-evolution and natural selection, possibly as a mechanism to increase system resilience (Lotka et al., 1922). Furthermore, existence of such physically based organizing principles, valid on a range of timescales as indicated in the work of Wang et al. (2015), can pave the way for a new generation of *behavioral* models (Schaefli et al., 2011). This is a promising development for two reasons. On the one hand, it can help constrain model structures inferred from rainfall–runoff data, and on the other hand, it can constrain the parameter combinations allowed for otherwise physically based models that are based on Newtonian mechanics (Li et al., 2014). A unique advantage of the approach adopted by Wang et al. (2015) is that it involved connecting the dots between empirical methods that have been widely used in engineering hydrology for over 50 years, and exploiting the simplicity in these estimation methods that arise from the very co-evolution we are trying to capture in our models (e.g., flood frequency, Guo et al., 2014). Potentially, if continued further, it will have the salutary effect of contributing to fundamental advances while maintaining coherence through avoiding fragmentation (Graham and Dayton, 2002; Blöschl et al., 2013).

7 Timescale interactions in the Anthropocene: from ecohydrology to socio-hydrology

The focus of the discussion in the last section was on timescale interactions and the co-evolution of hydrologic and other physical and biologic processes occurring in landscapes. In the otherwise natural catchments considered so far, the dynamic nature of catchment co-evolution was centered on vegetation, its adaptation of itself and its environment through feedbacks between hydrologic and ecologic processes across multiple timescales. This perspective was introduced to assist with developing transferable understanding between places in the course of data-based inference and top–down modeling under otherwise stationary conditions.

As one begins to look to the future, one recognizes the expanding human footprint, including land use and land cover (e.g., vegetation) changes and human interferences in the hydrologic cycle (e.g., water extraction from rivers or groundwater aquifers) and the inevitable acceleration of the timescale interactions. One can no longer count on stationarity for making predictions of catchment responses under these circumstances (Milly et al., 2008; Wagener et al., 2010). Furthermore, over longer periods of time, the timescale interactions must include two-way feedbacks between hydrological (and other Earth system processes) and human-social processes, and the emergent dynamics that result from these. Thus, as we transition from our focus on ecohydrology so far (Eagleson, 2002; Rodriguez-Iturbe, 2000) to the new field of socio-hydrology (Sivapalan et al., 2012), we can recognize both similarities as well as differences between the behaviors of vegetation and humans and, where possible, benefit from lessons learned from the practice of ecohydrology in the past 2 decades.

The co-evolution of water and vegetation does not stop with vegetation adapting to the water and energy balances and vice versa. We noticed that over time vegetation not only adapts itself to the prevailing climate variability and water balance dynamics (even as it modifies it at the same time), but it does so through adapting (or engineering) the landscape or environment around it as well, such as through modifying the soils or topography. Work by Gao et al. (2014) drew an analogy between humans and vegetation and highlighted parallels in water consumption behavior between them. We know humans build storages to cater to periods of drought – in engineering hydrology we estimate the required storage using (for example) the so-called Rippl method (Rippl, 1883). Analogously, Gao et al. (2014) proposed a root storage design method for vegetation, similar to Rippl, where the required storage capacity (with an analogy to root depth) was the between-year maximum of the within-year (seasonal) fluctuation (peak to trough) of soil moisture storage. They extracted this information from the outcomes of the modeling work they had done on a large number of MOPEX catchments across the United States, from which they estimated the pattern of root depth variation between the catchments. They discovered strong correlations between the root depth distributions obtained from such water balance analyses and those independently obtained for the actual vegetation classes found in these catchments. This observation lends strong support to the argument that vegetation adapts itself optimally to the prevailing water balances (seasonal and annual) through adapting itself and adapting its environment so as to, in this case, make maximum use of the water available (see also Troch et al., 2009; Yang et al., 2016; Brantley et al., 2017). These studies confirm and build on the prescient insight of Robert E. Horton, an early pioneer of hydrology who wrote:

> Natural vegetation of a region tends to develop to such an extent that it can utilize the largest possible proportion of the available soil moisture supplied by infiltration. – Robert E. Horton (1933)

Note in passing the similarities between Robert Horton's foresight with modern explanatory theories in ecology such as *ecological niche theory* (Chase, 2011) and *niche construction theory* (Odling-Smee et al., 2003) that relate, respectively, to how organisms modify themselves and modify their environment.

Admittedly, the examples presented above very much represent a top–down treatment of vegetation adaptation, which I invoked only to make the connection to comparative hydrology, and to highlight how model structures and parameterizations may vary between places through such adaptations. In reality, these broad-scale patterns indicative of co-evolution require further refinement in the context of *predictive ecohydrologic theories* regarding vegetation adaptation to its environment. One example of this is the vegetation optimality model (VOM) proposed by Schymanski et al. (2009) based on the hypothesis that natural vegetation co-evolves with its environment and that over time natural selection leads to a species composition most suited for the given environmental conditions. In VOM this is represented by a trade-off between water loss and carbon gain formulated in terms of the costs associated with the maintenance of roots, water transport tissues and foliage, and the benefits related to the exchange of water for CO_2 with the atmosphere, driven by photosynthesis. In VOM the optimal vegetation is taken as the one that maximizes "net carbon profit", i.e., the difference between carbon acquired by photosynthesis and carbon spent on maintenance of organs involved in its uptake. Note however that VOM is just one of many optimality theories that have been proposed in ecohydrology, and is only presented here as an illustrative example. There is increasing availability of data on evaporative fluxes from a large number of flux towers located in different biomes of the world (Thompson et al., 2011a, b) and new remotely sensed datasets on both vegetation cover and evapotranspiration rates over regional scales (Cheng et al., 2011) to complement existing rainfall–runoff data at catchment scales. There is enormous scope for these new datasets to be used to test alternative theories of vegetation adaptation (Brantley et al., 2017) and to provide guidance for the choice of model structures in top–down models, and thus help bridge the current impasse between top–down and bottom–up models.

7.1 Human-engineered landscapes and catchment water balances

We now know, at a minimum, how vegetation adapts itself to the water balance and adapts the water balance through its control. Increasingly, humans too interact with catchment water balances, in some ways similar to how vegetation interacts with the natural water balance. Humans extract water directly from rivers or groundwater aquifers to meet their needs, or might build dams across rivers to store water when nature's supply is low, such as during low flow or drought periods. Just as vegetation engineers the landscape to gain and maintain access to water, humans too engineer the landscape (e.g., catchments) to gain and maintain access to water, and to serve other human functions.

Previously hydrologists may only have been interested in the short-term (and local) effects of land use changes and human interferences in the hydrologic cycle to satisfy human needs or functions. With the expansion of the human footprint, they are increasingly concerned with longer-term hydrological changes brought about by human actions and how the changed hydrology then feeds back to generate secondary human reactions. Consequently, it is no longer sufficient, as in the past, to model the hydrology of pristine catchments and add human effects at the end (Wagener et al., 2010). For long-term planning and strategic decision-making purposes, humans must increasingly be treated as an intrinsic part of the catchment's water cycle. This also means that traditional definitions of catchments as topographically defined must be replaced or refined to account for administrative (or governance) units. One can thus imagine how the explicit inclusion of human behavior in hydrologic systems might over long time periods give rise to a co-evolution of water and people. For example, many ancient civilizations established themselves around sources of water, and their success and eventual collapse or dispersal were emergent outcomes of co-evolutionary feedbacks between humans and water (e.g., the Tarim basin in western China, Liu et al., 2014). Under these changing circumstances, one thus moves from ecohydrology, the study of two-way feedbacks between vegetation response and catchment water balance, to socio-hydrology, the new science dealing with two-way feedbacks between human behavior (e.g., water management) and catchment responses (e.g., water balance, flooding dynamics) (Sivapalan et al., 2012, 2014). This extends our earlier Earth system science perspective now to include human-social processes as well.

How far can we push the analogy between vegetation and humans? Unlike vegetation, humans need water not just for their physiological needs (i.e., for drinking), but also for sanitation, and even more so for use in industries, and for food and energy production. This introduces an economic purpose to the use of water, and excessive water extraction for economic uses impacts the environment in adverse ways, and impacts the ability of the environment to produce the ecosystem services (including the delivery of water) that humans also depend on. This introduces the notion of the water–food–energy–environment nexus, and the requirement to make tradeoffs between these various human needs. Two additional factors that enter the management of the water–food–energy–environment nexus are advances in technology (e.g., infrastructure that enables extraction and transfers of real water between places) and trade (e.g., trade of food and other commodities that depend on water for their production, also known as virtual water trade). Another compounding factor is the role of institutional system of administration, legislation and regulation of water (Sivapalan and Blöschl, 2015). Under these circumstances, human involvement in hydrologic systems in the past has been studied extensively in the context of water resource management, where the focus has been on optimal management (Loucks et al., 2005). Much of this work has worked on the belief that humans are optimizers, assumed to make rational decisions about allocation of resources between various needs, and opting to maximize economic livelihood. However, with the advent of socio-hydrology, hydrologic science has broadened to include the practices and outcomes of water resource management, in the presence of the aforementioned complexities, as themselves subjects for explanation and deeper understanding (Thompson et al., 2013).

A major complexity in understanding and eventually predicting human behavior in the context of water management is human agency (Sanderson et al., 2017). Unlike natural systems, e.g., vegetation, humans can choose among different modes of thinking and courses of action that impact the natural systems within which they are embedded. A major challenge is that human decision-making shapes, and is shaped by, the cultural contexts within which human societies are embedded. Humans draw upon commonly held values, beliefs and norms to guide their actions, and in turn can collectively change their cultures in the long term (Caldas et al., 2015). Human–water interactions are heavily influenced by differences and/or dynamic changes in human values and norms in respect of their economic livelihood and the environment. Unlike natural systems whose long-term co-evolutionary behavior can be expressed in terms of organizing principles (e.g., vegetation optimality, maximum entropy production), humans can appear to make decisions away from any optimality principles and to violate their own laws (Fig. 7, Sivapalan and Blöschl, 2015). For example, they may extract water from a well, even if it is illegal to do so; or they may choose to live with pollution, away from optimality reasons, giving the impression that unlike vegetation, humans are irrational. The net effects of such human decisions on water systems may manifest as puzzles, paradoxes, and other unintended consequences, nevertheless exhibiting similarities and differences that reflect distinct hydro-climatic, eco-environmental, and socio-economic features (Sivapalan, 2015; Pande and Sivapalan, 2017). These factors make human–water interactions much more complex and thus add to the already difficult prediction challenges in hydrology (Westerberg et al., 2017). Note, however, that differences highlighted above between natural and human-impacted systems can at least partly be attributed to the timescales of processes relative to the length of available data records. It is tempting to think that, given long enough records, co-evolutionary behavior in respect of coupled human–water system dynamics may indeed be governed by some (as yet unknown) organizing principles. In the end, then, it is a pragmatic choice whether one sees or assumes well-organized patterns within the available data records (Sivapalan and Blöschl, 2005).

It has been suggested that changes in culture and associated values and preferences may be more predictive of human behavior in the context of sustainability than rationality, or utility maximization (Caldas et al., 2015). Examples of human behavior different from utility maximization include (Sivapalan et al., 2014): the peaking in water resource availability as basins develop (peak water paradox, Kandasamy et al., 2014); increasing levee heights in urban environments at the expense of increased flood risk (levee effect, Di Baldassarre et al., 2013); increasing agriculture water consumption in spite of irrigation efficiency improvements (irrigation efficiency paradox) (Scott, 2011); and over-exploitation of coastal aquifers at the risk of causing saltwater intrusion (Chang and Clement, 2012). An example of non-optimal behavior in the context of coupled human–water system dynamics is lock-in or path dependence, which is one explanation offered for the collapse of the Maya civilization (Kuil et al., 2016). A more recent example of lock-in is the self-reinforcing action of the people of the Netherlands since the 11th century to drain their naturally peaty soils, leading to land subsidence, which ultimately led to the invention of the famous Gouda cheese, since the resulting low-lying areas that these actions produced were only suitable for dairy farming (Erkens et al., 2016). The science of socio-hydrology aims to explain these kinds of phenomena arising from two-way feedbacks associated with coupled human–water system dynamics, and to develop a generalized understanding that can connect diverse phenomena across many places and times (Sivapalan and Blöschl, 2015). There is still room for predictions in this new era of socio-hydrology: but given the difficulties of dealing with human agency and complex human behavior, the focus is not on predicting what will happen at a future prescribed time, as it is in traditional hydrology, but on mapping out the possibility space of future trajectories of system co-evolution (see Fig. 8, Srinivasan et al., 2017).

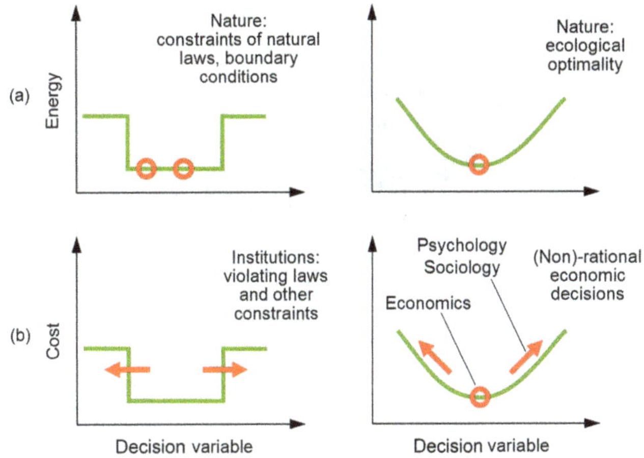

Figure 7. **(a)** Natural systems follow constraints and optimality laws; **(b)** Human systems may violate constraints and make irrational decisions away from optimality (taken from Sivapalan and Blöschl, 2015).

8 Engineering hydrology to Earth system science: from Newton to Darwin to Wegener

There is no question that there has been a fundamental transformation of hydrology in the past 50 years. My goal in this narrative has been to connect the dots and draw up a continuous thread through what I thought were major milestones in the evolution of the science. Hydrology (for me) started as engineering hydrology, addressing rather simple, well-defined problems (e.g., what is the 100-year flood?). Over time, problems became increasingly complex, such as the need to predict the effects of land use changes or climate change (e.g., on 100-year floods), which required more process-based approaches invoking Newtonian mechanics. Even as the ability to make predictions using such physi-

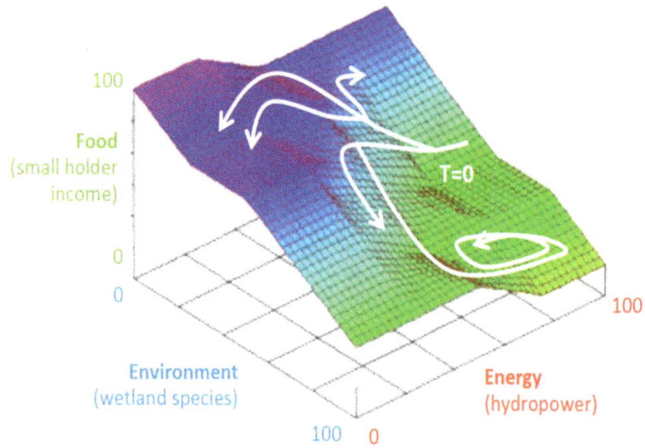

Figure 8. Trajectories illustrating the future possibility space that may be predicted by a socio-hydrological model addressed to management of the water–food–energy–environment nexus (taken from Srinivasan et al., 2017).

cally based models advanced, they encountered limitations due to unknown (and unknowable) heterogeneity of landscapes, and the lack of understanding of process interactions in the time domain, including how catchments functioned as whole ecosystems. Conceptual or functional models based on data-based inference partially helped to overcome these limitations, but they in turn suffered from heavy dependence on data, and the absence of holistic theories of catchment responses and ecosystem functioning.

The seeming impasse between the two approaches to predictions, mechanistic and functional, has motivated hydrologists to broaden the research from the traditional focus on estimation and prediction towards understanding, explanation and discovery. The shift towards understanding and explanation does not signify a lack of commitment to applications of hydrology, however. The goal of hydrology as use-inspired basic science (Stokes, 1997; Thompson et al., 2013) remains, which is to generate the basic understanding necessary to address the complex water sustainability or management problems that appear on the horizon. The nature and role of predictions have also evolved as problems become more complex and highly inter-disciplinary. The change in focus towards understanding and explanation is contributing to two major shifts in hydrology research. The first is increased attention to phenomena, emergent patterns that arise through process interactions across timescales and space scales. Numerical models are increasingly now used to test hypotheses about how these phenomena may have arisen and thus contribute towards accumulation of knowledge and understanding. The second major shift is towards comparative hydrology, i.e., comparative studies across gradients of climate, geology and human impacts. This is a Darwinian approach, which seeks to develop new insights and theories about catchments, as consisting of components that have co-evolved together, and to learn from the common history of their co-evolution.

Given this co-evolutionary perspective, hydrological problems needed to be framed broadly as inter-disciplinary problems that extend beyond the study of water flows in landscapes. In natural landscapes we are increasingly concerned with water flow processes interacting with and feeding back on land-forming (e.g., pedogenesis, soil erosion and deposition) and life-sustaining (e.g., biogeochemical, ecological) processes. With the expanding human footprint, hydrology also has to deal with vastly complex social processes. An example (one of many) of a new problem that hydrologists have to deal with in this context is the nature of interactions between people and droughts that might lead to exacerbation of water shortages (Apurv et al., 2017), conflicts, human migration, and even collapse (Kuil et al., 2016). This places hydrology squarely within the realm of Earth system science, with an alternate set of problems, questions, and study tools. With the broadening of the science there is increasing attention to a diversity of emergent phenomena that arise in the context of co-evolution of climate, soils, vegetation, and now humans.

Exploring phenomena of all kinds – defined as interesting catchment responses or ecosystem functioning with no obvious or immediate explanation – and asking how they came about, which is the scientific method practiced and perfected over centuries, is increasingly the main focus of most hydrological investigations. Bronowski (1956) describes the majesty of the scientific method in these terms:

> All science is the search for unity in hidden likenesses... The progress of science is the discovery at each step of a new order which gives unity to what had long seemed unlike... For order does not display itself of itself; if it can be said to be there at all, it is not there for the mere looking... order must be discovered and, in a deep sense, it must be created. What we see, as we see it, is mere disorder.
> – Bronowski (1956)

Increasingly, in a co-evolutionary context, one can adopt three different kinds of approaches to generate and study phenomena of interest to us. These may be framed as historical hydrology, comparative hydrology and process hydrology. Historical hydrology generates phenomena in the form of emergent dynamics in the time domain arising from timescale interactions. Comparative hydrology generates patterns of catchment responses between different catchments or places. Process hydrology generates patterns at a single place, resulting from process interactions in space (or space–time).

The evolution of the science and the growth of hydrological understanding over the past 50 years have benefited from the use of a combination of Newtonian and Darwinian approaches to address and explore increasingly complex prediction problems and emergent phenomena. The Newtonian

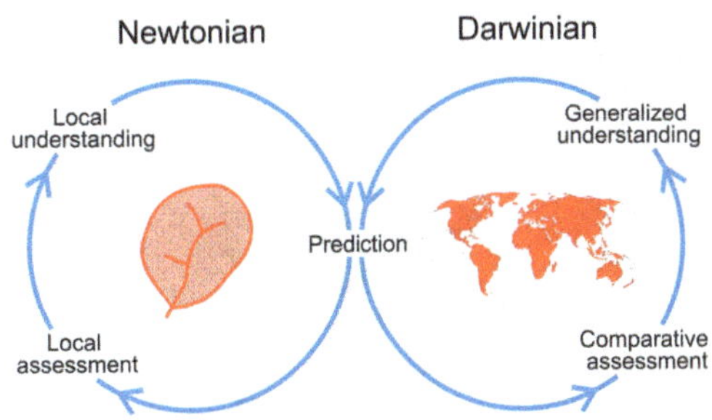

Figure 9. Newtonian–Darwinian synthesis: relative contributions of Newtonian (learning from individual catchments through detailed studies) and Darwinian (learning from a population of catchments through comparative studies) approaches (taken from Blöschl et al., 2013).

approach benefits from the fact that it is based on universal laws of mass, momentum and energy balances, with clear causality. It gains predictive power through advances in observing and understanding processes, and approaches to account for landscape heterogeneity and the resulting complexity of flow pathways and residence times. Yet it loses effectiveness in making general predictions by not being able to fully account for process interactions and feedbacks and parameter co-dependency that manifest in catchments through co-evolutionary processes. The legacy of past co-evolutionary processes in landscapes and their interconnections over long time periods lead to complex emergent spatial patterns and temporal dynamics. So an alternative approach to studying patterns and dynamics arising from such co-evolution is to study many catchments comparatively, treating them as legacies of past co-evolution, which is the Darwinian way. In short, the Newtonian approach generalizes by discovering universal laws governing particular processes through experimentation, inductive reasoning and through mathematical derivations, whereas the Darwinian approach generalizes by recognizing emergent phenomena through comparative analyses across places and then seeking explanations for how they came about (Guo et al., 2014). Of course, Newtonian and Darwinian methods are complementary to each other, and progress in terms of new theories of hydrology and improvements in our ability make future predictions everywhere will require a synthesis of these two approaches (Fig. 9) (Harte, 2002; Blöschl et al., 2013).

8.1 Crystal balling the future: from comparative hydrology to regional process hydrology

One of the benefits of developing a historical narrative, as I have done here, is that it provides a perspective view (rightly or wrongly) of where we have come from and where we may be (or should be) headed. I will therefore be remiss if I do not use the "crystal ball" that I have created in my mind based on this review to say something about what is in store for hydrology in the coming decade(s). In doing so I stand to benefit from a perspective that has helped me most to organize my narrative: this is the scale perspective. The issue of scale has been at the heart of most of the difficulties hydrologists have had in terms of developing new theories and advancing predictions. If at all we have made any progress in hydrology, this may be attributed to advances we have made in addressing scale issues (Blöschl and Sivapalan, 1995; Blöschl et al., 2013). So now, going over the progress we have made in hydrology over the last 50 years (Sivapalan and Blöschl, 2017), one cannot but recognize the change in focus from *time* during the empirical era (engineering hydrology), to *space* (Newtonian mechanics) during the process era, and back to *time* (Darwinian comparative hydrology) in the current co-evolutionary era. By induction, this symmetry tells me that even while remaining within the co-evolutionary framework, the next major change in focus is likely to be (or even should be) to go back to *space*, and to *space–time*.

It occurs to me that for far too long we hydrologists (or far too many of us) have focused on catchments as isolated objects, with an exclusive focus on their rainfall–runoff (input–output) transformations. Even in the context of comparative hydrology, we have continued to treat them as spatially distinct and independent objects, even while acknowledging or exploiting their common or similar co-evolutionary history. Clearly this is limiting, in terms of the phenomena that we can explore, and the kinds of predictions we increasingly have to make at larger spatial scales. If we extend the size of the study domain from isolated catchments to a whole region, then in many cases the catchments present within the region have spatial connectivity, as well as a common history. Catchments within a region can be connected to each other in three ways: downwind movement of water in the atmosphere, downstream water movement through surface pathways (e.g., rivers), and subsurface water movement (e.g., regional aquifers) from recharge areas in the mountains towards the ocean, leading to groundwater outcropping in streams, lakes and wetlands, and to estuaries and deltas in coastal regions. In this way there is a continual transformation across the region in the nature of interactions between atmospheric, surface and subsurface pathways of water, producing unique waterscapes. This inexorably takes us into a new field that we might as well call *regional process hydrology*. Just as a single catchment is viewed as a co-evolved ecosystem, owing to the co-evolution of different components (i.e., mountain range, headwater catchments, river network and floodplains, wetlands and estuaries), the *region* too needs to be looked at and managed as a whole ecosystem. An example of such a region satisfying these features is the arid Kaidu–Kongqi River basin in western China, which is supplied by snowmelt and glacier melt in its headwaters, feeding an extensive oasis and the terminal Bosten Lake, both of which support a thriving agricultural community in the midst of a vast desert (Fig. 10).

The Kaidu–Konqi River basin, and many others like it around the world (e.g., the Indus River basin in present-

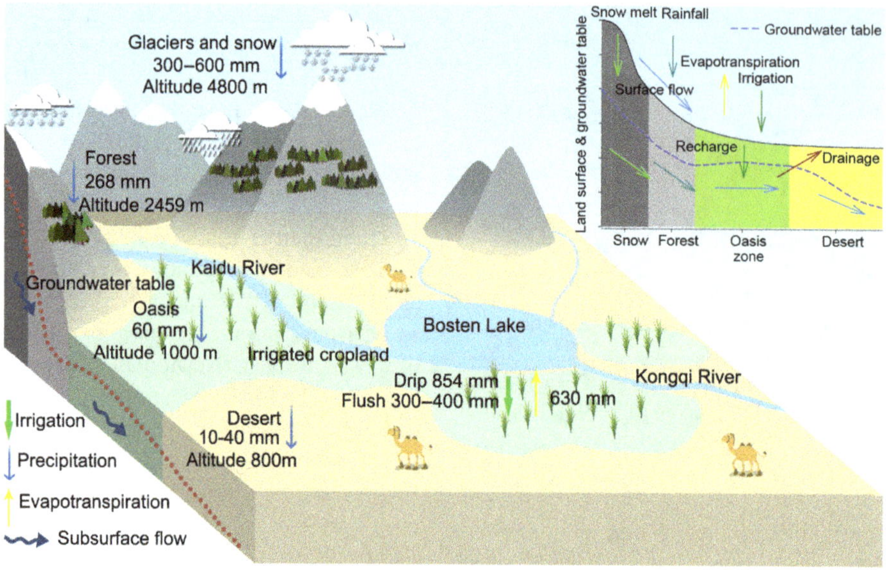

Figure 10. Regional process hydrology: schematic of the regional hydrologic cycle and waterscape in the Kaidu–Kongqi River basin, part of the much larger Tarim River basin, Xingjiang Province, in western China (taken from Zhang et al., 2014).

day Pakistan, the Cauvery River basin in southern India), have been cradles of human civilization that have waxed and waned as a result of long-term, large-scale climatic changes, as well as human alterations of their waterscapes and overexploitation of land and water resources (Liu et al., 2014). The complexity of water balances and long-term water management problems across such regions cannot be tackled by studying a few isolated catchments. As we begin to look at a region as a whole, the proposed regional outlook opens up new possibilities to advance hydrology, with altogether new emergent phenomena that arise through interactions of the atmosphere, land surface and regional groundwater aquifers, including the waterscapes that form as a result. This regional hydrology focus can also bring about the overdue unification of surface (catchment) hydrology with groundwater hydrology, and extensions of ecohydrology and socio-hydrology, to explore interesting, regional-scale phenomena.

I am not the first one to make the case for regional process hydrology, however. Tom Dunne, in his famous Abel Wolman Lecture (Dunne, 1998), called for a renewed focus on research targets at large spatial scales (i.e., regional) that are "of broad significance and might attract sustained interest from scientists in other fields and society at large". There is already much activity in regional process hydrology that is beginning to bring surface and/or groundwater hydrologists together with atmospheric scientists, with a focus on modeling, large-scale synergistic observations and data–model syntheses (Fan, 2015; Hipsey et al., 2015; Maxwell et al., 2015; Blöschl et al., 2017). Echoing Dunne (1998) himself, my main argument is to go beyond mere descriptions or narratives as we may have done in the past, and to reinforce the focus on phenomena, now regional-scale phenomena, and on process explanation and discovery, as a prelude to eventual predictions at these scales. The relevant processes of interest occur at all space scales and timescales, and are not limited to just small scales, and emergent phenomena arise out of these multi-scale (space, space–time) process interactions. Both Newtonian (bottom–up reductionist) and Darwinian (top–down functional) approaches and their synthesis retain currency, including the use of coarse-grained models to explore observed phenomena. So this is regional process hydrology that requires a whole system synergistic perspective.

Within the spirit of use-inspired basic science (Stokes, 1997; Thompson et al., 2013) the move towards regional hydrology also coincides with the evolution of the nature of water management problems that will be faced in the emergent Anthropocene. Thompson et al. (2013) presented several examples of anticipated century-scale trends in the main hydrologic drivers that will impact humanity in the future. These emerging future challenges include changing demography (Wagener et al., 2010), warming climate (Vörösmarty et al., 2000), rapid urbanization (Di Baldassarre et al., 2010), changing rainfall variability (Blöschl et al., 2017), loss of freshwater resources due to rising sea levels and saltwater intrusion (Chang et al., 2011), and globalization of water systems through real and virtual water trade (Konar et al., 2016). One can anticipate a range of water-related problems and secondary effects that will emerge in response to these drivers. Increasingly these problems will manifest at regional (and global) scales and can no longer be addressed through the study of individual, isolated catchments. Instead, we need a regional, increasingly global, approach to the study of these problems. Each of these problems is likely to draw attention

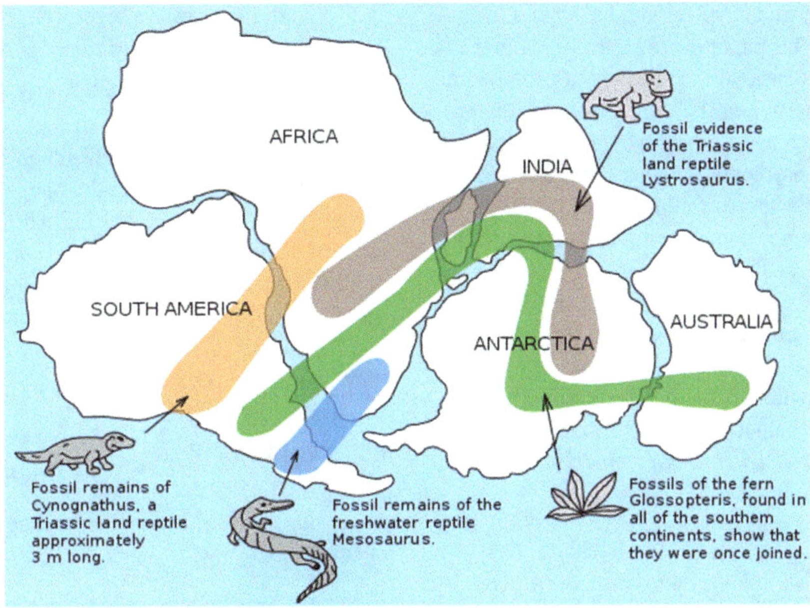

Figure 11. Connecting the dots: schematic describing the fossil patterns across continents that were the basis upon which Alfred Wegener proposed his continental drift theory: https://en.wikipedia.org/wiki/Alfred_Wegener (accessed on 28 October 2017).

to an array of emergent phenomena that manifest at the regional scale, and associated science questions, which could be the basis for the next major exciting phase of research in scientific hydrology.

The extension of the co-evolutionary perspective in space towards the regional scale takes us further from even Darwinian comparative hydrology (of isolated catchments) to the kind of approach adopted by Alfred Wegener, which eventually led to the discovery of plate tectonics. During his extensive journeys around the world, Wegener evaluated multiple variables (e.g., fossil records, rock types) across several continental regions with which he drew coherent spatial connections across places to support his reasoning on the continental drift (Fig. 11). Such an approach becomes feasible in hydrology too with the extraction of spatial patterns of multiple variables (e.g., climate, hydrology, vegetation, animals, people) across large regions, which could then serve as the basis for breakthroughs in understanding of regional water balance, and the co-evolutionary dynamics of climate, soils, vegetation, and human settlements that contribute to the formation of regional waterscapes. Much can be learned from studying spatial teleconnections (especially upstream–downstream) through remote sensing techniques and regional models (Guan et al., 2014), and how they impact and in turn are impacted by human alterations to landscapes and watercourses. Even though much of this is necessarily place-based, unique to particular regions, there is also much that can be transferred to other regions.

As the world becomes inter-connected through improvements in communications technology, trade, and the movement of people and ideas, along with global teleconnections in the climate system, it is increasingly necessary to go even further and treat the world as a *single* inter-connected system (Eagleson, 1986; Lall, 2014). Within the one-world system, anticipated global changes are likely to give rise to yet more complex phenomena manifesting as new and challenging water-related problems at both regional and global scales. This will require truly global and long-term vision and passion, the type of passion that led Alfred Wegener to discover continental drift and to open up the debate that ultimately led to the development of the theory of plate tectonics. Achieving this grand vision in hydrology will probably take us into what we have called the globalization era (Sivapalan and Blöschl, 2017), with a grand synthesis of global hydrology integrated with economics and climate science, as also concluded in an earlier review by Bierkens (2015). These are exciting times to be a hydrologist.

8.2 Lessons from Newton, Darwin, and Wegener: connecting the dots

This paper being an outgrowth of my Wegener Medal Lecture, in conclusion, I feel duty bound to reflect on the lessons we as hydrologists can learn from the methods adopted by not only Wegener, but also Newton and Darwin, and their contributions to science and the scientific pursuit. The ultimate goal in science is to "connect the dots", i.e., "to find order in disorder" (Bronowski, 1956), which should be the ultimate goal in hydrologic science too. It is in this spirit that Dooge (1986) exhorted us "to look for hydrological laws" and Klemeš (1986) warned against the "dilettantism" that results from failure to distinguish between science and applications of the science, as between hydrograph fitting and hydrological understanding. In his paper on "a hydrologic perspective", Klemeš (1988) emphasized the primacy of process understanding over techniques, i.e., the unravelling of the puzzles of the water cycle over the kinds of technologies that are adopted to solve societal problems, and warned against defining catchments merely by the different techniques that we might use from time to time to analyze them as engineering hydrologists. Regardless of the different methods they may have used, Newton, Darwin and Wegener, were ultimately giants in the art of *connecting the dots*, and we stand to benefit from the legacy they have left behind. Methods may come and go, but success depends on us not becoming slaves to the methods but treating them as necessary tools for a higher purpose, i.e., *advancing science through connecting the dots*.

> The method-oriented (wo)man is shackled; the problem-oriented (wo)man is at least reaching freely toward what is most important... willing repeatedly to put aside his(her) last methods and teach him(her)self new ones. – Adapted from Platt (1964).

Acknowledgements. This paper is an outgrowth from the Alfred Wegener Medal Lecture I gave at the General Assembly of the European Geosciences Union in April 2017. I thank the EGU for bestowing this honor on me and for giving me the opportunity to publish this paper in HESS. My thoughts on the evolution of hydrology came about through interactions with many students and colleagues over the years. They gained potency through my involvement in three community activities: the Predictions in Ungauged Basins (PUB) initiative, the University of Illinois Hydrologic Synthesis activity, and the Predictions under Change activity, which morphed into the Socio-hydrology movement, and both contributed to the launch of the IAHS Panta Rhei initiative. I am proud of the global hydrology community I served and learned from.

The paper has benefited from suggestions and comments I received from Keith Beven, Marc Bierkens, Günter Blöschl, Prabhakar Clement, John Ding, Markus Hrachowitz, Hongyi Li, Saket Pande, Jose Salinas, Majid Shafiee-Jood, Stan Schymanski, Richard Silberstein, Kevin Wallington, Dingbao Wang and Erwin Zehe. I am extremely grateful for their inputs, but in the final analysis, this is a personal account, and I am ultimately responsible for any errors of omission and commission that remain.

In closing, I bring up a line from *Kondraiventhan*, an anthology of poems written by legendary Tamil poetess Avvaiyar, who lived in southern India c. 1000 years ago. It reads:

எண்ணும் எழுத்தும் கண் எனத் தகும்

Numbers (*mathematics*) and letters (*language*)
are like our twin eyes
(i.e., *our windows to the world*)

These immortal words of Avvaiyar help me to celebrate the memory of two teachers I had at Hartley College, Point Pedro, Sri Lanka, the high school from which I entered university. S. Ratnasabapathy taught me Mathematics and S. S. Manuelpillai taught English. They both also shaped my intellectual outlook and the way to approach personal and professional challenges. It is not an exaggeration to say that the discipline, persistence and pursuit of excellence they instilled in me continue to guide me more than 50 years after they had taught me.

Edited by: Erwin Zehe

Competing interests. The author declares that he has no conflict of interest.

References

Abbott, M. B., Bathurst, J. C., Cunge, J. A., O'Connell, P. E., and Rasmussen, J.: An introduction to the European Hydrological System–Systeme Hydrologique Europeen, "SHE", 1: history and philosophy of a physically-based, distributed modelling system, J. Hydrol., 87, 45–59, 1986.

Apurv, T., Sivapalan, M., and Cai, X.: Understanding the role of climate characteristics in drought propagation, Water Resour. Res., 53, 9304–9329, https://doi.org/10.1002/2017WR021445, 2017.

Atkinson, S., Woods, R. A., and Sivapalan, M.: Climate and landscape controls on water balance model complexity over changing time scales, Water Resour. Res., 38, 1314, https://doi.org/10.1029/2002WR001487, 2002.

Band, L. and Wood, E. F.: Strategies for large-scale, distributed hydrologic simulation, Appl. Math. Comput., 27, 23–37, 1988.

Bear, J.: Dynamics of Fluids in Porous Media, Elsevier, Amsterdam, 764 pp., 1972.

Berghuijs, W. R., Sivapalan, M., Woods, R. A., and Savenije, H. H. G.: Patterns of similarity of seasonal water balance: A window into streamflow variability over a range of timescales, Water Resour. Res., 50, 5638–5661, https://doi.org/10.1002/2014WR015692, 2014.

Bergström, S.: Development and Application of a Conceptual Runoff Model for Scandinavian Catchments, Norrköping: SMHI, Report No. RHO 7, 1976.

Beven, K.: Kinematic subsurface stormflow, Water Resour. Res., 17, 1419–1424, 1981.

Beven, K.: Searching for the Holy Grail of scientific hydrology: $Qt = (S, R, \Delta t)A$ as closure, Hydrol. Earth Syst. Sci., 10, 609–618, https://doi.org/10.5194/hess-10-609-2006, 2006.

Beven, K. and Germann, P.: Macropores and water flow in soils, Water Resour. Res., 18, 1311–1325, 1982.

Beven, K. J.: Changing ideas in hydrology: the case of physically based models, J. Hydrol., 105, 157–172, 1989.

Beven, K. J.: Uniqueness of place and process representations in hydrological modelling, Hydrol. Earth Syst. Sci., 4, 203–213, https://doi.org/10.5194/hess-4-203-2000, 2000.

Beven, K. J.: Rainfall-Runoff Modelling: The Primer, 360 pp., John Wiley, Chichester, UK, 2012.
Beven, K. J.: What we see now: event-persistence and the predictability of hydro-eco-geomorphological systems, Ecol. Model., 298, 4–15, https://doi.org/10.1016/j.ecolmodel.2014.07.019, 2015.
Beven, K. J.: Facets of uncertainty: epistemic uncertainty, non-stationarity, likelihood, hypothesis testing, and communication, Hydrol. Sci. J., 61, 1652–1665, 2016.
Beven, K. J. and Binley, A. M.: The future of distributed models: model calibration and uncertainty prediction, Hydrol. Process., 6, 279–298, 1992.
Beven, K. J. and Cloke, H. L.: Comment on "Hyperresolution global land surface modeling: Meeting a grand challenge for monitoring Earth's terrestrial water" by E. F. Wood et al., Water Resour. Res., 48, W01801, https://doi.org/10.1029/2011WR010982, 2012.
Beven, K. J. and Kirkby, M. J.: A physically-based, variable contributing area model of basin hydrology, Hydrol. Sci. Bull., 24, 43–69, 1979.
Bierkens, M. F. P.: Global hydrology: State, trends, and directions, Water Resour. Res., 51, 4923–4947, https://doi.org/10.1002/2015WR017173, 2015.
Black, P. E.: Watershed functions, J. Am. Water Resour. Assoc., 33, 1–11, https://doi.org/10.1111/j.1752-1688.1997.tb04077.x, 1997.
Blöschl, G. and Sivapalan, M.: Scale issues in hydrological modelling – A review, Hydrol. Process., 9, 251–290, 1995.
Blöschl, G., Grayson, R. B., and Sivapalan, M.: On the representative elementary area (REA) concept and its utility for rainfall-runoff modelling, Hydrol. Process., 9, 313–330, 1995.
Blöschl, G., Sivapalan, M., Wagener, T., Viglione, A., and Savenije, H. H. G.: Runoff Prediction in Ungauged Basins – Synthesis across Processes, Places and Scales, Cambridge University Press, Cambridge, UK, 500 pp., 2013.
Blöschl, G., Blaschke, A. P., Broer, M., Bucher, C., Carr, G., Chen, X., Eder, A., Exner-Kittridge, M., Farnleitner, A., Flores-Orozo, A., Haas, P., Hogan, P., Kazemi Amiri, A., Oismüller, M., Parajka, J., Silasari, R., Stadler, P., Strauss, P., Vreugdenhil, M., Wagner, W., and Zessner, M.: The Hydrological Open Air Laboratory (HOAL) in Petzenkirchen: a hypothesis-driven observatory, Hydrol. Earth Syst. Sci., 20, 227–255, https://doi.org/10.5194/hess-20-227-2016, 2016.
Blöschl, G., Hall, J., Parajka, J., Perdigão, R. A. P., Merz, B., Arheimer, B., and Živković, N.: Changing climate shifts timing of European floods, Science, 357, 588–590, doi.org/10.1126/science.aan2506, 2017.
Brantley, S. L., Eissenstat, D. M., Marshall, J. A., Godsey, S. E., Balogh-Brunstad, Z., Karwan, D. L., Papuga, S. A., Roering, J., Dawson, T. E., Evaristo, J., Chadwick, O., McDonnell, J. J., and Weathers, K. C.: Reviews and syntheses: On the roles trees play in building and plumbing the critical zone, Biogeosciences, 14, 5115–5142, https://doi.org/10.5194/bg-14-5115-2017, 2017.
Bronowski, J.: Science and Human Values, 94 pp., Julian Messner Inc., New York, 1956.
Brunner, P. and Simmons, C. T.: HydroGeoSphere: a fully integrated, physically based hydrological model, Groundwater, 50, 170–176, 2012.
Budyko, M. I.: Climate and Life, 508 pp., Academic, NY, 1974.
Burt, T. P. and McDonnell, J. J.: Whither field hydrology? The need for discovery science and outrageous hydrological hypotheses, Water Resour. Res., 51, 5919–5928, 2015.
Caldas, M., Sanderson, M. R., Mather, M., Daniels, M. D., Bergtold, J. S., Aistrup, J., Heier Stamm, J. L., Haukos, D., Douglas-Mankin, K., Sheshukov, A. Y., and Lopez-Carr, D.: Endogenizing culture in sustainability science research and policy, P. Natl. Acad. Sci. USA, 112, 8157–8159, 2015.
Chang, S. W., Clement, T. P., Simpson, M. J., and Lee, K. K.: Does sea-level rise have an impact on saltwater intrusion?, Adv. Water Resour., 34, 1283–1291, https://doi.org/10.1016/j.advwatres.2011.06.006, 2011.
Chang, S. W. and Clement, T. P.: Experimental and numerical investigation of saltwater intrusion dynamics in flux controlled groundwater systems, Water Resour. Res., 48, W09527, https://doi.org/10.1029/2012WR012134, 2012.
Chase, J. M.: Ecological niche theory, in: The Theory of Ecology, edited by: Scheiner, S. M. and Willig, M. R., University of Chicago Press, Chicago, 93–108, 2011.
Cheng, L., Xu, Z., Wang, D., and Cai, X.: Assessing interannual variability of evapotranspiration at the catchment scale using satellite-based evapotranspiration data sets, Water Resour. Res., 47, W09509, https://doi.org/10.1029/2011WR010636, 2011.
Chow, V. T.: Handbook of Applied Hydrology, McGraw-Hill, New York, 1376 pp., 1964.
Chow, V. T., Maidment, D. R., and Mays, L.: Applied Hydrology, McGraw-Hill, New York, 572 pp., 1988.
Clark, M. P., Nijssen, B., Lundquist, J., Kavetski, D., Rupp, D., Woods, R., Gutmann, E., Wood, A., Brekke, L., Arnold, J., Gochis, D., and Rasmussen, R.: A unified approach to process-based hydrologic modeling. Part 1: Modeling concept, Water Resour. Res., 51, 2498–2514, https://doi.org/10.1002/2015WR017198, 2015.
Clark, M. P., Bierkens, M. F. P., Samaniego, L., Woods, R. A., Uijlenhoet, R., Bennett, K. E., Pauwels, V. R. N., Cai, X., Wood, A. W., and Peters-Lidard, C. D.: The evolution of process-based hydrologic models: historical challenges and the collective quest for physical realism, Hydrol. Earth Syst. Sci., 21, 3427–3440, doi.10.5194/hess-21-3427-2017, 2017.
Clement, T. P.: Complexities in hindcasting models – when should we say enough is enough?, Ground Water, 49, 620–629, https://doi.org/10.1111/j.1745-6584.2010.00765.x, 2011.
Crawford, N. H. and Linsley, R. K.: Digital Simulation in Hydrology: Stanford Watershed Model IV, Technical Report No. 39, Department of Civil Engineering, Stanford University, 210 pp., 1966.
Davies, P. C. W.: Why is the physical world so comprehensible?, CTNS Bulletin 12, 16–21, 1992.
Di Baldassarre, G., Montanari, A., Lins, H., Koutsoyiannis, D., Brandimarte, L., and Blöschl, G.: Flood fatalities in Africa: From diagnosis to mitigation, Geophys. Res. Lett., 37, L22402, https://doi.org/10.1029/2010GL045467, 2010.
Di Baldassarre, G., Viglione, A., Carr, G., Kuil, L., Salinas, J. L., and Blöschl, G.: Socio-hydrology: Conceptualising human-flood interactions, Hydrol. Earth Syst. Sci., 17, 3295–3303, https://doi.org/10.5194/hess-17-3295-2013, 2013.
Didszun, J. and Uhlenbrook, S.: Scaling of dominant runoff generation processes: Nested catchments approach us-

ing multiple tracers, Water Resour. Res., 44, W02410, https://doi.org/10.1029/2006WR005242, 2008.

Dooge, J. C. I.: Looking for hydrologic laws, Water Resour. Res., 22, 46S–58S, 1986.

Dooge, J. C. I.: History of Hydrology Film Interviews: J. C. I. Dooge Interviewed by D. R. Dawdy, Amer. Geophys. Union, available at: https://hydrology.agu.org/resources/history-of-hydrology-film-interviews, (last access: 1 March 2018) 1995.

Dunne, T.: Field studies of hillslope flow processes, Chapter 7, in: Hillslope Hydrology, edited by: Kirkby, M. J., Wiley, London, 227–293, 1978.

Dunne, T.: Wolman Lecture: Hydrologic science in landscapes on a planet in the future, in: Hydrologic Sciences: Taking Stock and Looking Ahead, Water Science and Technology Board, National Research Council, 10–43, Natl. Acad. Press, Washington, D.C., 1998.

Dunne, T. and Black, R. D.: Partial area contributions to storm runoff in a small New England watershed, Water Resour. Res., 6, 1296–1311 1970.

Eagleson, P. S.: Dynamic Hydrology, McGraw-Hill, New York, 462 pp., 1970.

Eagleson, P. S.: The emergence of global-scale hydrology, Water Resour. Res., 22, 6S–14S, 1986.

Eagleson, P. S.: Ecohydrology: Darwinian Expression of Vegetation Form and Function, Cambridge Univ. Press, Cambridge, UK, 345 pp., https://doi.org/10.1017/CBO9780511535680, 2002.

Erkens, G., van der Meulen, M. J., and Middelkoop, H.: Double trouble: subsidence and CO_2 respiration due to 1,000 years of Dutch coastal peatlands cultivation, Hydrogeol. J., 24, 551–568, https://doi.org/10.1007/s10040-016-1380-4, 2016.

Falkenmark, M. and Chapman, T.: Comparative Hydrology, UNESCO, Paris, 479 pp., 1989.

Fan, Y.: Groundwater in the Earth's critical zone: Relevance to large-scale patterns and processes, Water Resour. Res., 51, 3052–3069, https://doi.org/10.1002/2015WR017037, 2015.

Fan, Y. and Bras, R. L.: On the concept of a representative elementary area in catchment runoff, Hydrol. Process., 9, 821–832, 1995.

Fang, K. and Shen, C.: Full-flow-regime storage-streamflow correlation patterns provide insights into hydrologic functioning over the continental US, Water Resour. Res., 53, 8064–8083, https://doi.org/10.1002/2016WR020283, 2017.

Farmer, D., Sivapalan, M., and Jothityangkoon, C.: Climate, soil and vegetation controls upon the variability of water balance in temperate and semi-arid landscapes: Downward approach to hydrological prediction, Water Resour. Res., 39, 1035, https://doi.org/10.1029/2001WR000328, 2003.

Fenicia, F., Kavetski, D., and Savenije, H. H. G.: Elements of a flexible approach for conceptual hydrological modeling: 1. Motivation and theoretical development, Water Resour. Res., 47, W11510, https://doi.org/10.1029/2010wr010174, 2011.

Freeze, R. A. and Harlan, R. L.: Blueprint for a physically-based, digitally-simulated hydrologic response model, J. Hydrol., 9, 237–258, 1969.

Freeze, R. A.: Streamflow generation, Rev. Geophys. Space Phys., 12, 627–647, 1974.

Freeze, R. A.: A stochastic conceptual analysis of rainfall-runoff processes on a hillslope, Water Resour. Res., 6, 391–408, 1980.

Gaál, L., Szolgay, J., Kohnová, S., Parajka, J., Merz, R., Viglione, A., and Blöschl, G.: Flood timescales: Understanding the interplay of climate and catchment processes through comparative hydrology, Water Resour. Res., 48, W04511, https://doi.org/10.1029/2011WR011509, 2012.

Gao, H., Hrachowitz, M., Schymanski, S. J., Fenicia, F., Sriwongsitanon, N., and Savenije, H. H. G.: Climate controls how ecosystems size the root zone storage capacity at catchment scale, Geophys. Res. Lett., 41, 7916–7923, https://doi.org/10.1002/2014GL061668, 2014.

Gelhar, L. W. and Axness, C. L.: Three-dimensional stochastic analysis of macrodispersion in aquifers, Water Resour. Res., 19, 161–180, https://doi.org/10.1029/WR019i001p00161, 1983.

Gharari, S., Hrachowitz, M., Fenicia, F., and Savenije, H. H. G.: Hydrological landscape classification: investigating the performance of HAND based landscape classifications in a central European meso-scale catchment, Hydrol. Earth Syst. Sci., 15, 3275–3291, https://doi.org/10.5194/hess-15-3275-2011, 2011.

Graham, M. H. and Dayton, P. K.: On the evolution of ecological ideas: Paradigms and scientific progress, Ecology, 83, 1481–1489, 2002.

Grayson, R. B., Moore, I. D., McMahon, T. A.: Physically based hydrologic modeling: 2. Is the concept realistic?, Water Resour. Res., 28, 2659–2666, https://doi.org/10.1029/92WR01259, 1992.

Guan, K., Wood, E. F., Medvigy, D., Kimball, J., Pan, M., Caylor, K. K., Sheffield, J., Xu, X., and Jones, M. O.: Terrestrial hydrological controls on land surface phenology of African savannas and woodlands. J. Geophys. Res.-Biogeosci., 119, 1652–1669, https://doi.org/10.1002/2013JG002572, 2014.

Guo, J., Li, H.-Y., Leung, L. R., Guo, S., Liu, P., and Sivapalan, M.: Links between flood frequency and annual water balance behaviors: A basis for similarity and regionalization, Water Resour. Res., 50(, 937–953, https://doi.org/10.1002/2013WR014374, 2014.

Harman, C. J. and Sivapalan, M.: Similarity framework to assess controls on subsurface flow dynamics in hillslopes, Water Resour. Res., 45, W01417, https://doi.org/10.1029/2008WR007067, 2009.

Harman, C. J., Rao, P. S. C., Basu, N. B., McGrath, G. S., Kumar, P., and, Sivapalan, M.: Climate, soil and vegetation controls on the temporal variability of vadose zone transport, Water Resour. Res., 47, W00J13, https://doi.org/10.1029/2010WR010194, 2011.

Harman, C. and Troch, P. A.: What makes Darwinian hydrology "Darwinian"? Asking a different kind of question about landscapes, Hydrol. Earth Syst. Sci., 18, 417–433, https://doi.org/10.5194/hess-18-417-2014, 2014.

Harman, C. J., Lohse, K. A., Troch, P. A., and Sivapalan, M.: Spatial patterns of vegetation, soils and microtopography from terrestrial laser scanning on two semi-arid hillslopes of contrasting lithology, J. Geophys. Res.-Biogeosci., 119, 163–180, https://doi.org/10.1002/2013JG002507, 2014.

Harte, J.: Toward a synthesis of the Newtonian and Darwinian worldviews, Phys. Today, 55, 29–34, 2002.

Hassanizadeh, S. M. and Gray, W. G.: General conservation equations for multiphase systems: 1. Averaging procedure, Adv. Water Resour., 2, 131–144, 1979.

Hassanizadeh, S. M., Celia, M. A., and Dahle, H. K.: Dynamic effects in the capillary pressure-saturation relationship and their impacts on unsaturated flow, Vadose Z. J., 1, 38–57, 2002.

Hewlett, J. D. and Hibbert, A. R.: Factors affecting the response of small watersheds to precipitation in humid regions, in: Forest Hydrology, edited by: Sopper, W. E. and Lull, H. W., Pergamon Press, Oxford, 275–290, 1967.

Hipsey, M. R., Hamilton, D. P., Hanson, P. C., Carey, C. C., Coletti, J. Z., Read, J. S., Ibelings, B. W., Valesini, F. J., and Brookes, J. D.: Predicting the resilience and recovery of aquatic systems: A framework for model evolution within environmental observatories, Water Resour. Res., 51, 7023–7043, https://doi.org/10.1002/2015WR017175, 2015.

Horton, R. E.: The role of infiltration in the hydrologic cycle, Trans. Amer. Geophys. Union, 14, 446–460, 1933.

Hrachowitz, M., Savenije, H. H. G., Blöschl, G., McDonnell, J. J., Sivapalan, M., Pomeroy, J. W., Arheimer, B., Blume, T., Clark, M. P., Ehret, U., Fenicia, F., Freer, J. E., Gelfan, A., Gupta, H. V., Hughes, D. A., Hut, R. W., Montanari, A., Pande, S., Tetzlaff, D., Troch, P. A., Uhlenbrook, S., Wagener, T., Winsemius, H. C., Woods, R. A., Zehe, E., and Cudennec, C.: A decade of Predictions in Ungauged Basins (PUB) – a review, Hydrol. Sci. J., 58, 1–58, https://doi.org/10.1080/02626667.2013.803183, 2013.

Hrachowitz, M. and Clark, M. P.: HESS Opinions: The complementary merits of competing modelling philosophies in hydrology, Hydrol. Earth Syst. Sci., 21, 3953–3973, https://doi.org/10.5194/hess-21-3953-2017, 2017.

IACWD (Interagency Advisory Committee on Water Data): Guidelines for determining flood-flow frequency: Bulletin 17B of the Hydrology Subcommittee, Office of Water Data Coordination, U.S. Geological Survey, Reston, VA, 183 pp., available at: http://water.usgs.gov/osw/bulletin17b/bulletin_17B.html (last access: 1 March 2018), 1982.

Jakeman, A. J. and Hornberger, G. M.: How much complexity is warranted in a rainfall-runoff model?, Water Resour. Res., 29, 2637–2649, 1993.

Jothityangkoon, C., Sivapalan, M., and Farmer, D.: Process controls of water balance variability in a large semi-arid catchment: Downward approach to hydrological model development, J. Hydrol., 254, 174–198, 2001.

Jothityangkoon, C. and Sivapalan, M.: Framework for exploration of climatic and landscape controls on catchment water balance, with emphasis on inter-annual variability, J. Hydrol., 371, 154–168, https://doi.org/10.1016/j.jhydrol.2009.03.030, 2009.

Kandasamy, J., Sounthararajah, D., Sivabalan, P., Chanan, A., Vigneswaran, S., and Sivapalan, M.: Socio-hydrologic drivers of the pendulum swing between agriculture development and environmental health: A case study from Murrumbidgee river basin, Australia, Hydrol. Earth Syst. Sci., 18, 1027–1041, https://doi.org/10.5194/hess-18-1027-2014, 2014.

Kirchner, J. W.: Catchments as simple dynamical systems: Catchment characterization, rainfall-runoff modeling, and doing hydrology backward, Water Resour. Res., 45, W02429, https://doi.org/10.1029/2008WR006912, 2009.

Kleidon, A. and Renner, M.: Thermodynamic limits of hydrologic cycling within the Earth system: concepts, estimates and implications, Hydrol. Earth Syst. Sci., 17, 2873–2892, https://doi.org/10.5194/hess-17-2873-2013, 2013.

Klemeš, V.: Conceptualization and scale in hydrology, J. Hydrol., 65, 1–23, 1983.

Klemeš, V.: Dilettantism in Hydrology: Transition or Destiny?, Water Resour. Res., 22, 177S–188S, 1986.

Klemeš, V.: A hydrological perspective, J. Hydrol., 100, 3–28, 1988.

Kollet, S. J. and Maxwell, R. M.: Capturing the influence of groundwater dynamics on land surface processes using an integrated, distributed watershed model, Water Resour. Res., 44, W02402, https://doi.org/10.1029/2007WR006004, 2008.

Konar, M., Evans, T. P., Levy, M., Scott, C. A., Troy, T. J., Vörösmarty, C. J., and Sivapalan, M.: Water sustainability in a globalizing world: who uses the water?, Hydrol. Process., 30, 3330–3336, https://doi.org/10.1002/hyp.10843, 2016.

Kuil, L., Carr, G., Viglione, A., Prskawetz, A., and Blöschl, G.: Conceptualizing socio-hydrological drought processes: The case of the Maya collapse, Water Resour. Res., 52, 6222–6242, https://doi.org/10.1002/2015WR018298, 2016.

Kumar, M., Duffy, C. J., and Salvage, K. M.: A second-order accurate, finite volume-based integrated hydrologic modeling framework (FIHM) for simulation of surface and subsurface flow, Vadose Z. J., https://doi.org/10.2136/vzj2009.0014, 2009.

Lall, U.: Debates – The future of hydrological sciences: A (common) path forward? One water. One world. Many climes. Many souls, Water Resour. Res., 50, 5335–5341, https://doi.org/10.1002/2014WR015402, 2014.

Larsen, J. E., Sivapalan, M., Coles, N. A., and Linnet, P. E.: Similarity analysis of runoff generation processes in real-world catchments, Water Resour. Res., 30, 1641–1652, 1994.

Lee, H., Zehe, E., and Sivapalan, M.: Predictions of rainfall-runoff response and soil moisture dynamics in a microscale catchment using the CREW model, Hydrol. Earth Syst. Sci., 11, 819–849, https://doi.org/10.5194/hess-11-819-2007, 2007.

Levin, S. A.: The problem of pattern and scale in ecology, Ecology, 73, 1943–1967, 1992.

Li, H.-Y., Sivapalan, M., and Tian, F.: A comparative diagnostic analysis of runoff generation mechanisms in Oklahoma DMIP2 basins: The Blue River and the Illinois River, J. Hydrol., 418–419, 90–109, 2012.

Li, H.-Y., Sivapalan, M., Tian, F., and Harman, C. J.: Functional approach to exploring climatic and landscape controls on runoff generation. 1. Behavioral constraints on runoff volume. Water Resour. Res., 50, 9300–9322, https://doi.org/10.1002/2014WR016307, 2014.

Liu, Y., Tian, F., Hu, H., and Sivapalan, M.: Socio-hydrologic perspectives of the co-evolution of humans and water in the Tarim River Basin, Western China: the Taiji-Tire Model, Hydrol. Earth Syst. Sci., 18, 1289–1303, https://doi.org/10.5194/hess-18-1289-2014, 2014.

Linsley, R. K., Kohler, M. A., and Paulhus, J. L. H.: Hydrology for Engineers, McGraw-Hill, New York, 340 pp., 1958.

Loritz, R., Hassler, S. K., Jackisch, C., Allroggen, N., van Schaik, L., Wienhöfer, J., and Zehe, E.: Picturing and modeling catchments by representative hillslopes, Hydrol. Earth Syst. Sci., 21, 1225–1249, https://doi.org/10.5194/hess-21-1225-2017, 2017.

Loucks, D. P., Van Beek, E., Stedinger, J. R., Dijkman, J. P., and Villars, M. T.: Water Resources Systems Planning and Management: An Introduction to Methods, Models and Applications, 680 pp., UNESCO, Paris, 2015.

Lotka, A.: Natural selection as a physical principle, P. Natl. Acad. Sci. USA, 8, 151–154, 1922.

Lu, H., Moran, C. J., and Sivapalan, M.: A theoretical exploration of catchment-scale sediment delivery, Water Resour. Res., 41, W09415, https://doi.org/10.1029/2005WR004018, 2005.

L'vovich, M. I.: World Water Resources and Their Future, 415 pp., AGU, Washington, D.C., 1979.

Manabe, S.: Climate and the ocean circulation. 1. The atmospheric circulation and the hydrology of the earth's surface, Mon. Weather Rev., 97, 739–774, 1969.

Maxwell, R. M., Condon, L. E., and Kollet, S. J.: A high-resolution simulation of groundwater and surface water over most of the continental US with the integrated hydrologic model ParFlow v3, Geosci. Model Dev., 8, 923–937, https://doi.org/10.5194/gmd-8-923-2015, 2015.

McDonnell, J. J.: A rationale for old water discharge through macropores in a steep, humid catchment, Water Resour. Res., 26, 2821–2832, 1990.

McDonnell, J. J., Sivapalan, M., Vaché, K., Dunn, S., Grant, G., Haggerty, R., Hinz, C., Hooper, R. P., Kirchner, J. W., Roderick, M. L., Selker, J., and Weiler, M.: Moving beyond heterogeneity and process complexity: A new vision for watershed hydrology, Water Resour. Res., 43, W07301, https://doi.org/10.1029/2006WR005467, 2007.

McGrath, G. S., Hinz, C., and Sivapalan, M. The effect of rainfall regional variability on rapid pesticide risk leaching potential, J. Contam. Hydrol., 113, 56–65, 2010.

Mein, R. G. and Larson, C. L.: Modeling infiltration during steady rain, Water Resour. Res., 9, 384–394, 1973.

Milly, P. C. D., Betancourt, J., Falkenmark, M., Hirsch, R. M., Kundzewicz, Z. W., Lettenmaier, D. P., and Stouffer, R. J.: Stationarity is dead: Whither water management?, Science, 319, 573–574, https://doi.org/10.1126/science.1151915, 2008.

Minshall, N. E.: Predicting storm runoff on small experimental watersheds, Proc. Am. Soc. Civ. Eng., J. Hydraul. Div., 86, 17–38, 1960.

Mockus, V.: Estimation of total (and peak rates of) surface runoff for individual storms. Exhibit A in Appendix B, Interim Survey Report (Neosho) River Watershed, Washington, DC: Agricultural Research Service, United States Department of Agriculture, 61 pp., 1949.

Odling-Smee, F. J., Laland, K. N., and Feldman, M. W.: Niche Construction: The Neglected Process in Evolution, Princeton University Press, Princeton, 476 pp., 2003.

Oleson, K. W., Lawrence, D. W., Bonan, G. B., Drewniak, B., Huang, M., Koven, C. D., Levis, S., Li, F., Riley, W. J., Subin, Z. M., Swenson, S. C., Thornton, P. E.: Technical Description of Version 4.5 of the Community Land Model (CLM). NCAR Technical Report, NCAR/TN-503+STR, 420 pp, doi:10.5065/D6RR1W7M, http://opensky.ucar.edu/islandora/object/technotes%3A515/datastream/PDF/view (last access: last access: 1 March 2018.), 2013.

Or, D., Lehmann, P., and Assouline, S.: Natural length scales define the range of applicability of the Richards equation for capillary flows, Water Resour. Res., 51, 7130–7144, https://doi.org/10.1002/2015WR017034, 2015.

Pande, S. and Sivapalan, M.: Progress in socio-hydrology: a meta-analysis of challenges and opportunities, WIREs Water, 4, e1193, https://doi.org/10.1002/wat2.1193, 2017.

Patil, S., Sivapalan, M., Hassan, M. A., Ye, S., Harman, C. J., and Xu, X.: A network model for prediction and diagnosis of sediment dynamics at the watershed scale, J. Geophys. Res.-Earth Surf., 117, F00A04, https://doi.org/10.1029/2012JF002400, 2012.

Peters-Lidard, C. D., Clark, M., Samaniego, L., Verhoest, N. E. C., van Emmerik, T., Uijlenhoet, R., Achieng, K., Franz, T. E., and Woods, R.: Scaling, similarity, and the fourth paradigm for hydrology, Hydrol. Earth Syst. Sci., 21, 3701–3713, https://doi.org/10.5194/hess-21-3701-2017, 2017.

Platt, J. R.: Strong inference, Science, 146, 347–353, 1964.

Reggiani, P.: A Unifying Framework for Watershed Thermodynamics, PhD Dissertation, Centre for Water Research, Univ. West. Aust., 238 pp., 1999.

Reggiani, P., Sivapalan, M., and Hassanizadeh, S. M.: A unifying framework for watershed thermodynamics: Balance equations for mass, momentum, energy, entropy and the 2nd law of thermodynamics, Adv. Water Resour., 22, 367–398, 1998.

Reggiani, P., Hassanizadeh, S. M., Sivapalan, M., and Gray, W. G.: A unifying framework for watershed thermodynamics. Constitutive relationships. Adv. Water Resour., 23, 15–39, 1999.

Rippl, W.: The capacity of storage reservoirs for water supply, Minutes, Proc. Inst. Civil Eng., 71, 270–278, 1883.

Robinson, J. S., Sivapalan, M., and Snell, J. D.: On the relative roles of hillslope processes, channel routing and network geomorphology in the hydrological response of natural catchments, Water Resour. Res., 31, 3089–3101, 1995.

Robinson, J. S. and Sivapalan, M.: Temporal scales and hydrological regimes: Implications for flood frequency scaling, Water Resour. Res., 33, 2981–2999, 1997.

Robinson, T. M., La Pierre, K. J., Vadeboncoeur, M. A., Byrne, K. M., Thomey, M. L., and Colby, S. E.: Seasonal, not annual precipitation drives community productivity across ecosystems, Oikos, 122, 727–738, https://doi.org/10.1111/j.1600-0706.2012.20655.x, 2012.

Roderick, M. L. and Farquhar, G. D.: The pan evaporation paradox, in: Global Change and the Earth System: A Planet Under Pressure, Steffen, W., Sanderson, A., Tyson, P., Jäger, J., Matson, P., Moore III, B., Oldfield, F., Richardson, K., Schellnhuber, H.-J., Turner II, B. L., and Wasson, R., Springer-Verlag, p. 167, 2004.

Rodriguez-Iturbe, I. and Valdes, J. B.: The geomorphologic structure of hydrologic response, Water Resour. Res., 15, 1409–1420, 1979.

Rodriguez-Iturbe, I.: Ecohydrology: A hydrologic perspective of climate-soil-vegetation dynamics, Water Resour. Res., 36, 3–9, 2000.

Saco, P. M. and Moreno-de las Heras, M.: Ecogeomorphic coevolution of semiarid hillslopes: Emergence of banded and striped vegetation patterns through interaction of biotic and abiotic processes, Water Resour. Res., 49, 115–126, https://doi.org/10.1029/2012WR012001, 2013.

Sanderson, M. R., Bergtold, J. S., Heier Stamm, J. L., Caldas, M. M., and Ramsey, S. M.: Bringing the "social" into sociohydrology: Conservation policy support in the Central Great Plains of Kansas, USA, Water Resour. Res., 53, 6725–6743, https://doi.org/10.1002/2017WR020659, 2017.

Savenije, H. H. G.: Opinion paper: Linking Darcy's equation to the linear reservoir, Hydrol. Earth Syst. Sci. Discuss., https://doi.org/10.5194/hess-2017-580, in review, 2017.

Savenije, H. H. G. and Hrachowitz, M.: HESS Opinions "Catchments as meta-organisms – a new blueprint for hydrological modelling", Hydrol. Earth Syst. Sci., 21, 1107–1116, https://doi.org/10.5194/hess-21-1107-2017, 2017.

Schaefli, B., Harman, C. J., Sivapalan, M., and Schymanski, S. J.: HESS Opinions: Hydrologic predictions in a changing environment: behavioral modeling, Hydrol. Earth Syst. Sci., 15, 635–646, https://doi.org/10.5194/hess-15-635-2011, 2011.

Schymanski, S. J.: Transpiration as the Leak in the Carbon Factory: A Model of Self-Optimising Vegetation, PhD Dissertation, Sch. Env. Syst. Eng., Univ. of West. Aust., Perth, 244 pp., 2007.

Schymanski, S. J., Sivapalan, M., Roderick, M. L., Hutley, L., and Beringer, J.: An optimality-based model of the dynamic feedbacks between natural vegetation and the water balance, Water Resour. Res., 45, W01412, https://doi.org/10.1029/2008WR006841, 2009.

Scott, C. A.: The water-energy-climate nexus: resources and policy outlook for aquifers in Mexico, Water Resour. Res., 47, W00L04, https://doi.org/10.1029/2011WR010805, 2011.

Senatore, A., Mendicino, G., Gochis, D. J., Yu, W., Yates, D. N., and Kunstmann, H.: Fully coupled atmosphere-hydrology simulations for the central Mediterranean: Impact of enhanced hydrological parameterization for short and long time scales, J. Adv. Model. Earth Syst., 7, 1693–1715, https://doi.org/10.1002/2015MS000510, 2015.

Shen, C. and Phanikumar, M. S.: A process-based, distributed hydrologic model based on a large-scale method for surface-subsurface coupling, Adv. Water Resour., 33, 1524–1541, https://doi.org/10.1016/j.advwatres.2009.07.001, 2010.

Silberstein, R. P., Held, A., Hatton, T. J., Viney, N. R., and Sivapalan, M.: Energy balance of a natural jarrah (Eucalyptus marginata) forest in Western Australia. Measurements in spring and summer, Agric. Forest Meteorol., 109, 79–104, 2001.

Sivapalan, M., Jeevaraj, C. G., and Viney, N. R.: Catchment-scale water balance modeling to predict the effects of land use changes in forested catchments. 1. Small catchment water balance model, Hydrol. Process., 10, 393–411, 1996.

Sivapalan, M.: Process complexity at hillslope scale, process simplicity at the watershed scale: Is there a connection?, Hydrol. Process., 17, 1037–1041, https://doi.org/10.1002/hyp.5109, 2003.

Sivapalan, M., Blöschl, G., Zhang, L., and Vertessy, R.: Downward approach to hydrological prediction, Hydrol. Process., 17, 2101–2111, https://doi.org/10.1002/hyp.1425, 2003.

Sivapalan, M., Takeuchi, K., Franks, S. W., Gupta, V. K., Karambiri, H., Lakshmi, V., Liang, X., McDonnell, J. J., Mendiondo, E. M., O'Connell, P. E., Oki, T., Pomeroy, J. W., Schertzer, D., Uhlenbrook, S., and Zehe, E.: IAHS Decade on Predictions in Ungauged Basins (PUB), 2003–2012: Shaping an exciting future for the hydrological sciences, Hydrol. Sci. J., 48, 857–880, 2003.

Sivapalan, M.: Pattern, Process and Function: Elements of a New Unified Hydrologic Theory at the Catchment Scale, Contribution to: Encyclopaedia of Hydrologic Sciences, edited by: Anderson, M. G., 13, 193–219, John Wiley & Sons, 2005.

Sivapalan, M. and Young, P. C.: Downward approach to hydrological model development, in: Encyclopaedia of Hydrologic Sciences, edited by: Anderson, M. G., 134, 2081–2098, John Wiley & Sons, 2005.

Sivapalan, M. The secret to "doing better hydrological science": Change the question!, Hydrol. Process., 23, 1391–1396, https://doi.org/10.1002/hyp.7242, 2009.

Sivapalan, M., Savenije, H. H. G., and Blöschl, G.: Socio-hydrology: A new science of people and water, Hydrol. Process., 26, 1270–1276, https://doi.org/10.1002/hyp.8426, 2012.

Sivapalan, M., Konar, M., Srinivasan, V., Chhatre, A., Wutich, A., Scott, C. A., Wescoat, J. L., and Rodriguez-Iturbe, I.: Socio-hydrology: Use-inspired water sustainability science for the Anthropocene, Earth's Future, 2, 225–230, https://doi.org/10.1002/2013EF000164, 2014.

Sivapalan, M.: Debates – Perspectives in Socio-hydrology: Changing water systems and the "tyranny of small problems" – Socio-hydrology, Water Resour. Res., 51, 4795–4805, https://doi.org/10.1002/2015WR017080, 2015.

Sivapalan, M. and Blöschl, G.: Time scale interactions and the coevolution of humans and water, Water Resour. Res., 51, 6988–7022, https://doi.org/10.1002/2015WR017896, 2015.

Sivapalan, M. and Blöschl, G.: The growth of hydrological understanding: Technologies, ideas and societal needs shape the field, Water Resour. Res., 53, 8137–8146 https://doi.org/10.1002/2017WR021396, 2017.

Smith, J. A.: Representation of basin scale in flood peak distributions, Water Resour. Res., 28, 2993–2999, 1992.

Srinivasan, V., Sanderson, M., Garcia, M., Konar, M., Blöschl, G., and Sivapalan, M.: Prediction in a socio-hydrological world, Hydrol. Sci. J., 62, 338–345, https://doi.org/10.1080/02626667.2016.1253844, 2017.

Stephenson, G. R. and Freeze, R. A.: Mathematical simulation of subsurface flow contributions to snowmelt runoff, Reynolds Creek Watershed, Idaho, Water Resour. Res., 10, 284–294, 1974.

Stephenson, N. L.: Climatic control of vegetation distribution: The role of the water balance, Am. Nat., 135, 649–670, 1990.

Stokes, D. E.: Pasteur's Quadrant: Basic Science and Technological Innovation, Brookings Institution Press, Washington, D.C., 1997.

Sugawara, M.: The flood forecasting by a series storage type model, Proc. Int. Symp. on Floods and their Computation, Leningrad, USSR, 1–6, 1967.

Thomas, H. A.: Improved methods for national water assessment, water resource contract WR15249270, USGS unnumbered series, Harvard Water Resources Group, available at: http://pubs.er.usgs.gov/publication/70046351 (last access: 1 March 2018), 1981.

Thompson, S. E., Harman, C. J., Konings, A. G., Sivapalan, M., and Troch, P. A.: Comparative hydrology across Ameriflux sites: The variable roles of climate, vegetation and groundwater, Water Resour. Res., 47, W00J07, https://doi.org/10.1029/2010WR009797, 2011a.

Thompson, S. E., Harman, C. J., Troch, P. A., Brooks, P. D., and Sivapalan, M.: Scaling of ecohydrologically mediated water balance partitioning: A synthesis framework for catchment ecohydrology, Water Resour. Res., 47, W00J03, https://doi.org/10.1029/2010WR009998, 2011b.

Thompson, S. E., Sivapalan, M., Harman, C. J., Srinivasan, V., Hipsey, M., Reed, P., Montanari, A., and Blöschl, G.: Developing predictive insight into changing water systems: use-inspired hydrologic science for the Anthropocene, Hydrol. Earth Syst. Sci., 17, 5013–5039, https://doi.org/10.5194/hess-17-5013-2013, 2013.

Tian, F., Hu, H., Lei, Z., and Sivapalan, M.: Extension of the Representative Elementary Watershed approach for cold regions via explicit treatment of energy related processes, Hydrol. Earth Syst. Sci., 10, 619–644, https://doi.org/10.5194/hess-10-619-2006, 2006.

Tian, F., Li, H.-Y., and Sivapalan, M.: Model diagnostic analysis of seasonal switching of runoff generation mechanisms in the Blue River basin, Oklahoma, J. Hydrol., 418–419, 136–149, 2012.

Troch, P. A., Martinez, G. F., Pauwels, V. R. N., Durcik, M., Sivapalan, M., Harman, C. J., Brooks, P. D., Gupta, H. V., and Huxman, T. E.: Climate and vegetation water-use efficiency at catchment scales. Hydrol. Process., 23, 2409–2414, https://doi.org/10.1002/hyp.7358, 2009.

Vörösmarty, C. J., Green, P., Salisbury, J., and Lammers, R. B.: Global water resources: Vulnerability from climate change and population growth, Science, 289, 284–288, 2000.

Wagener, T., Sivapalan, M., Troch, P. A., and Woods, R. A.: Catchment classification and hydrologic similarity, Geog. Compass, 1, 901–931, https://doi.org/10.1111/j.1749-8198.2007.00039.x, 2007.

Wagener, T., Sivapalan, M., Troch, P. A., McGlynn, B. L., Harman, C. J., Gupta, H. V., Kumar, P., Rao, P. S. C., Basu, N. B., and Wilson, J. S.: The future of hydrology: An evolving science for a changing world, Water Resour. Res., 46, W05301, https://doi.org/10.1029/2009WR008906, 2010,

Wang, D. and Tang, Y.: A one-parameter Budyko model for water balance captures emergent behavior in Darwinian hydrologic models, Geophys. Res. Lett., 41, 4569–4577, https://doi.org/10.1002/2014GL060509, 2014.

Wang, D., Zhao, J., Tang, Y., and Sivapalan, M.: A thermodynamic interpretation of Budyko and L'vovich formulations of annual water balance: proportionality hypothesis and maximum entropy production, Water Resour. Res., 51, 3007–3016, https://doi.org/10.1002/2014WR016857, 2015.

Weingartner, R. and Aschwanden, H.: Discharge regime–the basis for the estimation of average flows, in: Hydrological Atlas of Switzerland, Map 5.2, Swiss Federal Office for the Environment, Bern, Switzerland, 1992.

Westerberg, I. K., Di Baldassarre, G., Beven, K. J., Coxon, G., and Krueger, T.: Perceptual models of uncertainty for sociohydrological systems: a flood risk change example, Hydrol. Sci. J., 62, 1705–1713, doi10.1080/02626667.2017.1356926, 2017.

Westhoff, M., Zehe, E., Archambeau, P., and Dewals, B.: Does the Budyko curve reflect a maximum-power state of hydrological systems? A backward analysis, Hydrol. Earth Syst. Sci., 20, 479–486, https://doi.org/10.5194/hess-20-479-2016, 2016.

Wing, O. E. J., Bates, P. D., Sampson, C. C., Smith, A. M., Johnson, K. A., and Erickson, T. A.: Validation of a 30 m resolution flood hazard model of the conterminous United States, Water Resour. Res., 53, 7968–7986, https://doi.org/10.1002/2017WR020917, 2017.

Wittenberg, H. and Sivapalan, M.: Watershed groundwater balance estimation using streamflow recession analysis and baseflow separation, J. Hydrol., 219, 20–33, 1999.

Wood, E. F., Sivapalan, M., Beven, K. J., and Band, L. E.: Effects of spatial variability and scale with implications to hydrologic modeling, J. Hydrol., 102, 29–47, 1988.

Wood, E. F., Roundy, J. K., Troy, T. J., van Beek, R., Bierkens, M. F. P., Blyth, E. M., de Roo, A., Döll, P., Ek, M., Famiglietti, J. S., Gochis, D., van de Giesen, N., Houser, P., Jaffe, P., Kollet, S., Lehner, B., Lettenmaier, D. P., Peters-Lidard, C., Sivapalan, M., Sheffield, J., Wade, A., and Whitehead, P.: Hyper-resolution global land surface modeling: Meeting a grand challenge for monitoring Earth's terrestrial water, Water Resour. Res., 47, W05301, https://doi.org/10.1029/2010WR010090, 2011.

Wood, E. F., Roundy, J. K., Troy, T. J., van Beek, R., Bierkens, M. F. P., Blyth, E. M., de Roo, A., Döll, P., Ek, M., Famiglietti, J. S., Gochis, D., van de Giesen, N., Houser, P., Jaffe, P., Kollet, S., Lehner, B., Lettenmaier, D. P., Peters-Lidard, C., Sivapalan, M., Sheffield, J., Wade, A., and Whitehead, P.: Reply to comment by Keith J. Beven and Hannah L. Cloke on "Hyper-resolution global land surface modeling: Meeting a grand challenge for monitoring Earth's terrestrial water", Water Resour. Res., 48, W01802, https://doi.org/10.1029/2011WR011202, 2012.

Woods, R. A., Sivapalan, M., and Duncan, M. J.: Investigating the Representative Elementary Area concept – An approach based on field data, Hydrol. Process., 9, 291–312, 1995.

Woolhiser, D. A.: Hydrologic and watershed modeling – state of the art, Trans. Amer. Soc. Agric. Engrs., 16, 553–559, 1973.

Yang, Y., Donohue, R. J., and McVicar, T. R.: Global estimation of effective plant rooting depth: Implications for hydrological modeling, Water Resour. Res., 52, 8260–8276, https://doi.org/10.1002/2016WR019392, 2016.

Ye, S., Covino, T. P., Sivapalan, M., Basu, N. B., Li, H.-Y., and Wang, S.-W.: Dissolved nutrient retention dynamics in river networks: A modeling investigation of transient flows and scale effects, Water Resour. Res., 48, W00J17, https://doi.org/10.1029/2011WR010508, 2012.

Yevjevich, V.: Misconceptions in hydrology and their consequences, Water Resour. Res., 4, 225–232, 1968.

Young, P. C.: Top-down and data-based mechanistic modelling of rainfall–flow dynamics at the catchment scale, Hydrol. Process., 17, 2195–2217, https://doi.org/10.1002/hyp.1328, 2003.

Young, P. C. and Beven, K. J.: Data-based mechanistic modelling and the rainfall-flow nonlinearity, Environmetrics, 5, 335–363, 1994.

Zehe, E. and Sivapalan, M.: Threshold behaviour in hydrological systems as (human) geo-ecosystems: manifestations, controls, implications, Hydrol. Earth Syst. Sci., 13, 1273–1297, https://doi.org/10.5194/hess-13-1273-2009, 2009.

Zehe, E., Lee, H., and Sivapalan, M.: Dynamical process upscaling for deriving catchment scale state variables and constitutive relations for meso-scale process models, Hydrol. Earth Syst. Sci., 10, 981–996, https://doi.org/10.5194/hess-10-981-2006, 2006.

Zehe, E., Ehret, U., Pfister, L., Blume, T., Schröder, B., Westhoff, M., Jackisch, C., Schymanski, S. J., Weiler, M., Schulz, K., Allroggen, N., Tronicke, J., van Schaik, L., Dietrich, P, Scherer, U., Eccard, J., Wulfmeyer, V., and Kleidon, A.: HESS Opinions: From response units to functional units: a thermodynamic reinterpretation of the HRU concept to link spatial organization and functioning of intermediate scale catchments, Hydrol. Earth Syst. Sci., 18, 4635–4655, https://doi.org/10.5194/hess-18-4635-2014, 2014.

Zhang, Z., Tian, F., and Sivapalan, M.: Groundwater dynamics under water saving irrigation and implications for sustainable water management in an oasis: Tarim River Basin, China, Hydrol. Earth Syst. Sci., 18, 3951–3967, https://doi.org/10.5194/hess-18-3951-2014, 2014.

7

Development and verification of a real-time stochastic precipitation nowcasting system for urban hydrology in Belgium

L. Foresti[1], M. Reyniers[1], A. Seed[2], and L. Delobbe[1]

[1]Royal Meteorological Institute of Belgium, Brussels, Belgium
[2]Bureau of Meteorology, Melbourne, Australia

Correspondence to: L. Foresti (loris.foresti@gmail.com)

Abstract. The Short-Term Ensemble Prediction System (STEPS) is implemented in real-time at the Royal Meteorological Institute (RMI) of Belgium. The main idea behind STEPS is to quantify the forecast uncertainty by adding stochastic perturbations to the deterministic Lagrangian extrapolation of radar images. The stochastic perturbations are designed to account for the unpredictable precipitation growth and decay processes and to reproduce the dynamic scaling of precipitation fields, i.e., the observation that large-scale rainfall structures are more persistent and predictable than small-scale convective cells. This paper presents the development, adaptation and verification of the STEPS system for Belgium (STEPS-BE). STEPS-BE provides in real-time 20-member ensemble precipitation nowcasts at 1 km and 5 min resolutions up to 2 h lead time using a 4 C-band radar composite as input. In the context of the PLURISK project, STEPS forecasts were generated to be used as input in sewer system hydraulic models for nowcasting urban inundations in the cities of Ghent and Leuven. Comprehensive forecast verification was performed in order to detect systematic biases over the given urban areas and to analyze the reliability of probabilistic forecasts for a set of case studies in 2013 and 2014. The forecast biases over the cities of Leuven and Ghent were found to be small, which is encouraging for future integration of STEPS nowcasts into the hydraulic models. Probabilistic forecasts of exceeding 0.5 mm h^{-1} are reliable up to 60–90 min lead time, while the ones of exceeding 5.0 mm h^{-1} are only reliable up to 30 min. The STEPS ensembles are slightly under-dispersive and represent only 75–90 % of the forecast errors.

1 Introduction

The use of radar measurements for urban hydrological applications has substantially increased during the last years (e.g., Berne et al., 2004; Einfalt et al., 2004; Bruni et al., 2015). Given the fast response time of urban catchments and sewer systems, radar-based very short-term precipitation forecasting (nowcasting) has the potential to extend the lead time of hydrological and hydraulic flow predictions.

Nowcasting concerns the accurate description of the current weather situation together with very short-term forecasts obtained by extrapolating the real-time observations. Quantitative precipitation nowcasting (QPN) is traditionally done by estimating the apparent movement of radar precipitation fields using optical flow or variational echo tracking techniques and extrapolating the last observed precipitation field into the future (e.g., Germann and Zawadzki, 2002; Bowler et al., 2004a). During recent years there has been significant progress in NWP modeling with radar data assimilation techniques (see a review in Sun et al., 2014), which reduces the useful lead time of extrapolation-based nowcasts compared with NWP forecasts. The development of seamless forecasting systems that optimally blend the extrapolation nowcast with the output of NWP models makes the definition of the nowcasting time range even fuzzier (see, e.g., Pierce et al., 2010).

Due to the lack of predictability of rainfall growth and decay processes at small spatial scales (Radhakrishna et al., 2012), it is very important to provide together with a forecast an estimation of its uncertainty. The established method to represent the forecast uncertainty in Numerical Weather Prediction (NWP) is to generate an ensemble of forecasts by perturbing the initial conditions of the model in the directions exhibiting the largest error growth, which amplify more the spread of the obtained ensemble. However, in the nowcasting range the computation of large NWP ensembles (50–100 members) that resolve features at the scales of 1 km and are updated every 5 min is still impossible to achieve. Consequently, the efforts in nowcasting research have recently focused on developing heuristic techniques for probabilistic precipitation nowcasting, which was the topic of the *Heuristic Probabilistic Forecasting Workshop* that was organized in Munich, Germany (Foresti et al., 2014).

Probabilistic QPN methods can be categorized into three main classes: analog, local Lagrangian and stochastic approaches. The analog-based approach derives the forecast probability density function (pdf) by retrieving a set of similar situations from an archive of precipitation events (Panziera et al., 2011; Foresti et al., 2015), the local Lagrangian approach derives the pdf by collecting the precipitation values in a neighborhood of a given grid point in Lagrangian frame of reference (Hohti et al., 2000; Germann and Zawadzki, 2004) and the stochastic approach exploits a random number generator to compute an ensemble of equally likely precipitation fields, for example by adding stochastic perturbations to a deterministic extrapolation nowcast (Pegram and Clothier, 2001a, b; Bowler et al., 2006; Metta et al., 2009; Berenguer et al., 2011; Seed et al., 2013; Atencia and Zawadzki, 2014; Dai et al., 2015). The stochastic approach is also extensively used to produce ensembles of precipitation fields that characterize the radar measurement uncertainty (e.g., Jordan et al., 2003; Germann et al., 2009) and for design storm studies (e.g., Willems, 2001a; Paschalis et al., 2013).

Uncertainty quantification is nowadays an integral part of both weather and hydrological forecasting (Pappenberger and Beven, 2006). Not surprisingly, an important part of hydro-meteorological research aims at understanding how to propagate the uncertainty of precipitation observations and forecasts into the hydrological models (e.g., Willems, 2001b; Cloke and Pappenberger, 2009; Collier, 2009; Zappa et al., 2010).

Several studies already analyzed the value of deterministic nowcasting systems for catchment hydrology (e.g., Berenguer et al., 2005) and for better control of urban drainage systems (e.g., Achleitner et al., 2009; Verworn et al., 2009; Thorndahl and Rasmussen, 2013). Since an important fraction of the uncertainty of hydrological predictions is due to the uncertainty of the input rainfall observations and forecasts, radar-based ensemble nowcasting systems are increasingly used as inputs for flood and sewer system modeling (e.g., Ehret et al., 2008; Silvestro and Rebora, 2012; Silvestro et al., 2013; Xuan et al., 2009, 2014). At longer forecast ranges, the NWP ensembles are also exploited for uncertainty propagation into hydrological models (see Roulin and Vannitsem, 2005; Thielen et al., 2009; Schellekens et al., 2011).

The Short-Term Ensemble Prediction System (STEPS) is a probabilistic nowcasting system developed at the Australian Bureau of Meteorology and the UK MetOffice (see the series of papers: Seed, 2003; Bowler et al., 2006; Seed et al., 2013). STEPS is operationally used at both weather services and provides short-term ensemble precipitation forecasts using both the extrapolation of radar images and the downscaled precipitation output of NWP models. The main idea behind STEPS is to represent the uncertainty due to the unpredictable precipitation growth and decay processes by adding stochastic perturbations to the deterministic extrapolation of radar images. The stochastic perturbations are designed to represent the scale-dependence of the predictability of precipitation and to reproduce the correct spatio-temporal correlation and growth of the forecast errors.

One of the first applications of STEPS in hydrology is presented in Pierce et al. (2005), who used the STEPS ensemble nowcasts to quantify the accuracy of flow predictions in a medium-sized catchment in the UK. The value of STEPS nowcasts for urban hydrology was extensively analyzed by Liguori and Rico-Ramirez (2012), Liguori et al. (2012), Liguori and Rico-Ramirez (2013) and Xuan et al. (2014). Liguori and Rico-Ramirez (2012) concluded that the performance of the radar-based extrapolation nowcast can be improved after 1 h lead time if blended with the output of a NWP model. They also found that, according to the Receiver Operating Characteristic (ROC) curve, the probabilistic nowcasts have more discrimination power than the deterministic ones. Liguori et al. (2012) integrated STEPS nowcasts as inputs into sewer system hydraulic models in an urban catchment in Yorkshire (UK). They concluded that the blending of radar and NWP forecasts has the potential to increase the lead time of flow predictions, but is strongly limited by the low accuracy of the NWP model in forecasting small-scale features. Liguori and Rico-Ramirez (2013) performed a detailed verification of the accuracy of flow predictions and concluded that the STEPS ensembles provide a similar performance as using a deterministic STEPS control forecast, but the ensembles lead to a slight underestimation of the flow predictions. Xuan et al. (2014) used ensemble STEPS nowcasts as inputs in a lumped hydrological model for a medium-sized catchment in the southwest of the UK. The hydrological model calibrated with rain gauges had lower RMSE than the one using radar data, but the ability of STEPS to account for the forecast uncertainty was useful in capturing some of the high flow peaks and extending the forecast lead time. However, the conclusions of the previous studies are strongly affected by the limited number of flood events analyzed. An extensive review of the usage of precipitation forecast systems for operational hydrological predictions in the UK from very short to long ranges (including STEPS) is provided in Lewis et al. (2015).

The goal of this paper is to present the development and verification of the STEPS system at the Royal Meteorological Institute of Belgium (RMI), referred to as STEPS-BE.

STEPS-BE provides in real-time 20-member ensemble precipitation nowcasts at 1 km and 5 min resolutions up to 2 h lead time on a 512×512 km domain using the Belgian 4 C-band radar composite as input. It was developed in the framework of the Belspo project PLURISK for better management of rainfall-induced risks in the urban environment. With respect to the original implementation of STEPS (Bowler et al., 2006), STEPS-BE includes two main improvements, which are designed to generate better STEPS nowcasts without NWP blending. The first one is related to the optical flow algorithm, which is extended with a kernel-based interpolation method to obtain smoother velocity fields. The second one concerns the generation of stochastic noise only within the advected radar composite. While the verification of STEPS nowcasts with NWP blending has already been extensive (Bowler et al., 2006; Seed et al., 2013), this paper analyzes the accuracy of STEPS ensemble nowcasts without NWP blending in the 0–2 h forecasting range.

Ensemble STEPS nowcasts are computed for a set of sewer overflow cases that affected the cities of Leuven and Ghent in 2013 and 2014. The accuracy of the ensemble mean forecast is verified using both continuous verification scores (multiplicative bias, RMSE) and categorical scores derived from the contingency table (probability of detection, false alarm ratio and Gilbert skill score). However, the most interesting part of this paper is the probabilistic and ensemble verification of STEPS nowcasts using both stratiform and convective rainfall events. Probabilistic nowcasts are verified using reliability diagrams and ROC curves. On the other hand, the dispersion of the nowcast ensembles is verified using rank histograms and by comparing the ensemble spread to the error of the ensemble mean.

The paper is structured as follows. Section 2 presents the radar data processing and case studies that are used to generate and verify the STEPS forecasts. Section 3 describes the STEPS nowcasting system and its extension and local implementation for Belgium (STEPS-BE). Section 4 illustrates the forecast verification results. Section 5 concludes the paper and discusses future perspectives.

2 Radar data and precipitation case studies

STEPS-BE integrates as input a composite image produced from the C-band radars of Wideumont (RMI, single-pol), Zaventem (Belgocontrol, single-pol), Jabbeke (RMI, dual-pol) and Avesnois (Meteo-France, dual-pol). The composite is produced on a 500 m resolution grid by combining single-radar pseudo Constant Altitude Plan Position Indicators (CAPPI) at a height of 1500 m a.s.l. The compositing algorithm takes the maximum reflectivity value from each radar at each grid point.

The radars have different hardware and scanning strategies, and are operated by different agencies (RMI, Belgocontrol and Meteo-France), which inevitably leads to differences in the data processing. The Wideumont and Zaventem radars eliminate the non-meteorological echoes using standard Doppler filtering. The Jabbeke radar includes an additional clutter filtering that uses a fuzzy logic algorithm based on the dual-polarization moments (essentially the co-polar correlation coefficient, the texture of the differential reflectivity and the texture of the specific differential phase shift). A static ground clutter map and a statistical filter are used by Meteo-France to remove the non-meteorological echoes of the Avensois radar. The French radar data processing chain is described in Tabary (2007) and in Figueras i Ventura and Tabary (2013).

Since the Zaventem radar is mainly used for aviation applications, its scanning strategy is optimized for the measurement of winds. Except for the lowest elevation scan, a dual PRF mode (1200/800 Hz) is used. The azimuths that are scanned with a high PRF (1200 Hz) only have a maximum range of 125 km and are more affected by the second trip echoes caused by convective cells located beyond the 125 km range.

All radars use the standard Marshall–Palmer relationship $Z = 200R^{1.6}$ to convert the measured reflectivity to rainfall rate. A composite image with more advanced radar-based quantitative precipitation estimation (QPE), that includes better ground clutter removal algorithms and also a correction for the bright band, was recently developed and the verification of the new product is ongoing.

STEPS forecasts were generated and verified for a set of sewer system overflow cases that affected the cities of Ghent and Leuven (see Table 1). The Ghent cases have a more stratiform character and occurred in late autumn and winter. On the other hand, the Leuven cases are more convective and occurred in summer months. A detailed climatology of convective storms in Belgium can be found in Goudenhoofdt and Delobbe (2009).

3 Short-Term Ensemble Prediction System (STEPS)

3.1 STEPS description

The Short-Term Ensemble Prediction System (STEPS) was jointly developed by the Australian Bureau of Meteorology (BOM) and the UK MetOffice (Bowler et al., 2006). STEPS forecasts are produced operationally at both weather services and are distributed to weather forecasters and a number of external users, in particular the hydrological services.

The key idea behind STEPS is to account for the unpredictable rainfall growth and decay processes by adding stochastic perturbations to the deterministic extrapolation of radar images (Seed, 2003). In order to be effective, the stochastic perturbations need to reproduce important statistical properties of both the precipitation fields and the forecast errors:

Table 1. List of precipitation events that caused sewer system floods in Ghent and Leuven.

City	Date	Event start (UTC)	Event end (UTC)	Duration	Predominant precipitation	Main wind direction
Ghent	10 Nov 2013	13:50	22:00	8:10 h	Stratiform	WNW → NNW
Ghent	3 Jan 2014	03:00	14:00	11 h	Stratiform	SW → WSW
Leuven	9–10 June 2014	06:30, 9th	15:30, 10th	33 h	Convective	SW
Leuven	19–20 July 2014	22:00, 19th	06:30, 20th	8:30 h	Convective	SSW

1. spatial scaling of precipitation fields,
2. dynamic scaling of precipitation fields,
3. spatial correlation of the forecast errors,
4. temporal correlation of the forecast errors.

The *spatial scaling* considers the precipitation field as arising from multiplicative cascade processes (Schertzer and Lovejoy, 1987; Seed, 2003). The presence of spatial scaling can be demonstrated by computing the 2-D Fourier power spectrum (PS) of a precipitation field. A 1-D PS can be obtained by radially averaging the 2-D PS. The precipitation field is said to be *scaling* if the 1-D PS draws a straight line on the log-log plot of the power against the spatial frequency (power law), which can be parametrized by one or two spectral exponents (see, e.g., Seed et al., 2013; Foresti and Seed, 2014). Within the multiplicative framework, a rainfall field is not represented as a collection of convective cells of a characteristic size but rather as a hierarchy of precipitation structures embedded in each other over a continuum of scales. STEPS considers the spatial scaling by decomposing the radar rainfall field into a multiplicative cascade using a fast Fourier transform (FFT) to isolate a set of eight spatial frequencies (Seed, 2003; Bowler et al., 2006; Seed et al., 2013). The top cascade levels (0, 1 and 2) represent the low spatial frequencies (large precipitation structures), while the bottom cascade levels (5, 6, 7) represent the high spatial frequencies (small precipitation structures). Another important behavior of rainfall fields is known as *dynamic scaling*, which is the empirical observation that the rate of temporal development of rainfall structures is a power law function of their spatial scale (Venugopal et al., 1999; Foresti and Seed, 2014). This means that large precipitation features are more persistent and hence predictable compared with small precipitation cells, which is closely related to the concept of scale-dependence of the predictability of precipitation (Germann and Zawadzki, 2002; Turner et al., 2004).

The stochastic perturbations should be able to reflect the properties of the forecast errors. Generating spatially and temporally correlated forecast errors is mandatory for hydrological applications, in particular when the correlation length of the errors is comparable or superior to the size and response time of the catchment. *Spatially correlated stochastic noise* can be constructed by applying a power law filter to a white noise field (Schertzer and Lovejoy, 1987). In practice it consists of three steps: computing the FFT of a white noise field, multiplying the obtained components in frequency domain by a given filter and applying the inverse FFT to return back to the spatial domain. The 1-D or 2-D power spectra of the rainfall field can be used as a filter to obtain noise fields that have the same scaling and spatial correlation of the rainfall field. The 1-D PS of the precipitation fields often appears to be a power law of the spatial frequency and explains why the procedure is also called power law filtering of white noise. In order to represent the anisotropies of the precipitation field, the 2-D PS can also be used as a filter. In the absence of a target precipitation field from which to derive the PS, the filter can be parametrized by using a climatological power law (see Seed et al., 2013). Finally, the *temporal correlations* are imposed by auto-regressive (AR) filtering. A hierarchy of AR processes defines the temporal evolution of the cascade levels. With the exception of forecast lead times beyond 2–3 h (Atencia and Zawadzki, 2014), an AR process of order 1 or 2 is already a good approximation to describe the temporal decorrelation of the forecast errors.

The practical implementation of STEPS to reproduce these important properties consists of the following steps (see Bowler et al., 2006; Foresti and Seed, 2014).

1. Estimation of the velocity field using optical flow on the last two radar rainfall images (Bowler et al., 2004a)
2. Decomposition of both rainfall fields into a multiplicative cascade using an FFT to isolate a set of eight spatial frequencies
3. Estimation of the rate of temporal evolution of rainfall features at each level of the cascade (Lagrangian autocorrelation)
4. Generation of a cascade of spatially correlated stochastic noise using as a filter the 1-D or 2-D power spectra of the last observed radar rainfall field. A Gaussian filter is used to isolate a given spatial frequency (see Foresti and Seed, 2014).
5. Stochastic perturbation of the rainfall cascade using the noise cascade (level by level)
6. Extrapolation of the cascade levels using a semi-Lagrangian advection scheme

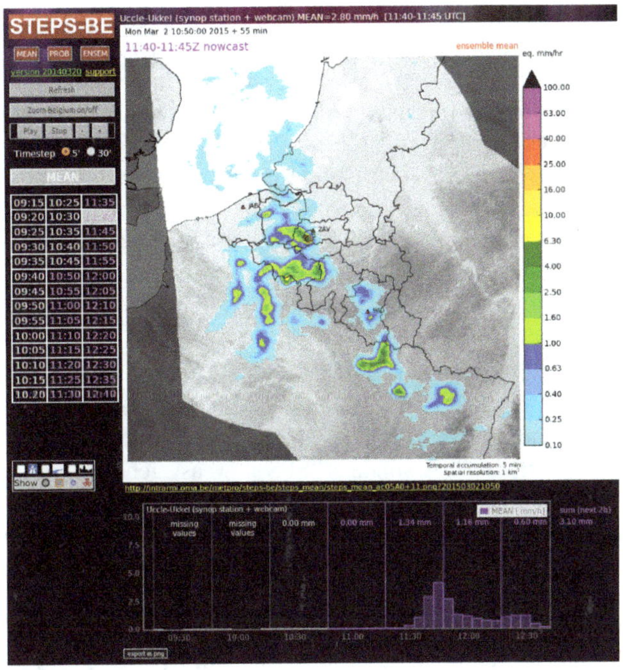

Figure 1. Web platform of STEPS-BE showing the ensemble mean forecast.

7. Application of the AR(1) or AR(2) model for the temporal update of the cascade levels at each forecast lead time using the Lagrangian auto-correlations estimated in step (3)

8. Recomposition of the cascade into a rainfall field

9. Probability matching of the forecast rainfall field with the original observed field (Ebert, 2001)

10. Computation of the forecast rainfall accumulations from the instant forecast rainfall rates. This procedure is known as advection correction and consists of advecting the instant rainfall rate forward over the 5 min period by discretizing the advection into smaller time steps.

3.2 STEPS implementation at RMI (STEPS-BE)

Bowler et al. (2006) introduced a general framework for blending a radar-based extrapolation nowcast with one or more outputs of downscaled NWP models (see also Pierce et al., 2010; Seed et al., 2013). Because of being designed for urban applications, the maximum lead time of STEPS-BE is restricted to 2 h. The operational NWP model of RMI (ALARO) runs only four times daily using a grid spacing of 4 km. Considering the model spin-up time and the absence of radar data assimilation, it is very unlikely that ALARO provides useful skill for blending its output with a radar-based extrapolation nowcast within the considered nowcasting range. It must also be remembered that the effective resolution of NWP models is much larger than the grid spacing. For instance, Grasso (2000) estimates the effective resolution to be at least 4 times the grid spacing, while Skamarock (2004) estimates it to be up to 7 times the grid spacing. ALARO would then only be able to resolve features that are greater than 20 km. For all these reasons, STEPS-BE only involves an extrapolation nowcast without NWP blending.

The STEPS-BE forecast domain is smaller than the extent of the 4 C-band radars composite (see Fig. 1). The radar field was upscaled from the original resolution of 500 m to 1 km and a sub-region of 512×512 grid points centered over Belgium was extracted. The forecast domain was extended by 32 pixels on each side to reduce the edge effects due to the FFT. This leads to an eight-level multiplicative cascade representing the following spatial scales (rounded to the nearest integer): *576*–*256*–114, 256–*114*–51, 114–*51*–23, 51–*23*–11, 23–*11*–4, 11–*4*–2, 4–*2*–1 and 2–*1* km. Italic characters mark the scales on which the Gaussian filter is centered in the frequency domain (see Foresti and Seed, 2014, for a more detailed explanation and visualization of the Gaussian FFT filter). The Gaussian filters of the largest and smallest spatial scales are truncated in order to preserve the power of the field. The top cascade level represents scales between the 576 and 256 km wavelengths and is not a perfect Gaussian filter. One can notice that the spatial scales are not exact multiples of 2. In fact, a multiplication factor of 2.246 was employed to match the enlarged STEPS-BE domain size.

STEPS-BE includes a couple of improvements compared with the original implementation of the BOM:

1. kernel interpolation of optical flow vectors;

2. generation of stochastic noise only within the advected radar mask.

The optical flow algorithm of Bowler et al. (2004a) estimates the velocity field by dividing the radar domain into a series of blocks within which the optical flow equation is solved. The minimization of the field divergence is only performed at the level of the block, which leaves sharp discontinuities in the velocity field between the blocks. In order to overcome this issue, a Gaussian kernel smoothing was applied to interpolate the velocity vectors located at the center of the blocks onto the fine radar grid. The bandwidth of the Gaussian kernel was chosen to be $\sigma = 24\,\text{km} = 0.4k$, where $k = 60$ grid points is the block size. This setting has the advantage of obtaining velocity fields that are less affected by the differential motion of small rainfall features and the presence of ground clutter. A too precise velocity field would provide increased predictability at very short lead times but worse forecasts at longer lead times due to excessive convergence and divergence of precipitation features during the advection. Smooth velocity fields could also be obtained by using a smaller block size and by compensating with a larger bandwidth of the smoothing kernel.

In STEPS-BE the 1-D power spectrum of the last observed rainfall field is used as a filter to generate the spatially correlated stochastic perturbations. The PS is parameterized using two spectral slopes to account for a scaling break that is often observed at the wavelength of 40 km (see Seed et al., 2013; Foresti and Seed, 2014). To simplify the computations, an auto-regressive model of order 1 (AR(1)) was employed for imposing the temporal correlations and to model the growth of forecast errors.

The original STEPS implementation (Bowler et al., 2006) was designed to blend the radar extrapolation nowcasts with the output of NWP models. However, the domain covered by the radars is smaller than the rectangular domain of the NWP model and small amounts of stochastic noise are generated by default also outside the radar composite. This setting was not adapted for radar-based nowcasts without NWP blending and needed some adaptation. In fact, when advecting the radar mask over several time steps, large areas with small amounts of stochastic rain appear outside the validity domain of the forecast and perturb the probability matching. In STEPS-BE the stochastic perturbations are only generated within the advected radar domain and set to zero elsewhere.

STEPS-BE can also account for the uncertainty in the estimation of the velocity field. The STEPS version that is implemented in the UK (Bowler et al., 2006) includes a detailed procedure to generate velocity perturbations that reproduce various statistical properties of the differences between the forecast velocity and the actual future diagnosed velocity (see details in Bowler et al., 2004b). In the BOM and RMI implementations a simpler procedure is applied. The diagnosed velocity field is multiplied by a single factor C that is drawn from the following distribution:

$$C = 10^{1.5N/10}, \quad (1)$$

where N is a normally distributed random variable with zero mean and unit variance. In other words, the velocity field is accelerated or decelerated by a single random factor without affecting the direction of the vectors. In fact, the uncertainty on the diagnosed speed was observed to be higher than that of the direction of movement (Bowler et al., 2006).

The BOM and RMI versions of STEPS also include a stochastic model for the radar measurement error, a broken-line model to account for the unknown future evolution of the mean areal rainfall and the possibility to use time-lagged ensembles. However, a nowcasting model with too many components is harder to calibrate and complicates the interpretation of the forecast fields. Because of these reasons, STEPS-BE only exploits the basic stochastic model for the velocity field and for the evolution of rainfall fields (due to growth and decay processes).

The core of STEPS and the STEPS-BE extensions are implemented in C/C^{++} and the production of figures in python. Bash scripts were written to call multiple STEPS instances and compute the ensemble members in parallel over several processors. Once all the ensemble members are computed, a separate script collects the corresponding netCDF files and calculates the forecast probabilities. Most of the computational cost of STEPS consists of filtering the white noise field with FFT, advecting and updating the radar cascade with the AR model. The re-calculation of optical flow fields on each processor takes less than 10 % of the total computational time.

The python matplotlib library is used to read the netCDF files, export the PNG figures and the time series of observed and forecast rainfall at the location of major cities and weather stations. A single STEPS nowcast generates more than 600 figures, which takes a significant fraction of the total computational time. In order to optimize the timing, a bash script monitors continuously the directory with incoming radar composites and triggers STEPS-BE once a field with a new time stamp is found. All these implementation details ensure that the user/forecaster can have access to an ensemble STEPS nowcast in less than 5 min after receiving the radar composite image.

The visualization system of STEPS-BE is very similar to the one of INCA-BE, the local Belgian implementation of the Integrated Nowcasting through Comprehensive Analysis system (INCA, Haiden et al., 2011) developed at the Austrian weather office (ZAMG). Figure 1 illustrates the web interface with an example of an ensemble mean nowcast. The user can highlight the major cities, weather stations and click to visualize the time series of observed and forecast precipitation/probability, which appears at the bottom of the web page. The navigation through the observations and forecast lead times is facilitated by the scroll wheel of the mouse. On the other hand, by clicking on the image it is possible to easily scroll through the various ensemble members or probability levels for a given lead time. Scrolling the ensemble members at different lead times is very instructive and can make the user aware of the forecast uncertainty. In fact, at a lead time of 5 min the ensemble members agree very well on the intensity and location of precipitation. This means that the ensemble spread is small and the probabilistic forecast is sharp; i.e., most of the forecast probabilities are close to 1 or 0 (see an explanation in Appendix A). On the other hand, at 1 or 2 h lead time the ensemble members disagree on the location and intensity of rainfall, which enhances the ensemble spread and decreases the sharpness of the probabilistic forecast. The web page includes extensive documentation to guide the user and a set of case studies to help understanding the strengths and limitations of STEPS. The visualization system was implemented with great attention to take full advantage of the multi-dimensional information content of probabilistic and ensemble forecasts.

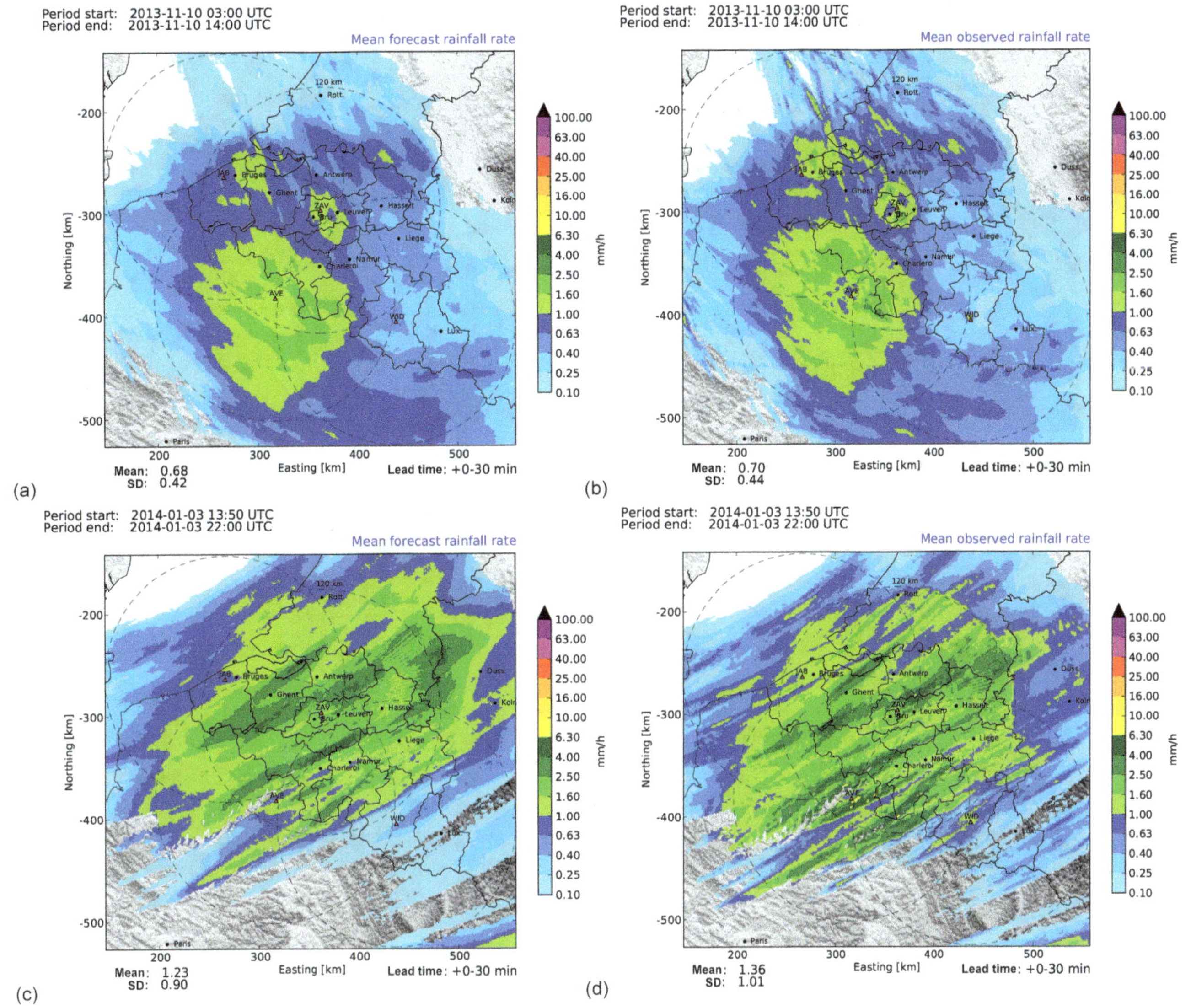

Figure 2. Average forecast and observed rainfall accumulations for the Ghent cases. **(a)** Forecast and **(b)** observed 0–30 min rainfall accumulations on 10 November 2013. **(c)** Forecast and **(d)** observed 0–30 min rainfall accumulations on 3 January 2014. The mean and standard deviation of the field within the 120 km range of the radars are shown on the bottom left corner. Field values are shown only if there are at least 10 samples for the computation of the mean. The red triangles denote the location of the Wideumont (WID, coordinates 438 km east/−405 km north), Zaventem (ZAV, 363/−296), Jabbeke (JAB, 266/−263) and Avesnois (AVE, 317/−382) radars. The 120 km range from the radar is displayed as a dashed circle. The mountain range of the Ardennes covers the three most southern districts of Belgium and Luxembourg (LUX).

4 Forecast verification

4.1 Verification set-up

This section presents the verification of STEPS-BE forecasts using a set of case studies (see Sect. 2). The accumulated radar observations were employed as reference for the verification. The rainfall rates are accumulated over the last 5 min by reversing the field vectors based on the observations and then performing the advection correction. The 30 min ensemble mean forecast was verified against the observed 30 min radar accumulations using both continuous and categorical verification scores. The deterministic verification procedure follows the one presented in Foresti and Seed (2015), which was designed to analyze the spatial distribution of the forecast errors. More details about the forecast verification setup and scores are given in Appendix A.

The continuous scores include the multiplicative bias and the root mean squared error (RMSE), while the categorical scores include the probability of detection (POD), false alarm ratio (FAR) and Gilbert skill score (GSS) derived from the contingency table for rainfall thresholds of 0.5 and

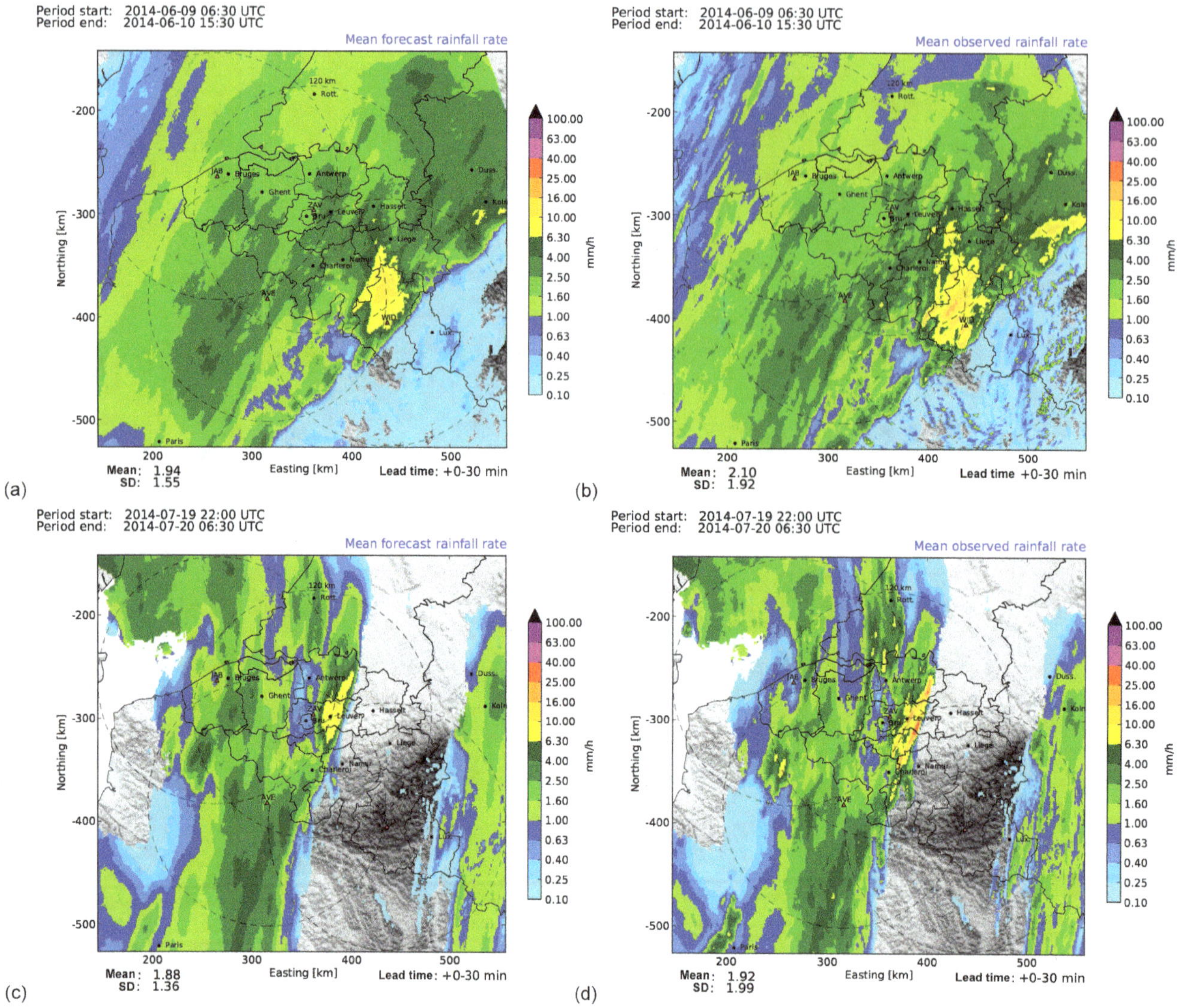

Figure 3. Average observed and forecast rainfall accumulations for the Leuven cases. **(a)** Forecast and **(b)** observed 0–30 min rainfall accumulations on 9–10 June 2014. **(c)** Forecast and **(d)** observed 0–30 min rainfall accumulations on 19–20 July 2014.

$5.0\,\text{mm}\,\text{h}^{-1}$. The rainfall thresholds are given in equivalent intensity independently of the forecast rainfall accumulation. Thus, a threshold of $5.0\,\text{mm}\,\text{h}^{-1}$ on a 30 min accumulation corresponds to 2.5 mm of rain. The multiplicative bias and the RMSE were evaluated only at the locations where the forecast or the verifying observations exceeded $0.1\,\text{mm}\,\text{h}^{-1}$, which can be referred to as a *weakly conditional verification*. The probabilistic forecast of exceeding 0.1, 0.5 and $5.0\,\text{mm}\,\text{h}^{-1}$ was verified using the reliability diagrams and ROC curves. Finally, the dispersion of the ensemble was analyzed by comparing the ensemble spread to the RMSE of the ensemble mean and by using rank histograms. The probabilistic and ensemble verification does not consider the spatial distribution of the errors and pools the data together in both space and time to derive the statistics.

4.2 Deterministic verification

Figures 2 and 3 show the average forecast and observed rainfall rates corresponding to the 0–30 min ensemble mean accumulation nowcast for the Ghent and Leuven cases, respectively. In other words, they represent the average forecast and observed rainfall rates over the duration of the precipitation event (for the 0–30 min lead time). The average was computed using the weak conditional principle explained above.

The average forecast and observed accumulations generally agree very well for the 0–30 min lead time forecast. The Ghent case on 10 November 2013 (Fig. 2a and b) is the only one with northwesterly flows and is characterized by the lowest average rainfall rates. The Avesnois radar demonstrates very well the range dependence of the average rainfall rates, which gradually decrease with increasing distance from the

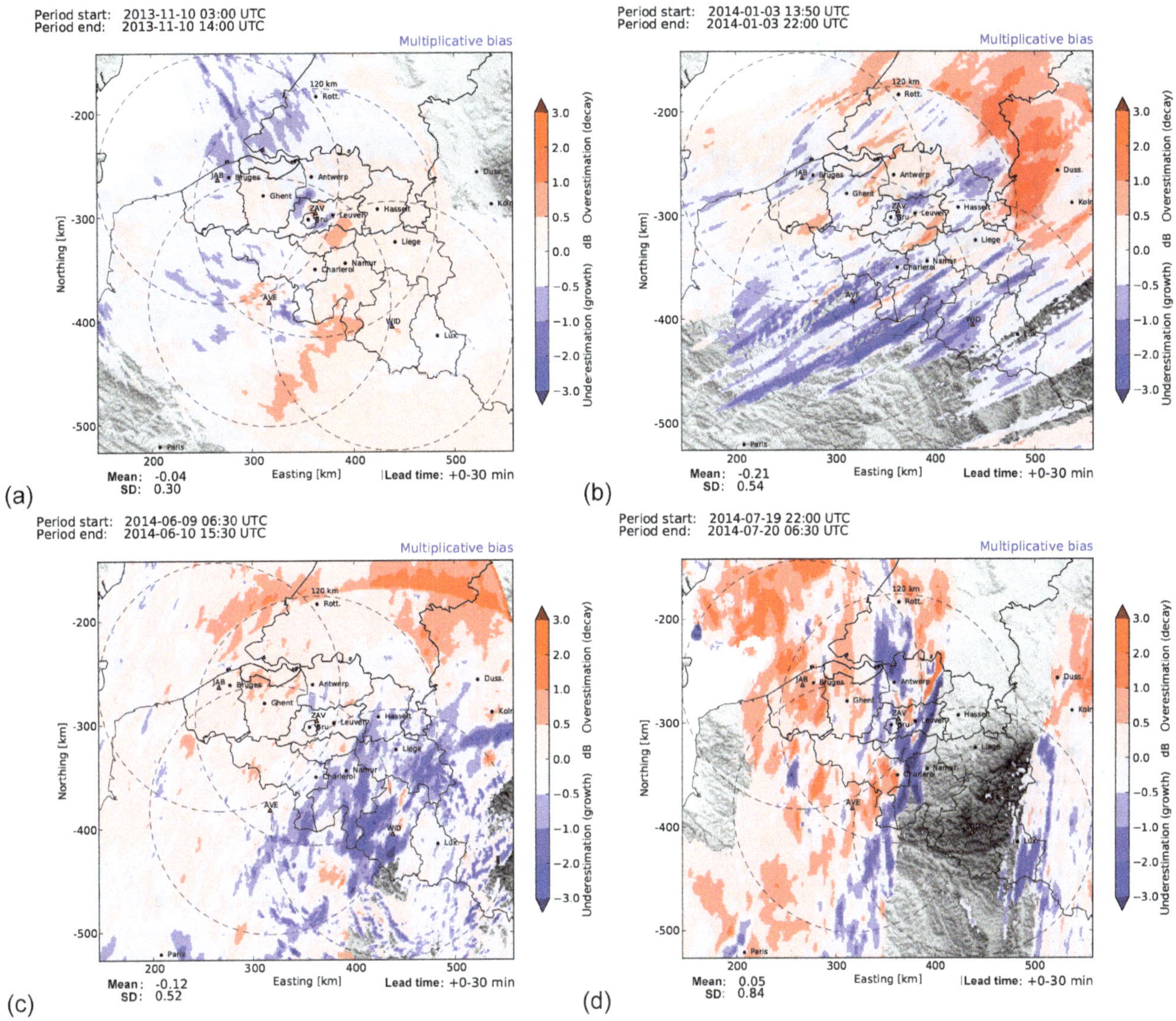

Figure 4. Average 0–30 min multiplicative forecast biases for the Ghent cases on **(a)** 10 November 2013 and on **(b)** 3 January 2014 and the Leuven cases on **(c)** 9–10 June 2014 and on **(d)** 19–20 July 2014. The interpretation of under- and over-estimations by STEPS as systematic rainfall growth and decay or simply as radar measurement biases is subject to interpretation as explained in the text.

radar. On the contrary, the smaller ring of high rainfall rates around the Zaventem radar is mostly due to the bright band (Fig. 2b).

The bright band effect influences the radar observations and hence the nowcasts based on their extrapolation. At longer lead times the larger rainfall estimates due to the bright band are extrapolated far from the location of the radar. The stochastic perturbations of STEPS can help to gradually dissolve the circular patterns introduced by the bright band effect. However, the bright band affects more the observations used for the verification, in particular when the rainfall is advected from upstream over the radar region. In such a case, the local larger rainfall estimates lead to a verification bias and the forecasts are wrongly supposed of rainfall underestimation. In spite of these issues, bright band effects might not be so important for urban hydrological applications. In fact, except for one stratiform case presented in this paper, pluvial floods mainly happen in summer with convective precipitation events, during which the bright band is absent or negligible.

The Ghent case on 3 January 2014 has higher rainfall rates and the elongated structures of precipitation areas demonstrate well the southwesterly flow regime (Fig. 2c and d). For this case the measurements of the Zaventem radar are also affected by second trip echoes, which appear as a set of radially oriented rainfall structures northwest of the radar. These alternating patterns are explained by the dual PRF mode of Zaventem (see Sect. 2).

The Leuven cases on 9 June and 19–20 July 2014 have an important convective activity (Fig. 3a–d). The maximum av-

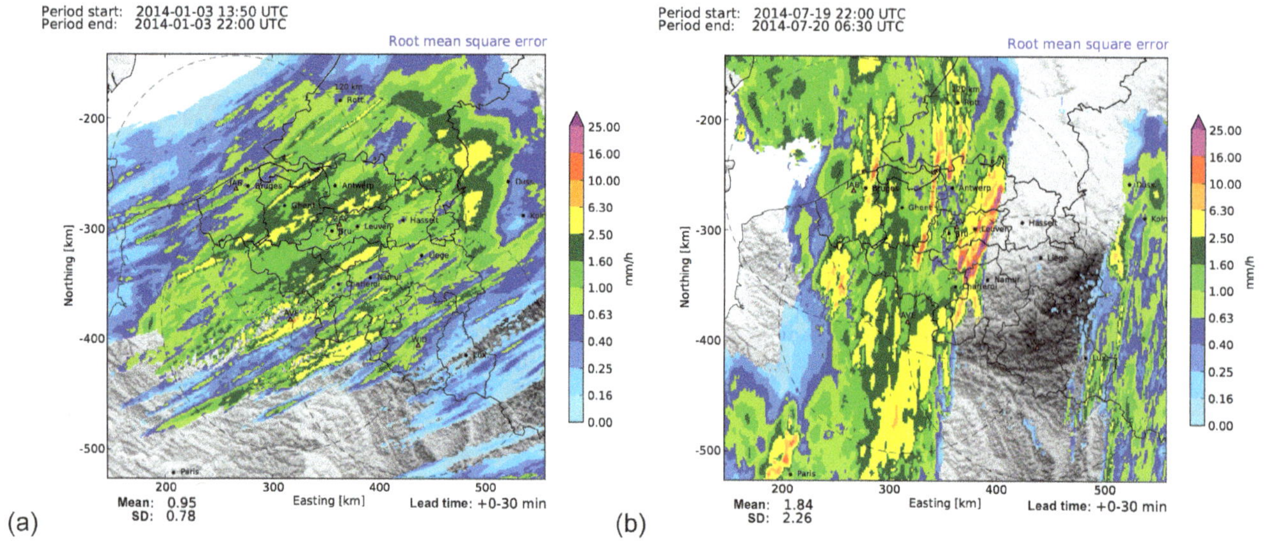

Figure 5. Average 0–30 min forecast RMSE for **(a)** the Ghent winter case on 3 January 2014 and **(b)** the Leuven summer case on 19–20 July 2014.

erage rainfall rates are located over the Ardennes mountain range and the city of Leuven, respectively. Since urban flash floods can be triggered by a single convective cell, the average rainfall rate over the duration of the event may not be as high in the considered city (e.g., Fig. 3b).

Figure 4 illustrates the multiplicative bias of the 0–30 min nowcast averaged over each of the four events. A detailed interpretation of such forecast biases using Australian radar data and their connection to orographic features is given in Foresti and Seed (2015), which point out that an important fraction of the forecast errors is caused by the biases of the verifying radar observations rather than systematic rainfall growth and decay processes due to orography. In Fig. 4a it is easy to notice the effect of bright band, which causes a series of systematic forecast biases around the Zaventem radar and perpendicularly oriented with respect to the prevailing flow direction (NW). Systematic rainfall underestimation occurs along the Belgian coast of the North Sea. One factor which contributes to this underestimation is the absence of visibility of the radar at longer ranges. The incoming precipitation is suddenly detected by the radar and therefore strongly underestimated by STEPS. The only situation where the range dependence of the rainfall estimation does not affect the forecast verification occurs when the velocity field is perfectly rotational and centered on the radar (assuming no beam blockage). All the other cases have to deal with the fact that the rainfall nowcast also extrapolates the biases of the radar observations! Contrary to expectation, on the upwind side of the Ardennes there is overestimation, which may depict a region of systematic rainfall decay. The bias over the city of Ghent is fortunately small and is included in the range from 0 to +0.5 dB (light overestimation, rainfall decay). Having small systematic biases over the cities of interest is very important for future integration of STEPS nowcasts as input in hydraulic models.

In Fig. 4b the systematic underestimation is also located upstream with respect to the prevailing winds (SW). The strong overestimations in Germany and the Netherlands are mostly due to the underestimation of rainfall by the verifying radar observations rather than caused by systematic rainfall decay. This is particularly visible after a range of 125 km from the Zaventem radar, which demonstrates again that discontinuities and biases in the radar observations lead to biases in the extrapolation forecast. Also in this case the bias over the city of Ghent is small but in the range from −0.5 to 0 dB (light underestimation, rainfall growth). Radar composite discontinuities are also visible in Fig. 4c but this time located at a range of 240 km north of the Wideumont radar when entering the area covered by the Jabbeke radar. This forecast bias is mainly explained by the negative calibration bias of the Jabbeke radar, which is known to slightly underestimate the rainfall rates with respect to the Wideumont radar. Strong underestimation occurs over the Ardennes due to the systematic initiation and growth of convection that cannot be predicted by STEPS (Fig. 4c). Fortunately the city of Leuven is located in a region with small biases in the range from −0.5 to +0.5 dB. Figure 4d is quite interesting since strong underestimations are located in front of the rain band (from Charleroi to Leuven and beyond) and overestimations at the rear of the rain band (west of the Jabbeke radar). The underestimations are due to systematic rainfall initiation in front of the rain band, while the overestimations are probably caused by a too slow extrapolation of rainfall, which tends to drag at the rear of the rain band. The two bands of underestimations south of Leuven are caused by two different thunderstorms. The first one passed over the city of Leuven and had

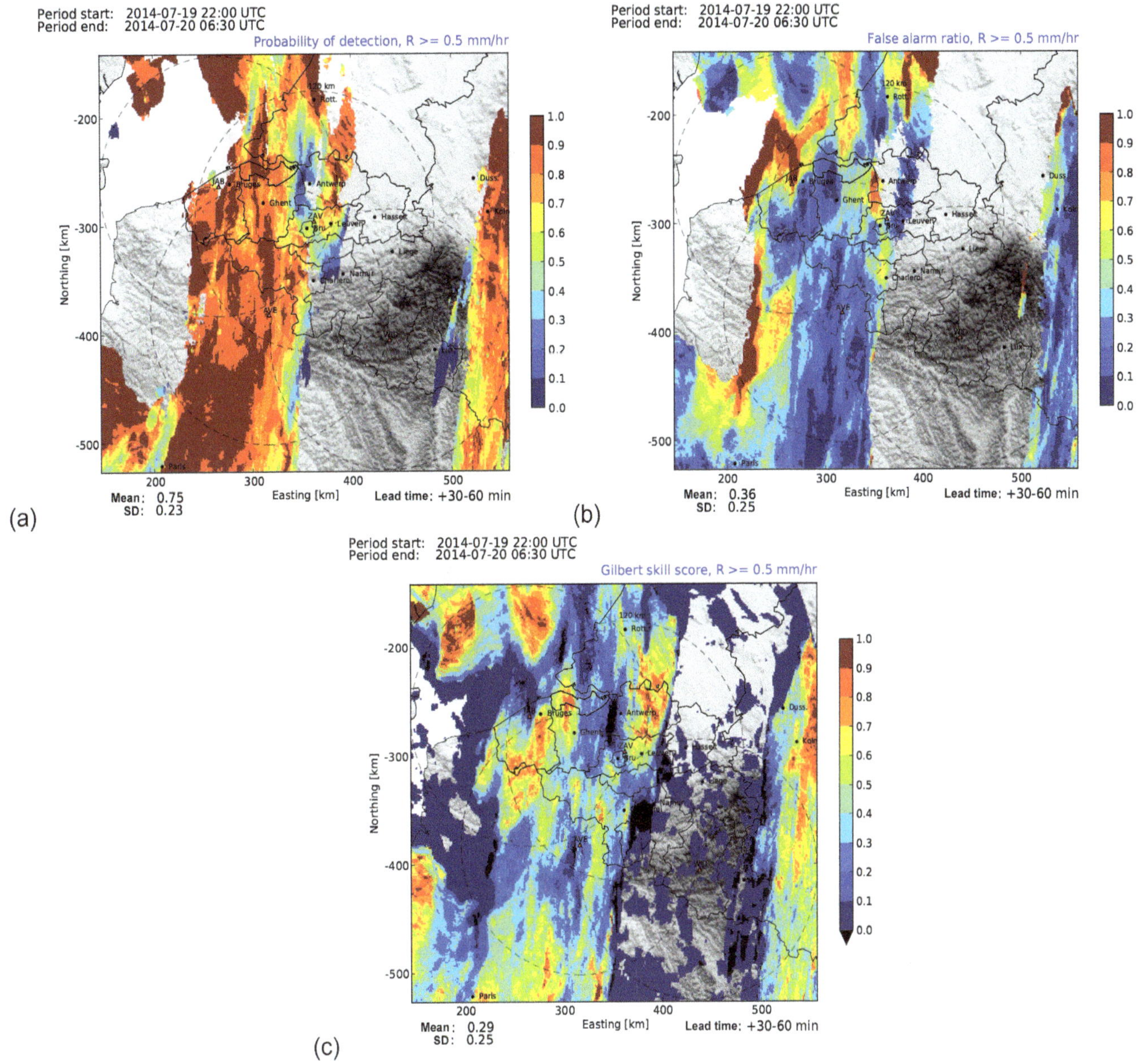

Figure 6. **(a)** POD, **(b)** FAR and **(c)** GSS of the 30–60 min ensemble mean forecast of exceeding $0.5\,\mathrm{mm\,h^{-1}}$ for the Leuven case on 19–20 July 2014.

a stronger westerly component with respect to the prevailing southerly flow. The second thunderstorm was weaker and had a stronger easterly component. When isolated convection does not follow the prevailing movement of the rainfall field, strong biases can appear in the nowcast during the first lead times.

Figure 5 shows the spatial distribution of the RMSE for the stratiform event on 3 January 2014 in Ghent and the convective event on 19–20 July 2014 in Leuven. If compared with Figs. 2d and 3d it is clear that the RMSE is strongly correlated with the regions having the highest mean rainfall accumulations (proportional effect). Thus, it is not surprising that the RMSE of the convective case (Fig. 5b) displays values exceeding $10\,\mathrm{mm\,h^{-1}}$ over the city of Leuven. The winter case only shows RMSE values below $2\,\mathrm{mm\,h^{-1}}$ over the city of Ghent.

Figure 6 illustrates an example of categorical verification of the 30–60 min ensemble mean forecast for the Leuven case on 19–20 July 2014 relative to the rainfall threshold of $0.5\,\mathrm{mm\,h^{-1}}$. The probability of detection is high everywhere (mean of 0.75) except in the neighborhood of Antwerp and south of Leuven, where the initiation of thunderstorms could not be predicted by STEPS (Fig. 6a). The false alarm ratio is quite low (mean of 0.36) and the regions with high values are

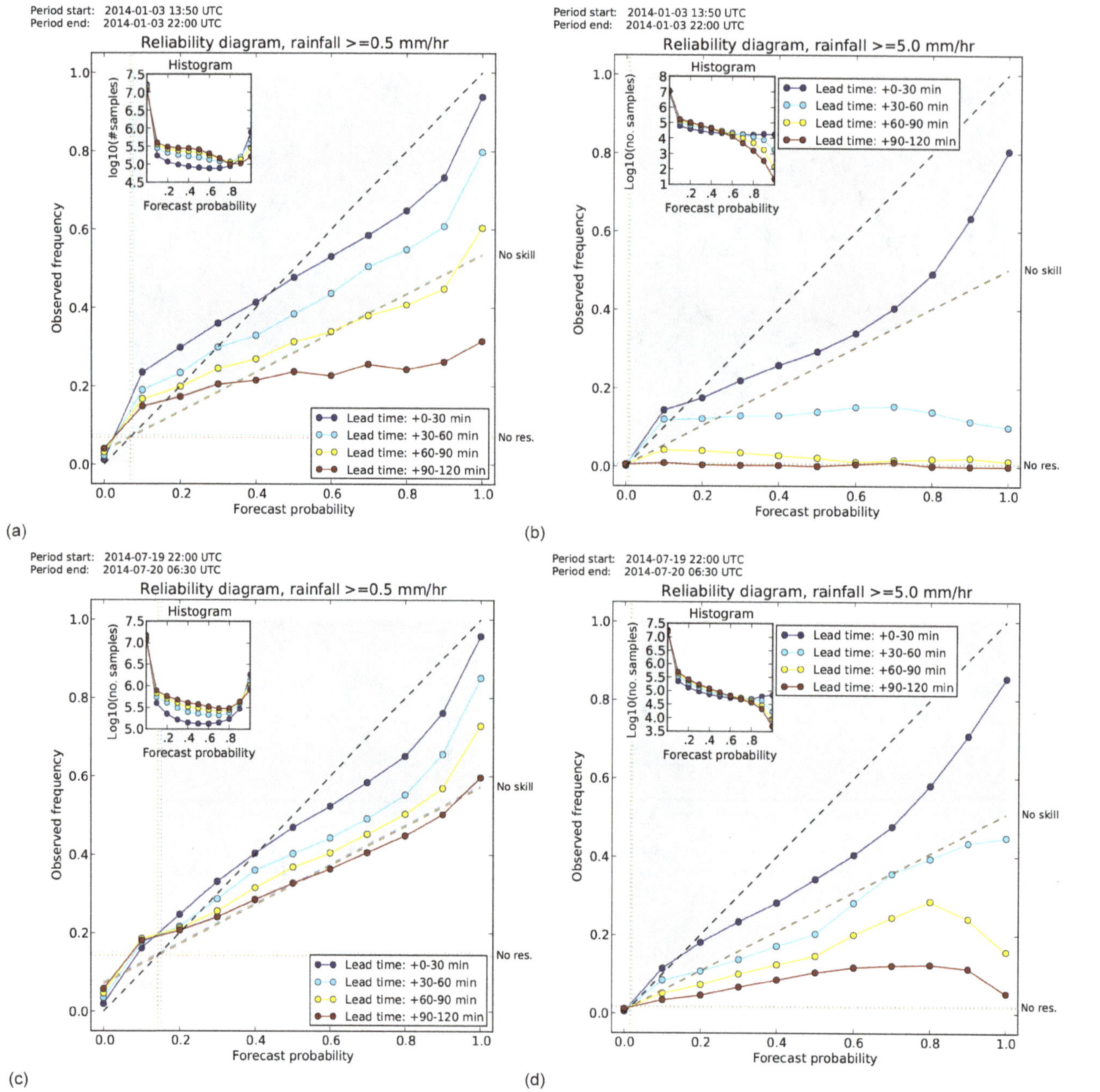

Figure 7. Reliability diagrams for the Ghent case on 3 January 2014 relative to the probabilistic forecast of exceeding **(a)** 0.5 mm h^{-1} and **(b)** 5.0 mm h^{-1}. **(c, d)** Same as **(a, b)** but for the Leuven case on 19–20 July 2014.

mainly located at the rear of the front where the rainfall is advected too slowly compared with the actual movement of the front (Fig. 6b). A high Gilbert skill score generally coincides with the regions with the highest rainfall accumulations and becomes lower at the edges of the rain areas (Fig. 6c). This finding can be explained conceptually if one thinks about the verification of the future path of a single convective cell. The regions with the highest uncertainty are located along the edges of the predicted thunderstorm path and the highest skill is obtained in the center of the predicted path.

4.3 Probabilistic verification

Figure 7 shows the reliability diagrams relative to the probabilistic forecast of exceeding the 0.5 and 5.0 mm h^{-1} rainfall thresholds for the Ghent case on 3 January 2014 (Fig. 7a and b) and the Leuven case on 19–20 July 2014 (Fig. 7c and d). The reference probabilistic forecast is taken as the climatological frequency of exceeding a given rainfall threshold during that precipitation event (horizontal dashed line). Unexpectedly, the forecasts of the stratiform case in Ghent

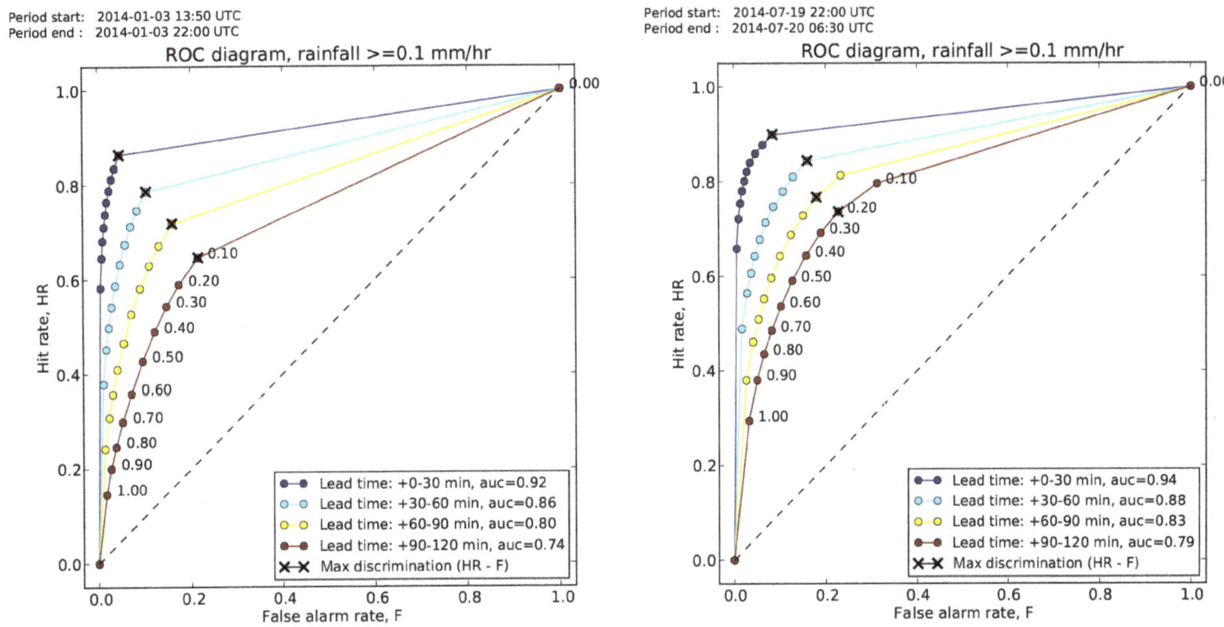

(a) (b)

Figure 8. ROC curves relative to the probabilistic forecast of exceeding 0.1 mm h^{-1} for **(a)** the Ghent case on 3 January 2014 and **(b)** the Leuven case on 19–20 July 2014.

are less reliable than the ones of the convective case in Leuven for both rainfall thresholds. Probabilistic forecasts of exceeding 0.5 mm h^{-1} for the Ghent case have a good reliability and positive Brier skill score (BSS) up to 60 min lead time (Fig. 7a). The higher rainfall threshold of 5.0 mm h^{-1} is harder to predict and there is skill only up to 30 min lead time (Fig. 7b). The convective case in Leuven is more predictable and the probabilistic forecast of exceeding 0.5 mm h^{-1} exhibits skill up to 90 min lead time (Fig. 7c). It is interesting to note that forecast probabilities that are close to the climatological frequency (intersection of lines around the probability 0.15) can easily fall outside the skillful region (Fig. 7c). In fact, a small systematic forecast bias is likely to be worse than the event climatology at those frequencies. The rainfall threshold of 5.0 mm h^{-1} shows again a limit of predictability of 30 min (Fig. 7d). Despite having a negative BSS, the following lead times (Fig. 7d) have higher resolution than the stratiform case in Ghent (Fig. 7b).

Figure 8 illustrates the ROC curves relative to the probabilistic forecast of exceeding 0.1 mm h^{-1} for the Ghent case on 3 January 2014 (Fig. 8a) and the Leuven case on 19–20 July 2014 (Fig. 8b). All the ROC curves are very far from the diagonal line of no skill. The probability level that is marked with a cross is the one that maximizes the difference between the hit rate (HR) and the false alarm rate (F) (not to be confused with the false alarm ratio, which is conditioned on the forecasts). This point is located within the probabilities 0.1 and 0.2, which means that an optimal forecast of the probability of rain is achieved when only 10–20 % of the ensemble members exceed the 0.1 mm h^{-1} threshold. A forecaster who is not scared of making false alarms would choose a lower probability level to increase the number of hits. On the contrary, an unconfident forecaster who would like to minimize the false alarms would choose a higher probability level, which has however the consequence of reducing the number of hits. As expected, the area under the ROC curves (AUC) decreases for increasing lead times. The discrimination skill for the convective event in Leuven is slightly higher than the one of the stratiform event in Ghent, which confirms the findings on the reliability diagrams (Fig. 7). This does not mean that small-scale features are easier to forecast than larger-scale features, which is known to be false (see Foresti and Seed, 2014). It means that the predictability of well-defined and organized convective systems is higher than the one of more moderate convection with shorter lifetime, at least for the cases analyzed in this paper.

4.4 Ensemble verification

Figure 9 compares the error of the ensemble mean (RMSE) and the ensemble spread for the Ghent case on 3 January 2014 and the Leuven case on 19–20 July 2014 (see interpretation of ensemble spread in Appendix A). In both cases the RMSE increases up to a lead time of 50–60 min and then starts a slow decrease, which can be counter-intuitive. However, it must be remembered that the ensemble mean forecast becomes smoother for increasing lead times, which reduces the double penalty error due to forecasting a thunderstorm at the wrong location. The ensemble spread also increases up to

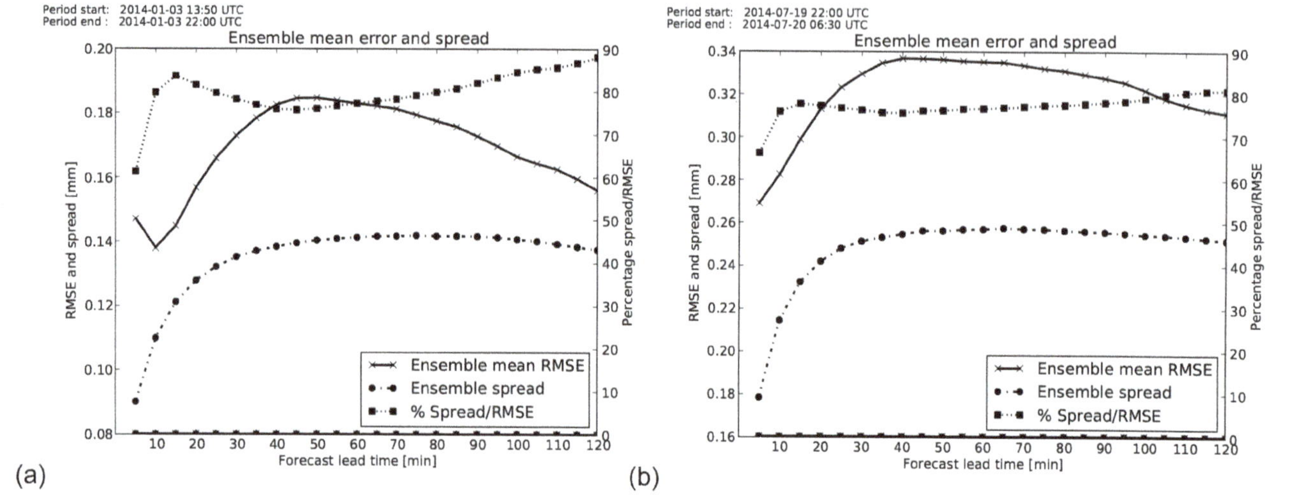

Figure 9. Comparison of ensemble spread and RMSE of the ensemble mean forecast at 5 min resolution for **(a)** the Ghent case on 3 January 2014 and **(b)** the Leuven case on 19–20 July 2014.

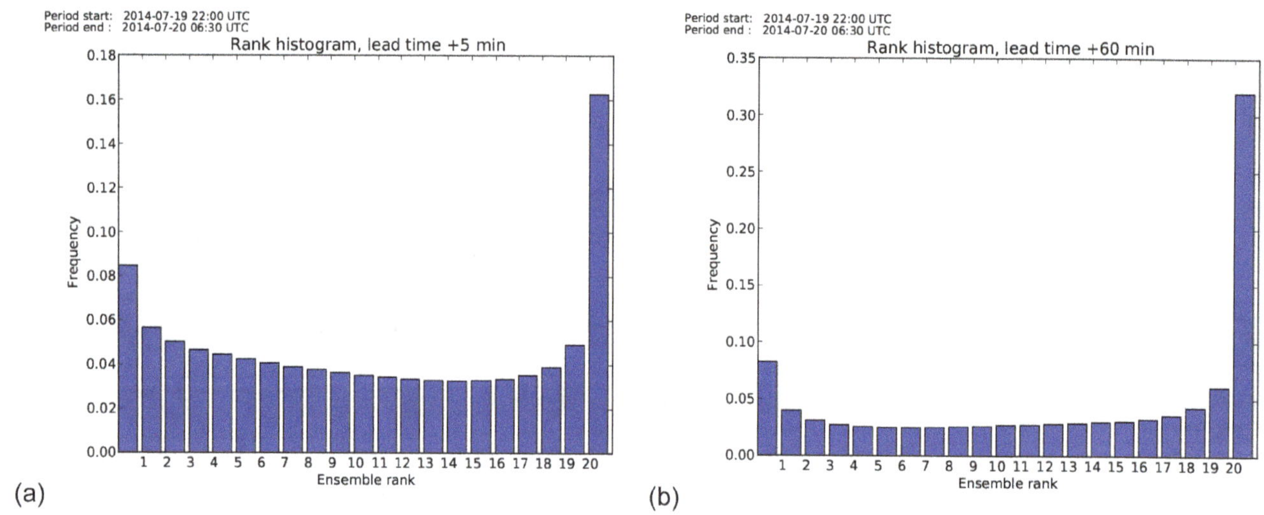

Figure 10. Rank histograms for the Leuven case on 19–20 July 2014 for a lead time of **(a)** 5 min and **(b)** 60 min.

50–60 min lead time and then slowly stabilizes. For both the Ghent and Leuven cases the ensemble spread is lower than the error of the ensemble mean at all lead times, which suggests that the ensemble forecasts are under-dispersive. The degree of under-dispersion is highest at a lead time of 5 min, with the spread values being equal to 60 % of the forecast error for the winter event in Ghent (Fig. 9a) and 70 % for the summer event in Leuven (Fig. 9b). Except for the 5 min lead time, the ensemble spread represents 75–90 % of the forecast error for the Ghent case (Fig. 9a) and 75–80 % for the Leuven case (Fig. 9b), which is a good result. It is not yet clear why the RMSE at a lead time of 5 min is higher than the one at 10 min for the winter case in Ghent (Fig. 9a).

The underestimation of the ensemble dispersion at the first lead time can be attributed to both the underestimation of the ensemble spread and the overestimation of the ensemble mean RMSE, but with different degrees according to the different causes. High RMSEs at the start of the nowcast can be due to using a very smooth velocity field for the advection (see Sect. 3.2), which does not exploit sufficiently the very short-term predictability of small-scale precipitation features, but is optimized for predictions at longer lead times. Another explanation for this underestimation of ensemble dispersion could be due to the space–time variability of the Z–R relationship. Spatial and temporal changes in the drop size distribution (DSD) can lead to changes in the estimated rainfall rate that is used for the verification. Therefore, there could be a mismatch between the "fixed" DSD of the forecasts and the variable DSD underlying the verifying observations. Another possible source of mismatch could be due to the advection correction with optical flow when computing the rainfall accumulations. The forecast accumulations are computed by advecting the previous rainfall field forward. On the other hand, the observed accumulations are computed by reversing the optical flow vectors and advecting the rainfall field backwards (see Sect. 4.1). This choice increases the differences when comparing the +0–5 min forecast accumulations (advection of the 0 min image forward) with the +0–5 min observed accumulations a posteriori (advection of the +5 min image backwards). The ideal approach would be to derive the accumulation by advecting both the previous image forward and the last image backwards. An optimal accumulation could be computed by a weighted average of the two advected images by discretizing the 5 min interval. However, such an approach is not very pragmatic and would require additional computational time in order to obtain a marginal improvement in the forecasts.

Figure 10 illustrates the rank histograms for the Leuven case on 19–20 July 2014 for lead times of 5 and 60 min. The U-shape of the rank histograms demonstrates again a certain degree of ensemble under-dispersion. In particular, all the ensemble members for the 5 min lead time are inferior to the observations in $\sim 16\,\%$ of the cases (Fig. 10a), while for the 60 min lead time it happens in more than $\sim 30\,\%$ of the cases (Fig. 10b). On the other hand, the fraction of observations falling below the value of the lowest ensemble member is only 8 % for both lead times. Despite the fact that STEPS is designed to reproduce the space–time variability of rainfall, it underestimates a certain fraction of the observed rainfall extremes. This underestimation grows with increasing lead time and depicts an increasing smoothness of the STEPS ensembles, which is probably due to the advection of the radar rainfall cascade (see Sect. 3, step 6). In fact, the small-scale rainfall features represented by the bottom cascade levels suffer more from numerical diffusion during the Lagrangian extrapolation, which is observed as a gradual loss of variability in the forecast ensembles.

4.5 Verification summary of the events

Table 2 provides a comparison of the verification scores for each event. The average standard deviation of the multiplicative biases of the 30 min lead time forecast is in the range 0.3–0.8. Except for the event on 19–20 July 2014 the biases remain well below 1 dB for all lead times, which is a positive result. Of course, these are average values, and locally they can even exceed 3 dB (see Fig. 4).

On the other hand, the RMSE values mark more the distinction between the two winter cases in Ghent and the two summer cases in Leuven. For the winter cases the RMSE values increase from 0.38–0.95 at a lead time of 30 min to 0.78–1.48 at 120 min, while for the summer cases from 1.84–2.45 to 2.52–3.38 mm h^{-1}. Thus, the RMSE of a 30 min lead time nowcast of the two convective cases is higher than the RMSE of a 120 min nowcast of the two stratiform cases, as might be expected. It is interesting to mention that linear verification scores such as the RMSE strongly depend on the variance of the data. Consequently, it would be difficult to compare the error of the STEPS ensemble mean nowcast with the one of a deterministic nowcast, for example computed by INCA-BE. In fact, the ensemble averaging process filters out the unpredictable precipitation features and is rewarded in terms of RMSE. Similar results were already observed in Foresti et al. (2015), who also pointed out the difficulty of comparing ensemble prediction systems having a different number of ensemble members.

The probability of detection relative to the 0.5 mm h^{-1} threshold decreases from 78–86 to 33–58 %, while the false alarm ratio increases from 10–17 to 46–65 %. The Gilbert skill score starts with values of 0.58–0.64 and 0.29–0.40 at the 30 and 60 min lead times, respectively, and decays to values of 0.08–0.20 at 120 min. Wang et al. (2009) reported a critical success index value of 0.45 for STEPS nowcasts of 0–60 accumulations relative to the 1 mm h^{-1} threshold. Considering that the GSS is the CSI corrected by random chance, this value is comparable with the ones of the 30–60 min accumulations obtained in this paper. The GSS values relative to the threshold of 5.0 mm h^{-1} are much lower. They oscillate between 0.15 and 0.44 for the first lead time and become very low and close to 0 afterwards. Thus, the predictability of rainfall structures exceeding 5.0 mm h^{-1} rarely exceeds 30 min according to the GSS.

The areas under the ROC curve values characterizing the potential discrimination power of the probabilistic forecast of exceeding 0.5 mm h^{-1} start at 0.92–0.95 at 30 min lead time and decrease to 0.69–0.79 at 120 min. For the probabilistic forecast of exceeding 5.0 mm h^{-1} they start at 0.88–0.90 and decrease to 0.62 for the convective cases and to 0.50 for the stratiform cases (no discrimination).

From all these results we can conclude that there is not much predictability beyond 2 h lead time by extrapolating the 4 C-band composite radar images in Belgium. Therefore, a maximum lead time of 2 h in STEPS-BE is a good choice. Extending this lead time requires blending the radar-extrapolation nowcast with the output of NWP models to increase the predictability of precipitation.

5 Conclusions

The Short-Term Ensemble Prediction System (STEPS) is a probabilistic nowcasting system based on the extrapolation of radar images developed at the Australian Bureau of Meteorology in collaboration with the UK MetOffice. The principle behind STEPS is to produce an ensemble forecast by perturbing a deterministic extrapolation nowcast with stochastic noise. The perturbations are designed to reproduce the spatial and temporal correlations of the forecast errors and the scale dependence of the predictability of precipitation.

This paper presented the local implementation, adaptation and verification of STEPS at the Royal Meteorological Institute of Belgium, referred to as STEPS-BE. STEPS-BE pro-

Table 2. Summary of the forecast verification scores of the next four 30 min accumulation forecasts for the precipitation events in Ghent and Leuven. The lead time shown is the end of the 30 min accumulation period (e.g., 60 min is relative to the 30–60 min accumulation). The bias values correspond to the standard deviation of the multiplicative bias, which is more interesting than its mean (often close to 0).

Event	Bias 30 min	Bias 60 min	Bias 90 min	Bias 120 min	RMSE 30 min	RMSE 60 min	RMSE 90 min	RMSE 120 min
	(dB)				(mm h^{-1})			
10/11/2013	0.30	0.49	0.61	0.70	0.38	0.59	0.71	0.78
03/01/2014	0.54	0.74	0.82	0.89	0.95	1.39	1.53	1.48
9–10/06/2014	0.52	0.63	0.66	0.69	2.45	3.26	3.40	3.38
19–20/07/2014	0.84	1.18	1.30	1.35	1.84	2.36	2.49	2.52
Event	POD 30 min	POD 60 min	POD 90 min	POD 120 min	FAR 30 min	FAR 60 min	FAR 90 min	FAR 120 min
	Forecast >= 0.5 mm h^{-1}				Forecast >= 0.5 mm h^{-1}			
10/11/2013	0.83	0.71	0.62	0.54	0.17	0.30	0.38	0.46
03/01/2014	0.80	0.63	0.49	0.33	0.10	0.25	0.45	0.65
9–10/06/2014	0.78	0.65	0.55	0.46	0.15	0.32	0.44	0.54
19–20/07/2014	0.86	0.75	0.66	0.58	0.17	0.36	0.50	0.61
Event	GSS 30 min	GSS 60 min	GSS 90 min	GSS 120 min	GSS 30 min	GSS 60 min	GSS 90 min	GSS 120 min
	Forecast >= 0.5 mm h^{-1}				Forecast >= 5.0 mm h^{-1}			
10/11/2013	0.58	0.38	0.27	0.20	0.15	0.02	0.0	0.0
03/01/2014	0.64	0.40	0.20	0.08	0.28	0.06	0.0	0.0
9–10/06/2014	0.59	0.38	0.26	0.17	0.44	0.20	0.09	0.04
19–20/07/2014	0.58	0.29	0.14	0.07	0.27	0.09	0.04	0.02
Event	AUC 30 min	AUC 60 min	AUC 90 min	AUC 120 min	AUC 30 min	AUC 60 min	AUC 90min	AUC 120 min
	Forecast >= 0.5 mm h^{-1}				Forecast >= 5.0 mm h^{-1}			
10/11/2013	0.95	0.89	0.84	0.79	0.88	0.67	0.56	0.50
03/01/2014	0.92	0.85	0.78	0.69	0.90	0.72	0.57	0.50
9–10/06/2014	0.93	0.86	0.81	0.76	0.89	0.77	0.68	0.62
19–20/07/2014	0.94	0.87	0.82	0.77	0.88	0.75	0.68	0.62

duces in real-time 20-member ensemble nowcasts at 1 km and 5 min resolutions up to 2 h lead time using the four C-band radar composite of Belgium. Compared with the original implementation, STEPS-BE includes a kernel-based interpolation of optical flow vectors to obtain smoother velocity fields and an improvement to generate stochastic noise only within the advected radar composite to respect the validity domain of the nowcasts.

The performance of STEPS-BE was verified using the radar observations as reference on four case studies that caused sewer system floods in the cities of Ghent and Leuven during the years 2013 and 2014. The ensemble mean forecast of the next four 30 min accumulations was verified using the multiplicative bias, the RMSE as well as some categorical scores derived from the contingency table: the probability of detection, false alarm ratio and Gilbert skill score (equitable skill score). The spatial distribution of multiplicative biases revealed regions of systematic over- and under-estimation by STEPS. The underestimations are often associated with the locations of convective initiation and thunderstorm growth, which cannot be predicted by STEPS. On the other hand, the regions of overestimation are mostly due to the underestimation of rainfall by the verifying observations rather than systematic rainfall decay (see Foresti and Seed, 2015, for a more detailed discussion). In order to disentangle the forecast and observation biases, detailed knowledge about the spatial distribution of the radar measurement errors for a given weather situation is needed. The multiplicative biases over the cities of Leuven and Ghent are very low (from -0.5 to $+0.5$ dB), which is a good starting point to integrate STEPS nowcasts as inputs into sewer system hydraulic models. The categorical forecast verification helped discovering the places with low probability of detection due to convective initiation at the front of the rain band and high false alarm ratio at the rear of the rain band, likely due to a too slow rainfall extrapolation by STEPS. Reliability diagrams demonstrated that probabilistic forecasts of exceeding 0.5 mm h^{-1} have skill up to 60–90 min lead time. On the contrary, convective features exceeding 5.0 mm h^{-1} are only predictable up to 30 min. In terms of reliability and discrimination ability, it was also observed that the forecasts of convective events have more skill than the ones of stratiform events. The STEPS ensembles are characterized by a certain degree of underestimation of the forecast uncertainty, with values of the ensemble spread close to 75–90 % of the forecast error.

The current contribution focused on the verification of STEPS-BE nowcasts using only four precipitation cases of different character. The deterministic and categorical verifications require many more cases to analyze the climatological distribution of the forecast errors, e.g., as done in Foresti and Seed (2015). On the other hand, the probabilistic and ensemble verification pools the data in both space and time and converges much faster to stable statistics.

From a research perspective, STEPS-BE could also be extended by including a stochastic model to account for the residual radar measurement errors, in particular to obtain more accurate estimations of the forecast uncertainty at short range. The STEPS framework also allows blending of the extrapolation nowcast with the output of NWP models, which is a necessary step to increase the predictability of precipitation for lead times beyond 2 h.

Table A1. Correspondence between the decibel scale and the power ratio.

dB	−6	−3	−1	−0.5	0	+0.5	+1	+3	+6
Power ratio (F/O)	0.251	0.501	0.794	0.891	1	1.122	1.259	1.995	3.981

Appendix A: Forecast verification scores

Forecast verification is an important aspect of a forecasting system. A forecast without an estimation of its accuracy is not very informative. For an in-depth description of forecast verification science and corresponding scores we refer to Jolliffe and Stephenson (2011) and the verification website maintained at the Bureau of Meteorology (http://www.cawcr.gov.au/projects/verification/).

The STEPS *ensemble mean forecast* was verified using the following scores.

- Multiplicative bias:

$$\text{bias} = \frac{1}{N}\sum_{i=1}^{N} 10\log_{10}\left(\frac{F_i + b}{O_i + b}\right), \tag{A1}$$

where F_i is the forecast rainfall at a given grid point, O_i is the observed rainfall at a given grid point, $b = 2\,\text{mm}\,\text{h}^{-1}$ is an offset to eliminate the division by zero and to reduce the contribution of the forecast errors at low rainfall intensities, and N is the number of samples. For the specific case of the verification of the spatial distribution of forecast biases, the summation is performed over time. Thus, N corresponds to the number of forecasts where either the forecast or the observed rainfall are greater than 0.1 mm h^{-1} at a given grid point (denoted as weak conditional verification). The bias is given in decibels (dB) in order to obtain a more symmetric distribution of the multiplicative errors centered at 0, which is not possible with the simple power ratio F/O. Table A1 summarizes the correspondence between the decibel scale and the power ratio. For example, a bias of +3 dB occurs when the forecast rainfall F is twice as much as the observed rainfall O.

- Root mean square error:

$$\text{RMSE} = \sqrt{\frac{1}{N}\sum_{i=1}^{N}(F_i - O_i)^2} \tag{A2}$$

- Contingency table of a dichotomous (yes/no) forecast, see Table A2, where the "hits" is the number of times that both the observation and the forecast exceed a given rainfall threshold (at a given grid point), the "false alarms" is the number of times that the forecast exceeds the threshold but the observation does not, the "misses" is the number of times that the forecast does not exceed the threshold but the observation does and the "correct negatives" is the number of times that both the observation and the forecast do not exceed the threshold.

Table A2. Contingency table of a categorical forecast.

		Observed		
		Yes	No	Total
Forecast	Yes	Hits	False alarms	Forecast yes
	No	Misses	Correct negatives	Forecast no
	Total	Observed yes	Observed no	Total

- Different scores can be derived from the contingency table to characterize a particular feature or skill of the forecasting system:

  - Probability of detection (hit rate):

$$\text{POD} = \frac{\text{hits}}{\text{hits} + \text{misses}} = \frac{\text{hits}}{\text{observed yes}}, \tag{A3}$$

The "POD" characterizes the fraction of observed events that were correctly forecast and is also known as the hit rate (HR).

- False alarm ratio:

$$\mathrm{FAR} = \frac{\text{false alarms}}{\text{hits} + \text{false alarms}} = \frac{\text{false alarms}}{\text{forecast yes}}, \quad \text{(A4)}$$

The "FAR" characterizes the fraction of forecast events that were wrongly forecast.

- False alarm rate:

$$F = \frac{\text{false alarms}}{\text{false alarms} + \text{correct negatives}} = \frac{\text{false alarms}}{\text{observed no}}. \quad \text{(A5)}$$

The false alarm rate F is conditioned on the observations, while the false alarm ratio FAR on the forecasts.

- Gilbert skill score (equitable threat score):

$$\mathrm{GSS} = \frac{\text{hits} - \text{hits}_{\text{random}}}{\text{hits} + \text{misses} + \text{false alarms} - \text{hits}_{\text{random}}}, \quad \text{(A6)}$$

where

$$\text{hits}_{\text{random}} = \frac{(\text{hits} + \text{misses})(\text{hits} + \text{false alarms})}{\text{total}} = \frac{(\text{observed yes})(\text{forecast yes})}{\text{total}} \quad \text{(A7)}$$

is the number of hits obtained by random chance, which is calculated by multiplying the marginal sums of the observed and forecast events (such as computing the theoretical frequencies for the Chi-squared test). The GSS characterizes the detection skill of the forecasting system with respect to random chance. In practice it corresponds to the critical success index (CSI) adjusted for the hits obtained by random chance.

The accuracy of probabilistic forecasts can be verified in various ways. In this paper we employ the reliability diagram and the Receiver Operating Characteristic (ROC) curve. The reliability diagram compares the forecast probability with the observed frequency. Reliability characterizes the agreement between the forecast probability and observed frequency. For a reliable forecasting system the two values should be the same, which happens for example when we observe rain 80 % of the time when it is forecast with 80 % probability (in average, diagonal line of Fig. 8). Unreliable forecasts exhibit departures from this optimum (bias). Resolution characterizes the ability of the forecasts to categorize the observed frequencies into distinct classes. The complete lack of resolution occurs when the forecast probabilities are completely unable to distinguish the observed frequencies, which generally corresponds to the climatological frequency of exceeding a given precipitation threshold (horizontal dashed line in Fig. 8). The Brier skill score (BSS) characterizes the relative accuracy of the probabilistic forecast compared to a reference system (see Jolliffe and Stephenson, 2011). Although the climatology or sample climatology of the event is often used as a reference, the BSS can also be computed against other reference forecasts, e.g., another probabilistic forecasting method or even a deterministic forecasting method treated as a probabilistic binary forecast. However, in such cases it is not possible to draw a unique horizontal line representing complete lack of skill in Fig. 8. The region where the probabilistic forecast has a positive BSS, i.e., it is better than the climatological frequency, is grayed out. In fact, the points located below the no skill line are closer to the climatological frequency and produce a negative BSS. Reliability diagrams usually contain the histogram of the forecast probabilities to analyze the sharpness of the forecasts (small inset in Fig. 8). Sharpness characterizes the ability to forecast probabilities that are different from the reference forecast. Sharp forecasting systems are "confident" about their predictions and give many probabilities around one and zero.

The ROC curve is used to analyze the discrimination power of a probabilistic forecast of exceeding a given rainfall threshold. It is constructed by plotting the hit rates and false alarm rates evaluated at increasing probability thresholds to make the binary decision whether it will rain or not. The ROC curve of a random probabilistic forecast system lies on the diagonal where the hit rate equals the false alarm rate (no skill): the forecast probabilities do not have discrimination power. When the false alarm rate is higher than the hit rate the forecast is worse than that obtained by random chance (below the diagonal). A skilled forecasting system is observed when the hit rates are higher than the false alarm rates, which draws a characteristic curve. The area under the ROC curve (AUC) measures the discrimination power of the probabilistic forecasts, with a maximum value of 1 (100 % of hits and 0 % of false alarms) and a minimum value of 0.5 for a random forecasting system. Values below 0.5 denote a forecasting system that performs worse than random chance. The AUC is computed by integrating over all the trapezoids that can be drawn below the ROC curve. The AUC is not sensitive to the forecast bias and the reliability of the forecast could still be improved through calibration. For this reason the AUC is only a measure of potential skill.

The ensemble forecasts are verified to detect whether there is over- or under-dispersion. It is common practice to compare the "skill" (error) of the ensemble mean with the ensemble spread (Whitaker and Loughe, 1998; Foresti et al., 2015):

$$\text{spread} = \frac{1}{N}\sum_{i=1}^{N}\sqrt{\frac{1}{M-1}\sum_{m=1}^{M}\left(F_{im} - \overline{F_i}\right)^2}, \quad \text{(A8)}$$

where M is the number of ensemble members (ensemble size), F_{im} is the forecast of a given ensemble member and $\overline{F_i}$ is the ensemble mean forecast (at a given grid point). Since we are not analyzing the spatial or temporal distribution of the ensemble spread, N corresponds to the total number of samples in space and time, which is the number of forecasts within a rainfall event multiplied by the number of grid points within a radar field. The weak conditional verification is also applied to the computation of the spread. The ensemble spread characterizes the variability of the ensemble members about the ensemble mean (standard deviation). For a good ensemble prediction system, the ensemble spread should be equal to the average variability of the observations about the ensemble mean, as measured by the RMSE of the ensemble mean (Eq. 2). If the spread is larger than the RMSE, the ensemble is overestimating the forecast uncertainty (over-dispersion), otherwise it is underestimating it (under-dispersion). It is interesting to mention that the ensemble mean RMSE and ensemble spread could also be computed starting from the logarithm of rainfall rates to account for the skewed distribution of precipitation (not used in this paper).

Another way to analyze the spread of ensemble forecasts is based on rank histograms (also known as a Talagrand diagram). First, the precipitation values of the ensemble members are ranked in increasing order. Then, the rank of the observation is evaluated by checking in which of the $M+1$ bins it falls. By repeating the operation for a large number of cases and forecasts it is possible to construct a histogram. A good ensemble prediction system displays a flat histogram; i.e., the observations are indistinguishable from the forecasts and each ensemble member is an equi-probable realization of the future state of the atmosphere. A bell-shaped histogram with a peak in the middle is observed in the case of ensemble over-dispersion. On the contrary, a U-shape histogram with peaks at the edges is observed in the case of ensemble under-dispersion, which is more common (in particular for NWP ensembles). In this case the values of the observations often fall below or above the lowest or highest value of the ranked ensemble, which is not dispersive enough to capture the extremes.

Acknowledgements. This research was funded by Belgian Science Policy Office (BelSPO) project PLURISK: Forecasting and management of rainfall-induced risks in the urban environment (SD/RI/01A). We thank Clive Pierce for the detailed discussions about the STEPS implementation and the guidance for various code improvements. We also acknowledge Meteo-France for providing the Avesnois data.

Edited by: R. Uijlenhoet

References

Achleitner, S., Fach, S., Einfalt, T., and Rauch, W.: Nowcasting of rainfall and of combined sewage flow in urban drainage systems, Water Sci. Technol., 59, 1145–51, 2009.

Atencia, A. and Zawadzki, I.: A comparison of two techniques for generating nowcasting ensembles – Part I: Lagrangian ensemble technique, Mon. Weather Rev., 142, 4036–4052, 2014.

Berenguer, M., Corral, C., Sánchez-Diezma, R., and Sempere-Torres, D.: Hydrological validation of a radar-based nowcasting technique, J. Hydrometeorol., 6, 532–549, 2005.

Berenguer, M., Sempere-Torres, D., and Pegram, G. G. S.: SBMcast – an ensemble nowcasting technique to assess the uncertainty in rainfall forecasts by Lagrangian extrapolation, J. Hydrol., 404, 226–240, 2011.

Berne, A., Delrieu, G., Creutin, J.-D., and Obled, C.: Temporal and spatial resolution of rainfall measurements required for urban hydrology, J. Hydrol., 299, 166–179, 2004.

Bowler, N. E. H., Pierce, C. E., and Seed, A. W.: Development of a precipitation nowcasting algorithm based upon optical flow techniques, J. Hydrol., 288, 74–91, 2004a.

Bowler, N. E. H., Pierce, C. E., and Seed, A. W.: STEPS: a probabilistic precipitation forecasting scheme which merges an extrapolation nowcast with downscaled NWP, Forecast Research Technical Report No. 433, MetOffice, Exeter, UK, 2004b.

Bowler, N. E. H., Pierce, C. E., and Seed, A. W.: STEPS: A probabilistic precipitation forecasting scheme which merges an extrapolation nowcast with downscaled NWP, Q. J. Roy. Meteor. Soc., 132, 2127–2155, 2006.

Bruni, G., Reinoso, R., van de Giesen, N. C., Clemens, F. H. L. R., and ten Veldhuis, J. A. E.: On the sensitivity of urban hydrodynamic modelling to rainfall spatial and temporal resolution, Hydrol. Earth Syst. Sci., 19, 691–709, doi:10.5194/hess-19-691-2015, 2015.

Cloke, H. L. and Pappenberger, F.: Ensemble flood forecasting: a review, J. Hydrol., 375, 613–626, 2009.

Collier, C. G.: On the propagation of uncertainty in weather radar estimates of rainfall through hydrological models, Meteorol. Appl., 16, 35–40, 2009.

Dai, Q., Rico-Ramirez, M. A., Han, D., Islam, T., and Liguori, S.: Probabilistic radar rainfall nowcasts using empirical and theoretical uncertainty models, Hydrol. Process., 29, 66–79, 2015.

Ebert, E. E.: Ability of a poor man's ensemble to predict the probability and distribution of precipitation, Mon. Weather Rev., 129, 2461–2480, 2001.

Ehret, U., Götzinger, J., Bárdossy, A., and Pegram, G. G. S.: Radar-based flood forecasting in small catchments, exemplified by the Goldersbach catchment, Germany, International Journal of River Basin Management, 6, 323–329, 2008.

Einfalt, T., Arnbjerg-Nielsen, K., Golz, C., Jensen, N. E., Quirmbachd, M., Vaes, G., and Vieux, B.: Towards a roadmap for use of radar rainfall data in urban drainage, J. Hydrol., 299, 186–202, 2004.

Figueras i Ventura, J. and Tabary, P.: The new French operational polarimetric radar rainfall rate product, J. Appl. Meteorol. Clim., 52, 1817–1835, 2013.

Foresti, L. and Seed, A.: The effect of flow and orography on the spatial distribution of the very short-term predictability of rainfall from composite radar images, Hydrol. Earth Syst. Sci., 18, 4671–4686, doi:10.5194/hess-18-4671-2014, 2014.

Foresti, L. and Seed, A.: On the spatial distribution of rainfall nowcasting errors due to orographic forcing, Meteorol. Appl., 22, 60–74, 2015.

Foresti, L., Seed, A., and Zawadzki, I.: Report of the Heuristic Probabilistic Forecasting Workshop, Munich, Germany, 13 pp., 30–31 August 2014, available at: https://sites.google.com/site/lorisforesti/projects/nowcasting/ScientificReport_HeuristicProbForecastingWorkshop_Munich_2014_121214.pdf, last access: 3 July 2015, 2014.

Foresti, L., Panziera, L., Mandapaka, P. V., Germann, U., and Seed, A.: Retrieval of analogue radar images for ensemble nowcasting of orographic rainfall, Meteorol. Appl., 22, 141–155, 2015.

Germann, U. and Zawadzki, I.: Scale-dependence of the predictability of precipitation from continental radar images – Part I: Methodology, Mon. Weather Rev., 130, 2859–2873, 2002.

Germann, U. and Zawadzki, I.: Scale-dependence of the predictability of precipitation from continental radar images – Part II: Probability forecasts, J. Appl. Meteorol., 43, 74–89, 2004.

Germann, U., Berenguer, M., Sempere-Torres, D., and Zappa, M.: REAL – ensemble radar precipitation estimation for hydrology in a mountainous region, Q. J. Roy. Meteor. Soc., 135, 445–456, 2009.

Goudenhoofdt, E. and Delobbe, L.: Evaluation of radar-gauge merging methods for quantitative precipitation estimates, Hydrol. Earth Syst. Sci., 13, 195–203, doi:10.5194/hess-13-195-2009, 2009.

Grasso, L. D.: The differentiation between grid spacing and resolution and their application to numerical modeling, B. Am. Meteorol. Soc., 81, 579–580, 2000.

Haiden, T., Kann, A., Wittmann, C., Pistotnik, G., Bica, B., and Gruber, C.: The Integrated Nowcasting through Comprehensive Analysis (INCA) system and its validation over the eastern Alpine region, Weather Forecast., 26, 166–183, 2011.

Hohti, H., Koistinen, J., Nurmi, P., Saltikoff, E., and Holmlund, K.: Precipitation nowcasting using radar-derived atmospheric motion vectors, in: Proc. of the 1st European Conf. on Radar in Meteorology and Hydrology (ERAD), Bologna, Italy, 4–8 September 2000, Physics and Chemistry of the Earth, Part B: Hydrology, Oceans and Atmosphere, 25, 1323–1327, 2000.

Jolliffe, I. T. and Stephenson, D. B.: Forecast Verification: a Practitioner's Guide in Atmospheric Science, 2nd edn., John Wiley and Sons, Chichester, 2011.

Jordan, P., Seed, A. W., and Weinnman, P. E.: A stochastic model of radar measurement errors in rainfall accumulations at catchment scale, J. Hydrometeorol., 4, 841–855, 2003.

Lewis, H., Mittermaier, M., Mylne, K., Norman, K., Scaife, A., Neal, R., Pierce, C., Harrison, D., Jewell, S., Kendon, M., Saunders, R., Brunet, G., Golding, B., Kitchen, M., Davies, P., and Pilling, C.: From months to minutes – exploring the value of high-resolution rainfall observation and prediction during the UK winter storms of 2013/2014, Meteorol. Appl., 22, 90–104, 2015.

Liguori, S. and Rico-Ramirez, M. A.: Quantitative assessment of short-term rainfall forecasts from radar nowcasts and MM5 forecasts, Hydrol. Process., 26, 3842–3857, 2012.

Liguori, S. and Rico-Ramirez, M. A.: A practical approach to the assessment of probabilistic flow predictions, Hydrol. Process., 27, 18–32, 2013.

Liguori, S., Rico-Ramirez, M. A., Schellart, A., and Saul, A.: Using probabilistic radar rainfall nowcasts and NWP forecasts for flow prediction in urban catchments, Atmos. Res., 103, 80–95, 2012.

Metta, S., Rebora, N., Ferraris, L., von Hardernberg, J., and Provenzale, A.: PHAST: a phase-diffusion model for stochastic nowcasting, J. Hydrometeorol., 10, 1285–1297, 2009.

Panziera, L., Germann, U., Gabella, M., and Mandapaka, P. V.: NORA – nowcasting of orographic rainfall by means of analogues, Q. J. Roy. Meteor. Soc., 137, 2106–2123, 2011.

Pappenberger, F. and Beven, K. J.: Ignorance is bliss: or seven reasons not to use uncertainty analysis, Water Resour. Res., 42, W05302, doi:10.1029/2005WR004820, 2006.

Paschalis, A., Molnar, P., Fatichi, S., and Burlando, P.: A stochastic model for high-resolution space–time precipitation simulation, Water Resour. Res., 49, 8400–8417, 2013.

Pegram, G. G. S. and Clothier, A. N.: High resolution space–time modelling of rainfall: the "String of Beads" model, J. Hydrol., 241, 26–41, 2001a.

Pegram, G. G. S. and Clothier, A. N.: Downscaling rainfields in space and time, using the String of Beads model in time series mode, Hydrol. Earth Syst. Sci., 5, 175–186, doi:10.5194/hess-5-175-2001, 2001b.

Pierce, C., Bowler, N., Seed, A., Jones, A., Jones, D., and Moore, R.: Use of a stochastic precipitation nowcast scheme for fluvial flood forecasting and warning, Atmos. Sci. Lett., 6, 78–83, 2005.

Pierce, C., Hirsch, T., and Bennett, A. C.: Formulation and evaluation of a post-processing algorithm for generating seamless, high resolution ensemble precipitation forecasts, Forecasting R&D Technical Report 550, MetOffice, Exeter, UK, 2010.

Radhakrishna, B., Zawadzki, I., and Fabry, F.: Predictability of precipitation from continental radar images. Part V: growth and decay, J. Atmos. Sci., 69, 3336–3349, 2012.

Roulin, E. and Vannitsem, S.: Skill of medium-range hydrological ensemble predictions, J. Hydrometeorol., 6, 729–744, 2005.

Schellekens, J., Weerts, A. H., Moore, R. J., Pierce, C. E., and Hildon, S.: The use of MOGREPS ensemble rainfall forecasts in operational flood forecasting systems across England and Wales, Adv. Geosci., 29, 77–84, doi:10.5194/adgeo-29-77-2011, 2011.

Schertzer, D. and Lovejoy, S.: Physical modelling and analysis of rain and clouds by anisotropic scaling multiplicative processes, J. Geophys. Res., 92, 9696–9714, 1987.

Seed, A.: A dynamic and spatial scaling approach to advection forecasting, J. Appl. Meteorol., 42, 381–388, 2003.

Seed, A. W., Pierce, C. E., and Norman, K.: Formulation and evaluation of a scale decomposition-based stochastic precipitation nowcast scheme, Water Resour. Res., 49, 6624–6641, 2013.

Silvestro, F. and Rebora, N.: Operational verification of a framework for the probabilistic nowcasting of river discharge in small and medium size basins, Nat. Hazards Earth Syst. Sci., 12, 763–776, doi:10.5194/nhess-12-763-2012, 2012.

Silvestro, F., Rebora, N., and Cummings, G.: An attempt to deal with flash floods using a probabilistic hydrological nowcasting chain: a case study, Nat. Hazards Earth Syst. Sci. Discuss., 1, 7497–7515, doi:10.5194/nhessd-1-7497-2013, 2013.

Sun, J., Xue, M., Wilson, J. W., Zawadzki, I., Ballard, S. P., Onvlee-Hooimeyer, J., Joe, P., Barker, D. M., Li, P.-W., Golding, B., Xu, M., and Pinto, J.: Use of NWP for nowcasting convective precipitation: recent progress and challenges, B. Am. Meteor. Soc., 95, 409–426, 2014.

Tabary, P.: The new French operational radar rainfall product – Part I: Methodology, Weather Forecast., 22, 393–408, 2007.

Thielen, J., Bartholmes, J., Ramos, M.-H., and de Roo, A.: The European Flood Alert System – Part 1: Concept and development, Hydrol. Earth Syst. Sci., 13, 125–140, doi:10.5194/hess-13-125-2009, 2009.

Thorndahl, S. and Rasmussen, M. R.: Short-term forecasting of urban storm water runoff in real-time using extrapolated radar rainfall data, J. Hydroinform., 15, 897–912, 2013.

Turner, B. J., Zawadzki, I., and Germann, U.: Predictability of precipitation from continental radar images. Part III: Operational nowcasting implementation (MAPLE), J. Appl. Meteorol., 43, 231–248, 2004.

Venugopal, V., Foufoula-Georgiou, E., and Sapozhnikov, V.: Evidence of dynamic scaling in space–time rainfall, J. Geophys. Res., 104, 31599–31610, 1999.

Verworn, H. R., Rico-Ramirez, M. A., Krämer, S., Cluckie, I., and Reichel, F.: Radar-based flood forecasting for river catchments, Water Manage., 162, 159–168, 2009.

Wang, J. Keenan, T., Joe, P., Wilson, J., Lai, E. S. T., Liang, F., Wang, Y., Ebert, E. E., Ye, Q., Bally, J., Seed, A., Chen, M., Xue, J., and Conway, B.: Overview of the Beijing 2008 Olympics Project. Part I: Forecast Demonstration Project, WMO World Weather Research Programme, Report, 133 pp., 2009.

Whitaker, J. S. and Loughe, A. F.: The relationship between ensemble spread and ensemble mean skill, Mon. Weather Rev., 26, 3292–3302, 1998

Willems, P.: A spatial rainfall generator for small spatial scales, J. Hydrol., 252, 126–144, 2001a.

Willems, P.: Stochastic description of the rainfall input errors in lumped hydrological models, Stoch. Env. Res. Risk A., 15, 132–152, 2001b.

Skamarock, W. C.: Evaluating mesoscale NWP models using kinetic energy spectra, Mon. Weather Rev., 132, 3019–3032, 2004.

Xuan, Y., Cluckie, I. D., and Wang, Y.: Uncertainty analysis of hydrological ensemble forecasts in a distributed model utilising short-range rainfall prediction, Hydrol. Earth Syst. Sci., 13, 293–303, doi:10.5194/hess-13-293-2009, 2009.

Xuan, Y., Zhu, D., Triballi, P., and Cluckie, I.: Forecast uncertainty of a lumped hydrological model coupled with the STEPS radar rainfall nowcasts, in: Int. Symp. Weather Radar and Hydrol., Washington DC, US, 7–10 April 2014, 9 pp., 2014.

Zappa, M., Beven, K., Bruen, M., Cofino, A., Kok, K., Martin, E., Nurmi, P., Orfila, B., Roulin, E., Seed, A., Schroter, K., Szturc, J., Vehvilainen, B., Germann, U., and Rossa, A.: Propagation of uncertainty from observing systems and NWP into hydrological models: COST-731 Working Group 2, Atmos. Sci. Lett., 11, 83–91, 2010.

8

Dual-polarized quantitative precipitation estimation as a function of range

Micheal J. Simpson and Neil I. Fox

University of Missouri, School of Natural Resources, Water Resources Program, Department of Soil, Environmental, and Atmospheric Sciences, 332 ABNR Building, Columbia, Missouri 65211, USA

Correspondence: Micheal J. Simpson (mjs5h7@mail.missouri.edu)

Abstract. Since the advent of dual-polarization radar technology, many studies have been conducted to determine the extent to which the differential reflectivity (ZDR) and specific differential phase shift (KDP) add benefits to estimating rain rates (R) compared to reflectivity (Z) alone. It has been previously noted that this new technology provides significant improvement to rain-rate estimation, primarily for ranges within 125 km of the radar. Beyond this range, it is unclear as to whether the National Weather Service (NWS) conventional $R(Z)$-convective algorithm is superior, as little research has investigated radar precipitation estimate performance at larger ranges. The current study investigates the performance of three radars – St. Louis (KLSX), Kansas City (KEAX), and Springfield (KSGF), MO – with 15 tipping bucket gauges serving as ground truth to the radars. With over 300 h of precipitation data being analyzed for the current study, it was found that, in general, performance degraded with range beyond, approximately, 150 km from each of the radars. Probability of detection (PoD) in addition to bias values decreased, while the false alarm rates increased as range increased. Bright-band contamination was observed to play a potential role as large increases in the absolute bias and overall error values near 120 km for the cool season and 150 km in the warm season. Furthermore, upwards of 60 % of the total error was due to precipitation being falsely estimated, while 20 % of the total error was due to missed precipitation. Correlation coefficient values increased by as much as 0.4 when these instances were removed from the analyses (i.e., hits only). Overall, due to the lowest normalized standard error (NSE) of less than 1.0, a National Severe Storms Laboratory (NSSL) $R(Z,\text{ZDR})$ equation was determined to be the most robust, while a $R(\text{ZDR,KDP})$ algorithm recorded NSE values as high as 5. The addition of dual-polarized technology was shown to estimate quantitative precipitation estimates (QPEs) better than the conventional equation. The analyses further our understanding of the strengths and limitations of the Next Generation Radar (NEXRAD) system overall and from a seasonal perspective.

1 Introduction

In 2012, the National Weather Service (NWS) began upgrading the Next Generation Radar (NEXRAD) system from single to dual polarization. The potential benefits of this upgrade were investigated by the National Severe Storms Laboratory (NSSL) and the Cooperative Institute for Mesoscale Meteorological Studies. These advantages include, but are not limited to, (1) significant improvement in radar rainfall estimation (Ryzhkov et al., 2005; Gourley et al., 2010) through better representation of precipitation shape (Brandes et al., 2002; Gorgucci et al., 2000, 2006; Berne and Uijlenhoet, 2005); (2) discrimination between solid and liquid precipitation (Zrnic and Ryzhkov, 1996), allowing for better distinction between areas of heavy rain and hail (Park et al., 2009; Giangrande and Ryzhkov, 2008; Cunha et al., 2013); (3) identifying the melting layer position in the radar field (Straka et al., 2000; Park et al., 2009); (4) hardware calibration (Holleman et al., 2010; Hubbert et al., 2017); and (5) calculating drop-size distributions retrieved from measurements of reflectivity (Z), differential reflectivity (ZDR), and specific differential phase shift (KDP) as opposed to using ground-based point-located disdrometers (Zhang et al., 2001; Brandes et al., 2004; Anagnostou et al., 2008).

Rain-rate retrieval by weather radars is an estimation based upon the dielectric properties of the hydrometeors encountered in the atmosphere. Therefore, there is no direct measurement of rainfall, and this inherently introduces error. However, dual-polarized radar technology allows for in-depth analyses on the microphysics of precipitation that single polarization was incapable of conducting. In spite of this technology, conflicting studies report the benefits for quantitative precipitation estimation (QPE). For example, Gourley et al. (2010) and Cunha et al. (2015) reported that conventional $R(Z)$ algorithms have significantly better bias than algorithms containing ZDR and/or KDP, while others (e.g., Ryzhkov et al., 2013; Simpson et al., 2016) report the opposite. This could be due, at least in part, to the fact that hydrometeor types (e.g., rain versus hail) vary on spatial scales that cannot be easily resolved by even densely gauged networks.

Multiple studies have found that the performance of radar rain-rate estimates decreases as range increases (Smith et al., 1996; Ryzhkov et al., 2003), which is caused, primarily, by degradation of beam quality with range. Furthermore, the researchers also discuss how the probability of detection (PoD) at larger ranges decreases, as the radar beam overshoots shallow, stratiform precipitation, especially winter precipitation (Kumjian, 2013b). Bright-banding can also play a crucial role in significantly increasing the amount of precipitation estimated by the radar, prompting many researchers to produce automated bright-band detection algorithms (e.g., Zhang et al., 2008; Zhang and Qi, 2010).

Despite these overall disadvantages, studies have shown that radar rain-rate algorithms seldom exceed absolute errors on the order of $10\,\mathrm{mm\,h^{-1}}$ (Berne and Uijlenhoet, 2013). However, many of these studies have looked at a small sample of rain events (on the order of 10–50 h) (Kitchen and Jackson, 1993; Smith et al., 1996; Ryzhkov et al., 2003; Gourley et al., 2010; Cunha et al., 2013). Long-term performances of weather radar are becoming more common in recent years as the availability of data becomes more abundant (e.g., Haylock et al., 2008; Goudenhoofdt and Delobbe, 2012, 2016; Fairman et al., 2012). Additionally, few studies (e.g., Smith et al., 1996; Cunha et al., 2015; Simpson et al., 2016) quantified QPE errors including the probability of detection and false alarm ratio. In order to gain a better understanding of the performance of weather radars on rain-rate estimates, more data must be collected over a broad range of precipitation regimes in addition to an overall broader region of interest.

The overarching objective of the current study was to assess the performance of three different radars within the state of Missouri at various ranges from the radar, using terrestrial-based tipping bucket gauges as ground-truth data. Radar rain-rate estimation algorithms include 55 algorithms encompassing standard $R(Z)$ relations as well as algorithms containing dual-polarization variables including differential reflectivity and the specific differential phase shift. A rain-rate echo classification algorithm was also tested for performance in correctly identifying the suitable rain-rate algorithm to choose based on the Z, ZDR, and KDP radar fields. The current work expands upon that of Simpson et al. (2016) such that a larger sample of data was analyzed (over 300 h of rainfall data from 46 separate days in 2014) to encompass multiple different precipitation regimes for both summer and winter, with several ground-truth tipping buckets to analyze the performance of three separate radars as a function of range, and further expanding upon the effects of erroneous precipitation estimates on the overall radar error. Objectives for this study included the following: (1) statistically analyze the performance of each radar at various ranges (compared against the gauges); (2) compute (a) the amount of precipitation incorrectly estimated by the radar (quantifying the probability of false detection, PoFD) and (b) the amount of precipitation incorrectly missed by the radar but measured by the rain gauge; (3) test the overall best radar rain-rate algorithm; and (4) perform objectives (1), (2), and (3) while the data are separated into warm and cool seasons, which have been shown to result in significantly different QPEs (Smith et al., 1996; Ryzhkov et al., 2003; Cunha et al., 2015).

2 Study area and methods

2.1 Study area

National Weather Service radars from St. Louis (KLSX), Kansas City (KEAX), and Springfield (KSGF), MO, are able to scan the majority of the state of Missouri. Because of this, the three aforementioned radars were used to assess overall performance in estimating precipitation for this study. Each radar covered a 200 km radius for which a different number of gauges were within their domains: KLSX, KEAX, and KSGF covered 9, 8, and 5 gauges, respectively (Fig. 1).

Missouri is characterized as a continental type of climate, marked by relatively strong seasonality. Furthermore, Missouri is subject to frequent changes in temperature, primarily due to its inland location and its lack of proximity to any large lakes. All of Missouri experiences below-freezing temperatures on a yearly basis. For example, the majority of the state typically registers 110 days with temperatures below freezing, while the Bootheel (i.e., southeast region) records, on average, 70 days of below-freezing day temperatures, emphasizing the typical northwest to southeast warming pattern of temperatures observed in the state. Because of the large variability in temperature, the warm and cool seasons were defined from an agronomic perspective, primarily taking probabilities of freezing into account. Based on the climatological averages of Missouri, from 1983 to 2013, November through April registered average minimum temperatures below freezing, which was considered the cool season, while May through October's minimum average temperatures were above freezing and constituted the warm season.

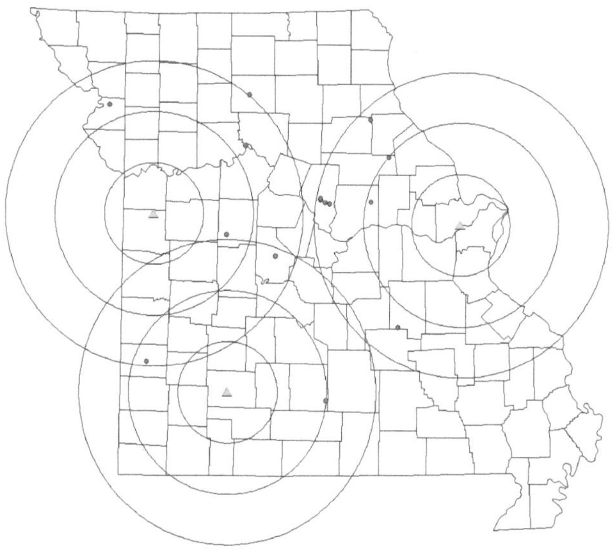

Figure 1. Study location (Missouri) with St. Louis (KLSX), Kansas City (KEAX), and Springfield (KSGF), MO, radars (triangles) overlaid with 50, 100, and 150 km range rings in addition to the 15 terrestrial-based precipitation gauges utilized as ground-truthed data.

2.2 Rainfall data

In order for the results to be comparable across the domains of the three radars it was necessary to select days on which rain was observed widely across the state. Although measurable rainfall occurs on more than 100 days of the year in Missouri with only 50 days typically recording greater than 0.254 mm, 2014 recorded 46 days with measurable rainfall throughout the state. Furthermore, occurrence of rain was defined as the observation of an amount greater than 0.5 mm (equivalent to two rain gauge tips) in an hour. This amounted to a total of approximately 300 h of rain across those 46 days. This represents a relatively standard year of rainfall for the state of Missouri. Furthermore, the days were chosen based on availability of data from the National Climate Data Center's (NCDC) Hierarchal Data Storage System (HDSS) for all three radars, in addition to error-free performance notes from each of the gauges used. The dates analyzed were split near evenly between the warm (May–October) and cool (November–April) seasons, therefore encompassing an overall performance of each of the radars throughout the year with no preferential bias towards rain or snow. Additionally, days were distributed evenly during the summer between convective and stratiform events with a threshold of 38 dBZ (Gamache and Houze, 1982).

Terrestrial-based precipitation gauge data were collected from 15 separate weather stations within the Missouri Mesonet, established by the Commercial Agriculture Program of University Extension (Table 1). All precipitation data were aggregated in hourly intervals to match the temporal resolution of the gauges. Observed precipitation data were collected using Campbell Scientific TE525 tipping buckets located at each of the locations for the study (Table 1). The precipitation gauges have a 15.4 cm orifice which funnels to a fulcrum which registers 0.254 mm of rainfall per tip. For the current study, none of the day's average temperature values fell outside of the gauge's maximum performance range of 0 to 50 °C. Accuracy in gauge measurements range between −1 and 1, −3 and 0, and −5 and 0 % for precipitation up to 25.4, 25.4 to 50.8, and 50.8 to 76.2 mm h^{-1}, respectively, which are, primarily, associated with local random errors and errors in tip-counting schemes (Kitchen and Blackall, 1992; Habib et al., 2001).

Each tipping bucket is located, approximately, 1 m above the ground in areas clear of buildings and properly maintained vegetation height to mitigate turbulence effects (Habib et al., 1999). Due to the well-maintained nature of the mesonet gauges, these errors were assumed negligible and, therefore, allowed for the gauges to be representative of the true rainfall rate. In spite of the non-homogeneous spacing of the gauges, unbiased statistics including the normalized mean bias (NMB) and normalized standard error (NSE) were utilized.

2.3 Radar data and radar rainfall algorithms

Next Generation Radar level-II data were retrieved from the NCDC's HDSS. Files were processed using the Warning Decision Support System – Integrated Information (WDSS-II) program (Lakshmanan et al., 2007a) to assess reflectivity (Z) in addition to dual-polarized radar variables including differential reflectivity and specific differential phase shift. Three other variables were also generated based on a KDP-based smoothing field (Ryzhkov et al., 2003) for reflectivity, differential reflectivity, and specific differential phase: DSMZ, DZDR, and DKDP, respectively. These were implemented to determine whether the additional KDP-smoothing fields tend to over- or underestimate QPEs (Simpson et al., 2016). A rain-rate echo classification variable (RREC) was also computed, which chooses whether an $R(Z)$, R(KDP), $R(Z,$ZDR), or R(ZDR, KDP) algorithm is implemented in estimating rain rates based on the radar fields of Z, ZDR, and KDP (Kessinger et al., 2003) to determine whether a multi-parameter algorithm is superior to a single algorithm.

All seven variables (Z, ZDR, KDP, DSMZ, DZDR, DKDP, and RREC) were converted from their native polar grid to 256 × 256 1 km Cartesian grids, where the lowest radar elevation scans (0.5°) were used to mitigate uncalculated effects from evaporation and wind drift (Ciach, 2002). An average of 5 min scans were used for each of the variables, which were aggregated to hourly totals to be compared to the hourly tipping-bucket accumulations. In spite of previous reports suggesting 5 min to hourly aggregates can have significant effects on QPE (e.g., Fabry et al., 1994), Shucksmith et al.'s (2011) criterion of present accumulation exceeding 26 % for a pixel size of 1 km was not reached.

Table 1. Terrestrial-based precipitation gauge locations used for the study in addition to the National Weather Service radars from Springfield, MO (KSGF), Kansas City, MO (KEAX), and St. Louis, MO (KLSX), used in conjunction with each gauge.

Gauge location	Latitude (° N)	Longitude (° W)	Radar(s) used
Bradford	38.897236	−92.218070	KLSX, KEAX
Brunswick	39.412667	−93.196500	KEAX
Capen Park	38.929237	−92.321297	KLSX, KEAX
Cook Station	37.797945	−91.429645	KLSX, KSGF
Green Ridge	38.621147	−93.416652	KEAX, KSGF
Jefferson Farm	38.906992	−92.269976	KLSX, KEAX
Lamar	37.493366	−94.318185	KSGF
Linneus	39.856919	−93.149726	KEAX
Monroe City	39.635314	−91.725370	KLSX
Mountain Grove	37.153865	−92.268831	KSGF
Sanborn Field	38.942301	−92.320395	KLSX, KEAX
St. Joseph	39.757821	−94.794567	KEAX
Vandalia	39.302300	−91.513000	KLSX
Versailles	38.434700	−92.853733	KEAX, KSGF
Williamsburg	38.907350	−91.734210	KLSX

Table 2. List of single- and dual-polarimetric algorithms used for radar rainfall estimates.

$R(Z) = aZ^b$			
Precipitation type	a	b	c
Stratiform	200	1.6	–
Convective	300	1.4	–
Tropical	250	1.2	–
$R(\mathrm{KDP}) = a\vert\mathrm{KDP}\vert^b \mathrm{sign}(\mathrm{KDP})$			
Algorithm number			
1	50.7	0.85	–
2	54.3	0.81	–
3	51.6	0.71	–
4	44	0.82	–
5	50.3	0.81	–
6	47.3	0.79	–
$R(Z, \mathrm{ZDR}) = aZ^b \mathrm{ZDR}^c$			
Algorithm number			
7	6.70×10^{-3}	0.927	−3.43
8	7.46×10^{-3}	0.945	−4.76
9	1.42×10^{-2}	0.77	−1.67
10	1.59×10^{-2}	0.737	−1.03
11	1.44×10^{-2}	0.761	−1.51
$R(\mathrm{ZDR}, \mathrm{KDP}) = a\vert\mathrm{KDP}\vert^b \mathrm{ZDR}^c \mathrm{sign}(\mathrm{KDP})$			
Algorithm number			
12	90.8	0.93	−1.69
13	136	0.968	−2.86
14	52.9	0.852	-0.53
15	63.3	0.851	−0.72

The latitude and longitude of each of the 15 gauges were matched with the radar pixel that corresponds to the Cartesian grid value of the seven radar variables which were then implemented in rain-rate calculations. These rain-rate calculations were calculated using the equations presented by Ryzhkov et al. (2005) (Table 2), which were gathered from multiple studies using disdrometers to derive a relationship between reflectivity, differential reflectivity, and specific differential phase (Bringi and Chandrasekar, 2001; Brandes et al., 2002; Illingworth and Blackman, 2002; Ryzhkov et al., 2003). Standard $R(Z)$ algorithms were also included to test whether the addition of dual-polarized technology improves QPEs.

With the use of both (i) Z, ZDR, and KDP and (ii) DSMZ, DZDR, and DKDP fields produced by WDSS-II, the number of algorithms tested was 55. This includes the three standard single-polarized algorithms (stratiform, convective, and tropical) which were calculated using reflectivity $R(Z)$ and then calculated as R(DSMZ), while algorithms 1–6 (R(KDP)) were also calculated as R(DKDP). Algorithms 7–11 ($R(Z$, ZDR)) were additionally calculated as $R(Z$, DZDR), R(DSMZ, ZDR), and R(DSMZ, DZDR), while the same four combinations of non- and KDP-smoothed fields were applied to the R(KDP, ZDR) algorithms (12–15). Quality controlling methods for the algorithms include mitigation of clutter, sun spikes, beam blockage, anomalous propagation, and removal of non-precipitation echoes (including biological and chaff returns) through w2qcnn the w2qcnndp algorithms (Lakshmanan et al., 2007b, 2010, 2014).

2.4 Statistical analyses

To test the performance of each algorithm, several statistical analyses were calculated. The average difference (Bias) was calculated as

$$\text{Bias} = \frac{\sum(R_i - G_i)}{N}, \tag{1}$$

where R_i is each hourly aggregated radar-estimated rainfall amount calculated from one of the 55 algorithms, G_i is the hourly aggregated gauge (observed) measurement, and N is the total number of observations which, for this study, was 300 h. A second statistical parameter, the normalized mean bias (NMB), was calculated as

$$\text{NMB} = \frac{1}{N}\frac{\sum(R_i - G_i)}{\sum G_i}. \tag{2}$$

The normalized mean bias is included in the analyses due to the fact that overestimations (i.e., radar estimates larger than gauge measurements) and underestimations (i.e., radar estimates smaller than gauge measurements) are treated proportionately. This is directly analogous to choosing the mean absolute error (MAE) opposed to the standard deviation as the MAE does not penalize smaller or larger errors, obscuring the overall results (Chai and Draxler, 2014). Bias measurements (Bias and NMB) were calculated to determine whether radar-derived rain rates were over- or underestimated in comparison to the gauges. However, to calculate the overall magnitude of error associated with the performance of the radars, the absolute values of Eqs. (1) and (2) were performed to yield the mean absolute error and normalized standard error, respectively.

Several other meteorological parameters were calculated, including the probability of detection (PoD), which was calculated as

$$\text{PoD} = \frac{\sum |R_i \cdot G_i > 0 \;\&\; R_i > 0|}{\sum |G_i|}, \tag{3}$$

where the bullet $(\cdot)$ indicates "if", to determine how accurate the radars were at correctly detecting precipitation. The probability of detection values range between 0.0 (radar did not detect any precipitation correctly) and 1.0 (radar detected the occurrence of all precipitation 100 % correctly). The probability of false detection (PoFD) takes into account the amount of precipitation the radars incorrectly estimated when the gauges recorded zero values, and it was calculated as

$$\text{PoFD} = \frac{\sum R_i \cdot (G_i = 0 \;\&\; R_i > 0)}{\sum G_i}. \tag{4}$$

Quantitative measures including the missed precipitation amount (MPA) and the false precipitation amount (FPA) were defined such that

$$\text{MPA} = \sum R_i \cdot (G_i > 0 \;\&\; R_i = 0), \tag{5}$$

$$\text{FPA} = \sum R_i \cdot (G_i = 0 \;\&\; R_i > 0), \tag{6}$$

which analyzes the total amount of precipitation due to misses and false alarms. The total precipitation error was also recorded to assess the overall error from each radar.

3 Results and discussion

3.1 Overall algorithm performance

To test the overall performance of each radar, it was necessary to determine the overall best algorithm for each statistical measure. The best algorithm from each grouping of equations was determined to have the lowest normalized standard error, indicating the best performance relative to the gauge-recorded precipitation amount (Ryzhkov et al., 2005). This reduces the impact of bias inherent within the dataset between the warm and cool season, as well as between stratiform and convective events, and allows for statistical measurements in spite of the (typical) non-Gaussian behavior of precipitation (Kleiber et al., 2012; Alaya et al., 2017).

From the results obtained, the three $R(Z)$, three R(DSMZ), and RREC algorithms displayed a particular bias in favor of the $R(Z)$-convective algorithm for all three

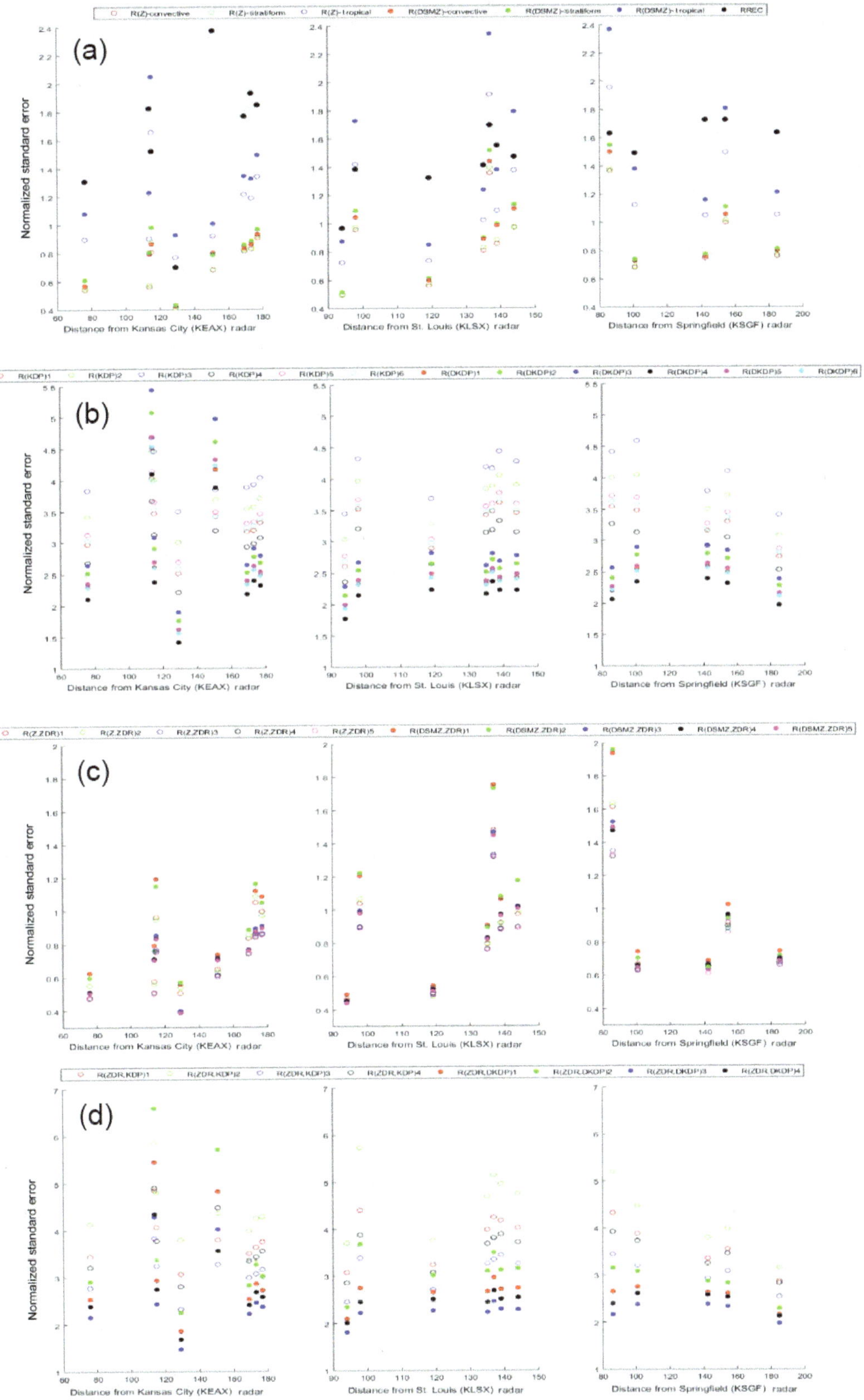

Figure 2. Normalized standard error values for the overall performance of the **(a)** 3 $R(Z)$, 3 R(DSMZ), and RREC algorithms; **(b)** 6 R(KDP) and 6 R(DKDP) algorithms (algorithms 1–6 from Table 2); **(c)** 5 R(Z,ZDR) and 5 R(DSMZ,ZDR) algorithms (Eqs. 7–11 from Table 2); and **(d)** 4 R(ZDR,KDP) and 4 R(ZDR,DKDP) algorithms (Eqs. 12–15 from Table 2) for the three radars utilized for the current study.

radars with $R(Z)$-stratiform displaying similar performance (Fig. 2a). This could be due, at least in part, to the near-equal stratiform and convective precipitation regimes throughout 2014. Although errors generally increased as range increased for KEAX and KLSX, the results were nebulous for KSGF. The lowest NSE values were, typically, closest to each of the radars (between 0.4 and 0.8), with the notable exception of the closest gauge to KSGF. In general, the RREC performed worst at the largest of ranges, potentially due to the algorithm's ability to incorrectly assess the hydrometeors present (Cifelli et al., 2010; Yang et al., 2016). Additionally, the poor performance by the R(DSMZ)-tropical equation is due to the lack of tropical precipitation within central Missouri. Overall, the KDP-smoothed reflectivity fields (DSMZ) performed worse than their counterparts, resulting in over-prediction of precipitation and, thus, larger errors (Simpson et al., 2016). Errors did not exceed 2.4 NSE units for any of these algorithms.

However, the performance of the KDP-smoothed KDP field (DKDP) performed better than the original specific differential phase shift field (Fig. 2b). For nearly all gauges for each of the three radars, R(DKDP)4 performed the best, with NSE values ranging from 1.4 to 4.1. The range of NSE values were largest at KEAX, while the spread was relatively small for KLSX and KSGF. In spite of this, the overall spread of the performance of the 12 KDP algorithms varied greatly (average of 2 NSE units), exhibiting the sensitivity of KDP estimates on QPE (Ryzhkov et al., 2005; Cunha et al., 2013). In general, the NSSL-derived R(KDP) equations (i.e., Eqs. 4–6) outperformed those from Bringi and Chandrasekar (2001, Eq. 1), Brandes et al. (2002, Eq. 2), and Illingworth and Blackman (2002, Eq. 3). Regardless, the magnitudes were all, approximately, more than 1 NSE unit than the performance of the $R(Z)$ algorithms.

The algorithms with the lowest NSE values were Eqs. (7)–(11). For example, the overall lowest NSE was at a distance of 130 km from KEAX (0.3), with no locations exceeding NSE values of 2.0 (Fig. 2c). The large values at the closest location for KSGF (85 km, 1.3–1.9 NSE units) and the fifth closest gauge to KLSX (135 km, 1.3–1.8 NSE units), Cook Station, were similar to the $R(Z)$ and R(DSMZ) results, indicating potential issues with reflectivity measurements. Additionally, these locations were the closest in performance to the R(KDP) and R(DKDP) NSE values. Observations from this gauge (Cook Station) indicated hail occurred during the evening of 1 August, for which KDP estimates would be more ideal than Z for QPE (Ryzhkov et al., 2005; Kumjian, 2013a; Cunha et al., 2015). In spite of this, the overall spread in performance of the $R(Z$,ZDR) equations was lower than the R(KDP) equations, demonstrating the robust performance of $R(Z$,ZDR) for QPE (Wang and Chandrasekar, 2010; Seo et al., 2015).

The R(ZDR,KDP) algorithms performed the worst, overall (Fig. 2d). In spite of the differential reflectivity being implemented, the overall NSE values increased in magnitude, exceeding six units for the second gauge analyzed by KEAX. Algorithms containing DKDP measurements performed better than simply KDP, demonstrating that, even with the scaling behavior of ZDR, DKDP is superior to KDP estimates. This provides a potential solution to the noisiness that tends to be exhibited in the KDP field (Ruzanski and Chandrasekar, 2012).

Due to the overall NSE values obtained, for the remainder of the analyses, Eq. (11) (i.e., $R(Z$,ZDR)5) and Eq. (13) (i.e., R(ZDR,KDP)2) will be utilized as the best and worst algorithms, respectively. Equations containing DZDR were not included in the following discussion due to the very large QPE errors for each radar.

3.2 KEAX

The overall bias showed that there was a positive bias, peaking near 5.5 mm h^{-1} at the second gauge for KEAX, approximately 115 km from the radar for both the best- and worst-performing algorithms (Fig. 3). This corresponds well with the spike in the falsely detected precipitation recorded, which is canceled by the maximum in missed precipitation at the second distance of, approximately, 150 km. The overall worst algorithm, Eq. (13), an R(ZDR,KDP) relationship, revealed a decreasing trend in bias as the distance from the radar increased. For example, a bias of 4 mm h^{-1} was observed at a distance of 75 km from the radar, whereas the bias reduced to 3 mm h^{-1} at distances near 175 km. This could be due, at least in part, to the algorithm's utilization of KDP, which performs poorly in frozen (especially light) precipitation (Zrnic and Ryzhkov, 1996; Kumjian, 2013a), causing the overestimation. Conversely, the algorithm with the lowest bias was an $R(Z$,ZDR) algorithm (Eq. 11). There was a maximum in the bias calculations while utilizing Eq. (11) near 120 km, similar to Eq. (13); however, there was a more pronounced minimum in the data near 150 km. Furthermore, it appears the data oscillates around a bias value of 0 mm h^{-1} when using Eq. (13). This could be due to ZDR's capability to respond to precipitation shape (Kumjian, 2013a), which helps to scale the reflectivity portion of the rainfall estimation algorithm to a more accurate value (Seo et al., 2015). In general, the cool season displayed a larger magnitude of error in terms of bias for both algorithms.

The normalized mean bias reveals the same trend in values for bias but with an overall decrease in magnitude. It is important to note, however, that the algorithms that tend to perform the worst (e.g., algorithms containing KDP) result in anomalous range responses which would be due, at least in part, to a stronger response to precipitation type (Kumjian, 2013c). This indicates that observations above the melting layer are dominant, for which QPEs tend not to be calculated (Cifelli et al., 2011; Seo et al., 2015), but are important for regions devoid of adequate radar coverage (Ryzhkov et al., 2003; Simpson et al., 2016).

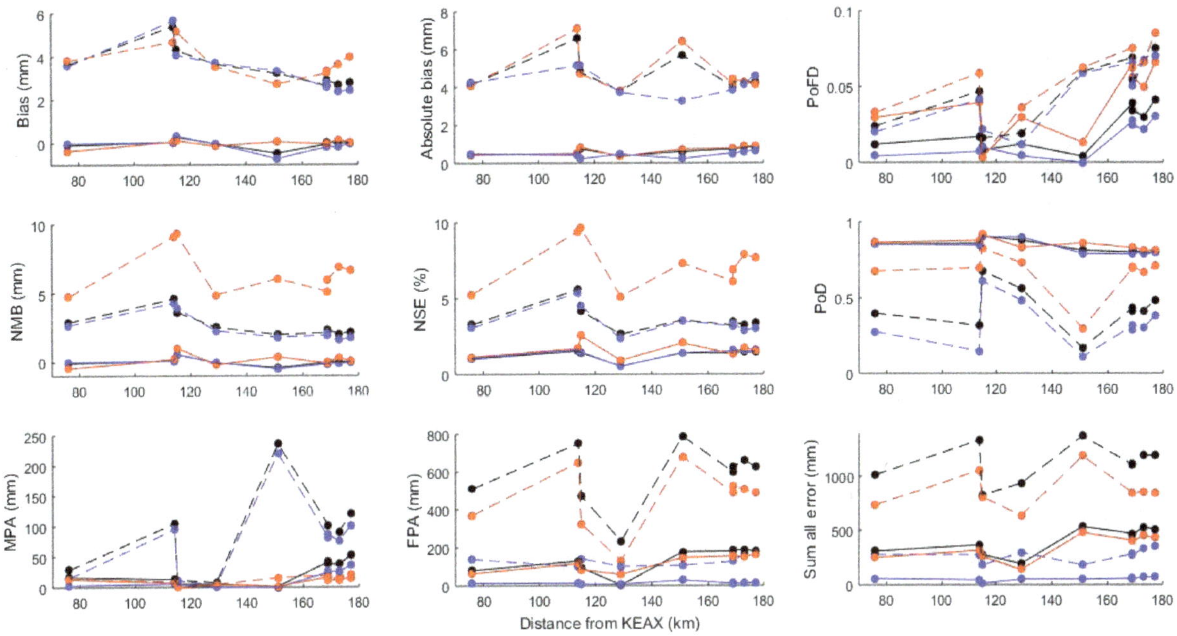

Figure 3. Values of analyses from the Kansas City radar. Dashed lines and points represent the analyses of the worst-performing algorithm (R(ZDR,KDP)), while the solid lines and points represent the analyses of the best-performing algorithm (R(Z,ZDR)). Red, blue, and black colors represent analyses conducted during the warm season, cool season, and overall, respectively.

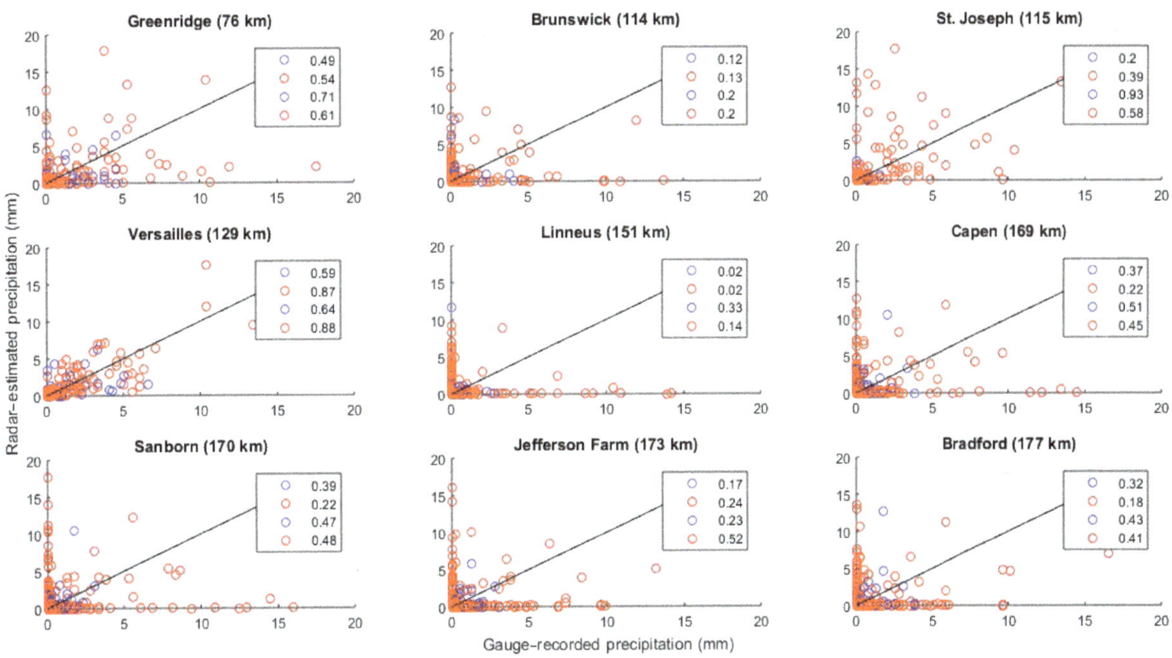

Figure 4. Correlation coefficient values for the nine locations analyzed by the Kansas City radar with the R(Z,ZDR) NSSL equation. Blue and red scatter points represent the cool and warm season data, respectively. The top two numbers on each plot indicate the overall R^2 value, whereas the bottom two numbers represent the R^2 when false alarms and misses are removed.

The absolute bias and normalized standard error shows the same maxima in the data at the second gauge (Brunswick) that was present in the bias data (6.2 mm h^{-1} and 5.6, respectively). However, a second maxima is located at the fifth gauge at, approximately, 150 km (Linneus) with values of 5.9 mm h^{-1} and 4.0, respectively. Bright-band issues are detected due, at least in part, to the increased missed precipitation amount (240 mm) at this particular distance for the

R(ZDR,KDP) equation (i.e., worst-performing algorithm). There was also a pronounced minimum in the absolute bias and NSE results at the fourth gauge for equations 11 and 13, 4.0 and 0.8 mm h^{-1}, and 2.8 and 0.8, respectively, potentially indicating an idealized range of QPE for KEAX. Furthermore, the historical records at this particular gauge showed less issues (e.g., clogging) than any of the others analyzed by the KEAX radar. This highlights the importance of choosing ground-truth data, in particular tipping buckets which are prone to numerous errors (Ciach and Krajewski, 1999b).The largest contributions to the NSE and NMB were due to the warm season.

The probability of detection results indicate a large difference in algorithm choice for correctly detecting precipitation. The low PoD at, approximately 150 km, indicates overshooting of the beam. This is further evidenced by the MPA results, as about 225 mm of precipitation was missed by the radar at 150 km, whereas only 100 mm of precipitation was missed by the radar at the second gauge at 120 km. Although Eq. (11), an $R(Z$,ZDR) algorithm, was superior in terms of the bias, the same algorithm with a KDP-smoothed reflectivity value, R(DSMZ,ZDR), revealed the overall least amount of the falsely missed precipitation (by 10 mm). However, the summation of the amount of precipitation falsely detected (PoFD) by KEAX showed a larger source of error than the MPA in terms of magnitude. For example, at the second (fifth) gauge, only 100 (225) mm of precipitation was missed by the radar, but over 700 (725) mm of precipitation was incorrectly estimated by the radar.

Correlation coefficient (CC) values for any of the nine stations analyzed by KEAX range from 0.02 (Linneus, 151 km) to 0.93 for the cool season (St. Joseph, 115 km). The lowest R^2 values were due to a combination of false alarms and misses. For example, the CC values for the warm seasons at Sanborn (170 km) and Jefferson Farm (173 km) were 0.22 and 0.24, respectively, whereas when the instances of false alarms and misses were removed the CC values increased to 0.48 and 0.52. Few locations (Brunswick, 114 km; and Versailles, 129 km) saw little improvement in the CC values when only hits were analyzed (less than 0.1 increase), indicating the mean absolute error (in terms of hits) contributed the largest portion of error.

3.3 KLSX

Unlike the KEAX data, the gauges used for analyses for the KLSX radar span between 90 and 150 km. Furthermore, five out of the eight gauges were located within a 10 km of range from one another, near 140 km from the radar, limiting the data available for analyses between 100 and 140 km (Fig. 5).

The bias and NMB both show a relatively modest peak in values near the second gauge of 5 mm, which decreases to approximately 3.6 mm at the third gauge, 120 km from the radar. The worst-performing algorithm, Eq. (13), had the same R(ZDR,KDP) relation as the worst KEAX bias and NMB data. Additionally, the overall trend of decreasing bias and NMB as distance from the radar increases was noted, presumably due to overshooting effects similar to the KEAX data. Furthermore, the overall non-biased results from the $R(Z$,ZDR) equation demonstrates its robust capabilities in QPE, in spite of its sensitivity to calibration (Zrnic et al., 2005; Bechini et al., 2009; Gorgucci et al., 1992).

The double maxima in the absolute bias graph are present as with the KEAX data, but are not as pronounced. For example, the absolute biases at 95 and 140 km from KLSX were 5.9 and 4.9 mm for Eq. 13, while the values were 1.1 and 1.4 mm for Eq. 11, accordingly. Additionally, the overall minima in the absolute bias for both KEAX and KLSX are at, approximately, 125 km from the radar (3.9 and 1.0 mm h^{-1}, respectively, for Eqs. 13 and 11). The relative distance from the radars are the same, where the two maxima for KEAX were at 115 and 150 km, while the maxima were at, approximately, 100 and 140 km for KLSX. The overall best- and worst-performing algorithms at KLSX for the absolute bias and NSE were Eqs. (11) and (13), the $R(Z$,ZDR) and R(ZDR,KDP) algorithms, respectively.

The magnitude of error in terms of absolute bias, normalized mean bias, and normalized standard error all showed a decreasing pattern as distance from KLSX increased. This was due, primarily, from a maximum in the false precipitation amount at 95 km from the radar. Historical notes at this location indicate frequent clogging of the rain gauge, either due to bugs or leaves. From a particular series of events spanning from 1 to 4 April and 1 to 3 August 2014, over 130 mm of precipitation occurred during each period which was not captured by the gauge, resulting in a large amount of overall error. These results indicate the importance of dual gauges in the same vicinity (Krajewski et al., 1998; Ciach and Krajewski, 1999a). Interestingly, the cool season displayed a larger NSE (5 % for R(ZDR,KDP)) potentially due to the very low probability of detection (0.2) at this range of 118 km.

One of the main differences between the KLSX and KEAX data was the decreased probability of detection at 120 km for KLSX, while there was an increased probability of detection for KEAX. In general, the PoD values were worse for KLSX when compared to KEAX. For example, Eq. (11) had no PoD values below 0.90, whereas no PoD values exceeded 0.84 for KLSX. There was also a slight trend of increasing PoD values as distance from the St. Louis radar increased and, at one point near 140 km, the best algorithm, R(DSMZ)-convective, and the worst algorithm, KDP1, were not significantly different ($p < 0.10$). Additionally, the maxima in the PoD while utilizing KDP1 correspond to a minima in the R(DSMZ) detection percentage, which is well correlated by the similarly valued MPA results.

The missed precipitation amount (MPA) showed that the cool season contributed the most, whereas the warm season contributed the most amount of the false precipitation amount. The $R(Z$,ZDR) equation only registered, on average, 25 mm of MPA and 160 mm of FPA, whereas the

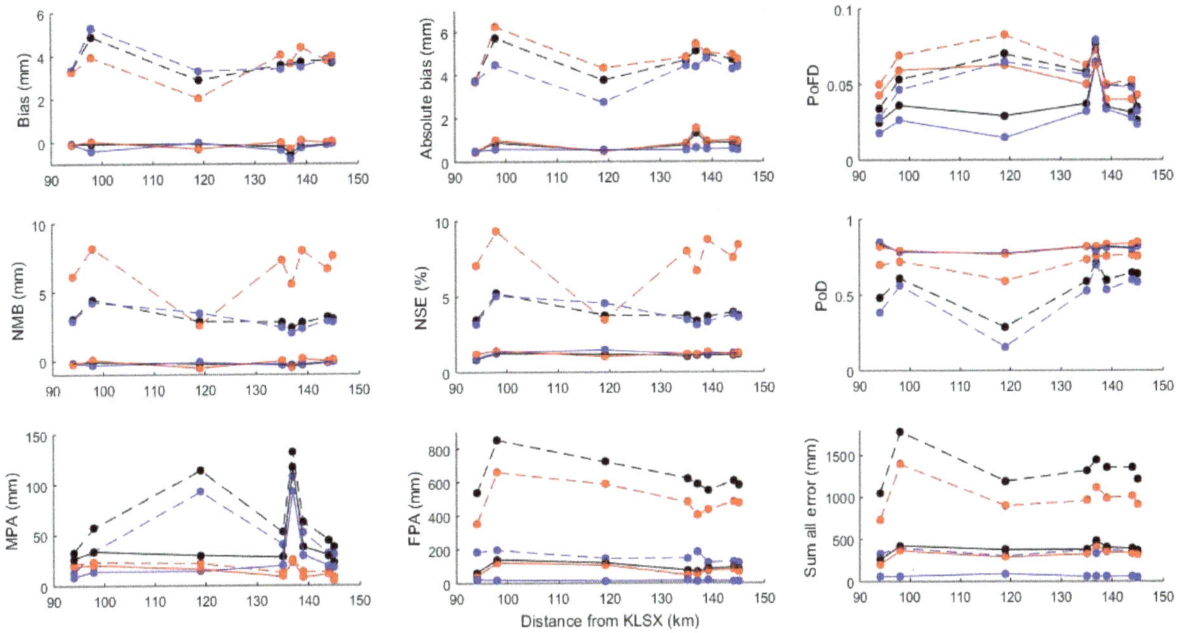

Figure 5. Values of analyses from the St. Louis radar. Dashed lines and points represent the analyses of the worst-performing algorithm (R(ZDR,KDP)), while the solid lines and points represent the analyses of the best-performing algorithm (R(Z,ZDR)). Red, blue, and black colors represent analyses conducted during the warm season, cool season, and overall, respectively.

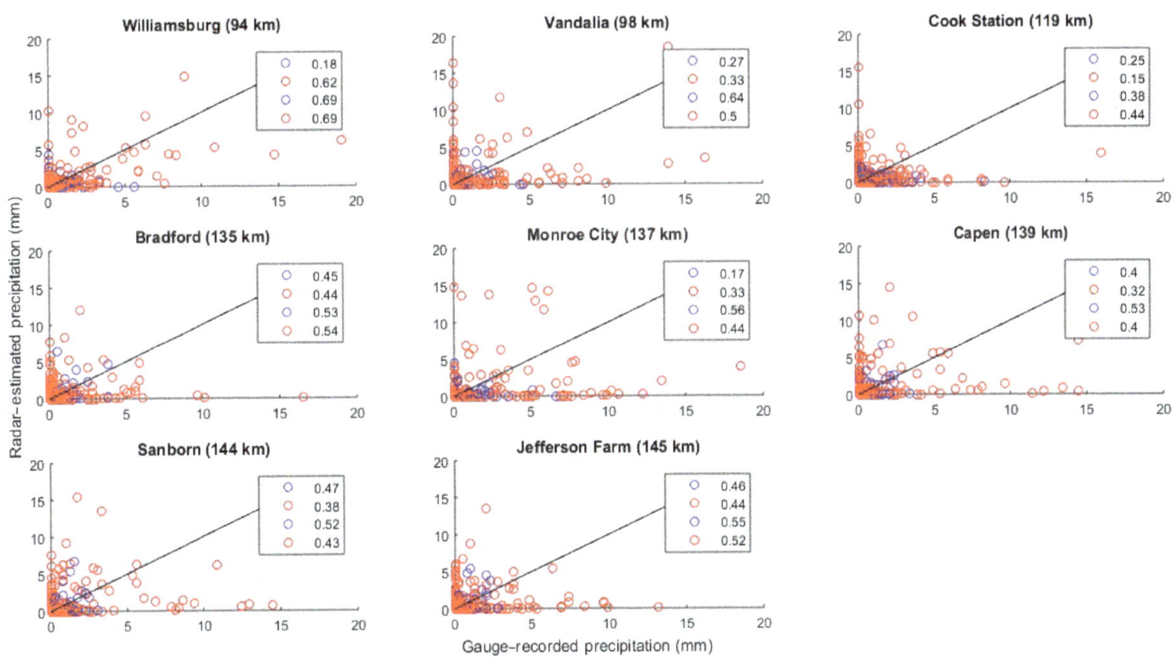

Figure 6. Correlation coefficient values for the eight locations analyzed by the St. Louis radar with the R(Z,ZDR) NSSL equation. Blue and red scatter points represent the cool and warm season data, respectively. The top two numbers on each plot indicate the overall R^2 value, whereas the bottom two numbers represent the R^2 when false alarms and misses are removed.

R(ZDR,KDP) equation was very dependent upon range. For example, the FPA from R(ZDR,KDP) decreased as range increased from the radar from a maximum of, approximately, 850 to 620 mm. However, the fifth-furthest gauge (137 km from KLSX) displayed a sharp increase in the MPA for both cool seasons (above 100 mm).

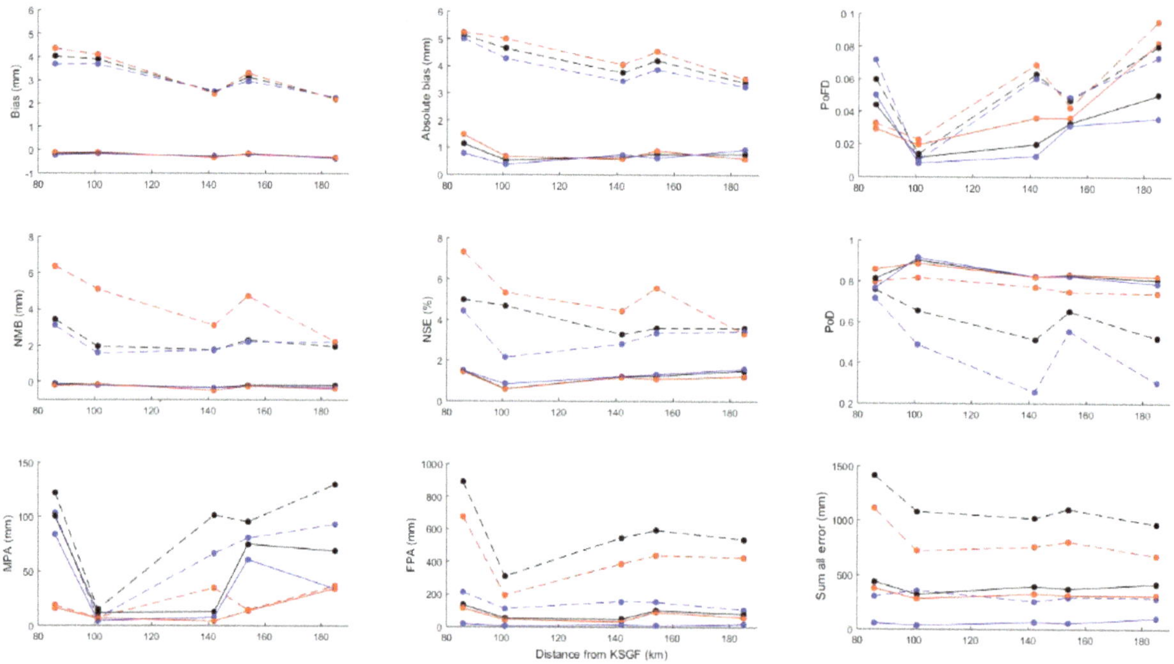

Figure 7. Values of analyses from the Springfield radar. Dashed lines and points represent the analyses of the worst-performing algorithm (R(ZDR,KDP)), while the solid lines and points represent the analyses of the best-performing algorithm ($R(Z$,ZDR)). Red, blue, and black colors represent analyses conducted during the warm season, cool season, and overall, respectively.

3.4 KSGF

In spite of the fact that the KLSX and KEAX data strongly suggest false precipitation errors near 100 km in addition to bright-banding near 150 km from the radars, the KSGF results reveal an overall smooth decrease (increase) of error with range (Fig. 7) for R(ZDR,KDP) and $R(Z$,ZDR), accordingly. One of the main reasons for this could be due to the fact that only five gauges were analyzed from KSGF (the fewest of the three radars analyzed), smoothing the overall trend lines.

The bias remained relatively constant near −0.3 mm for $R(Z$,ZDR), whereas the bias exhibited a sharp decrease from 4 to 2.7 mm over a distance of, approximately, 100 km. In general, the cool season displayed lower bias magnitudes when compared to the warm season, similar to the KEAX results. This may be due, at least in part, to the low PoFD values for the warm season close to the KSGF radar.

Similar to the bias, the absolute bias for $R(Z$,ZDR) was constant at all ranges (near 1 mm), whereas the R(ZDR,KDP) equation decreased from 5.2 to 3.8 mm. This is potentially due to the low cool season PoD values (below 0.6), while the warm season R(ZDR,KDP) values (near 0.8) remained constant. A larger contribution from more correctly detected precipitation in addition to the decreasing trends in the NMB and NSE would result in a lower absolute bias.

The closest location (90 km) typically displayed the largest errors for the R(ZDR,KDP) equation and then decreased in error magnitude as range increased. In spite of this, the PoFD results indicate both algorithms increased in PoFD values as range increased, with the warm season typically dominating, particularly due to the large convective clouds dominating in the warm season. False detection values as low as 0.01 for the cool season while utilizing $R(Z$,ZDR) were observed at distances near 100 and 140 km from the radar.

Normalized standard error values increased from 0.7 % at a distance of 105 km to 1.8 % at a distance of 185 km for $R(Z$,ZDR). Large NSE values for the warm season (7.5 %) were calculated for R(ZDR,KDP), which decreased to 3.8 % at 185 km from the radar. Furthermore, this was the only instance when the warm season was less than the cool season in terms of NSE. Otherwise, the overall NSE decreased from 5 to 3.9 % for R(ZDR,KDP). The NMB followed a similar trend for the KDP-containing algorithm, with a noticeable exception at the second gauge (105 km from KSGF), where the overall NSE was closer to the warm season than the cool season. This is due to the low PoFD values at this location, in addition to a smaller difference between the two algorithm's FPA measurements.

The MPA results, unlike for KEAX and KLSX, displayed a larger range of performance between seasons. However, the warm season still exhibited the overall best performance in terms of MPA, yet contributed the most to the FPA for both $R(Z$,ZDR) and R(ZDR,KDP). In spite of the MPA typically increasing as range increased, the FPA was more nebulous. For example, the second gauge (105 km from

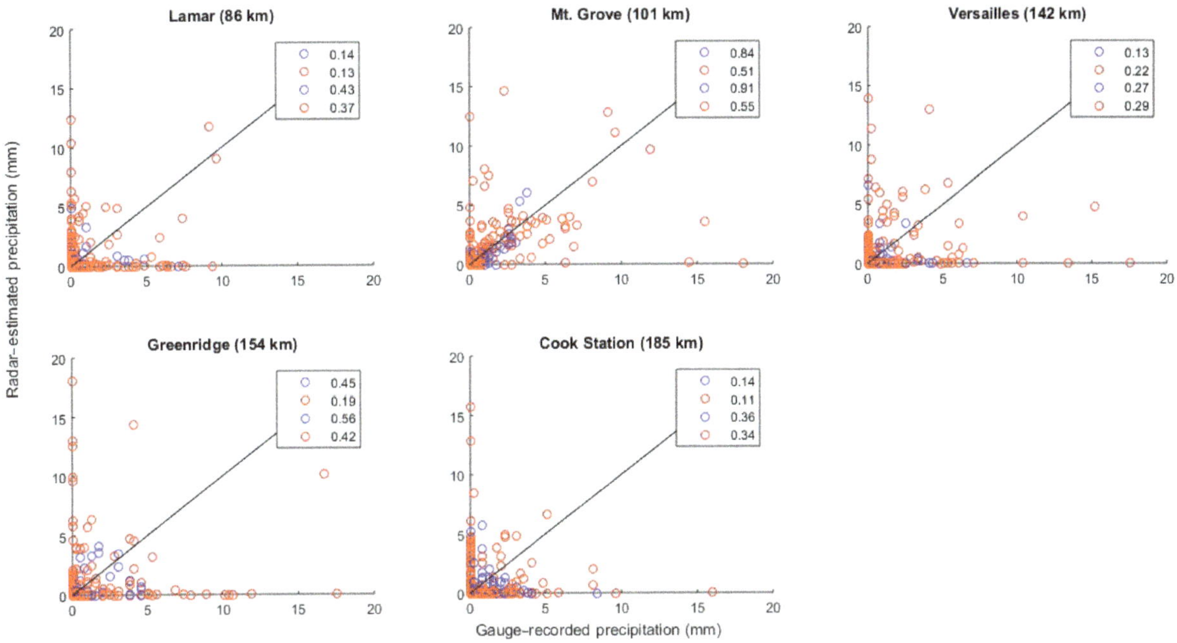

Figure 8. Correlation coefficient values for the five locations analyzed by the Springfield radar with the $R(Z,\mathrm{ZDR})$ NSSL equation. Blue and red scatter points represent the cool and warm season data, respectively. The top two numbers on each plot indicate the overall R^2 value, whereas the bottom two numbers represent the R^2 when false alarms and misses are removed.

KSGF) had the overall lowest NSE (0.8 %), MPA (15 mm), and FPA (95 mm) for $R(Z,\mathrm{ZDR})$. The third-furthest location (142 km) resulted in slightly larger errors, overall, while the fourth-furthest location had errors similar to the second gauge (105 km). Then, at the furthest tipping bucket location (185 km), NSE values increased, whereas FPA and MPA decreased. Therefore, the furthest location's errors are due, primarily, from discrepancies between precipitation magnitude between the gauge and radar.

Excluding Versailles (142 km from KSGF), the cool season exhibited larger R^2 values in comparison to the cool season (Fig. 8). Furthermore, CC values exceeded 0.9 when false alarms and misses were excluded from Mountain Grove (101 km) and were 0.84 when included. Otherwise, the other four stations analyzed by the Springfield radar displayed many counts of false alarms and misses, leading to low R^2 values.

Due to the relatively large ranges from the Springfield radar, most of the correlation coefficient values were low in comparison to either KLSX or KEAX. For the warm (cool) season without false alarms and misses, R^2 values ranged from 0.44 (0.38) to 0.34 (0.36) for KLSX and KSGF, respectively, at Cook Station (119 and 185 km). Similarly, the CC values ranged from 0.61 (0.71) to 0.42 (0.56) at Green Ridge (76 and 154 km) for KEAX and KSGF, accordingly.

4 Conclusions

Dual-polarization technology was implemented in the National Weather Service Next Generation Radar network in the spring of 2012 to, primarily, improve quantitative precipitation estimation and hydrometeor classification. The current study observed over 300 h of precipitation data with three separate radars in Missouri using 55 algorithms including the three conventional $R(Z)$ radar rain-rate estimation algorithms (stratiform, convective, and tropical) along with a myriad of $R(\mathrm{KDP})$, $R(Z,\mathrm{ZDR})$, and $R(\mathrm{ZDR},\mathrm{KDP})$ algorithms, which can be found in Ryzhkov et al. (2005). Additionally, a KDP-smoothing field of reflectivity, differential reflectivity, and the specific differential phase shift (DSMZ, DZDR, and DKDP, respectively) were measured and used for analyses. Unlike previous studies, the current work emphasizes the amount of precipitation correctly and incorrectly estimated by the radar in comparison to the terrestrial-based precipitation gauges through measurements of the missed and false precipitation amount.

For all three radars – Kansas City, St. Louis, and Springfield, MO – the majority of precipitation error (over 60 %) was contributed by the amount of precipitation falsely detected by the radar (up to 725 mm), while 20 % was due to the radar missing the precipitation (up to 225 mm) for KEAX. Similar magnitudes of error were reported for KLSX and KSGF, with an overall error in precipitation for each radar ranging between 250 mm for the best performing of the 55 algorithms, Eq. (11) (an $R(Z,\mathrm{ZDR})$ algorithm), and up to 2000 mm for the worst-performing algorithms, R(ZDR,KDP) Eq. (13). The R(Z,ZDR) equation (an NSSL algorithm) was determined to be the most robust due to it registering the lowest NSE. These values of false precipitation amount and missed precipitation amount generally increased as range from the radar increased.

Most algorithms showed a degradation in the normalized standard error with range. In particular, the KDP-smoothed equations displayed larger biases and NSE values than their non-KDP counterparts, with the exception of R(KDP) algorithms themselves. Some larger errors were recorded at gauge locations close to the radar, potentially due to bright-banding effects which were determined to be due to the large false precipitation amount analyzed at these locations.

The data were divided into summer (May–October) and winter (November–April; 59 and 41 % of the entire data, respectively). Despite the winter data contributing less than the summertime data, they accounted for 20 % of the overall MPA and 40 % to the overall PoFD. The R^2 values were lower during the winter in comparison to the warm season, primarily due to the smaller magnitude of precipitation that occurred. Furthermore, CC values increased by as much as 0.4 when instances of hits and misses were removed from the analyses, resulting in the warm season outperforming the cool season CC values at particularly short ranges from the radar.

These results aid in our understanding of the possibilities for hydrometeorological studies. Nearly 50 % of the 300 h where precipitation occurred analyzed for the study consisted of either falsely estimated precipitation by the radar, or missed by the radar. Furthermore, these errors accumulate between 500 and 2000 mm of precipitation depending on the algorithms chosen. Although the overall performance increased when false alarms and misses were removed, correlation coefficient values still, typically, remained below 0.50 at ranges beyond 130 km.

Furthermore, results demonstrate the issues with analyzing QPE from a single gauge, explaining why the Community Collaborative Rain, Hail, and Snow Network (Kelsch, 1998; Cifelli et al., 2005; Reges et al., 2016) or other densely gauged networks (e.g., the Hydrometeorological Automated Data System, HADS; and the Meteorological Assimilation Data Ingest System, MADIS) tend to be more utilized since results have shown that measurements or quality-controlled techniques made by these organizations, especially CoCoRaHS (Community Collaborative Rain, Hail and Snow Network), are significantly more accurate than rain gauges (Simpson et al., 2017), especially for convective events (Moon et al., 2009).

Author contributions. NF designed the experiment and provided feedback while MS carried out the calculations and wrote the manuscript.

Competing interests. The authors declare that they have no conflict of interest.

Acknowledgements. This material is based upon work supported by the National Science Foundation under award number IIA-1355406. Any opinions, findings, and conclusions or recommendations expressed in this material are those of the authors and do not necessarily reflect the views of the National Science Foundation.

Edited by: Remko Uijlenhoet

References

AgEBB (Agricultural Electronic Bulletin Board): Missouri Mesonet, available at: http://agebb.missouri.edu/weather/stations/, last access: February 2017.

Alaya, M. A., Ourda, T. B. M. J., and Chebana, F.: Non-Gaussian spatiotemporal simulation of multisite precipitation: Downscaling framework, Clim. Dynam., 50, 1–15, https://doi.org/10.1007/s00382-017-3578-0, 2017.

Anagnostou, M. N., Anagnostou, E. N., Vulpiani, G., Montopoli, M., Marzano, F. S., and Vivekanandan, J.: Evaluation of X-band polarimetric-radar estimates of drop-size distributions from coincident S-band polarimetric estimated and measured raindrop spectra, IEEE T. Geosci. Remote, 46, 3067–3075, 2008.

Bechini, R., Baldini, L., Cremonini, R., and Gorgucci, E.: Differential reflectivity calibration for operational radars, J. Atmos. Ocean. Tech., 25, 1542–1555, 2009.

Berne, A. and Krajewski, W. F.: Radar for hydrology: Unfulfilled promise or unrecognized potential?, Adv. Water Resour., 51, 357–366, 2013.

Berne, A. and Uijlenhoet, R.: A stochastic model of range profiles of raindrop size distributions: application to radar attenuation correction, Geophys. Res. Lett., 32, https://doi.org/10.1029/2004GL021899, 2005.

Brandes, E. A., Zhang, G., and Vivekanandan, J.: Experiments in rainfall estimation with a polarimetric radar in a subtropical environment, J. Appl. Meteorol., 41, 674–685, 2002.

Brandes, E. A., Zhang, G., and Vivekanandan, J.: Drop size distribution retrieval with polarimetric radar: model and application, J. Appl. Meteorol., 43, 461–475, 2004.

Bringi, V. N. and Chandrasekar, V.: Polarimetric Doppler weather radar, principles and applications, Cambridge University Press: Cambridge, UK, 636 pp., 2001.

Chai, T. and Draxler, R. R.: Root mean square error (RMSE) or mean absolute error (MAE)? – Arguments against avoiding RMSE in the literature, Geosci. Model Dev., 7, 1247–1250, https://doi.org/10.5194/gmd-7-1247-2014, 2014.

Ciach, G. J. and Krajewski, W. F.: On the estimation of radar rainfall error variance, Adv. Water Resour., 22, 585–595, 1999a.

Ciach, G. J. and Krajewski, W. F.: Radar-raingage comparisons under observational uncertainties, J. Appl. Meteorol., 38, 1519–1525, 1999b.

Ciach, G. J.: Local random errors in tipping-bucket rain gauge measurements, J. Atmos. Ocean. Tech., 20, 752–759, 2002.

Cifelli, R., Doesken, N., Kennedy, P., Carey, L., Rutledge, S. A., Gimmestad, C., and Depue, T.: The community collaborative rain, hail, and snow network: Informal education for scientists and citizens, B. Am. Meteorol. Soc., 86, 1069–1077, 2005.

Cifelli, R., Chandrasekar, V., Lim, S., Kennedy, P. C., Wang, Y., and Rutledge, S. A.: A new dual-polarization radar rainfall algorithm: Application in Colorado precipitation events, J. Atmos. Ocean. Tech., 28, 352–364, 2010.

Cunha, L. K., Smith, J. A., Baeck, M. L., and Krajewski, W. F.: An early performance of the NEXRAD dual-polarization radar rainfall estimates for urban flood applications, Weather Forecast., 28, 1478–1497, 2013.

Cunha, L. K., Smith, J. A., Krajewski, W. F., Baeck, M. L., and Seo, B.: NEXRAD NWS polarimetric precipitation product evaluation for IFloods, J. Hydrometeorol., 16, 1676–1699, 2015.

Fabry, F., Bellon, A., Duncan, M. R., and Austin, G. L.: High resolution rainfall measurements by radar for very small basins: the sampling problem reexamined, J. Hydrol., 161, 415–428, 1994.

Fairman, J. G., Schultz, D. M., Kirschbaum, D. J., Gray, S. L., and Barrett, A. I.: A radar-based rainfall climatology of Great Britain and Ireland, Weather, 70, 153–158, https://doi.org/10.1002/wea.2486, 2012.

Gamache, J. F. and Houze, R. A.: Mesoscale air motions associated with a tropical squall line, Mon. Weather Rev., 110, 118–135, 1982.

Giangrande, S. E. and Ryzhkov, A. V.: Estimation of rainfall based on the results of polarimetric echo classification, J. Appl. Meteorol., 47, 2445–2460, 2008.

Gorgucci, E., Scarchilli, G., and Chandrasekar, V.: Calibration of radars using polarimetric techniques, IEEE T. Geosci. Remote, 30, 853–858, 1992.

Gorgucci, E., Scarschilli, G., Chandrasekar, V., and Bringi, V. N.: Measurement of mean raindrop shape from polarimetric radar observations, J. Atmos. Sci., 57, 3406–3413, 2000.

Gorgucci, E., Baldini, L., and Chandrasekar, V.: What is the shape of a raindrop? An answer from radar measurements, J. Atmos. Sci., 63, 3033–3044, 2006.

Goudenhoofdt, E. and Delobbe, L.: Long-term evaluation of radar QPE using VPR correction and radar-gauge merging, International Association of Hydrological Sciences Publications, 351, 249–254, 2012.

Goudenhoofdt, E. and Delobbe, L.: Generation and verification of rainfall estimates from 10-yr volumetric weather radar measurements, J. Hydrometeorol., 133, 1191–1204, 2016.

Gourley, J. J., Giangrande, S. E., Hong, Y., Flamig, Z., Schuur, T., and Vrugt, J.: Impacts of polarimetric radar observations on hydrologic simulation, J. Hydrometeorol., 11, 781–796, 2010.

Habib, E., Krajewski, W. F., Nespor, V., and Kruger, A.: Numerical simulation studies of rain gauge data correction due to wind effect, J. Geophys. Res., 104, 723–734, 1999.

Habib, E., Krajewski, W. F., and Kruger, A.: Sampling errors of tipping-bucket rain gauge measurements, J. Hydrol. Eng., 6, 159–166, 2001.

Haylock, M. R., Hofstra, N., Klein Tank, A. M. G., Klok, E. J., Jones, P. D., and New, M.: A European daily high-resolution gridded data set of surface temperature and precipitation for 1950–2006, J. Geophys. Res., 113, https://doi.org/10.1029/2008JD010201, 2008.

Holleman, I., Huuskonen, A., Gill, R., and Tabary, P.: Operational monitoring of radar differential reflectivity using the sun, J. Atmos. Ocean. Tech., 27, 881–887, 2010.

Hubbert, J. C.: Differential reflectivity calibration and antenna temperature, J. Atmos. Ocean. Tech., 34, 1885–1906, 2017.

Illingworth, A. and Blackman, T. A.: The need to represent raindrop size spectra as normalized gamma distributions for the interpretation of polarization radar observations, J. Appl. Meteorol., 41, 286–297, 2002.

Kelsch, M.: The Fort Collins flash flood: Exceptional rainfall and urban runoff, Preprints, 19th Conference on severe local storms, Minneapolis, MN, American Meteorological Society, 404–407, 1998.

Kessinger, C., Ellis, S., and Van Andel, J.: The radar echo classifier: a fuzzy logic algorithm for theWSR-88D, 19th Conf. on Inter. Inf. Proc. Sys. (IIPS) for Meteor., Ocean., and Hydr., Amer. Meteor. Soc., Long Beach, CA, 2003.

Kitchen, M. and Blackall, M.: Representativeness errors in comparisons between radar and gauge measurements of rainfall, J. Hydrol., 134, 13–33, 1992.

Kitchen, M. and Jackson, P. M.: Weather radar performance at long range – simulated and observed, J. Appl. Meteorol., 32, 975–985, 1993.

Kleiber, W., Katz, R. W., and Rajagopalan, B.: Daily spatiotemporal precipitation simulation using latent and transformed Gaussian processes, Water Resour. Res., 48, W01523, https://doi.org/10.1029/2011WR011105, 2012.

Krajewski, W. F., Kruger, A., and Nespor, V.: Experimental and numerical studies of small-scale rainfall measurements and variability, Water Sci. Technol., 37, 131–138, 1998.

Kumjian, M. R.: Principles and applications of dual-poarization weather radar, Part 1: Description of the polarimetric radar variables, Journal of Operational Meteorology, 1, 226–242, 2013a.

Kumjian, M. R.: Principles and applications of dual-poarization weather radar, Part 2: Warm and cold season applications, Journal of Operational Meteorology, 1, 243–264, 2013b.

Kumjian, M. R.: Principles and applications of dual-poarization weather radar, Part 3: Artifacts. Journal of Operational Meteorology, 1, 265–274, 2013c.

Lakshmanan, V., Smith, T., Stumpf, G., and Hondl, K.: The warning decision support system – integrated information, Weather Forecast., 22, 596–612, 2007a.

Lakshmanan, V., Fritz, A., Smith, T., Hondl, K., and Stumpf, G.: An automated technique to quality control radar reflectivity data, J. Appl. Meteorol. Clim., 46, 288–305, 2007b.

Lakshmanan, V., Zhang, J., and Howard, K.: A technique to censor biological echoes in radar reflectivity data. J. Appl. Meteorol. Clim., 49, 453–462, 2010.

Lakshmanan, V., Karstens, C., Krause, J., and Tang, L.: Quality control of weather radar data using polarimetric variables, J. Atmos. Ocean. Tech., 31, 1234–1249, 2014.

Moon, J. T., Guinan, P. E., Snider, D. J., and Lupo, A. R.: CoCoRaHS in Missouri: Four years later, the importance of observations, Transactions of the Missouri Academy of Science, 43, 7–18, 2009.

NOAA: Hierarchical Data Storage System, available at: https://www.ncdc.noaa.gov/has/has.dsselect, last access: February 2017.

Park, H. S., Ryzhkov, A. V., and Zrnic, D. S.: The hydrometeor classification algorithm for the polarimetric WSR-88DL Description and application to an MCS, Weather Forecast., 24, 730-748, 2009.

Reges, H. W., Doesken, N., Turner, J., Newman, N., Bergantino, A., and Schwalbe, Z.: CoCoRaHS: The evolution and accomplishments of a volunteer rain gauge network, B. Am. Meteorol. Soc., 97, 1831–1846, 2016.

Ruzanski, E. and Chandrasekar, V.: Nowcasting rainfall fields derived from specific differential phase, J. Appl. Meteorol. Clim., 51, 1950–1959, 2012.

Ryzhkov, A., Kumjian, M., Ganson, S., and Zhang, P.: Polarimetric characteristics of melting hail. Part II: Practical implications, J. Appl. Meteorol. Clim., 52, 2871–2886, 2013.

Ryzhkov, A. V., Giangrande, S., and Schurr, T.: Rainfall measurements with the polarimetric WSR-88D radar, National Severe Storms Laboratory Rep., Norman, OK, 98, 2003.

Ryzhkov, A. V., Giangrande, S., and Schurr, T.: Rainfall estimation with a polarimetric prototype of WSR-88D, J. Appl. Meteorol., 44, 502–515, 2005.

Shucksmith, P. E., Sutherland-Stacey, L., and Austin, G. L.: The spatial and temporal sampling errors inherent in low resolution radar estimates of rainfall, Meteorol. Appl., 18, 354–360, 2011.

Simpson, M. J., Hubbart, J. A., and Fox, N. I.: Ground truthed performance of single and dual-polarized radar rain rates at large ranges, Hydrol. Process., 30, 3692–3703, 2016.

Simpson, M. J., Hirsch, A., Grempler, K., and Lupo, A. R.: The importance of choosing precipitation datasets, Hydrol. Process., 31, 1–13, https://doi.org/10.1002/hyp.11381, 2017.

Seo, B.-C., Dolan, B., Krajewski, W., Rutledge, S. A., and Petersen, W.: Comparison of single- and dual-polarization-based rainfall estimates using NEXRAD data for the NASA Iowa Flood Studies project, J. Hydrometeorol., 16, 1658–1675, 2015.

Smith, J., Seo, D. J., Baeck, M. L., and Hudlow, M. D.: An intercomparison study of NEXRAD precipitation estimates, Water Resour. Res., 32, 2035–2045, 1996.

Straka, J. M., Zrnic, D. S., and Ryzhkov, A. V.: Bulk hydrometeor classification and quantification using polarimetric radar data: Synthesis of relations, J. Appl. Meteorol., 39, 1341–1372, 2000.

Wang, Y. and Chandrasekar, V.: Quantitative precipitation estimation in the CASA X-band dual-polarization radar network, J. Atmos. Ocean Tech., 27, 1665–1676, 2010.

Yang, L., Yang, Y., Liu, P., and Wang, L.: Radar-derived quantitative precipitation estimation based on precipitation classification, Adv. Meteorol., 2016, 2457489, https://doi.org/10.1155/2016/2457489, 2016.

Zhang, G., Vivekanandan, J., and Brandes, E. A.: A method for estimating rain rate and drop size distribution from polarimetric radar measurements, IEEE T. Geosci. Remote, 39, 830–841, 2001.

Zhang, J. and Qi, Y.: A real-time algorithm for the correction of brightband effects in radar-derived QPE, J. Hydrometeorol., 11, 1157–1171, 2010.

Zhang, J., Langston, C., and Howard, K.: Brightband identification based on vertical profiles of reflectivity from the WSR-88D, J. Atmos. Ocean. Tech., 25, 1859–1872, 2008.

Zrnic, D. S. and Ryzhkov, A. V.: Advantages of rain measurements using specific differential phase, J. Atmos. Ocean. Tech., 13, 454–464, 1996.

Zrnic, D. S., Melknikov, V. M., and Carter, J. K.: Calibrating differential reflectivity on the WSR-88D, J. Atmos. Ocean. Tech., 23, 944–951, 2005.

9

Numerical modeling and sensitivity analysis of seawater intrusion in a dual-permeability coastal karst aquifer with conduit networks

Zexuan Xu[1], **Bill X. Hu**[2], **and Ming Ye**[3]

[1]Climate and Ecosystem Sciences Division, Lawrence Berkeley National Laboratory, Berkeley, California 94720, USA
[2]Institute of Groundwater and Earth Sciences, Jinan University, Guangzhou, Guangdong, 510632, China
[3]Department of Earth, Ocean and Atmospheric Science, Florida State University, Tallahassee, Florida 32306, USA

Correspondence: Zexuan Xu (xuzexuan@gmail.com) and Bill Hu (bill.x.hu@gmail.com)

Abstract. Long-distance seawater intrusion has been widely observed through the subsurface conduit system in coastal karst aquifers as a source of groundwater contaminant. In this study, seawater intrusion in a dual-permeability karst aquifer with conduit networks is studied by the two-dimensional density-dependent flow and transport SEAWAT model. Local and global sensitivity analyses are used to evaluate the impacts of boundary conditions and hydrological characteristics on modeling seawater intrusion in a karst aquifer, including hydraulic conductivity, effective porosity, specific storage, and dispersivity of the conduit network and of the porous medium. The local sensitivity analysis evaluates the parameters' sensitivities for modeling seawater intrusion, specifically in the Woodville Karst Plain (WKP). A more comprehensive interpretation of parameter sensitivities, including the nonlinear relationship between simulations and parameters, and/or parameter interactions, is addressed in the global sensitivity analysis. The conduit parameters and boundary conditions are important to the simulations in the porous medium because of the dynamical exchanges between the two systems. The sensitivity study indicates that salinity and head simulations in the karst features, such as the conduit system and submarine springs, are critical for understanding seawater intrusion in a coastal karst aquifer. The evaluation of hydraulic conductivity sensitivity in the continuum SEAWAT model may be biased since the conduit flow velocity is not accurately calculated by Darcy's equation as a function of head difference and hydraulic conductivity. In addition, dispersivity is no longer an important parameter in an advection-dominated karst aquifer with a conduit system, compared to the sensitivity results in a porous medium aquifer. In the end, the extents of seawater intrusion are quantitatively evaluated and measured under different scenarios with the variabilities of important parameters identified from sensitivity results, including salinity at the submarine spring with rainfall recharge, sea level rise, and a longer simulation time under an extended low rainfall period.

1 Introduction

Many serious environmental issues have been caused by seawater intrusion in coastal regions, such as soil salinization, marine and estuarine ecological changes, and groundwater contamination (Bear, 1999). Werner et al. (2013) pointed out that climate variations, groundwater pumping, and fluctuating sea levels are important factors in the mixing of seawater and freshwater in the aquifer. Custodio (1987) and Shoemaker (2004) summarized the control factors of seawater intrusion into a coastal aquifer, including the geologic and lithological heterogeneity, localized surface recharge, paleohydrogeological conditions, and anthropogenic influences. Particularly, seawater intrusion in a coastal aquifer is significantly impacted by sea level rise, which has been recognized as a serious environmental threat in the 21st century (Voss and Souza, 1987; Bear, 1999; IPCC, 2007). In the Ghyben–Herzberg relationship, a small rise of sea level causes extended seawater intrusion and significantly moves the mixing interface position further landward in a coastal aquifer (Werner and Simmons, 2009). For example, Essink et al. (2010) systematically studied the exacerbated seawater intrusion under sea level rise and global climate change.

Likewise, high tides associated with hurricanes or tropical storms have been found to temporarily affect the extent of seawater intrusion in a coastal aquifer (Moore and Wilson, 2005; Wilson et al., 2011).

Modeling seawater intrusion in a coastal aquifer requires a coupled density-dependent flow and salt transport groundwater model. The simulated salinity is computed by the groundwater velocity field from flow modeling, and salinity in turn determines water density and affects the simulation of flow field. Several variable-density numerical models have been developed and widely used to study seawater intrusion, including SUTRA (Voss and Provost, 1984) and FEFLOW (Diersch, 2002). SEAWAT is a widely used density-dependent model, which solves flow equations with a finite difference method and transport equations with three major classes of numerical techniques (Guo and Langevin, 2002; Langevin et al., 2003). Generally speaking, most variable-density models are numerically complicated and computationally expensive. These models require a smaller timestep and an implicit procedure for solving flow and transport equations iteratively many times in each timestep (Werner et al., 2013).

In addition, a karst aquifer is particularly vulnerable to groundwater contamination including seawater intrusion in a coastal region since sinkholes and karst windows are usually connected by well-developed subsurface conduit networks. Some karst caves are found open to the sea and become submarine springs below sea level, connected with the conduit network as natural pathways for seawater intrusion. Fleury et al. (2007) reviewed the studies of freshwater discharge and seawater intrusion through karst conduits and submarine springs in coastal karst aquifers, and they summarized the important control factors, including the hydraulic gradient of the equivalent freshwater head, hydraulic conductivity, and seasonal precipitation variation. For example, seawater intrudes through the conduit network as preferential flow and contaminates the fresh groundwater resources in a coastal karst aquifer (Calvache and Pulido-Bosch, 1997). As an indicator of rainfall and regional freshwater recharges, salinity at the outlet of a conduit system is diluted by freshwater discharge during a rainfall season, but it remains constant as saline water during a low rainfall period (Martin and Dean, 2001; Martin et al., 2012).

Modeling groundwater flow in a dual-permeability karst aquifer is a challenging issue since groundwater flow in a karst conduit system is often non-laminar (Davis, 1996; Shoemaker et al., 2008; Gallegos et al., 2013). Several discrete-continuum numerical models, such as MODFLOW-CFPM1 (Shoemaker et al., 2008) and CFPv2 (Reimann et al., 2014, 2013; Xu et al., 2015a, b), have been developed to simultaneously solve the non-laminar flow in the conduit, the Darcian flow in a porous medium, and the exchanges between the two systems. However, these constant-density karst models have limitations in simulating the density-dependent seawater intrusion processes in a coastal aquifer. The VDFST-CFP, developed by Xu and Hu (2017), is based on a density-dependent discrete-continuum modeling approach to study seawater intrusion in a coastal karst aquifer with conduits. However, VDFST-CFP is not able to simulate the seawater intrusion processes addressed in this study due to the computational constraints and the numerical method limitations associated with the aquifer geometry and the domain scale. Therefore, the variable-density SEAWAT model is still applied in this study, in which Darcy's equation is used to compute flow not only in the porous medium, but also in the conduit with large values of hydraulic conductivity and effective porosity.

Since simulating seawater intrusion in karst aquifers is challenging, sensitivity analysis is important to provide guidelines for understanding the hydrology model, data collection, and groundwater resources management. Several sensitivity studies have evaluated the parameters in karst aquifers. Kaufmann and Braun (2000) reported that boundary conditions and sink recharges are important to the preferential flow path in a karst aquifer. Scanlon et al. (2003) also confirmed that recharge is important to karst spring discharge. Regional sensitivity analysis has been widely used to show this relationship of karst spring discharge with different hydrological processes in a local karst catchment (Chang et al., 2017). Chen et al. (2017) and Hartmann et al. (2015) applied Sobol's global sensitivity method to evaluate parameters using different objective functions under different hydrodynamic conditions. However, very few studies have addressed the parameters' sensitivities for modeling seawater intrusion in a coastal karst aquifer. Shoemaker (2004) performed a sensitivity analysis of the SEAWAT model for seawater intrusion in a homogeneous porous aquifer, concluding that dispersivity is an important parameter in the head, salinity, and groundwater flow simulations and observations in the transition zone. Shoemaker (2004) also concluded that salinity observations are more effective than head observations, and head and salinity simulations and observations are more sensitive to parameters at the "toe" of the transition zone. The sensitivity results in this study confirm some conclusions in Shoemaker (2004) and highlight the significance of a conduit network for seawater intrusion in a coastal karst aquifer with interaction between a karst conduit and a porous medium.

The parameter sensitivities are evaluated to address the impacts of the two major challenges in this study, the density-dependent flow and transport coupled seawater intrusion processes and the dual-permeability karst system. This study aims to strengthen the understanding of the roles of model parameters and boundary conditions in simulating seawater intrusion in the coastal karst region. To the best of our knowledge, this is the first attempt to assess the parameters' sensitivities for modeling seawater intrusion in a vulnerable dual-permeability karst aquifer. The rest of the paper is arranged as follows: the details of local and global sensitivity analysis methods are introduced in Sect. 2. The model setup, hydrological conditions, model discretization, and initial and boundary conditions are discussed in Sect. 3. The results of local and global sensitivity analysis are discussed in Sect. 4. The scenarios of seawater intrusion simulation with different boundary conditions and simulation time are presented in Sect. 5. Conclusions are made in Sect. 6.

2 Methods

The governing equations used in the SEAWAT model can be found in Guo and Langevin (2002), including the variable-density flow equation with additional density terms and the advection-dispersion solute transport equation. The local and global sensitivity methods used in this study are briefly introduced below. Note that the sensitivity analysis does not necessarily need field observations; it only evaluates the model simulations with respect to parameters instead. Field observational data, especially head and salinity measurements in the conduit, are seldom available considering the difficulties of sensor installation in the deep subsurface conduit network. Model calibration is beyond the scope of this study, due to the lack of observational data in the Woodville Karst Plain (WKP).

2.1 Local sensitivity analysis

In this study, UCODE_2005 (Poeter and Hill, 1998) is used in the local sensitivity analysis to evaluate the derivatives of model simulations with respect to parameters at the specified values (Hill and Tiedeman, 2006). The forward difference approximation of sensitivity is calculated as the derivative of the ith simulation with respect to the jth model parameters:

$$\left.\frac{\partial y'_i}{\partial x_j}\right|_b \approx \frac{y'_i(\boldsymbol{x}+\Delta\boldsymbol{x}) - y'_i(\boldsymbol{x})}{\Delta x_j}, \tag{1}$$

where y'_i is the value of the ith simulation; x_j is the jth estimated parameter; $\boldsymbol{x}$ is a vector of the specified values of estimated parameter; $\Delta\boldsymbol{x}$ is a vector of zeros except that the jth parameter equals Δx_j.

Since parameters can have different units, scaled sensitivities are used to compare the parameter sensitivities. In UCODE_2005, a scaling method is used to calculate the dimensionless scaled sensitivity (DSS) values with the following equation:

$$\mathrm{dss}_{ij} = \left.\left(\frac{\partial y'_i}{\partial x_j}\right)\right|_x |x_j|\,\omega_{ii}^{1/2}, \tag{2}$$

where dss_{ij} is the dimensionless scaled sensitivity of the ith simulation with respect to the jth parameter; and ω_{ii} is the weight of the ith simulation, which is computed by the inverse of error variances (square of error standard deviation) as $\omega_{ii} = 1/\sigma^2$. The values of error standard deviation used in this study are referenced from Shoemaker (2004). The head measurement error was assumed to be normally distributed with a mean of zero and a standard deviation of ~ 0.003 m. The standard deviation was based on standard error estimates for water levels measured in wells by the USGS in Florida. The salinity measurement error was assumed to be normally distributed with a mean of zero and a standard deviation of ~ 0.1 PSU (Practical Salinity Unit). Shoemaker et al. (2004) believed that a 0.1 PSU standard deviation was thought to be appropriate for a salinity range from 3 to 35 PSU. In this study, the eleven evaluated locations (column no. 25 to no. 75 with an interval of five cells) have the same weights for each simulation, in terms of the head and salinity in this study. The DSS values of different simulations with respect to each parameter are accumulated as the composite scaled sensitivity (CSS) values, which reflect the total amount of information provided by the simulation for the estimation of one parameter. The CSS of the jth parameter is evaluated via

$$\mathrm{css}_j = \sum_{i=1}^{\mathrm{ND}} \left[\left.(\mathrm{dss}_{ij})^2\right|_x / \mathrm{ND} \right]^{1/2}, \tag{3}$$

where ND is the number of simulated quantities, in terms of the head and salinity simulations in this study.

2.2 Morris method for global sensitivity analysis

The local sensitivity analysis is conceptually straightforward and easy to perform without expensive computational cost; however, it only calculates the parameter sensitivities at one specified value for each parameter instead of the ranges. In addition, the local sensitivity indices are based on the first-order derivative only, assuming a linear relationship of simulated quantities with respect to parameters.

The global sensitivity analysis evaluates the nonlinear relationship of parameters with simulations, and/or parameters involved in interaction with other factors. The Morris method is applied in this study to evaluate the global parameter sensitivities (Morris, 1991). The design of the Morris method is made up of an individually randomized "one-step-at-a-time" (OAT) experiment, which perturbs only one input parameter and computes a new simulated output in each run. The Morris method is composed of a number r of local changes at different points of the possible range values. For each parameter, a discrete number of values called levels are chosen within the parameter range.

In the Morris method, the k-dimensional vector $\boldsymbol{x}$ of the model parameters has components x_i to be divided into p uniform intervals. The global parameter sensitivity is evaluated from the difference of simulation results by changing one parameter at a time, which is called an elementary effect (EE), d_i, defined as

$$d_i = \frac{1}{\tau_y}\frac{\left[y\left(x_1^*,\ldots,x_{i-1}^*,x_i^*+\Delta,x_{i+1}^*,\ldots,x_k^*\right) - y(x_1^*,\ldots,x_k^*)\right]}{\Delta}, \tag{4}$$

where Δ is the relative distance in the parameter coordinate; τ_y is the output scaling factor; $\{x_i^*\}$ is the parameter set selected in a sampling method.

To compute the EE for the k parameters, $(k+1)$ simulations will run with the perturbation of each parameter, which is called one "path" (Saltelli et al., 2004). An ensemble of EEs is generated with multiple paths of a parameter set. The total number of model runs is $r(k+1)$, where r is the number of paths.

Two sensitivity measures are proposed by the Morris method to approximate parameter sensitivities: the mean μ estimates the overall influence of the factor on the output, and the standard deviation σ estimates the nonlinear effect between input and output and/or the parameter interactions (Saltelli et al., 2004). The mean μ and standard deviation σ of the EEs are evaluated with the r-independent paths in the Morris method:

$$\mu = \sum_{i=1}^{r} d_i / r \tag{5}$$

$$\sigma = \sqrt{\sum_{i=1}^{r} (d_i - \mu)^2 / r}. \tag{6}$$

In this study, the EEs for the Morris method are not generated by Monte Carlo random sampling, which usually needs an extremely large number (> 250) of paths for the eleven parameters in this study and takes a very long time to complete sensitivity computation without parallelization. To save the running time and computational cost, a more efficient trajectory sampling is developed by Saltelli et al. (2004), which has become a widely used method to generate the ensemble of EEs in the Morris method but ensure the confidence of global sensitivity results. In the trajectory method, the choice of parameter p is usually even, and Δ is equal to $\pm p/[2(p-1)]$, either positive or negative. The trajectory method starts by randomly selecting a "seed" value x^* for the vector $\boldsymbol{x}$. Each component x_i of x^* is randomly sampled from the set $(0, 1/(p-1), 2/(p-1), \ldots, 1)$. The randomly selected vector $\boldsymbol{x}^*$ is used to generate the other sampling points but not one of them, which means that the model is never evaluated at vector $\boldsymbol{x}^*$. The first sampling point, $x^{(1)}$, is obtained by changing one or more components of x^* by Δ. The second sampling point, $x^{(2)}$, is generated from x^* but differs from $x^{(1)}$ in its ith component that has been either increased or decreased by Δ, but conditioned on the domain, and the index i is randomly selected in the set $\{1, 2, \ldots, k\}$. In other word, $x^{(2)} = (x_1^{(1)}, \ldots, x_{i-1}^{(1)}, x_i^{(1)} \pm \Delta, x_{i+1}^{(1)}, \ldots, x_k^{(1)})$. The third sampling point, $x^{(3)}$, differs from $x^{(2)}$ for only one component j; for any $j \neq i$, it will be $x_j^{(3)} = x_j^{(2)} \pm \Delta$. A succession of $(k+1)$ sampling points $x^{(1)}, x^{(2)}, \ldots, x^{(k+1)}$ is produced in the input parameters' space called a trajectory, with the key characteristic that two consecutive points differ in only one component. Note that the choice of components x^* to be increased or decreased has the condition that x_i is still within the domain. In the trajectory sampling, any component i of the "base" vector $\boldsymbol{x}^*$ has been selected at least once by Δ in order to calculate one EE for each parameter.

Once a trajectory has been constructed and evaluated by the Morris method, an EE for each parameter $i, i = 1, \ldots, k$, can be computed. If $x^{(l)}$ and $x^{(l+1)}$, with l in the set in $(1, \ldots, k)$, are two sampling points differing in their ith component, the EEs associated with the parameter i is computed as

$$d_i\left(x^{(l)}\right) = \frac{\left[y\left(x^{(l+1)}\right) - y\left(x^{(l)}\right)\right]}{\Delta}. \tag{7}$$

A random ensemble of r EEs is preselected at the beginning of sampling, but the starting point of each trajectory sampling is also randomly generated. In other words, the points belonging to the same trajectory are not independent, but the r points sampled from each distribution belonging to different trajectories are independent.

3 Model development

3.1 Study site

The numerical model developed in this paper is based on the parameter values of a porous medium and conduit measured in the aquifer at the Woodville Karst Plain (WKP). The Spring Creek Springs (SCS) is a system consisting of 14 submarine springs located in the Gulf of Mexico (Fig. 1). SCS is an outlet of the subsurface conduit network and the entrance of seawater intrusion, exactly located at the shoreline beneath the sea level. Davis and Verdi (2014) described a groundwater cycling conceptual model to explain the hydrogeological conditions in the WKP. In this conceptual model of seawater and freshwater interaction, seawater intrudes through subsurface conduit networks during low precipitation periods, while rainfall recharge dilutes and pushes the intruded seawater out from the submarine spring during high rainfall periods, usually after a heavy storm event. Later on, the conceptual model was quantitatively simulated by a constant-density CFPv2 numerical model in Xu et al. (2015b). Tracer test studies and cave diving investigations indicate that the conduit system starts from the submarine spring and extends 18 km landward, connecting with an inland spring called Wakulla Spring, although the exact locations of the subsurface conduits are unknown and difficult to explore (Kernagis et al., 2008; Kincaid and Werner, 2008). Evidence shows that seawater intrusion has been observed through the subsurface conduit system for more than 18 km in the WKP (Xu et al., 2016). In addition, Davis and Verdi (2014) also point out that sea level rise in the Gulf of Mexico in the 20th century could be a reason for increasing discharge at an inland karst spring (Wakulla Spring) and decreasing discharge at SCS, when the hydraulic gradient between the two springs is directed towards the gulf.

In this study, a two-dimensional SEAWAT model is set up to simulate seawater intrusion via the SCS through the major subsurface conduit network in the WKP (Fig. 1). Figure 2

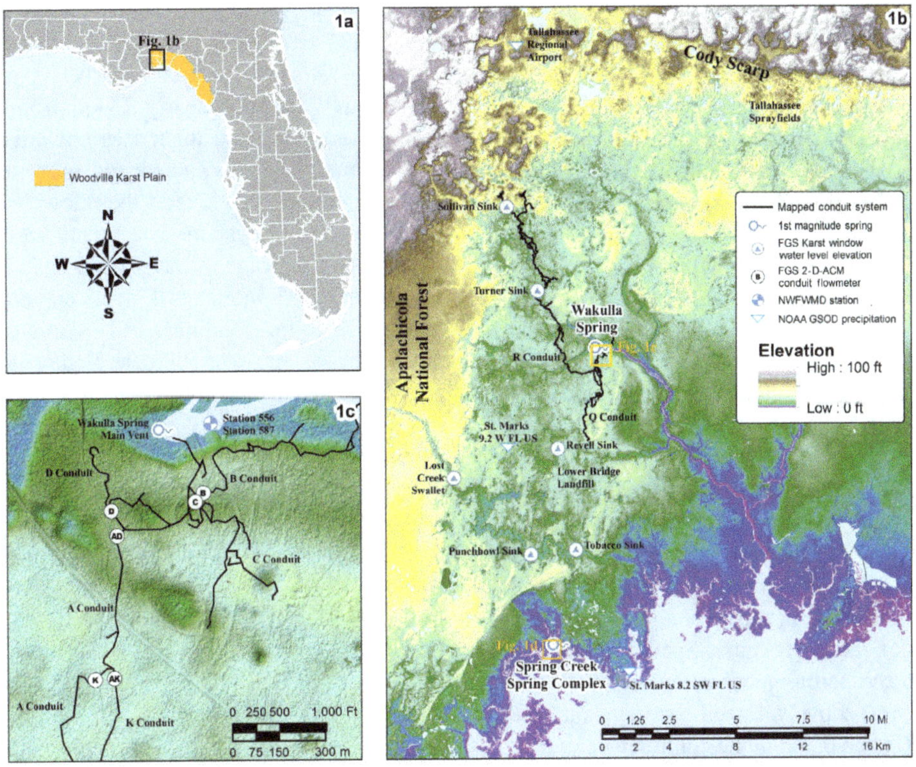

Figure 1. **(a)** Locations of the Woodville Karst Plain (WKP) and the study site. **(b)** The map of the Woodville Karst Plain showing the locations of features of note in the study. **(c)** Details of the cave system near Wakulla Spring. Modified from Xu et al. (2016).

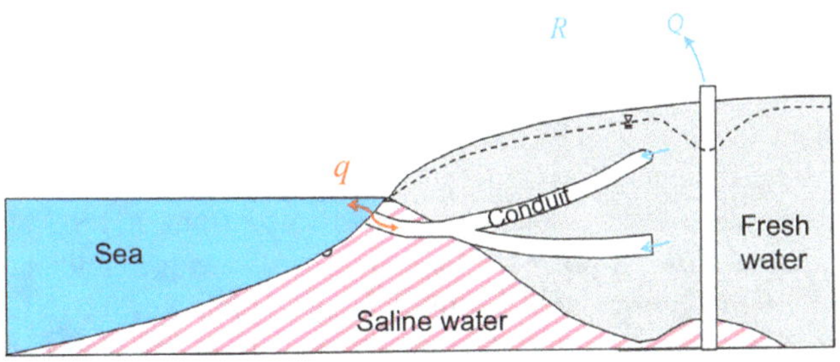

Figure 2. Schematic figure of a coastal karst aquifer with conduit networks and a submarine spring opening to the sea in a cross section. Flow direction q would be seaward when sea level drops, pumping rate Q is low, and precipitation recharge R is large; however, reversal flow occurs when sea level rises, pumping rate Q is high, or precipitation recharge R is small.

presents the cross section schematic figure in a coastal karst aquifer with a conduit network and a submarine spring opening to the sea. The model spatial domain is not a straight line from the SCS to Wakulla Spring, but it is a cross section along the major conduit pathway of seawater intrusion between the two springs. The conduit geometry in the model is set as 18 km long and 91 m deep, with a height of 10 m in the horizontal part and a width of 50 m in the vertical part.

The two-dimensional model has some limitations on simulating seawater intrusion in the entire aquifer, usually assuming that the quantities are constant parallel to the shoreline. The simulation of seawater intrusion in a perpendicular direction to the cross section as well as three-dimensional flow and transport in the porous matrix are ignored and beyond the scope of this study. However, most SEAWAT models are set up for a two-dimensional cross section with finer resolution vertical discretization. Note that this study only aims to evaluate the parameter sensitivities and simulate seawater intrusion within the vertical cross section of a karst conduit in the aquifer, including flow and transport through the conduit network, the salinity plume in the porous medium, and the exchanges between the two systems. In addition, the two-dimensional assumption is reasonable in the study site since relatively large hydraulic conductivity layers are found at nearly the same depth as the conduit network (Werner, 2001). The permeable layers indicate the possibility of a large extension of the conduit network parallel to the shoreline, although no direct evidence has been found.

3.2 Hydrological parameters

Table 1 presents the hydrological parameter values of the Upper Floridan aquifer (UFA) in the WKP and boundary conditions used in the model. These parameters were calibrated in the regional-scale groundwater flow and solute transport models by Davis et al. (2010) and have been applied in many previous modeling studies (Gallegos et al., 2013; Xu et al., 2015a, b). It should be pointed out that model calibration has not been conducted in this study since the head and salinity observational field data are insufficient, particularly in the conduit, considering the difficulties of monitoring devices' installation in the subsurface conduit. The parameter values in Table 1 are evaluated in the following local sensitivity analysis and then applied in the seawater intrusion scenarios in Sect. 5.

The values of hydrological parameters (hydraulic conductivity, specific storage, and effective porosity) in the conduit are generally greater than those of the surrounding porous medium. Hydraulic conductivity of the porous medium is assigned as 2286 m day^{-1} and as large as 610 000 m day^{-1} for the conduit system. Note that even the hydraulic conductivity of the porous medium in the study region is larger than most alluvial aquifers, due to numerous small fractures and relatively large pores existing in the karst aquifer associated with the dissolution of carbonate rocks. Specific storage and effective porosity in the porous medium are assumed to be 5×10^{-7} and 0.003, respectively. Specific storage and effective porosity are 0.005 and 0.300 in the conduit layer, respectively. The longitudinal dispersivity is estimated to be 10 m in the porous medium, but it is assumed to be a very small value (0.3 m) in the conduit because advection is dominating and dispersion is negligible in the solution of transport in the conduit.

3.3 Spatial and temporal discretization

The grid discretization and boundary conditions of the two-dimensional SEAWAT numerical model are shown in Fig. 3, with 140 columns and 37 layers in the cross section. Guo and Langevin (2002) and Werner et al. (2013) pointed out that a fine-resolution vertical grid is required for accurately modeling the density-dependent flow and solute transport. The vertical thickness of each grid cell is set uniformly as 3.048 m (10 ft) in this study, significantly smaller than the large thickness of 152 m in many previous constant-density modeling studies in the WKP, for example, Davis and Katz (2007); Davis et al. (2010); Xu et al. (2015a, b); and Gallegos et al. (2013).

Based on the field scale, the horizontal discretization for each cell is set uniformly at 152 m, except columns no. 22 and no. 139, which are 15.2 m in the vertical conduit network connecting the submarine spring (SCS) and inland spring (Wakulla Spring), respectively. The sizes of spring outlets and the conduit network are based on the observational field data and the calibrated values from the previous modeling studies (Gallegos et al., 2013). For model simplicity, the size of horizontal conduit network is assumed constant in this study. The outlet of vertical conduit system is the submarine spring (SCS) located at the shoreline at column no. 22. The conduit system starts from the submarine spring, descends downward to layer no. 29 (nearly 100 m below sea level), horizontally extends nearly 18 km from column no. 22 to column no. 139, and then rises upward to the top through column no. 139. Seawater intrudes at the SCS on the first layer of column no. 22 and then flows vertically downward into the conduit system. The inland spring is simulated by the DRAIN package as the general head boundary condition in the SEAWAT model. All layers are simulated as confined aquifer since the conduit is fully saturated, which is consistent with the previous numerical models used in Davis et al. (2010) and Xu et al. (2015a, b) in the WKP.

A transient 7-day stress in the SEAWAT model is evaluated throughout this study, except the scenarios of longer simulation time for evaluating seawater intrusion under an extended low rainfall period in Sect. 5.4. The timestep of flow model is set as 0.1 days, and the timestep of transport model is determined by SEAWAT automatically.

3.4 Initial and boundary conditions

The initial condition of head is constant within each layer, set as 0.0 m as the present-day sea level for the cells from the boundary on the left (column no. 1) to the shoreline (column no. 22), and it gradually rises to 1.52 m at inland boundary on the right, determined by the elevation of Wakulla Spring. Note that the head values are written in the input files of the SEAWAT model instead of the equivalent freshwater head. The initial conditions of salinity are assumed to be a constant value of 35.0 PSU, assuming no freshwater dilution at the sea boundary and the leftmost 10 columns. The seawater–freshwater mixing zone is assumed to be from 35 PSU at column no. 11 to 0 PSU at column no. 45, with a gradient of 1.0 PSU per column. Salinity is set uniformly as 0.0 PSU from column no. 46 to the inland boundary on the right, as uncontaminated freshwater before seawater intrudes. Several test cases have been made to confirm that the initial conditions do not significantly affect the modeling results.

The boundary conditions are also presented in Fig. 3. The less permeable confining unit of the UFA base is simulated at the bottom of the model domain as the no-flow boundary condition. The constant head and concentration inland boundary condition on the right is 1.5 m as the elevation of inland spring and 0.0 PSU as uncontaminated freshwater. The seawater boundary on the left is 3.38 km away from the shoreline, set as 0.0 m constant head as the present-day sea level and 35.0 PSU constant concentration as seawater without mixing. The boundary conditions of head and salinity at the submarine spring (column no. 22, layer no. 1) are adjusted and evaluated in the scenarios of different sea level, salinity, and rainfall conditions in Sect. 5.

Table 1. The symbols and definitions of parameters used in this study, the specified evaluated values in local sensitivity study, and evaluation ranges (the lower and upper constraints) of each parameter in global sensitivity analysis.

Parameter	Definitions	Lower	Upper	Evaluated value	Unit
HY_P	Hydraulic conductivity (porous medium)	1.524	4.572	2.286	($\times 10^3$) m day−1
HY_C	Hydraulic conductivity (conduit)	3.048	9.144	6.096	($\times 10^5$) m day−1
SS_P	Specific storage (porous medium)	4.00	6.00	5.00	($\times 10^{-7}$) dimensionless
SS_C	Specific storage (conduit)	0.03	0.07	0.05	dimensionless
RCH	Recharge rate on the surface	0.00	0.03	0.01	m day−1
H_SL	Sea level at the submarine spring	−0.305	0.914	0.305	meters
PO_P	Porosity (porous medium)	0.001	0.005	0.003	dimensionless
PO_C	Porosity (conduit)	0.200	0.400	0.300	dimensionless
SC	Salinity at the submarine spring	0.0	35.0	35.0	PSU
DISP_P	Longitudinal dispersivity (porous medium)	6.10	12.20	10.00	meters
DISP_C	Longitudinal dispersivity (conduit)	0.15	0.60	0.30	meters

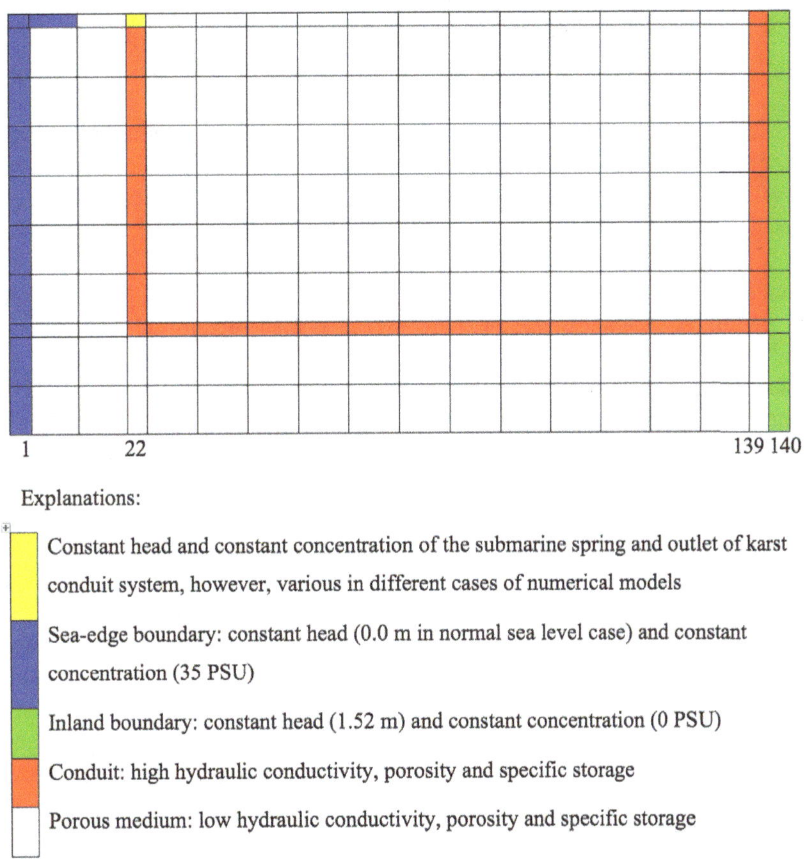

Figure 3. Schematic figure of finite difference grid discretization and boundary conditions applied in the SEAWAT model. Every cell represents 10 horizontal cells and 4 vertical cells, except the boundary and conduit layer in color with smaller width. The submarine spring is located at column no. 22, layer no. 1, and the inland spring is located at column no. 139, layer no. 1. The conduit system starts from the top of column no. 22, descends downward to layer no. 29, horizontally extends to column no. 139, and then rises upward to the top through column no. 139.

4 Sensitivity analysis

Sensitivity analysis evaluates the uncertainties of salinity and head simulations with respect to eleven parameters, and it helps to understand the effects of variations and interactions of aquifer parameters and boundary conditions on simulations. The symbols and definitions of the eleven parameters are listed in Table 1, as well as the values computed in the local sensitivity analysis and the parameter ranges evaluated in the global sensitivity analysis (Table 1). There are six parameters in the groundwater flow model, including hydraulic conductivity (HY_P and HY_C), specific storage (SS_P and SS_C) of the conduit and of the porous medium, recharge rate (RCH), and the sea level at the submarine spring (H_SL). The other five parameters, including effective porosity (PO_P and PO_C), dispersivity (DISP_P and DISP_C) of the conduit and the porous medium, and the salinity at the submarine spring (SC), are in the solute transport model.

4.1 Local sensitivity analysis

In the local sensitivity analysis, the CSS values of parameters with respect to head and salinity simulations are calculated at several locations along the conduit network and the porous medium, respectively. The CSS values are computed for the parameter values in the maximum seawater intrusion benchmark case in Sect. 5.1, which is developed to quantitatively evaluate the extent of seawater intrusion specifically in the WKP after a 7-day low precipitation period. The parameters to be adjusted and evaluated in the scenarios are also determined based on the local sensitivity result.

Parameter sensitivities are computed at several locations, from column no. 25 to column no. 75 with an interval of five cells along the horizontal conduit (layer no. 29), where column no. 25 is close to the shoreline as fully contaminated by seawater, and column no. 75 is assumed to be the uncontaminated freshwater aquifer. The parameter sensitivities of simulations in a porous medium are evaluated at layer no. 24, 15.2 m, or five layers above the conduit layer, from column no. 25 to column no. 75, with an interval of five cells along the horizontal direction.

4.2 Local sensitivity analysis of simulations in the conduit

Figure 4 shows the arithmetic mean of CSS values computed in the evaluated locations along the conduit layer. The largest CSS value indicates that salinity at the submarine spring (SC) is the most important parameter in both salinity and head simulations. Hydraulic conductivity, specific storage, and effective porosity of the conduit (HY_C, SS_C, and PO_C), as well as the sea level at the submarine spring (H_SL), are also important parameters. Simulations are not sensitive to hydraulic conductivity, specific storage, and effective porosity of the porous medium (HY_P, SS_P, and PO_P), recharge rate (RCH), and dispersivity (DISP_C and DISP_P). Generally speaking, the parameter sensitivities with respect to head simulations are similar and consistent with salinity simulations.

The boundary conditions of the conduit system, including salinity and sea level at the submarine spring (SC and H_SL), are important in modeling seawater intrusion in the WKP. Seawater enters the conduit system at the submarine spring and intrudes landward through the subsurface conduit system. The most important parameter is identified as the salinity at the submarine spring (SC), which affects the equivalent freshwater head in terms of water density at the inlet of the conduit system and affects flow simulation within the conduit system. The salinity at the submarine spring (SC) is determined by freshwater mixing and dilution from the conduit network; in other words, it is controlled by rainfall recharge and freshwater discharge from the aquifer to the sea. In this study, rainfall recharge is represented by salinity at the submarine spring with freshwater dilution instead of the recharge flux on the surface (RCH), which is not an important parameter and not applicable to represent the total rainfall recharge in the two-dimensional SEAWAT model. On the other hand, the sea level at the submarine spring (H_SL) has an intermediate CSS, indicating that it is also important in flow field and salinity transport simulations. However, sea level is not as important as the salinity at the submarine spring (SC). In other words, the extent of seawater intrusion in the conduit is more sensitive to rainfall recharge and freshwater discharge represented by the parameter SC, rather than the sea level and/or tide level variations.

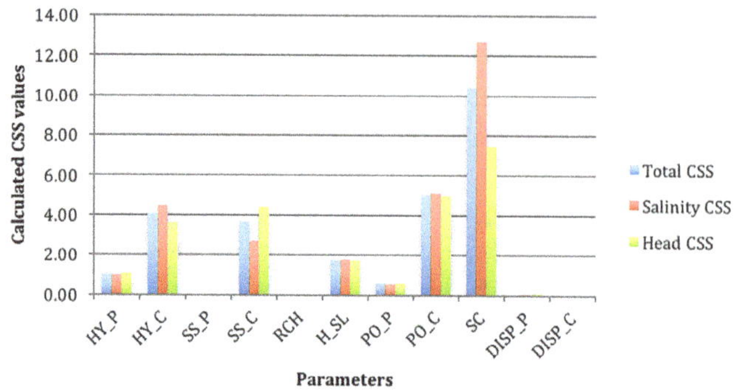

Figure 4. The CSS (composite scaled sensitivity) values of all parameters with respect to simulations in the conduit (layer no. 29) in the local sensitivity analysis.

Dispersivity is usually an important parameter in the sensitivity analysis of transport modeling in a porous medium aquifer (Shoemaker et al., 2004). However, the conduit and porous medium dispersivities (DISP_C and DISP_P) are not evaluated as important parameters in the dual-permeability model in this study. Advection dominates in the transport of seawater in the high permeability conduit network, while dispersion is negligible in such a high-velocity flow condition. Moreover, the dispersion solution and dispersivity sensitivities in the conduit are inaccurately calculated when conduit flow becomes turbulent. On the other hand, the numerical dispersion is significantly greater than the physical dispersion in the conduit. The Peclet number can be as great as 2500, far beyond the theoretical criterion (< 4) for solving the advection dispersion transport equation using the finite

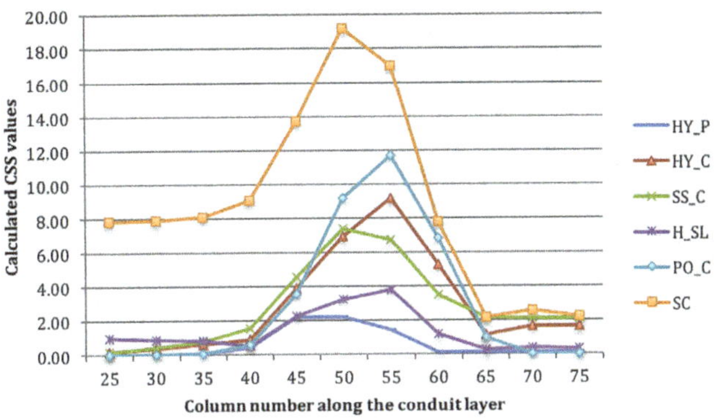

Figure 5. The CSS (composite scaled sensitivity) values of selected parameters at different locations along the conduit layer (from column no. 25 to column no. 75) in the local sensitivity analysis.

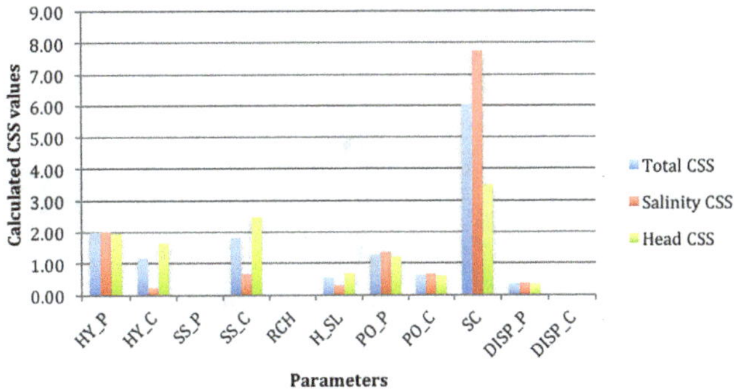

Figure 6. The CSS (composite scaled sensitivity) values of all parameters with respect to simulations in the porous medium (layer no. 24) in the local sensitivity analysis.

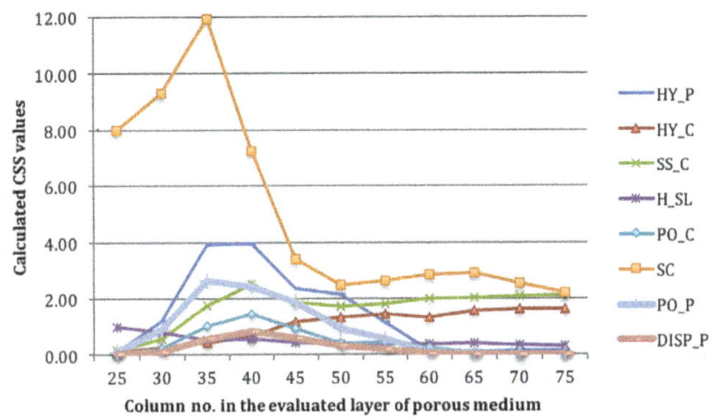

Figure 7. The CSS (composite scaled sensitivity) values at different locations in the porous medium (from column no. 25 to column no. 75 at layer no. 24) in the local sensitivity analysis.

difference method (Zheng and Bennett, 2002). Dispersivity sensitivities have large uncertainty in this study, indicating that the continuum SEAWAT model is not applicable to accurately compute the salinity dispersion in the conduit. An experiment of deactivating the DSP (dispersion) package in SEAWAT confirms that dispersion is negligible within the conduit network in this study. Instead of the dispersion computed by dispersivity, numerical dispersion is the major reason for the range of mixing interface shown in this study.

The parameters with the six largest CSS values are presented in Fig. 5, with respect to the combination of head and salinity simulations in the evaluated locations along the conduit network, from column no. 25 to column no. 75. The largest CSS values are found at either column no. 50 or no. 55 within the conduit, which matches with the position of the seawater–freshwater mixing zone along the conduit network in the maximum seawater intrusion case (Sect. 5.1). The largest CSS values are found at the mixing zone more than anywhere else for all parameters because head and salinity simulations only change significantly near the mixing zone but remain constant in other locations.

4.3 Local sensitivity analysis of simulations in the porous medium

Figure 6 shows the arithmetic mean of CSS values computed in the evaluated locations in the porous medium (layer no. 24). The largest CSS value indicates that salinity at the submarine spring (SC) is also the most important parameter with respect to simulations in the porous medium, although it is a boundary condition of the conduit system. However, some parameter sensitivities exhibit different patterns compared to the results of simulations in the conduit. The hydraulic conductivity and effective porosity of both the conduit and porous medium (HY_ C, HY_P, PO_C and PO_P), specific storage of the conduit (SS_C), and dispersivity of the porous medium (DISP_P), have intermediate CSS values. The CSS values at different evaluated locations along the layer of the porous medium are plotted in Fig. 7, except the three unimportant parameters. Similar to the sensitivity analysis of simulations along the conduit, the largest CSS values are found at either column no. 35 or no. 40, which is the mixing zone position in the porous medium in the maximum seawater intrusion case (Sect. 5.1).

The important rules of the boundary condition and hydrological parameters of the conduit system on simulations in the porous medium are highlighted in the local sensitivity analysis. Salinity at the submarine spring (SC) remains the most important parameter and determines the seawater intrusion plume in the porous medium. The conduit parameters, such as hydraulic conductivity, effective porosity, and specific storage (HY_C, PO_C, and SS_C), are also important to the simulations in the porous medium. The CSS values of conduit parameters indicate that groundwater flow and seawater transport through the conduit system have a significant impact on the simulations in the surrounding porous medium. In summary, simulations in the porous medium are sensitive to both the conduit and porous medium parameters, highlighting the interaction between the two domains in simulating seawater intrusion in the dual-permeability WKP coastal

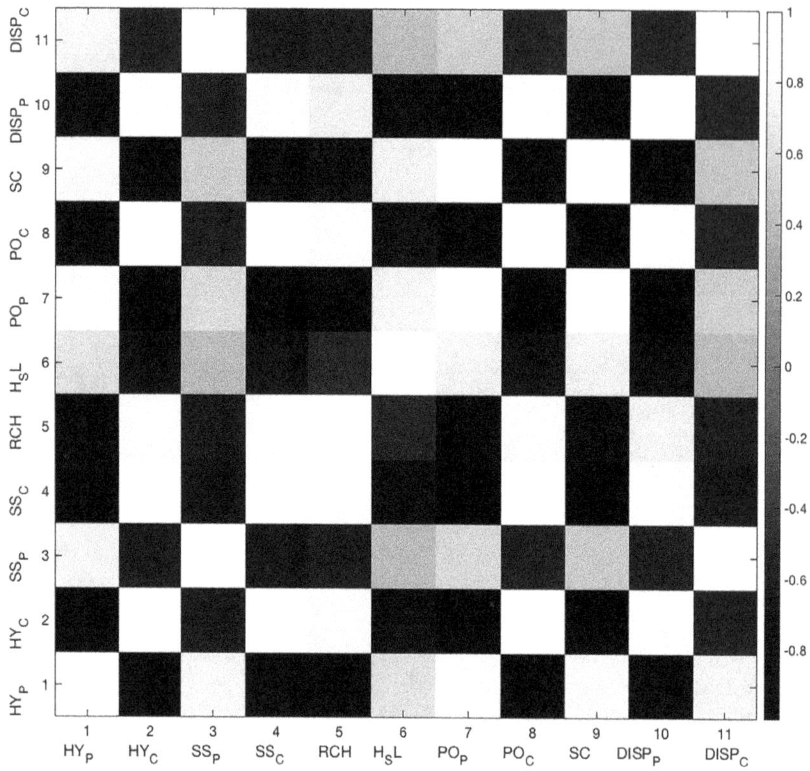

Figure 8. The Pearson pattern correlation coefficient matrix for all eleven parameters.

karst aquifer. As a result, simulations and observations of salinity and head in the conduits and other karst features have significance for calibrating numerical models and values for understanding seawater intrusion.

4.3.1 Parameter correlations

The correlation coefficients and covariance matrix of all parameters are calculated and presented in Fig. 8. The white and black colors represent positive and negative parameter correlations, respectively. Generally speaking, hydrological parameters of the porous medium are positively correlated with the other parameters of the porous medium but negatively correlated with conduit parameters, and vice versa. On the other hand, hydraulic conductivity, specific storage, and porosity have a similar correlation pattern among all evaluated parameters, while the correlation of dispersion is different to others. For example, hydraulic conductivity (HY_P) has a strong positive correlation with specific storage (SS_P) and porosity (PO_P); however, it has a negative correlation with dispersivity (DISP_P). The correlations of conduit parameters exhibit a similar relationship as well. The results can be explained such that a larger hydraulic conductivity would result in higher seepage velocity in either the conduit or the porous medium by Darcy's law; therefore, salt transport from the submarine springs also results in higher salinity in both the conduit porous medium domains. However, larger dispersivity could decrease the peak values of salinity concentration but enlarge contaminant plumes due to stronger dispersion and diffusion.

4.4 Global sensitivity analysis

The local sensitivity analysis analyzes the parameter sensitivities specifically for the seawater intrusion in the WKP, as the maximum seawater intrusion case in Sect. 5.1. However, the local sensitivity result is not representative of the entire parameter range and higher order derivatives of simulations. The global sensitivity analysis is essential to provide a comprehensive understanding of the relationship between simulations and parameters for modeling seawater intrusion in a coastal karst aquifer.

The derivatives of simulations with respect to the selected parameters in Fig. 9 clearly indicate that local sensitivity results are not representative of the entire parameter range. For example, both head and salinity simulations in the conduit are nearly constant in relation to the variation of an unimportant parameter (DISP_P) in the local sensitivity study. However, simulations are nonlinear to salinity at the submarine spring (SC). Parameter SC is identified as the most important parameter in the local sensitivity analysis, partially because the CSS value is computed at the largest derivative value where salinity is 35 PSU.

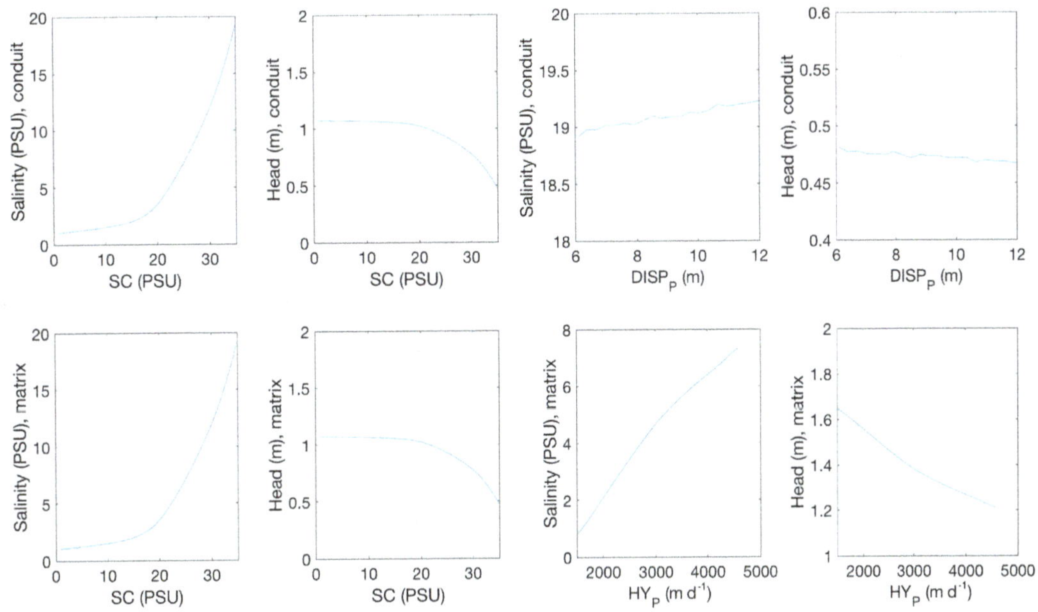

Figure 9. The nonlinear relationship between head and salinity simulations with respect to parameters SC, DISP_P, and HY_P. (Note that the scale for each plot is different.)

The locations in the conduit and porous medium systems with the largest CSS values from the local sensitivity analysis are evaluated in the global sensitivity analysis. Parameter sensitivities are computed at the locations with largest CSS values in the previous local sensitivity analysis, specifically, column no. 50, layer no. 29 in the conduit and column no. 35, layer no. 24 in the porous medium, respectively. The trajectory sampling method developed by Saltelli et al. (2004) is introduced in Sect. 2.2 and applied in the global sensitivity analysis, with the recommended choice of $p = 4$ and $r = 10$ by Saltelli et al. (2004).

4.4.1 Global sensitivity analysis of simulations in the conduit

In the global sensitivity analysis, the mean and standard deviation of the EEs for salinity simulation in the conduit (column no. 50, layer no. 29) are presented in Fig. 10a. Consistent with the local sensitivity analysis, the largest mean value of EEs indicates that parameter SC is the most important parameter in salinity simulations. Parameter SC also has the largest standard deviation of the EEs due to the nonlinear relationship between the salinity simulation and parameter SC shown in Fig. 9, in which the derivatives vary with different parameter values. The hydraulic conductivity and effective porosity of the conduit (HY_C and PO_C), as well as sea level (H_SL), are all important to salinity simulation with relatively large mean and standard deviation values of EEs. Generally speaking, the local and global sensitivity study results for salinity simulation in the conduit are similar; however, the standard deviation of EEs provides additional information of parameter sensitivities in the global sensitivity study.

The global sensitivities for head simulations with respect to parameters are more complicated than salinity simulations (Fig. 9b). The mean and standard deviation of EEs for head simulations are smaller than those for salinity simulations, consistent with the conclusion of Shoemaker (2004) that salinity simulations are more effective than head simulations. The two largest mean values of EEs show that the specific storage (SS_C) and effective porosity (PO_C) of the conduit are the two most important parameters. As mentioned in the local sensitivity analysis, parameters in the transport model are also important to the head simulation in a coupled density-dependent flow and transport model. For example, effective porosity is important in head simulation since the solution of salinity transport in turn determines the density and impact flow calculation in the model, particularly in the study of density-dependent seawater intrusion. In addition, head simulations are also sensitive to the boundary conditions of salinity in the transport model since equivalent freshwater head is a function of density in terms of salinity in the coupled variable-density flow and transport model for simulating seawater intrusion. Different from salinity simulation, salinity at the submarine spring (SC) no longer has the largest mean of EEs. However, the standard deviation of EEs for parameter SC is still the largest due to the nonlinear relationship with head simulation shown in Fig. 9.

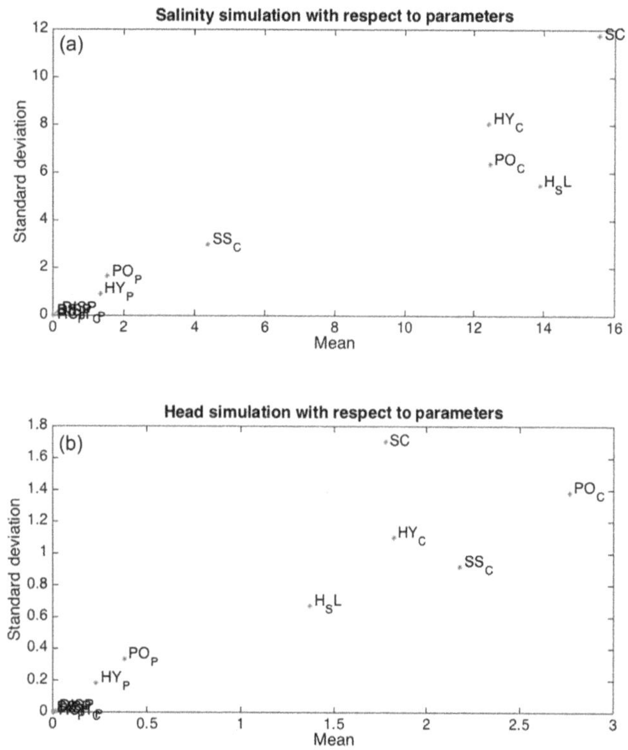

Figure 10. Mean and standard deviation of the EEs (elementary effects) of parameters with respect to simulations in the conduit (column no. 50, layer no. 29) in the global sensitivity analysis by the Morris method: **(a)** salinity simulation (top); **(b)** head simulation (bottom).

One of the major findings in the global sensitivity analysis is that the hydraulic conductivity of the conduit (HY_C) has smaller means and standard deviations of EEs than the other two parameters (PO_C and SS_C), and it no longer becomes the most important parameter as shown in the previous local sensitivity analysis. This is different from common knowledge and empirical experience in hydrogeological modeling but is actually reasonable in the karst aquifer with a non-laminar conduit flow. In the SEAWAT model, the Darcy equation is used to calculate the flow velocity in the whole model domain including the conduit system; however, it is only accurate for laminar seepage flow in the porous medium. Groundwater flow is usually non-laminar and even turbulent in the conduit system when the conduit flow rate is nonlinear to the head gradient and hydraulic conductivity. The simulation of conduit flow is beyond the applicability of the Darcy equation in the SEAWAT model, with relatively large error and uncertainty in the relationship between hydraulic conductivity and head simulation. Then, the uncertainty of hydraulic conductivity sensitivities can be too large and difficult to be accurately measured.

4.4.2 Global sensitivity analysis of simulations in the porous medium

The hydraulic conductivity of the porous medium (HY_P) and salinity at the submarine spring (SC) are identified as the two most important parameters for salinity simulations in the porous medium (Fig. 11a). Compared to parameter HY_P, parameter SC has a much larger CSS value at 35.0 PSU with the largest derivative in the local sensitivity analysis (Fig. 6) and also a larger standard deviation of EEs in the global sensitivity analysis. Local sensitivity analysis overestimates the sensitivity of parameter SC within the range, and global sensitivity analysis provides a more comprehensive understanding of the physical meaning of parameter SC, for example, variability of rainfall recharge and freshwater discharge. As the boundary condition of the conduit system, salinity at the submarine spring (SC) determines the equivalent freshwater head at the inlet of seawater intrusion and affects simulations in the conduit and also the surrounding porous medium via exchanges between the two systems. The global sensitivity results highlight the significance of conduit and porous medium interactions in a dual-permeability aquifer. Similar to salinity at the submarine spring (SC), dynamic interactions between the conduit and the porous medium in this study are clearly shown in the relatively large mean of EEs for sea level (H_SL), effective porosity, and specific storage of the conduit (PO_C and SS_C). Effective porosity is important for head simulations in this study since the density-dependent flow and transport models are coupled for simulating seawater intrusion.

On the other hand, parameter sensitivities for simulations in the porous medium are different to the sensitivities for simulations in the conduit. The porous medium hydraulic conductivity (HY_P) is an important term in the flow equation for solving head and advective velocity for the transport equation (Fig. 11b), similar to most sensitivity results of hydrological modeling for flow in a porous medium. For the simulations in the conduit, effective porosity and specific storage of the conduit (PO_C and SS_C) are more important than hydraulic conductivity (HY_C) because of the large uncertainty in conduit flow computation by Darcy's equation in the continuum SEAWAT model.

5 Seawater intrusion scenarios

In this section, the extents of seawater intrusion are quantitatively measured and evaluated under different scenarios of boundary conditions, which are identified as the important parameters in the local sensitivity analysis. In each scenario, only one parameter is adjusted and others are constant, as in the maximum seawater intrusion benchmark case in Sect. 5.1.

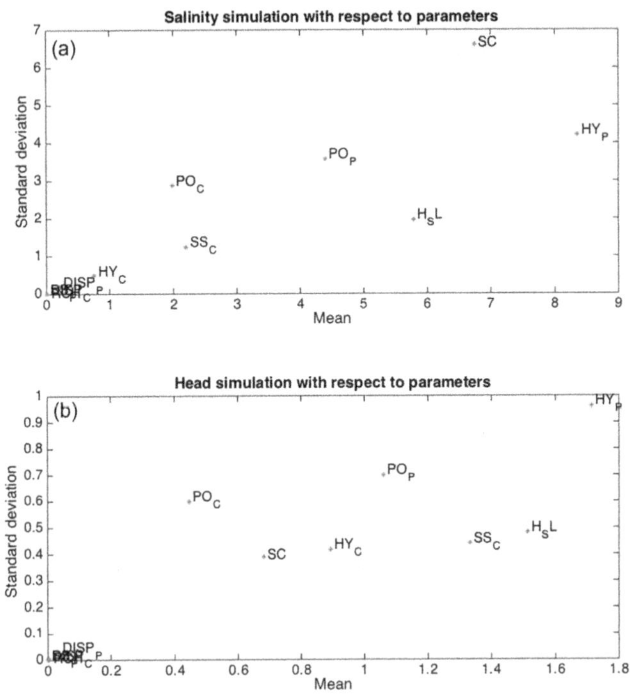

Figure 11. Mean and standard deviation of the EEs (elementary effects) of parameters with respect to simulations in the porous medium (column no. 35, layer no. 24) in the global sensitivity analysis by the Morris method: **(a)** salinity simulation (top); **(b)** head simulation (bottom).

5.1 The maximum seawater intrusion benchmark case

The local sensitivity analysis computes the sensitivities of parameter values in the maximum seawater intrusion benchmark case, which assumes the head and salinity boundary conditions are 0.0 m as the present-day sea level, and 35.0 PSU as seawater without dilution at the conduit system outlet, respectively. Salinity and sea level at the submarine spring (SC and H_SL) are identified as two important parameters and then adjusted in the following two scenarios. In this case, the longest distance of seawater intrusion is simulated with the assumption that freshwater recharge is negligible, and the outlet of conduit system is filled with undiluted seawater. Figure 12 presents the simulated salinity and head profile in the cross section after a 7-day simulation.

According to the Ghyben–Herzberg relationship, landward seawater intrusion is on the bottom of the aquifer beneath the seaward freshwater on the top. The equivalent freshwater head at the submarine spring is calculated as 2.29 m when salinity is 35.0 PSU at the submarine spring, and undiluted seawater is filled within the 91 m deep submarine cave and conduit network. The equivalent freshwater head at the submarine spring is higher than the 1.52 m constant head at the inland spring, diverts the hydraulic gradient landward, and causes seawater to intrude into the aquifer. Seawater intrudes further landward through the highly permeable conduit network; it also contaminates the surrounding porous medium via exchange on the conduit wall. The seawater–freshwater mixing zone in the deep porous medium beneath the conduit is only slightly behind the seawater front in the conduit because high-density saline water easily descends from the conduit and flows downward. The area with relatively smaller salinity to the left of the vertical conduit network nearshore is due to the freshwater discharge dilution from the aquifer to the sea since the equivalent freshwater head is only 2.29 m at the submarine spring but remains 0 m in other areas. The mixing zone position in the conduit, defined as the location with a salinity of 5.0 PSU, is measured at nearly 5.80 km landward from the shoreline. The width of mixing interfaces, defined as the distance between the locations with a salinity of 1.0 and 25.0 PSU, is roughly the same as seven grid cells or 1.13 km in both the conduit and porous medium.

5.2 Salinity variation at the submarine spring (SC)

Sensitivity analysis indicates that the salinity at the submarine spring (SC) is generally the most important parameter for simulations in both the conduit and the porous medium. Salinity at the submarine spring is diluted by a large amount of rainfall recharge and freshwater discharge after a significant precipitation event, but it remains highly saline after an extended low rainfall period, as shown in the maximum seawater intrusion benchmark case in Sect. 5.1. The equivalent freshwater head at the submarine spring is 2.29 m when salinity is 35.0 PSU, and it proportionally decreases to 0.0 m when salinity is 0.0 PSU and freshwater is filled within the conduit system. The impact of freshwater recharge on seawater intrusion is evaluated in four scenarios with salinity levels of 0.0, 10.0, 20.0, and 30.0 PSU at the submarine spring (Fig. 13). The mixing zone in both the conduit and porous medium is measured at 4.0 (4.5) km away from the shoreline in the case of a salinity of 10.0 (20.0) PSU at the submarine spring. Compared to the maximum seawater intrusion benchmark case, rainfall recharge and freshwater discharge dilute seawater intrusion and move the interface significantly seaward. The mixing zone is very close to the shoreline when salinity is 0.0 PSU at the submarine spring and seawater intrusion is blocked by a large amount of freshwater dilution. The shape of the mixing interface is similar to the maximum seawater intrusion benchmark, but the width of the mixing interface is wider due to the slower advective flow with a smaller or even reversed hydraulic gradient from the aquifer to the sea. In the scenarios of freshwater dilution, the solution of dispersion becomes more accurate and important in salinity transport with slower groundwater seepage flow. Generally speaking, a heavy rainfall event dilutes the intruded seawater and moves the mixing interface seaward.

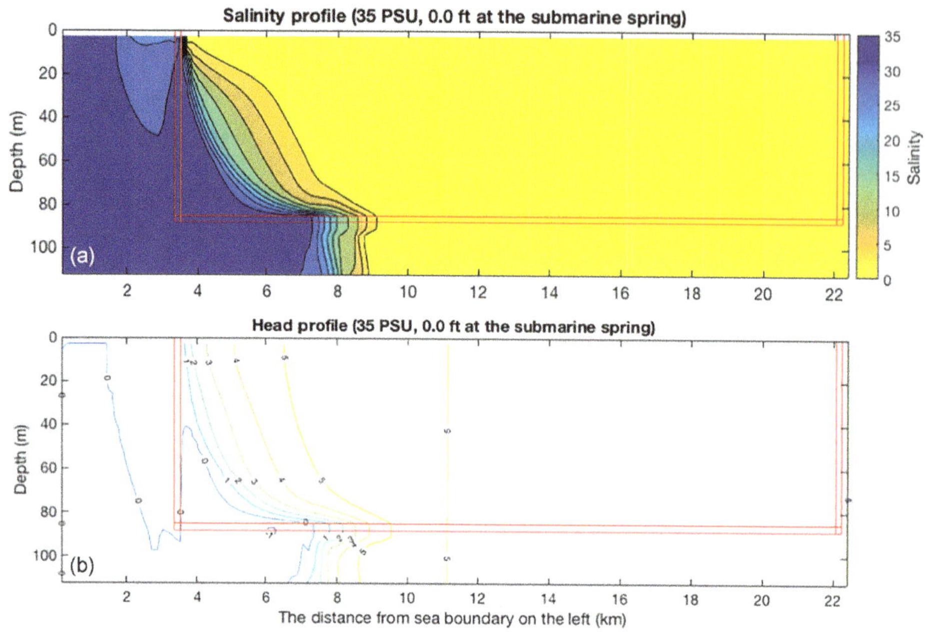

Figure 12. Salinity **(a)** and head **(b)** simulations of the maximum seawater intrusion benchmark case (35 PSU, 0.0 m at the submarine spring).

5.3 Sea level variation at the submarine spring (H_SL)

In addition to salinity, sensitivity analysis indicates that sea level at the submarine spring is also an important parameter. IPCC (2007) predicted an approximation of 1.0 m sea level rise by the end of 21st century, which has significant impacts on seawater intrusion in a coastal karst aquifer. The extents of seawater intrusion in the conduit and porous medium under 0.91 and 1.82 m sea level rise conditions are quantitatively evaluated in this study (Fig. 14). Salinity at the submarine spring remains 35.0 PSU, but the head at the submarine spring increases to simulate rising sea level. The simulated salinity profiles show that the width and shape of the mixing zone are similar to the results in the maximum seawater intrusion benchmark. However, the mixing zone is intruded landward along the conduit to almost 7.08 km from the shoreline with 0.91 m sea level rise, which is 1.28 km further inland than the simulation under present-day sea level. In the other extreme case of 1.82 m sea level rise, seawater intrudes an additional 0.97 km further inland along the conduit than the simulated result with 0.91 m sea level rise, or 2.25 km further landward than the simulation under present-day sea level. Compared to the porous alluvial aquifer, seawater intrudes further landward through the conduit network in the a dual-permeability karst aquifer under sea level rise. This scenario confirms the concerns of severe seawater intrusion in the coastal karst aquifer under sea level rise, and it also highlights the value of the conduit system as the major pathway for long-distance seawater intrusion. In addition, sea level rise might have great impacts on the regional flow field and hydrological conditions in a coastal aquifer. Davis and Verdi (2014) reported an increasing groundwater discharge at the inland Wakulla Spring in the WKP associated with the rising sea level in the past decades. The relationship between spring discharge and sea level was quantitatively simulated by a CFPv2 numerical model in Xu et al. (2015b); however, this is beyond the scope of this study.

5.4 Extended low rainfall period

The elapsed time in simulations is set at constant in the sensitivity analysis and the previous scenarios for consistent comparison purposes. However, extents of seawater intrusion under scenarios of extended low rainfall periods are presented in Fig. 15, with the extended simulated time of 14, 21, and 28 days. The boundary conditions of salinity and sea level at the submarine spring remain 35.0 PSU and 0.0 m, respectively, as the maximum seawater intrusion benchmark.

Seawater persistently intrudes through both the conduit and the porous medium domains during the extended low rainfall period since the 2.29 m equivalent freshwater head at the submarine spring is higher than the inland freshwater boundary. Compared to the maximum seawater intrusion benchmark with a stress period of 7-day elapsed time in the simulation, the mixing zone position moves an additional 1.29 km landward in the conduit and the surrounding porous medium in the 14-day simulation. In the predictions of 21-day (28-day) extended low rainfall period, the mixing zone finally arrives at 7.56 (7.89) km from the shoreline. Above all, seawater intrudes further inland through the conduit network during an extended low rainfall period, contaminates fresh groundwater resources in the aquifer, and becomes an environmental issue in coastal regions.

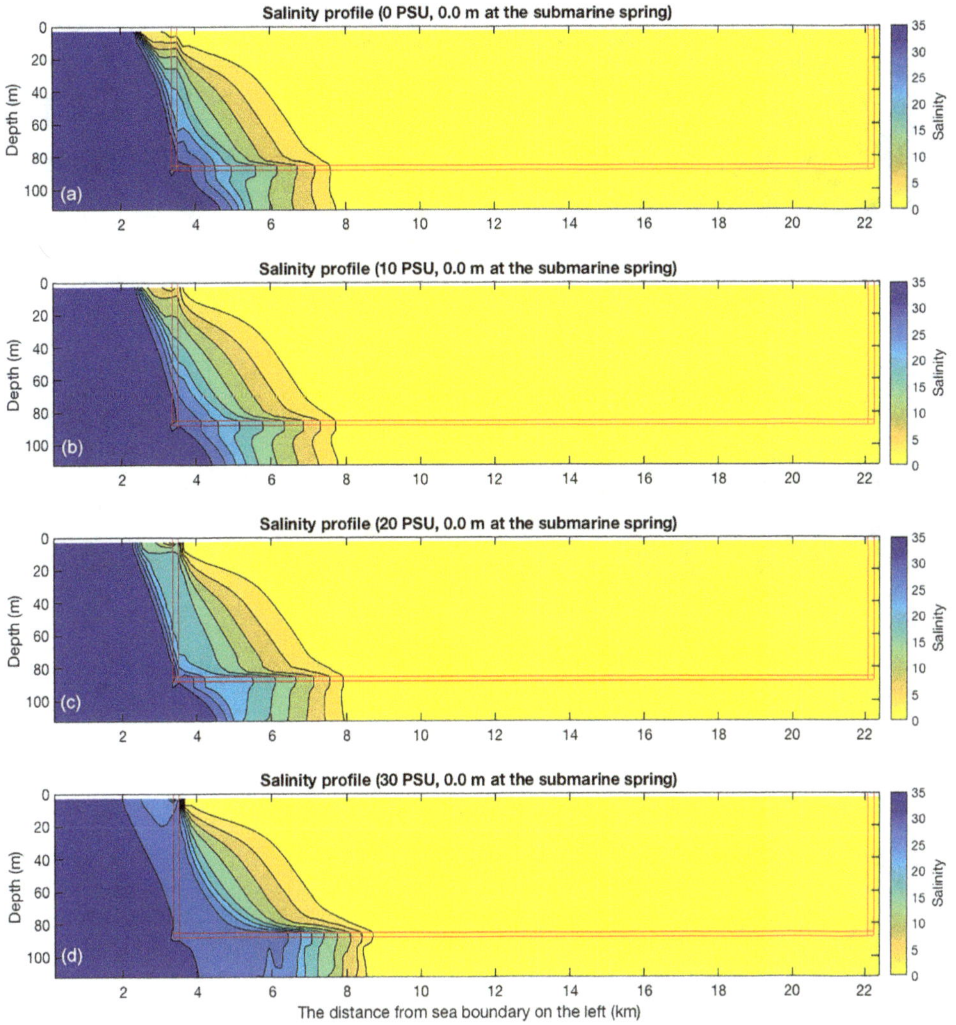

Figure 13. Salinity simulation of seawater intrusion with various salinities at the submarine spring, indicating different rainfall recharge and freshwater discharge conditions: **(a)** 0.0 PSU, 0.0 m at the submarine spring; **(b)** 10.0 PSU, 0.0 m at the submarine spring; **(c)** 20.0 PSU, 0.0 m at the submarine spring; **(d)** 30.0 PSU, 0.0 m at the submarine spring (from top to bottom).

6 Conclusions

In this study, a two-dimensional SEAWAT model is developed to study seawater intrusion in a dual-permeability coastal karst aquifer with a conduit network. Local and global sensitivity analyses are used to evaluate the parameter sensitivities and then understand the roles of hydrogeology conditions and karst features in seawater intrusion. Some major conclusions from the sensitivity analysis are summarized here.

1. Compared to the one specified parameter value computed in the local sensitivity analysis, the global sensitivity analysis is important to accurately estimate the parameter sensitivities in wide ranges, due to the parameter interactions and nonlinear relationship between simulations and parameters shown in Fig. 9. Different from constant-density karst studies, head simulations are sensitive to the boundary conditions and parameters of transport equation since the solution of salinity in terms of density determines the equivalent freshwater head in the coupled density-dependent flow and transport SEAWAT model.

2. Overall, salinity at the submarine spring (SC) is the most important parameter. The boundary conditions and hydrological parameters of the conduit system are not only important to the simulations in the conduit, but also to the porous medium via exchanges between the two systems. The submarine spring and conduit system are the major entrance and pathway, respectively, for seawater intrusion in the coastal karst aquifer. Sensitivity analysis indicates that the simulations of seawater intrusion through the conduit are important for un-

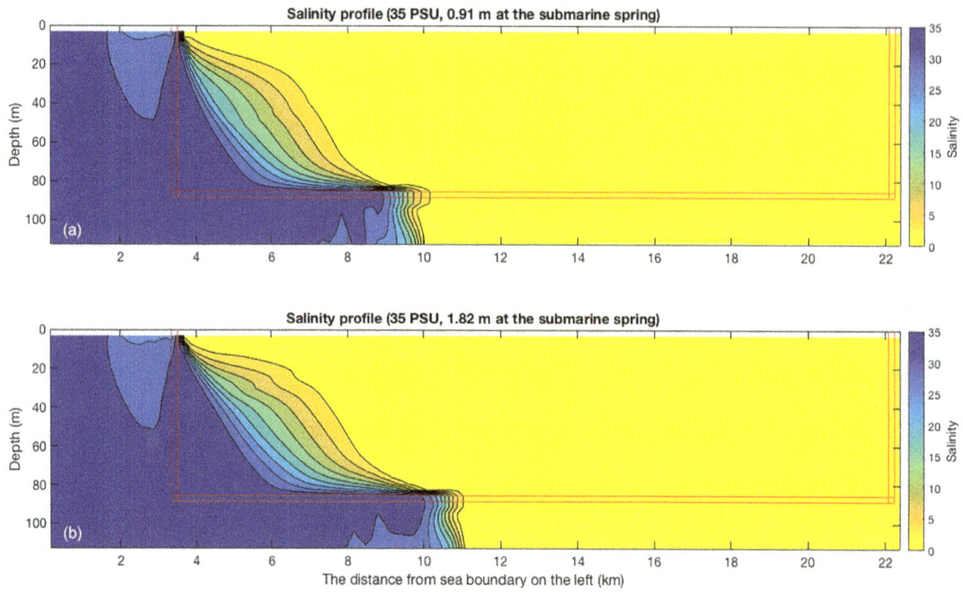

Figure 14. Salinity simulation of seawater intrusion with various sea level conditions: **(a)** 35.0 PSU, 0.91 m at the submarine spring; **(b)** 35.0 PSU, 1.82 m at the submarine spring (from top to bottom).

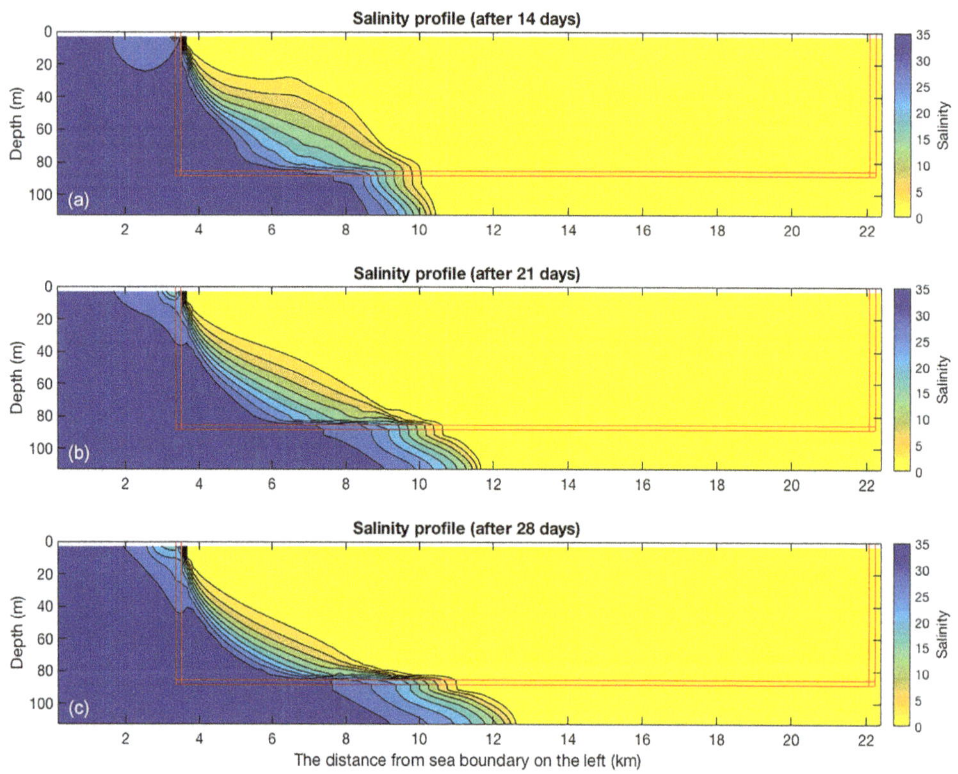

Figure 15. Salinity simulation of the maximum seawater intrusion benchmark case (35 PSU, 0.0 m at the submarine spring) with extended simulation time during a low rainfall period: **(a)** 14-day simulation period; **(b)** 21-day simulation period; **(c)** 28-day simulation period (from top to bottom).

derstanding the hydrogeological processes in the dual-permeability karst aquifer, and field observational data, particularly within the conduit system, are essential for the model calibration.

3. Different from the previous studies in Shoemaker (2004), dispersivity is no longer an important parameter for simulations in the conduit. Advection is dominant but dispersion is negligible in salinity transport under the conditions of turbulent flow in the conduit and also the relatively fast seepage flow in the surrounding porous medium. The interaction between the conduit and the porous medium significantly changes the flow field and affects the applicability of the transport model. In the simulated salinity profile, mixing process is mostly due to numerical dispersion instead of the solution of dispersion equation since the Peclet number is extremely large in the domain and beyond the criteria of solving transport equation by finite difference method.

4. Hydraulic conductivity is no longer an important parameter for simulations in the conduit. Conduit flow is usually non-laminar and beyond the applicability range of the Darcy equation used in the SEAWAT model, which assumes a linear relationship between specific discharge and head gradient. Therefore, the uncertainty and sensitivity of conduit permeability is difficult to be accurately evaluated by hydraulic conductivity in the continuum model.

The extents of seawater intrusion and width of mixing interface are quantitatively measured with a different salinity and sea level at the submarine spring, which are identified as important parameters in the sensitivity study. In the maximum seawater intrusion benchmark case with salinity and head as 35.0 PSU and 0.0 m at the submarine spring, respectively, the mixing zone in the conduit moves to 5.80 km from the shoreline with a width of 1.13 km after a 7-day low rainfall period. Rainfall and regional recharges dilute the salinity at the submarine spring (SC) and significantly shift the mixing zone position seaward to 4.0 (4.5) km away from the shoreline with a salinity of 10.0 (20.0) PSU. Compared to the benchmark, seawater intrudes an additional 1.29 (2.25) km further landward along the conduit under 0.91 (1.82) m sea level rise at the submarine spring (H_SL). In addition, the impacts of extended low rainfall on seawater intrusion through the conduit network are also quantitatively assessed with a longer elapsed time in simulations. The mixing zone moves to 7.56 (7.89) km from the shoreline after a 21 (28)-day low precipitation period.

In summary, the modeling and field observations of the karst features, including the subsurface conduit network, the submarine spring, and karst windows, are critical for understanding seawater intrusion in a coastal karst aquifer and important for model calibration. The discrete-continuum density-dependent flow and transport model, for example, the VDFST-CFP in Xu and Hu (2017), is important to accurately simulate seawater intrusion and assess parameter sensitivities in the coastal karst aquifer with conduit networks. Advanced numerical methods and/or high-performance computing are expected to solve the issue of Peclet number limitation and reduce the uncertainty of the dispersion solution with a finer spatial resolution in this study.

Competing interests. The authors declare that they have no conflict of interest.

Acknowledgements. The authors would like to thank associate editor Mauro Giudici and two anonymous reviewers for their constructive comments and suggestions in the review process. Bill X. Hu is partially supported by the National Natural Science Foundation of China (grant no. 41530316).

Edited by: Mauro Giudici

References

Bear, J.: Seawater intrusion in coastal aquifers, Springer Science & Business Media, 1999.

Calvache, M. and Pulido-Bosch, A.: Effects of geology and human activity on the dynamics of salt-water intrusion in three coastal aquifers in southern Spain, Environ. Geol., 30, 215–223, 1997.

Chang, Y., Wu, J., Jiang, G., and Kang, Z.: Identification of the dominant hydrological process and appropriate model structure of a karst catchment through stepwise simplification of a complex conceptual model, J. Hydrol., 548, 75–87, 2017.

Chen, Z., Hartmann, A., and Goldscheider, N.: A new approach to evaluate spatiotemporal dynamics of controlling parameters in distributed environmental models, Environ. Model. Softw., 87, 1–16, 2017.

Custodio, E.: Salt-fresh water interrelationships under natural conditions, Groundwater Problems in Coastal Areas, UNESCO Studies and Reports in Hydrology, 45, 14–96, 1987.

Davis, J. H.: Hydraulic investigation and simulation of groundwater flow in the Upper Floridan aquifer of north central Florida and southwestern Georgia and delineation of contributing areas for selected City of Tallahassee, Florida, Water-Supply Wells, U.S. Geological Survey, Tallahassee, Florida, 95-4296, 55 pp., 1996.

Davis, J. H. and Katz, B. G.: Hydrogeologic investigation, water chemistry analysis, and model delineation of contributing areas for City of Tallahassee public-supply wells, Tallahassee, Florida, Geological Survey (US)2328-0328, 2007.

Davis, J. H. and Verdi, R.: Groundwater Flow Cycling Between a Submarine Spring and an Inland Fresh Water Spring, Groundwater, 52, 705–716, 2014.

Davis, J. H., Katz, B. G., and Griffin, D. W.: Nitrate-N movement in groundwater from the land application of treated municipal wastewater and other sources in the Wakulla Springs Springshed, Leon and Wakulla counties, Florida, 1966–2018, US Geol Surv Sci Invest Rep, 5099, 90 pp., 2010.

Diersch, H.: FEFLOW reference manual, Institute for Water Resources Planning and Systems Research Ltd, 278, 2002.

Essink, G., Van Baaren, E., and De Louw, P.: Effects of climate change on coastal groundwater systems: a modeling study in the Netherlands, Water Resour. Res., 46, W00F04, https://doi.org/10.1029/2009WR008719, 2010.

Fleury, P., Bakalowicz, M., and de Marsily, G.: Submarine springs and coastal karst aquifers: a review, J. Hydrol., 339, 79–92, 2007.

Gallegos, J. J., Hu, B. X., and Davis, H.: Simulating flow in karst aquifers at laboratory and sub-regional scales using MODFLOW-CFP, Hydrogeol. J., 21, 1749–1760, 2013.

Guo, W. and Langevin, C.: User's guide to SEWAT: a computer program for simulation of three-dimensional variable-density ground-water flow, Water Resources Investigations Report. United States Geological Survey, 2002.

Hartmann, A., Gleeson, T., Rosolem, R., Pianosi, F., Wada, Y., and Wagener, T.: A large-scale simulation model to assess karstic groundwater recharge over Europe and the Mediterranean, Geosci. Model Dev., 8, 1729–1746, https://doi.org/10.5194/gmd-8-1729-2015, 2015.

Hill, M. C. and Tiedeman, C. R.: Effective groundwater model calibration: with analysis of data, sensitivities, predictions, and uncertainty, John Wiley & Sons, 2006.

IPCC: Contribution of Working Groups I, II and III to the Fourth Assessment Report of the Intergovernmental Panel on Climate Change, Geneva, Switzerland, 104 pp., 2007.

Kaufmann, G. and Braun, J.: Karst aquifer evolution in fractured, porous rocks, Water Resour. Res., 36, 1381–1391, 2000.

Kernagis, D. N., McKinlay, C., and Kincaid, T. R.: Dive Logistics of the Turner to Wakulla Cave Traverse, 2008.

Kincaid, T. R. and Werner, C. L.: Conduit Flow Paths and Conduit/Matrix Interactions Defined by Quantitative Groundwater Tracing in the Floridan Aquifer, Sinkholes and the Engineering and Environmental Impacts of Karst: Proceedings of the Eleventh Multidisciplinary Conference, Am. Soc. of Civ. Eng. Geotech. Spec. Publ., 288–302, 2008.

Langevin, C. D., Shoemaker, W. B., and Guo, W.: MODFLOW-2000, the US Geological Survey Modular Ground-Water Model–Documentation of the SEAWAT-2000 Version with the Variable-Density Flow Process (VDF) and the Integrated MT3DMS Transport Process (IMT), US Department of the Interior, US Geological Survey, 2003.

Martin, J. B. and Dean, R. W.: Exchange of water between conduits and matrix in the Floridan aquifer, Chem. Geol., 179, 145–165, 2001.

Martin, J. B., Gulley, J., and Spellman, P.: Tidal pumping of water between Bahamian blue holes, aquifers, and the ocean, J. Hydrol., 416, 28–38, 2012.

Moore, W. S. and Wilson, A. M.: Advective flow through the upper continental shelf driven by storms, buoyancy, and submarine groundwater discharge, Earth Planet. Sc. Lett., 235, 564–576, 2005.

Morris, M. D.: Factorial sampling plans for preliminary computational experiments, Technometrics, 33, 161–174, 1991.

Poeter, E. P. and Hill, M. C.: Documentation of UCODE, a computer code for universal inverse modeling, DIANE Publishing, 1998.

Reimann, T., Liedl, R., Giese, M., Geyer, T., Maréchal, J.-C., Dörfliger, N., Bauer, S., and Birk, S.: Addition and Enhancement of Flow and Transport processes to the MODFLOW-2005 Conduit Flow Process, 2013 NGWA Summit – The National and International Conference on Groundwater, 2013.

Reimann, T., Giese, M., Geyer, T., Liedl, R., Maréchal, J. C., and Shoemaker, W. B.: Representation of water abstraction from a karst conduit with numerical discrete-continuum models, Hydrol. Earth Syst. Sci., 18, 227–241, https://doi.org/10.5194/hess-18-227-2014, 2014.

Saltelli, A., Tarantola, S., Campolongo, F., and Ratto, M.: Sensitivity analysis in practice: a guide to assessing scientific models, John Wiley & Sons, 2004.

Scanlon, B. R., Mace, R. E., Barrett, M. E., and Smith, B.: Can we simulate regional groundwater flow in a karst system using equivalent porous media models? Case study, Barton Springs Edwards aquifer, USA, J. Hydrol., 276, 137–158, 2003.

Shoemaker, W. B.: Important observations and parameters for a salt water intrusion model, Ground Water, 42, 829–840, 2004.

Shoemaker, W. B., Kuniansky, E. L., Birk, S., Bauer, S., and Swain, E. D.: Documentation of a conduit flow process (CFP) for MODFLOW-2005, U.S. Geological Survey Techniques and Methods, Book 6, chap. A24, 50 pp., 2008.

Voss, C. I. and Provost, A. M.: SUTRA, US Geological Survey Water Resources Investigation Reports, 84-4369, 1984.

Voss, C. I. and Souza, W. R.: Variable density flow and solute transport simulation of regional aquifers containing a narrow freshwater–saltwater transition zone, Water Resour. Res., 23, 1851–1866, 1987.

Werner, A. D. and Simmons, C. T.: Impact of sea–level rise on sea water intrusion in coastal aquifers, Groundwater, 47, 197–204, 2009.

Werner, A. D., Bakker, M., Post, V. E., Vandenbohede, A., Lu, C., Ataie-Ashtiani, B., Simmons, C. T., and Barry, D. A.: Seawater intrusion processes, investigation and management: recent advances and future challenges, Adv. Water Resour., 51, 3–26, 2013.

Werner, C. L.: Preferential flow paths in soluble porous media and conduit system development in carbonates of the Woodville Karst Plain, Florida, Master of Science, Department of Earth, Ocean and Atmosphere Science, Florida State University, Tallahassee, FL, 2001.

Wilson, A. M., Moore, W. S., Joye, S. B., Anderson, J. L., and Schutte, C. A.: Storm-driven groundwater flow in a salt marsh, Water Resour. Res., 47, W02535, https://doi.org/10.1029/2010WR009496, 2011.

Xu, Z. and Hu, B. X.: Development of a discrete-continuum VDFST-CFP numerical model for simulating seawater intrusion to a coastal karst aquifer with a conduit system, Water Resour. Res., 53, 688–711, https://doi.org/10.1002/2016WR018758, 2017.

Xu, Z., Hu, B. X., Davis, H., and Cao, J.: Simulating long term nitrate-N contamination processes in the Woodville Karst Plain using CFPv2 with UMT3D, J. Hydrol., 524, 72–88, 2015a.

Xu, Z., Hu, B. X., Davis, H., and Kish, S.: Numerical study of groundwater flow cycling controlled by seawater/freshwater interaction in a coastal karst aquifer through conduit network using CFPv2, J. Contam. Hydrol., 182, 131–145, 2015b.

Xu, Z., Bassett, S. W., Hu, B., and Dyer, S. B.: Long distance seawater intrusion through a karst conduit network in the Woodville Karst Plain, Florida, Scientific Reports, 6, 2016.

Zheng, C. and Bennett, G. D.: Applied contaminant transport modeling, Wiley-Interscience New York, 2002.

10

Bathymetry observations of inland water bodies using a tethered single-beam sonar controlled by an unmanned aerial vehicle

Filippo Bandini[1], Daniel Olesen[2], Jakob Jakobsen[2], Cecile Marie Margaretha Kittel[1], Sheng Wang[1], Monica Garcia[1], and Peter Bauer-Gottwein[1]

[1]Department of Environmental Engineering, Technical University of Denmark, Kgs. Lyngby, Denmark
[2]National Space Institute, Technical University of Denmark, Kgs. Lyngby, 2800, Denmark

Correspondence: Filippo Bandini (fban@env.dtu.dk)

Abstract. High-quality bathymetric maps of inland water bodies are a common requirement for hydraulic engineering and hydrological science applications. Remote sensing methods, such as space-borne and airborne multispectral imaging or lidar, have been developed to estimate water depth, but are ineffective for most inland water bodies, because of the attenuation of electromagnetic radiation in water, especially under turbid conditions. Surveys conducted with boats equipped with sonars can retrieve accurate water depths, but are expensive, time-consuming, and unsuitable for unnavigable water bodies.

We develop and assess a novel approach to retrieve accurate and high-resolution bathymetry maps. We measured accurate water depths using a tethered floating sonar controlled by an unmanned aerial vehicle (UAV) in a lake and in two different rivers located in Denmark. The developed technique combines the advantages of remote sensing with the potential of bathymetric sonars. UAV surveys can be conducted also in unnavigable, inaccessible, or remote water bodies. The tethered sonar can measure bathymetry with an accuracy of $\sim 2.1\,\%$ of the actual depth for observations up to 35 m, without being significantly affected by water turbidity, bed form, or bed material.

1 Introduction

Accurate topographic data from the riverbed and floodplain areas are crucial elements in hydrodynamic models. Detailed bathymetry maps of inland water bodies are essential for simulating flow dynamics and forecasting flood hazard (Conner and Tonina, 2014; Gichamo et al., 2012; Schäppi et al., 2010), predicting sediment transport and streambed morphological evolution (Manley and Singer, 2008; Nitsche et al., 2007; Rovira et al., 2005; Snellen et al., 2011), and monitoring instream habitats (Brown and Blondel, 2009; Powers et al., 2015; Strayer et al., 2006; Walker and Alford, 2016). Whereas exposed floodplain areas can be directly monitored from aerial surveys, riverbed topography is not directly observable from airborne or space-borne methods (Alsdorf et al., 2007). Thus, there is a widespread global deficiency in bathymetry measurements of rivers and lakes.

Within the electromagnetic spectrum, visible wavelengths have the greatest atmospheric transmittance and the smallest attenuation in water. Therefore, remote sensing imagery from satellites, such as Landsat (Liceaga-Correa and Euan-Avila, 2002), QuickBird (Lyons et al., 2011), IKONOS (Stumpf et al., 2003), WorldView-2 (Hamylton et al., 2015; Lee et al., 2011), and aircrafts (Carbonneau et al., 2006; Marcus et al., 2003), has been used to monitor the bathymetry of inland water bodies. However, bathymetry can only be derived from optical imagery when water is very clear and shallow, the sediment is comparatively homogeneous, and atmospheric conditions are favorable (Legleiter et al., 2009; Lyzenga, 1981; Lyzenga et al., 2006; Overstreet and Legleiter, 2017).

Thus, passive remote sensing applications are limited to shallow gravel-bed rivers, in which water depth is on the order of the Secchi depth (depth at which a Secchi disk is no longer visible from the surface).

Similarly, airborne lidars operating with a green wavelength can be applied to retrieve bathymetry maps (Bailly et al., 2010; Hilldale and Raff, 2008; Legleiter, 2012), but also this method is limited by water turbidity, which severely restricts the maximum depth to generally 2–3 times the Secchi depth (Guenther, 2001; Guenther et al., 2000).

Because of satellite or aircraft remote sensing limitations, field surveys, which are expensive and labor intensive, are normally required to obtain accurate bathymetric cross sections of river channels. Some preliminary tests using a green wavelength ($\lambda = 532$ nm) terrestrial laser scanning (TLS) for surveying submerged areas were performed (Smith et al., 2012; Smith and Vericat, 2014). However, TLS suffers from similar limitations as lidar. Furthermore, the highly oblique scan angles of TLS make refraction effects more problematic (Woodget et al., 2015) and decrease returns from the bottom while increasing returns from the water surface (Bangen et al., 2014). Therefore, field surveys are normally performed using single-beam or multi-beam swath sonars transported on manned boats or more recently on unmanned vessels (e.g., Brown et al., 2010; Ferreira et al., 2009; Giordano et al., 2015) . However, boats cannot be employed along unnavigable rivers and require sufficient water depth for navigation.

Unmanned aerial vehicles (UAVs) offer the advantage of enabling a rapid characterization of water bodies in areas that may be difficult to access by human operators (Tauro et al., 2015b). Bathymetry studies using UAVs are so far restricted to (i) spectral signature-depth correlation based on passive optical imagery (Flener et al., 2013; Lejot et al., 2007) or (ii) DEM (digital elevation model) generation through stereoscopic techniques from through-water pictures, correcting for the refractive index of water (Bagheri et al., 2015; Dietrich, 2016; Tamminga et al., 2014; Woodget et al., 2015).

The high cost, size, and weight of bathymetric lidars severely limit their implementation on UAVs. An exception is the novel topo-bathymetric laser profiler, bathymetric depth finder BDF-1 (Mandlburger et al., 2016). This lidar profiler can retrieve measurements of up to 1–1.5 times the Secchi depth; thus, it is only suitable for shallow gravel-bed water bodies. The system weighs ~ 5.3 kg and requires a large UAV platform (e.g., multi-copters with a weight of ~ 25 kg).

To overcome these limitations, we assess a new operational method to estimate river bathymetry in deep and turbid rivers. This new technique involves deploying an off-the-shelf, floating sonar, tethered to and controlled by a UAV. With this technique we can combine (i) the advantages of UAVs in terms of the ability to survey remote, dangerous, unnavigable areas with (ii) the capability of bathymetric sonars to measure bathymetry in deep and turbid inland water bodies.

Figure 1. Pictures of the UAV and the tethered sonar. These pictures were retrieved in **(a)** Marrebæk Kanal, Denmark; and **(b)** Furesø lake, Sjælland, Denmark. In **(b)** the drone was flown a few hundred meters from the shore and the picture was retrieved using an optical camera onboard an auxiliary UAV (DJI Mavic Pro).

UAV measurements of water depth (i.e., elevation of the water surface above the bed) can enrich the set of available hydrological observations along with measurements of water surface elevation (WSE), i.e., elevation of the water surface above sea level (Bandini et al., 2017; Ridolfi and Manciola, 2018; Woodget et al., 2015), and surface water flow (Detert and Weitbrecht, 2015; Tauro et al., 2015a, 2016; Virili et al., 2015).

2 Materials and methods

The UAV used for this study was the off-the-shelf DJI hexacopter Spreading Wings S900 equipped with a DJI A-2 flight controller.

2.1 UAV payload

The UAV was equipped with a Global Navigation Satellite System (GNSS) receiver for retrieving accurate position, an inertial measurement unit (IMU) to retrieve angular and linear motion, and a radar system to measure the range to water surface. A picture of the UAV and the tethered sonar is shown in Fig. 1.

The onboard GNSS system is a NovAtel receiver (OEM628 board) with an Antcom (3G0XX16A4-XT-1-4-Cert) dual frequency GPS and GLONASS flight antenna. The UAV horizontal and vertical position is estimated with $\sim$ 2–3 cm accuracy in carrier-phase differential GPS mode. The onboard inertial measurement unit is an Xsense MTi-10 series. The optical camera is a SONY RX-100 camera. The radar is an ARS 30X radar developed by Continental. The radar and GNSS systems are the same instrumentation as described in Bandini et al. (2017), in which WSE was measured by subtracting the range measured by the radar (range between the UAV and the water surface) from the altitude observed by the GNSS instrumentation (i.e., altitude above ref-

Figure 2. Deeper sonar is connected to a UAV with a wire winch.

erence ellipsoid, convertible into altitude above geoid level). In this research, the radar and GNSS instrumentation are used to (i) retrieve WSE and (ii) observe the accurate position of the tethered sonar.

2.2 Sonar instrumentation

The sonar used for this study was the Deeper Smart Sensor PRO+ manufactured by the company Deeper, UAB (Vilnius, Lithuania). It costs ~ USD 240 and weighs ~ 100 g.

The sonar is tethered to the UAV with a physical wire connection as shown in Fig. 2. For specific applications, the sonar can be lowered or raised using a remotely controlled lightweight wire winch, as shown in Fig. 2. The maximum extension of the wire was ~ 5 m. A remotely controlled emergency hook can be installed to release the sonar in case of emergency (e.g., if the wire is caught in obstacles).

This sonar is a single-beam echo sounder with two frequencies: 290 and 90 kHz, with 15 and 55° beam angles, respectively. The 90 kHz frequency is developed to locate fish with a large scanning angle, while the narrow field of view of the 290 kHz frequency gives the highest bathymetric accuracy. For this reason, the 290 kHz frequency is used for observing the bottom structure. The sonar is capable of measuring depths up to 80 m and has a minimum measuring depth of 0.3–0.5 m depending on the substrate material. The 15° beam angle of the 290 kHz frequency results in a ground footprint of ~ 26 cm at 1 m water depth. This footprint is not optimal for resolving small-scale features at large water depths.

The observations retrieved by the sonar include time, approximate geographical coordinates of the sonar, sonar depth measurements (including waveform shape), size and depth of identified fish, and water temperature. It is essential to analyze multiple echo returns to identify the actual water depth, especially in shallow water. Indeed, when a sound pulse returns from the bottom, only a very small part of the echo hits the receiving transducer. The major portion hits the wa-

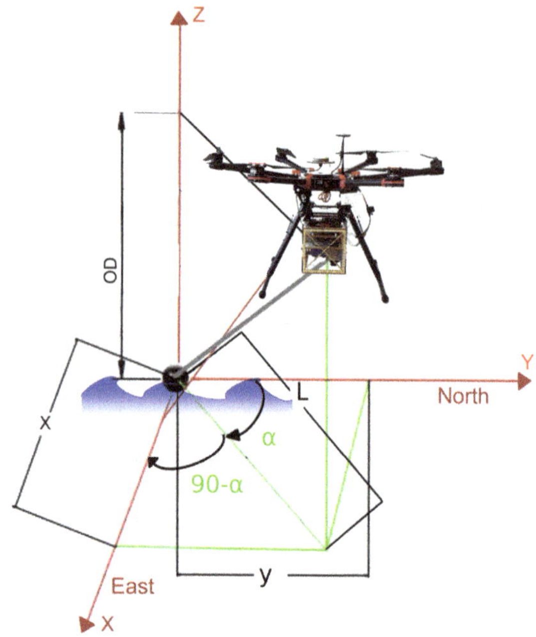

Figure 3. Sonar is the center of the reference system X, Y, Z. The horizontal displacement between the sonar and the drone is computed along the X and Y directions, while the vertical displacement is computed along the Z axis (object distance – OD). The angle α is the azimuth, i.e., the angle between the Y axis pointing north and the vector between the drone and the sonar, projected onto the horizontal plane (in green). The azimuth angle is measured clockwise from north (i.e., α is positive in the figure).

ter surface and is reflected back to the bottom of the water body. From the bottom, it is reflected upwards again and hits the receiving transducer a second time. In shallow water, this double-path reflection is strong enough to generate a second echo that must be filtered out.

The sonar has a built-in GPS receiver to identify its approximate location. However, the accuracy of this GPS is several meters (up to 30 m). The large error of this single frequency GPS receiver is related to many different factors, including disturbance of the GNSS signal by water beneath the sonar, the drone, and the topography surrounding the water body. The accuracy of either GPS option is suboptimal for the generation of bathymetry maps; thus, more accurate measurements of the sonar position are necessary. The drone absolute position is accurately known through the differential GNSS system described in Bandini et al. (2017). In order to estimate the relative position of the sonar with respect to the drone, the payload system measures the offset and orientation of the sonar. This concept is described in Fig. 3.

The displacement between the sonar and the principal point of the onboard camera sensor is denoted by the variables x and y, in which x measures the displacement along the east direction and y along the north direction. As shown in Fig. 3, the azimuth angle is necessary to compute the sonar displacement in Cartesian coordinates.

The horizontal displacement between the sonar and the onboard camera can be estimated with the observations from the different sensors comprising the drone payload: (i) the GNSS system (to measure drone absolute coordinates), (ii) optical camera (to measure displacement of the sonar with respect to the drone in pixel units), (iii) radar (to convert the displacement from pixels to metric units), and (iv) IMU (to project this displacement into the east and north direction). In this framework, the optical SONY camera continuously captures pictures (with focus set to infinity) of the underlying water surface to estimate the sonar position. Lens distortion needs to be corrected for because the SONY RX-100 camera is not a metric camera. Numerous methods have been discussed in the literature to correct for lens distortion (e.g., Brown, 1971; Clarke and Fryer, 1998; Faig, 1975; Weng et al., 1992). In this research the software PTLens was used to remove lens radial distortion because the lens parameters of the SONY RX-100 camera are included in the software database. The displacement of the sonar with respect to the camera principal point can be measured in pixels along the vertical and horizontal axis of the image. This displacement in pixels is converted into metric units through Eqs. (1)–(4). A representation of the variables contained in these equations is given in Fig. 4. Application of Eqs. (1) and (2) requires the following input parameters: the sensor width (Wsens) and sensor height (Hsens), the focal length (F), and the object distance (OD). OD is the vertical range to the water surface and is measured by the radar. Equations (1) and (2) compute the width (WFOV) and height (HFOV) of the field of view.

$$\mathrm{WFOV} = \mathrm{Wsens} \cdot \frac{\mathrm{OD}}{F} \tag{1}$$

$$\mathrm{HFOV} = \mathrm{Hsens} \cdot \frac{\mathrm{OD}}{F} \tag{2}$$

Equations (3) and (4) compute the displacement, in metric unit, between the sonar and the center of the camera sensor along the horizontal (Lw) and vertical (Lh) axis of the picture. Application of Eqs. (3) and (4) requires the following input parameters: the width (WFOV) and height (HFOV) of the field of view, the sensor resolution in pixels along the horizontal ($n_{\mathrm{pix_w}}$) and the vertical ($n_{\mathrm{pix_h}}$) direction, and the measured distance in pixels between the sonar and the center of the image along the horizontal ($\mathrm{pix_w}$) and vertical ($\mathrm{pix_h}$) image axis.

$$\mathrm{Lw} = \frac{\mathrm{WFOV}}{n_{\mathrm{pix_w}}} \cdot \mathrm{pix_w} \tag{3}$$

$$\mathrm{Lh} = \frac{\mathrm{HFOV}}{n_{\mathrm{pix_h}}} \cdot \mathrm{pix_h} \tag{4}$$

The magnitude of the displacement vector, $\boldsymbol{L}$, between the sonar and the camera principal point and the angle φ (angle between the camera vertical axis and the displacement vector) are computed through Eqs. (5) and (6). Figure 5 shows a picture retrieved by the camera. In the current payload setup, the vertical axis of the camera is aligned with the drone nose (heading).

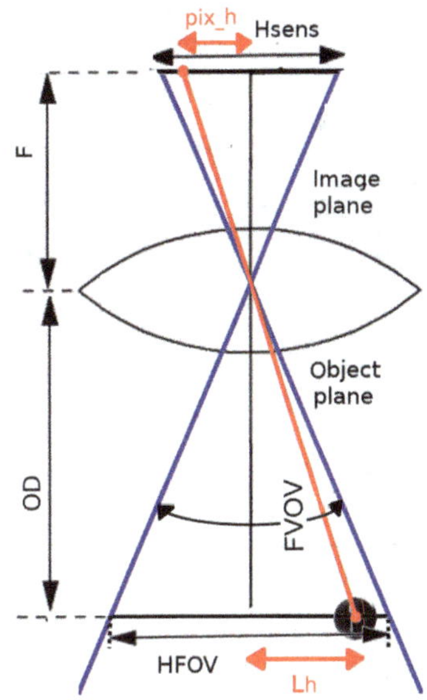

Figure 4. Relationship between FOV (field of view in degrees), HFOV (height of the field of view, in metric unit), OD (object distance), F (focal length), $\mathrm{pix_h}$ (distance in pixels between center of the image and the object in the image, along the vertical axis of the image captured by the camera), Hsens (sensor height), and Lh (distance in metric units between the object and center of the sensor along the vertical axis of the image). The drawing is valid under the assumption that the image distance (distance from the rear nodal point of the lens to the image plane) corresponds to the focal length.

$$\boldsymbol{L} = \sqrt{\mathrm{Lw}^2 + \mathrm{Lh}^2} \tag{5}$$

$$\varphi = \tan^{-1}\frac{\mathrm{Lw}}{\mathrm{Lh}} \tag{6}$$

The azimuth angle α of the sonar is computed through Eq. (7), which requires φ and β as inputs. The symbol β denotes the drone heading (angle between the drone's nose and the direction of the true north, measured clockwise from north). This heading angle is measured by the onboard IMU system.

$$\alpha = \beta + \varphi \tag{7}$$

Equations (8) and (9) compute the variables x and y, which represent the displacement of the sonar with respect to the principal point of the onboard camera sensor along the east and north direction, respectively.

$$x = \cos(90 - \alpha) \cdot \boldsymbol{L} \tag{8}$$

$$y = \sin(90 - \alpha) \cdot \boldsymbol{L} \tag{9}$$

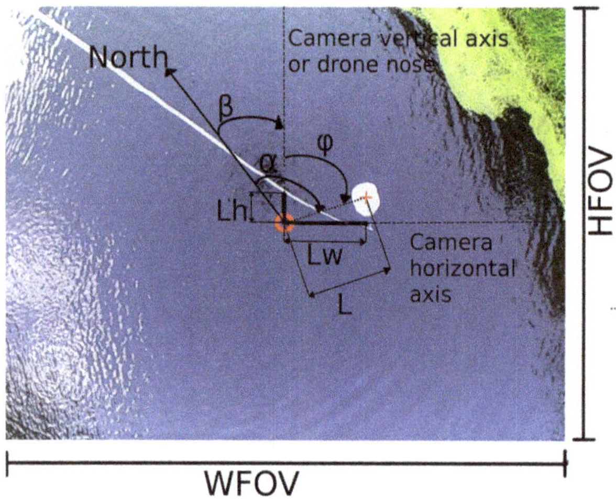

Figure 5. UAV-borne picture of the tethered sonar. WFOV and HFOV are the width and height of the field of view. The tethered sonar is located below the white polyester board floating on the water surface. The red circle indicates the center of the image, while the red cross indicates the exact position of the sonar. The north direction is retrieved by the IMU. The vertical axis of each image captured by the camera coincides with the drone heading. β is the angle between the drone heading and the north. φ is the angle measured clockwise from the camera vertical axis to the vector ($\boldsymbol{L}$), which is the vector on the horizontal plane connecting the sonar to the image center. Lh and Lw are the vertical and horizontal components of the vector $\boldsymbol{L}$. α is the azimuth angle measured clockwise from the north direction to $\boldsymbol{L}$. Angles and vectors highlighted in this figure are on the horizontal plane, i.e., on the water surface.

The absolute position of the drone is simultaneously retrieved by the GNSS antenna installed on the top of the drone. The offset between the sensor of the camera onboard the drone and the phase center of the GNSS antenna position is constant and known a priori. This offset vector also needs to be converted to spatial real-world coordinates at each time increment accounting for the drone heading. Using this framework, the absolute sonar position can be computed in Cartesian coordinates by summing the relative displacement x and y to the camera absolute position.

2.3 Case studies

First, the accuracy of the water depth measured with the sonar was assessed against measurements obtained by the survey boat. Secondly, UAV surveys were conducted to evaluate the accuracy of the depth measured by the sonar and the accuracy of the sonar position.

2.3.1 On boat accuracy evaluation

A bathymetric survey was conducted on a boat in the Furesø lake, Denmark. A second reference sonar, the Airmar EchoRange SS510 Smart Sensor (developed by Airmar, Milford, USA), was deployed to assess the accuracy of the Deeper sonar. According to the technical data sheet, the SS510 Smart Sensor weighs around 1.3 kg, has a resolution of 3 cm, a 9° beam angle, a measuring range from 0.4 to 200 m, and nominal accuracy 0.25 % in depth measurements at full range. The horizontal positions of the sonars during the surveys were acquired with a real-time kinematic (RTK) GNSS rover installed on the boat.

During this survey, ground truth depth measurements were retrieved in selected locations to validate the observations of the two sonars. Ground truth measurements were retrieved using a measuring system consisting of a heavy weight ($\sim$ 5 kg) attached to an accurate measuring tape. This reference system has an accuracy of $\sim$ 10–15 cm in water depth up to 40 m.

2.3.2 UAV-borne measurements

Flights were conducted in Denmark (DK) above Furesø lake (Sjælland, DK), above Marrebæk Kanal (Falster, DK), and Åmose Å (Sjælland, DK).

The flights above Furesø demonstrate the potential of the airborne technology for retrieving measurements at a line-of-sight distance of a few hundred meters from the shore. The flight above Marrebæk Kanal demonstrates the possibility of retrieving accurate river cross sections, which can potentially be used to inform hydrodynamic river models. The flight above Åmose Å shows the possibility to retrieve observations with high spatial resolution, enabling the construction of bathymetric maps of entire river stretches. The accuracy of the observed river cross sections is evaluated by comparison with ground truth observations. Ground truth observations of the river cross sections were obtained by a manual operator wading into the river and taking measurements with a RTK GNSS rover of (i) the orthometric height of the river bottom and (ii) the WSE. Ground truth depth was then computed by subtracting the orthometric height of the bottom from the WSE measurements.

3 Results

3.1 On boat sonar accuracy

Figure 6 shows the measurements retrieved by the two sonars in the lake. The background map is from Google Earth. WSE retrieved by the RTK GNSS station was 20.40 ± 0.05 m a.s.l. (above sea level) during this survey.

The maximum water depth retrieved during the survey is $\sim$ 36 m. In Fig. 7, we report the difference between the observations retrieved by the SS510 and the Deeper sonar.

Figure 7 shows high consistency between the two sonars. However, littoral areas with dense submerged vegetation show larger errors. While in the deepest area ($\sim$ 30 m deep) the Deeper sonar observed multiple returns of the sound wave caused by suspended sediments, the analysis of the waveform was more complicated and subject to errors. In

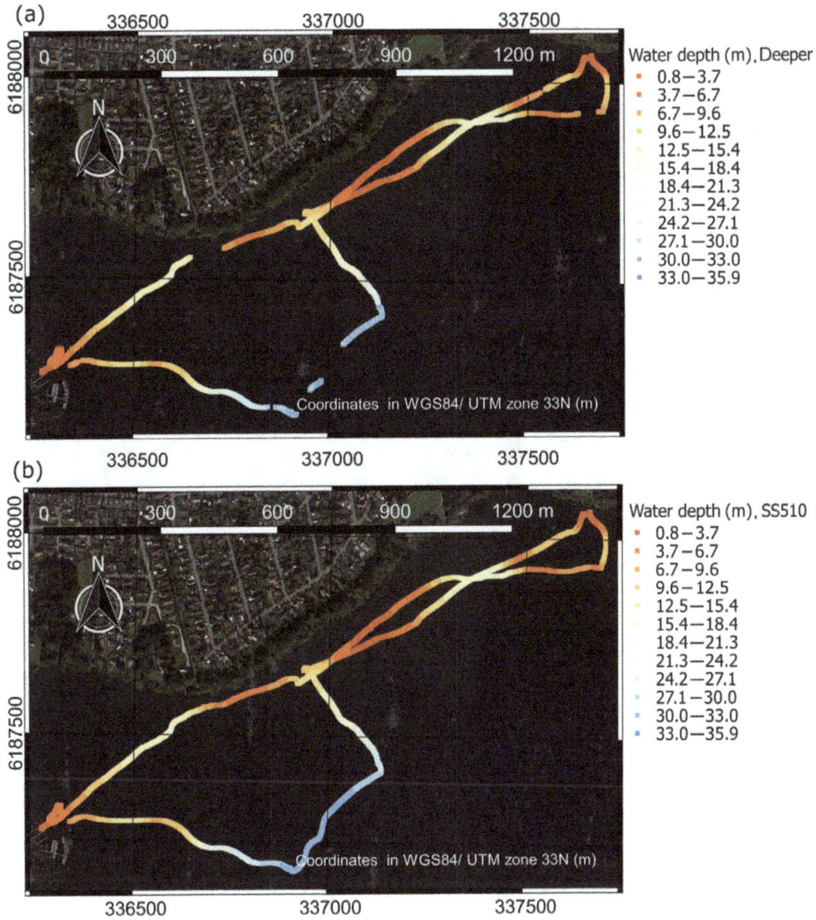

Figure 6. Water depth measurements retrieved in Furesø by the two sonars: **(a)** observations with Deeper sonar; **(b)** observations with SS510 sonar.

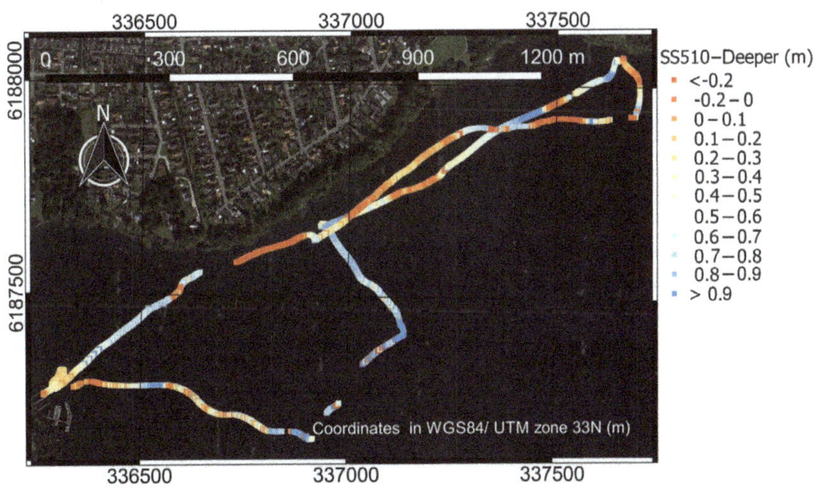

Figure 7. Difference between water depth measured by SS510 sonar and the Deeper sonar.

Table 1. Statistics comparing the Deeper sonar, SS510 sonar, and ground truth observations.

Statistics	Sample size	Root mean square error (RMSE) (m)	Mean absolute error (m)	Mean bias error (MBE) (m)	Relative error (%)
SS510 sonar – Deeper sonar*	57 528	0.38	0.32	0.27	3.70 %
Before bias correction					
Deeper sonar – ground truth	5	0.58	0.52	0.48	3.80 %
SS510 sonar – ground truth	5	0.675	0.56	0.56	3.65 %
After bias correction					
Deeper sonar – ground truth	5	0.12	0.11	5×10^{-4}	2.10 %
SS510 sonar – ground truth	5	0.052	0.047	−0.01	0.57 %

* Statistics computed after removing outliers (above the 95th percentile and below the 5th percentile).

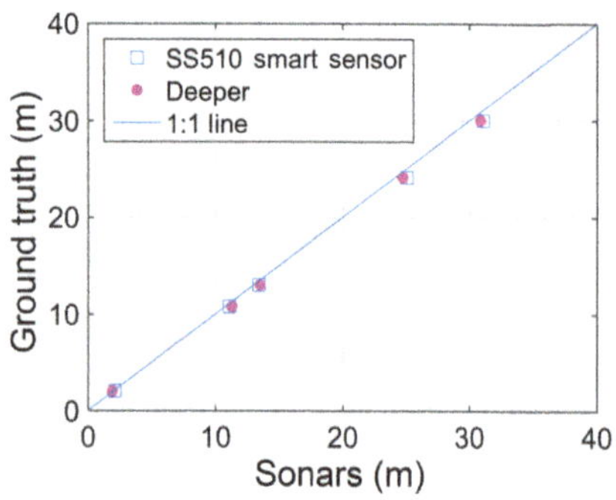

Figure 8. Relationship between measurements of two sonars and ground truth.

this area, the Deeper sonar fails to retrieve some water depth observations, where the waveform analysis does not show a well-defined strong return echo.

The observations retrieved by the two sonars are compared with ground truth observations in Fig. 8.

Figure 8 depicts a systematic overestimation of water depth by both sensors. The relationship between the observations of the two sonar sensors (x) and ground truth (y) can be described with a linear regression of the form shown in Eq. (10), in which β_0 is the offset (y intercept), β_1 is the slope, and ε is a random error term:

$$y = \beta_0 + \beta_1 x + \varepsilon. \tag{10}$$

This survey showed an offset of zero. Thus, the bias between the ground truth observations and the sonar observations can be corrected by multiplying the sonar observations by β_1. Linear regression lines can be fitted to the observations shown in Fig. 8 with a R^2 of ~ 0.99. Appendix A shortly describes how physical variables (such as depth, salinity, and temperature) can affect water depth observations using sonars.

Table 1 shows comparative statistics between the Deeper, the SS510 sonar, and the ground truth observations.

Table 1 shows a difference of ~ 30 cm between the measurements of the two sonars, with the Deeper sonar generally underestimating water depth. This can be due to the wider scanning angle of the Deeper (15°) compared to the SS510 sonar (9°). The Deeper sonar is more affected by steep slopes, in which the depth tends to represent the most shallow point in the beam because of the larger scanning angle. The Deeper and SS510 observations can be corrected multiplying by the slope β_1 (~ 0.97 for the Deeper and ~ 0.96 for the SS510 sonar). The correction factor is site specific as it depends on the bed form and material, as well as on the water properties (temperature, salinity, and pressure). Therefore, the acquisition of a sample of ground control points is required.

3.2 UAV-borne measurements

In Fig. 9 we show the observations of the UAV-borne survey above Furesø.

Figure 10 depicts the UAV observations of four different cross sections of Marrebæk Kanal.

The accuracy of ground truth observations depends on both (i) the accuracy of the GNSS observations and (ii) the accuracy in positioning the GNSS pole in contact with the river bed. A vertical accuracy of ~ 5–7 cm and a horizontal accuracy of ~ 2–3 cm are estimated for the RTK GNSS ground truth observations, while the accuracy of the UAV-borne river-cross-section observations depends on (i) the error in absolute position of the sonar and (ii) the sonar's accuracy in measuring depth. The Deeper sonar shows a systematic overestimation of water depth in Fig. 10, which can be corrected by multiplying by the slope coefficient ($\beta 1 \sim 0.95$ for this specific survey). Figure 11 shows the observations after correction for the measurement bias.

UAV-borne bathymetric surveys provide high spatial resolution. Surveys can be interpolated to obtain bathymetric maps of entire river stretches. Figure 12 shows UAV observations in Åmose Å retrieved with the Deeper sonar at a resolu-

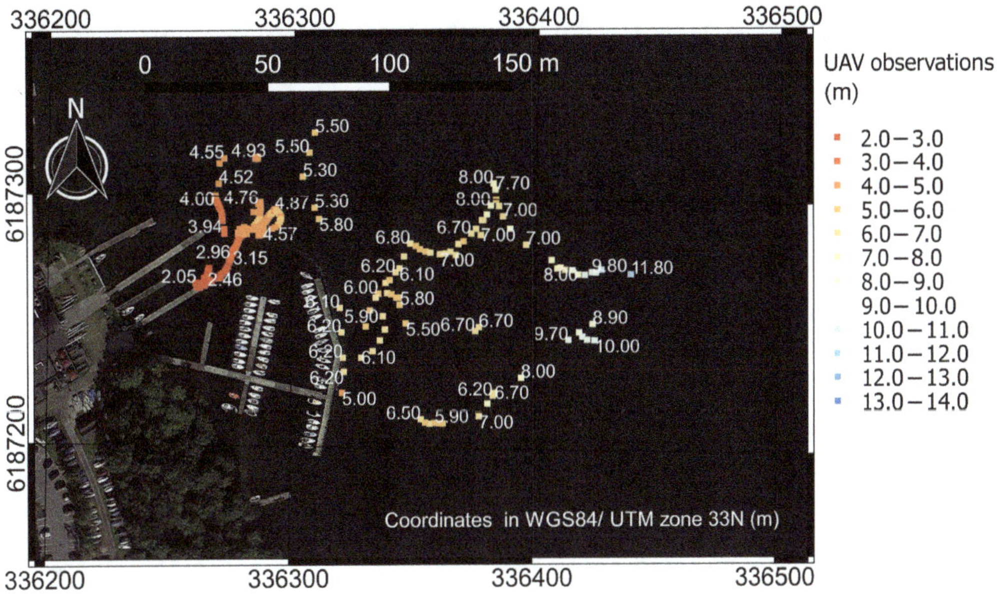

Figure 9. Water depth (m) observations retrieved in Furesø with the the sonar tethered to the UAV.

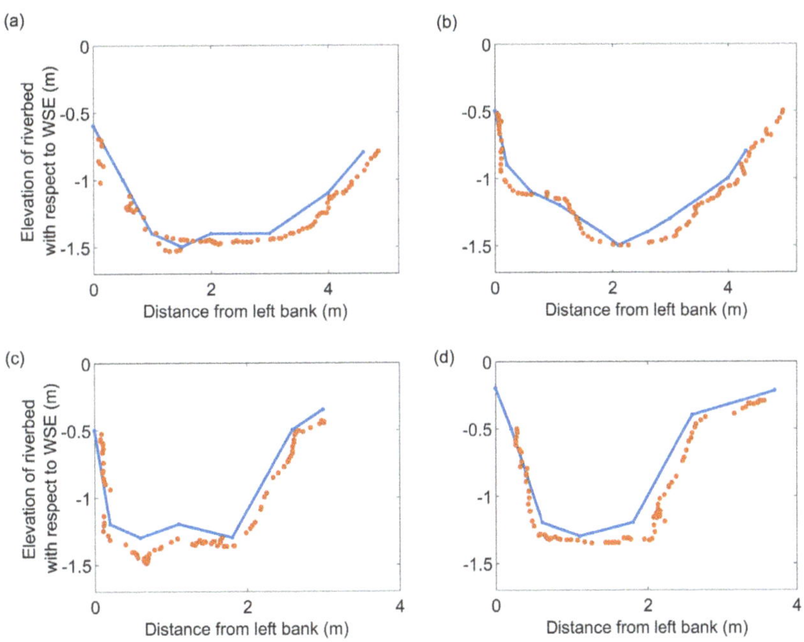

Figure 10. River cross sections retrieved at different locations along Marrebæk Kanal. The y axis shows the difference between riverbed elevation and WSE (opposite sign of water depth). Red points are retrieved with UAV-borne observations and blue lines are the ground truth observations. The latitude and longitude coordinates of the left bank of the river cross sections are **(a)** 54.676300, 11.913296°; **(b)** 54.675507, 11.913628°; **(c)** 54.682117, 11.911957°; and **(d)** 54.681779, 11.910723° (WGS84 reference system).

tion of ~ 0.5 m. These observations were interpolated using the triangulated irregular network method. Two ground truth cross sections were retrieved with the RTK GNSS rover. The investigated stretch of Åmose Å has a length of ~ 85 m and a maximum water depth of ~ 1.15 m.

Figure 12 shows that the minimum depth restriction is a significant limitation of the Deeper sonar in small rivers and streams. Water depth values smaller than ~ 0.5 m are generally not measured by the Deeper sonar. Furthermore, the soft sediment and the submerged vegetation cause significant errors in the Deeper observations when compared to ground truth cross sections. In this survey, it was not possible to identify a systematic error and thus correct for the bias of the UAV-borne observations.

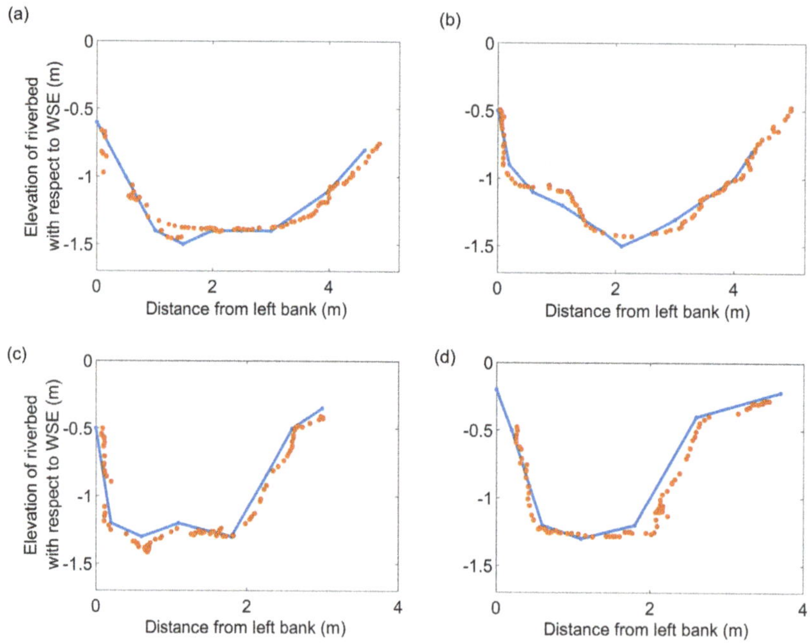

Figure 11. River-cross-section observations retrieved from Marrebæk Kanal at the locations shown in Fig. 10 after bias correction of the Deeper sonar observations. Red points are retrieved with UAV-borne observations and blue lines are the ground truth observations.

3.3 Accuracy of the Deeper sonar position

The accuracy of the absolute position of the Deeper sonar depends on the accuracy of (i) the drone horizontal position, (ii) the drone heading, and (iii) the relative position of the sonar with respect to the drone. The accuracies of these observations are reported in Table 2.

The accuracy of the relative position of the sonar depends on the image analysis procedure implemented to convert an offset from pixel into metric units. This procedure is also affected by the accuracy of the radar-derived WSE, because OD is an input to Eqs. (1) and (2). Tests were conducted in static mode using a checkerboard, placed at a series of known distances between 1 and 4 m, to evaluate the accuracy of measuring true distances in the image. These experiments showed that the offset between the camera and the sensor could be determined with an accuracy of 3 % of its actual value. The error in the conversion from image units to true distance units is mainly due to the (i) uncorrected lens distortion and (ii) assumption, made in Eqs. (1) and (2), that focal length is precisely known and that the distance between the rear nodal point of the lens and the image plane is exactly equal to focal length.

An error propagation study evaluated the overall accuracy of the absolute position of the sonar in real-world horizontal coordinates. For detailed information, see the data in the Supplement. The uncertainties of β, Lw, and Lh have the larger impact on the overall accuracy, compared to other error sources, such as OD and drone horizontal position. Since the offset (L) between the center of the camera and the sonar typically assumes values between 0 and 2 m, the overall accuracy of the Deeper sonar position is generally better than 20 cm. This accuracy is acceptable for most bathymetric surveys, particularly in light of the spatial resolution (15° beam divergence) of the Deeper sonar measurements.

Table 2. Accuracy of the different sensors used to measure the absolute position of the sonar.

Sensor	Observation	Accuracy
IMU	β (drone heading)	3°
GNSS	Drone horizontal position	2 cm at twice the standard deviation (Bandini et al., 2017)
Radar	OD (range to water surface)	0.5 % of the actual range (Bandini et al., 2017)
Camera	Lw, Lh (offset between sonar and camera center along horizontal and vertical axis of the picture)	≈ 3 % of the actual value

4 Discussion

Bathymetry can be measured with both in situ and remote sensing methods. In situ methods generally deploy bathymetric sonars installed on vessels. Remote sensing methods include (i) lidar techniques, (ii) methods evaluating

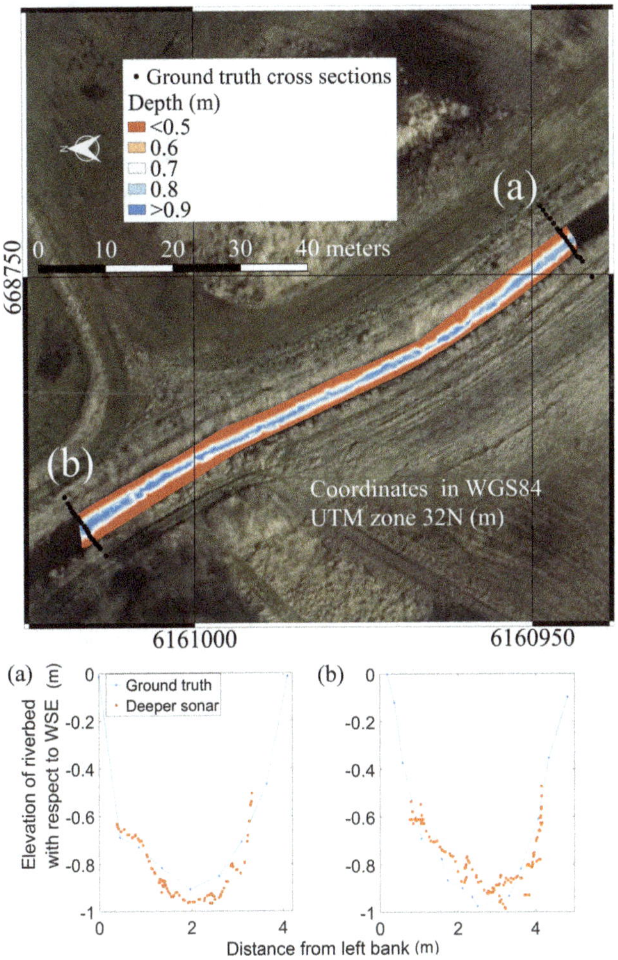

Figure 12. Bathymetry observations in Åmose Å. Top panel shows the surveyed river stretch (north direction pointing towards the left side of the map as indicated by the north arrow). Background map is an airborne orthophoto provided by the Danish Styrelsen for Dataforsyning og Effektivisering (https://kortforsyningen.dk/, last access: 6 September 2017). Raster foreground map shows UAV-borne observations interpolated with the triangulated irregular network method. Two ground truth cross sections were retrieved, which are shown in the bottom panels: **(a)** upstream and **(b)** downstream cross section. In the cross section plots, the x axis shows the distance from left bank (west bank), and the y axis shows the difference between riverbed elevation and WSE (opposite sign of water depth).

the relationship between spectral signature and depth, and (iii) through-water photogrammetry. Remote sensing methods generally allow for larger spatial coverage than in situ methods, but only shallow and clear water bodies can be surveyed. Table 3 shows a comparison of the different remote sensing and in situ techniques. UAV-borne sonar depth measurements bridge the gap between ground surveys and remote sensing techniques. The deployed Deeper sonar can measure deep and turbid water, and reach remote and dangerous areas, including unnavigable streams, when it is tethered to UAV. For depths up to ~ 30 m, the 2.1 % accuracy complies with the first accuracy level established by the International Hydrographic Organization (IHO) for accurate bathymetric surveys. Indeed, for depths of 30 m, the accuracy of the tethered sonar is ~ 0.630 m, while the first IHO level standard requires an accuracy better than 0.634 m. Conversely, for depths greater than 30 m, the UAV-borne sonar measurements comply with the second IHO level. Because of the large beam angle of the Deeper sonar, small-scale bathymetric features at greater depth cannot be resolved. However, a large beam angle (e.g., 8–30°) is an intrinsic limitation of single-beam sonar systems. For these reasons, when detection of small-scale features is required, surveys are generally performed with vessels equipped with multi-beam swath systems or side-scan imaging sonars. These systems are significantly more expensive, heavier, and larger than single-beam sonars, which makes integration with UAV platforms difficult.

Table 3 does not include methods requiring the operator to wade into a river, e.g., measurements taken with a RTK GNSS rover (e.g., Bangen et al., 2014). To take measurements with a GNSS rover, the operator must submerge the antenna pole until it reaches the river bed surface. Therefore, this method can only be used for local observations. Furthermore, innovative approaches such as using a ground penetrating radar (GPR) are not included because they are still at the level of local proof-of-concept applications (Costa et al., 2000; Spicer et al., 1997) and generally require cableways to suspend instrumentation a few decimeters above the water surface.

In order to obtain reliable measurements and ensure effective post-processing of the data, the techniques shown in Table 3 require initial expenditure and expertise from multiple fields, e.g., electric and software engineers (for technology development and data analysis), pilots (e.g., UAVs and manned aircrafts), experts in river navigation (for boats), surveyors (e.g., for GNSS rovers, photogrammetry), hydrologists, and geologists. In Appendix B, the typical survey expenditures for the different techniques are shown.

Future research

UAV-borne measurements of water depth have the potential to enrich the set of available hydrological observations. Their advantages compared to airborne, satellite, and manned boat measurements were demonstrated in this study. The competitiveness of UAVs in measuring water depth, compared to the capabilities of unmanned aquatic vessels equipped with sonar and RTK GNSS systems, is currently limited to water bodies that do not allow navigation of unmanned aquatic vessels, e.g., because of high water currents, slopes, or obstacles. The full potential of UAV-borne hydrological observations will be exploited only with flight operations beyond visual line of sight. The new generation of waterproof rotary wing UAVs equipped with visual navigation sensors and automatic

Table 3. Comparison of different approaches for measuring river bathymetry.

Technique	Platform	Spatial resolution (m)	Max. water depth	Typical error	Applicability (e.g., water clarity)	References
Spectral signature	Satellite	High-resolution commercial satellites[a]: ≈ 2 m; medium-resolution satellites[b]: typically > 30 m	1–1.5 m	0.10–0.20 m	≈ 1–1.5 times the Secchi depth	Fonstad and Marcus (2005), Legleiter and Overstreet (2012).
	Manned aircraft	Typically 0.5–4 m				Carbonneau et al. (2006), Legleiter and Roberts (2005), Winterbottom and Gilvear (1997).
	UAV	0.05–0.20 m				Flener et al. (2013), Lejot et al. (2007).
Through-water photogramme-try	Manned aircraft	Typically 0.1–0.5 m	0.6–1.5 m	0.08–0.2 m	≈ Secchi depth	Feurer et al. (2008), Lane et al. (2010), Westaway et al. (2001).
	UAV	Typically 0.01–0.1 m				Bagheri et al. (2015), Dietrich (2016), Tamminga et al. (2014), Woodget et al. (2015).
Lidar	UAV	≈ 0.020 m at 20 m range	1–1.5 m	≈ 0.10 m with standard deviation of 0.13 m	≈ 1–1.5 times the Secchi depth	Mandlburger et al. (2016).
	Manned aircraft	Few dm-several m	6 m	0.05–0.3 m	≈ 2–3 times the Secchi depth	Bailly et al. (2012, 2010), Charlton et al. (2003), Hilldale and Raff (2008), Kinzel et al. (2007).
TLS[c]	Banks of the water body	Typically ≈ 0.05 m	0.5 m, but typically ≈ 0.1 m	0.005–0.1 m	Clear water	Bangen et al. (2014), Heritage and Hetherington (2007), Smith et al. (2012), Smith and Vericat (2014).
Single-or multi-beam swath sonars	Manned/ unmanned vessels	Depending on the instrumentation and water depth	Sonars have minimum depth requirements (min 0.2–1 m)	Variable	Navigable streams	Widely known methodology
Sonar tethered to UAV	UAV	Depending on the water depth[d]	0.5–80 m	≈ 3.8 %[e] ≈ 2.1 %[f] of actual depth	All water conditions	Methodology described in this paper

[a] Multispectral bands: IKONOS, QuickBird, and WorldView-2. [b] Landsat. [c] Terrestrial laser scanner (TLS). [d] The divergence of the sonar cone beam is 15°. [e] Before bias correction. [f] After bias correction.

pilot systems will make it possible to collect hyperspatial observations in remote or dangerous locations, without requiring the operator to access the area.

5 Conclusions

UAVs are flexible and low-cost platforms. UAVs allow operators to retrieve hyperspatial hydrological observations with high spatial and temporal resolution. Automatic flight, together with computer vision navigation, allows UAVs to monitor dangerous or remote areas, including unnavigable streams.

This study shows how water depths can be retrieved by a tethered sonar controlled by UAVs. In particular, we highlighted the following:

- The accuracy of the measured water depth is not significantly affected by bottom structure and water turbidity if the sound waveform is correctly processed. However, submerged vegetation and soft sediments can affect sonar observations.

- Observations were retrieved for water depths ranging from 0.5 up to 35 m. Accuracy can be improved from $\sim 3.8\,\%$ to $\sim 2.1\,\%$ after correction of the observational bias, which can be identified and quantified by acquiring a representative sample of ground truth observations. The observational bias, which was observed in most experiments, can be caused by the dependence of the sound wave speed on temperature, salinity, and pressure. The relatively wide beam angle (15°) of the UAV-tethered sonar implies coarse spatial resolution, especially at large water depths, and limits the detection of small-scale differences in depth.

- The accuracy and maximum survey depth achieved in this study exceed those of any other remote sensing techniques and are comparable with bathymetric sonars transported by manned or unmanned aquatic vessels.

Code and data availability. Datasets used in the study are available online in the repository archived in Zenodo.org, https://doi.org/10.5281/zenodo.1309416 (Bandini et al., 2018). The repository contains MATLAB datasets, scripts, together with vector and raster files that can be used to replicate the figures of this paper.

Appendix A

In Fig. 8 the measurements of the two different sonars lie along a line with a nearly constant slope (not coincident with the 1 : 1 line) with respect to the ground truth observations.

The equation presented by Chen and Millero (1977) is the international standard algorithm, often known as the UNESCO algorithm, that computes the speed of sound (c) in water as a complex function of temperature (T), salinity (S), and pressure (P).

This equation has a range of validity: temperature 0 to 40 °C, salinity 0 to 40 parts per thousand, pressure 0 to 1000 bar (Wong and Zhu, 1995). Measurements were conducted in the Furesø lake, which has a salinity of less than 0.5 ‰, a recorded surface temperature between 12 and 19°, and a depth up to ~ 35 m. A sensitivity analysis with one factor varying at the time was applied to the Chen and Millero equation to estimate the range of variability of the speed of sound at different temperature, salinity, and depth (or pressure) values, as shown in Fig. A1.

As shown Fig. A1, temperature has the largest influence on speed of sound. Thus, the slope of linear regression between sonar and ground truth measurements is mainly determined by the temperature profiles and only to a lesser extent by the salinity and depth. Indeed, although the two sonars measure the surface temperature of water, no internal compensation is performed for the vertical temperature profile.

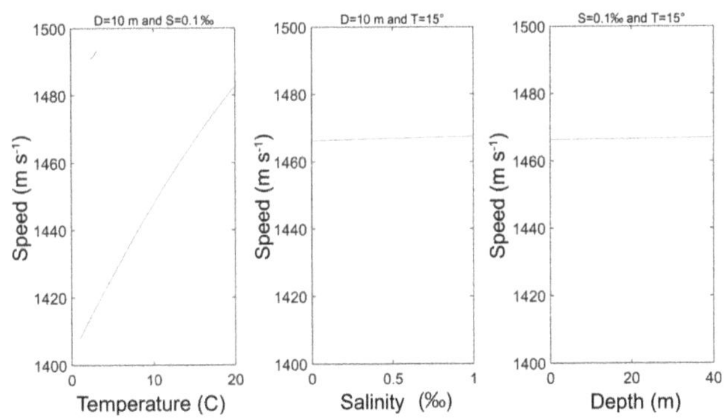

Figure A1. Sound speed for varying temperature, salinity, and depth.

Appendix B

Costs related to the individual approaches to measure bathymetry are difficult to estimate and compare. Costs include an initial expenditure and additional expenses depending on the nature of each survey. These typically depend on the duration of the survey, on the size of the area to be surveyed, on the needed accuracy and resolution, on the cost of labor, and on the water body characteristics. Table B1 compares the approximate costs for the techniques that are most commonly used to retrieve water depth.

Table B1. Cost comparison for different techniques.

Technique	Platform	Cost of instrumentation (currency: US Dollars)	Costs per survey (currency: US dollars)	Reference
Spectral signature	Satellite	Costs sustained by space agencies	High resolution: USD 10–30 km^{-2} With minimum order image size: 25–100 km^2 Medium resolution (e.g., Landsat): open access	http://www.landinfo.com/satellite-imagery-pricing.html, last access: 2 September 2017
	Manned aircraft	Multispectral cameras: USD 15 000–200 000	Minimum survey cost: $\sim$ USD 15 000–20 000 Rate km^{-2}: USD 300–800 km^{-2}	Online data collection
	UAV	Multispectral cameras: USD 15 000–200 000 Medium-size UAV: USD 3000–30 000	Minimum survey cost: $\sim$ USD 100–300 h^{-1} of survey	Online data collection
Through-water photogrammetry	Manned aircraft	Cameras: USD 1000–30 000	Minimum survey cost: $\sim$ USD 15 000–20 000 Rate km^{-2}: USD 300–800 km^{-2}	Online data collection
	UAV	Cameras: USD 500–10 000 Medium-size UAV: USD 3000–30 000	Minimum survey cost: $\sim$ USD 100–300 h^{-1} of survey	Online data collection
Lidar	UAV	Lidar: $\approx$ USD 120 000 Large-size UAV: USD 15 000–30 000	Minimum survey cost: $\sim$ USD 100–300 h^{-1} of survey	RIEGL Laser Measurement Systems GmbH, personal communication, 2017
	Manned aircraft	Lidar: USD 100 000–2 500 000 (price range available on the market)	Minimum survey cost: $\sim$ USD 15 000–20 000 Rate km^{-2}: USD 300–800 km^{-2} Post-processing: additional USD 150–300 km^{-2}	Bangen et al. (2014)
TLS	In situ	TLS: USD 65 000–225 000	Minimum survey cost: $\sim$ USD 60–100 h^{-1} Survey efficiency: 1.4–1.9 h/scan	Bangen et al. (2014)
Single-beam and multi-beam swath sonar	Manned Boat	USD 200–2000 (single-beam sonar) USD 20 000–100 000 (multi-beam sonar)	Minimum survey cost: $\sim$ USD 100–500 h^{-1} of survey	Online data collection
Sonar tethered to UAV	UAV	Sonar: USD 240 Radar, camera, IMU, and GNSS: USD 6000–10 000 Medium-size UAV: USD 3000–30 000	Minimum survey cost: $\sim$ USD 100–300 h^{-1} of survey. Survey efficiency: average flight speed of $\sim$ 0.5 $m\,s^{-1}$	This paper

Competing interests. The authors declare that they have no conflict of interest.

Acknowledgements. Ole Smith, Thyge Bjerregaard Pedersen, Karsten Stæhr Hansen, and Mikkel Lund Schmedes from Orbicon A/S provided help and technical support during the bathymetric surveys.

The Innovation Fund Denmark is acknowledged for providing funding for this study via the project Smart UAV (125-2013-5).

Edited by: Anas Ghadouani

References

Alsdorf, D. E., Rodriguez, E., and Lettenmaier, D. P.: Measuring surface water from space, Rev. Geophys., 45, 1–24, https://doi.org/10.1029/2006RG000197, 2007.

Bagheri, O., Ghodsian, M., and Saadatseresht, M.: Reach scale application of UAV+SfM method in shallow rivers hyperspatial bathymetry, Int. Arch. Photogramm., 40, 77–81, 2015.

Bailly, J.-S., Kinzel, P. J., Allouis, T., Feurer, D., and Le Coarer, Y.: Airborne LiDAR Methods Applied to Riverine Environments, in: Fluvial Remote Sensing for Science and Management, edited by: Carbonneau, P. E. and Piégay, H., 141–161, 2012.

Bailly, J. S., le Coarer, Y., Languille, P., Stigermark, C. J., and Allouis, T.: Geostatistical estimations of bathymetric LiDAR errors on rivers, Earth Surf. Proc. Land., 35, 1199–1210, https://doi.org/10.1002/esp.1991, 2010.

Bandini, F., Jakobsen, J., Olesen, D., Reyna-Gutierrez, J. A., and Bauer-Gottwein, P.: Measuring water level in rivers and lakes from lightweight Unmanned Aerial Vehicles, J. Hydrol., 548, 237–250, https://doi.org/10.1016/j.jhydrol.2017.02.038, 2017.

Bandini, F., Olesen, D., Jakobsen, J., Kittel, C. M. M., Wang, S., Garcia, M., and Bauer-Gottwein, P.: Dataset used in "Bathymetry observations of inland water bodies using a tethered single-beam sonar controlled by an Unmanned Aerial Vehicle" (https://doi.org/10.5194/hess-2017-625), https://doi.org/10.5281/ZENODO.1309417, 2018.

Bangen, S. G., Wheaton, J. M., Bouwes, N., Bouwes, B., and Jordan, C.: A methodological intercomparison of topographic survey techniques for characterizing wadeable streams and rivers, Geomorphology, 206, 343–361, https://doi.org/10.1016/j.geomorph.2013.10.010, 2014.

Brown, C. J. and Blondel, P.: Developments in the application of multibeam sonar backscatter for seafloor habitat mapping, Appl. Acoust., 70, 1242–1247, https://doi.org/10.1016/j.apacoust.2008.08.004, 2009.

Brown, D. C.: Close-range camera calibration, Photogramm. Eng., 37, 855–866, 1971.

Brown, H. C., Jenkins, L. K., Meadows, G. A., and Shuchman, R. A.: BathyBoat: An autonomous surface vessel for stand-alone survey and underwater vehicle network supervision, Mar. Technol. Soc. J., 44, 20–29, 2010.

Carbonneau, P. E., Lane, S. N., and Bergeron, N.: Feature based image processing methods applied to bathymetric measurements from airborne remote sensing in fluvial environments, Earth Surf. Proc. Land., 31, 1413–1423, https://doi.org/10.1002/esp.1341, 2006.

Charlton, M. E., Large, A. R. G., and Fuller, I. C.: Application of airborne lidar in river environments: The River Coquet, Northumberland, UK, Earth Surf. Proc. Land., 28, 299–306, https://doi.org/10.1002/esp.482, 2003.

Chen, C. and Millero, F. J.: Speed of sound in seawater at high pressures, J. Acoust. Soc. Am., 62, 1129–1135, https://doi.org/10.1121/1.381646, 1977.

Clarke, T. and Fryer, J.: The development of camera calibration methods and models, Photogramm. Rec., 16, 51–66, https://doi.org/10.1111/0031-868X.00113, 1998.

Conner, J. T. and Tonina, D.: Effect of cross-section interpolated bathymetry on 2D hydrodynamic model results in a large river, Earth Surf. Proc. Land., 39, 463–475, https://doi.org/10.1002/esp.3458, 2014.

Costa, J. E., Spicer, K. R., Cheng, R. T., Haeni, F. P., Melcher, N. B., Thurman, E. M., Plant, W. J., and Keller, W. C.: Measuring stream discharge by non-contact methods: A proof-of-concept experiment, Geophys. Res. Lett., 27, 553–556, https://doi.org/10.1029/1999GL006087, 2000.

Detert, M. and Weitbrecht, V.: A low-cost airborne velocimetry system: proof of concept, J. Hydraul. Res., 53, 532–539, https://doi.org/10.1080/00221686.2015.1054322, 2015.

Dietrich, J. T.: Bathymetric Structure from Motion: Extracting shallow stream bathymetry from multi-view stereo photogrammetry, Earth Surf. Proc. Land., 42, 355–364, https://doi.org/10.1002/esp.4060, 2016.

Faig, W.: Calibration of close-range photogrammetry systems: Mathematical formulation, Photogramm. Eng. Rem. S., 41, 1479–1486, 1975.

Ferreira, H., Almeida, C., Martins, A., Almeida, J., Dias, N., Dias, A., and Silva, E.: Autonomous bathymetry for risk assessment with ROAZ robotic surface vehicle, in OCEANS '09 IEEE, Bremen, 2009.

Feurer, D., Bailly, J.-S., Puech, C., Le Coarer, Y., and Viau, A. A.: Very-high-resolution mapping of river-immersed topography by remote sensing, Prog. Phys. Geogr., 32, 403–419, https://doi.org/10.1177/0309133308096030, 2008.

Flener, C., Vaaja, M., Jaakkola, A., Krooks, A., Kaartinen, H., Kukko, A., Kasvi, E., Hyyppä, H., Hyyppä, J., and Alho, P.: Seamless mapping of river channels at high resolution using mobile liDAR and UAV-photography, Remote Sens., 5, 6382–6407, https://doi.org/10.3390/rs5126382, 2013.

Fonstad, M. A. and Marcus, W. A.: Remote sensing of stream depths with hydraulically assisted bathymetry (HAB) models, Geomorphology, 72, 320–339, https://doi.org/10.1016/j.geomorph.2005.06.005, 2005.

Gichamo, T. Z., Popescu, I., Jonoski, A., and Solomatine, D.: River cross-section extraction from the ASTER global DEM for flood modeling, Environ. Model. Softw., 31, 37–46, https://doi.org/10.1016/j.envsoft.2011.12.003, 2012.

Giordano, F., Mattei, G., Parente, C., Peluso, F., and Santamaria, R.: Integrating sensors into a marine drone for bathymetric 3D surveys in shallow waters, Sensors, 16, 41, https://doi.org/10.3390/s16010041, 2015.

Guenther, G. C.: Airborne Lidar Bathymetry, in Digital Elevation Model Technologies and Applications, The DEM Users Manual, 253–320, 8401 Arlington Blvd., 2001.

Guenther, G. C., Cunningham, A. G., Larocque, P. E., Reid, D. J., Service, N. O., Highway, E., and Spring, S.: Meeting the Accuracy Challenge in Airborne Lidar Bathymetry, EARSeL eProceedings, 1, 1–27, 2000.

Hamylton, S., Hedley, J., and Beaman, R.: Derivation of High-Resolution Bathymetry from Multispectral Satellite Imagery: A

Comparison of Empirical and Optimisation Methods through Geographical Error Analysis, Remote Sens., 7, 16257–16273, https://doi.org/10.3390/rs71215829, 2015.

Heritage, G. L. and Hetherington, D.: Towards a protocol for laser scanning in fluvial geomorphology, Earth Surf. Proc. Land., 32, 66–74, https://doi.org/10.1002/esp.1375, 2007.

Hilldale, R. C. and Raff, D.: Assessing the ability of airborne LiDAR to map river bathymetry, Earth Surf. Proc. Land., 33, 773–783, https://doi.org/10.1002/esp.1575, 2008.

Kinzel, P. J., Wright, C. W., Nelson, J. M., and Burman, A. R.: Evaluation of an Experimental LiDAR for Surveying a Shallow, Braided, Sand-Bedded River, J. Hydraul. Eng., 133, 838–842, https://doi.org/10.1061/(ASCE)0733-9429(2007)133:7(838), 2007.

Lane, S. N., Widdison, P. E., Thomas, R. E., Ashworth, P. J., Best, J. L., Lunt, I. A., Sambrook Smith, G. H., and Simpson, C. J.: Quantification of braided river channel change using archival digital image analysis, Earth Surf. Proc. Land., 35, 971–985, https://doi.org/10.1002/esp.2015, 2010.

Lee, K. R., Kim, A. M., Olsen;, R. C., and Kruse, F. A.: Using WorldView-2 to determine bottom-type and bathymetry, Proc. SPIE 8030, Ocean Sensing and Monitoring III, , 80300D (5 May 2011), vol. 2, 2011.

Legleiter, C. J.: Remote measurement of river morphology via fusion of LiDAR topography and spectrally based bathymetry, Earth Surf. Proc. Land., 37, 499–518, https://doi.org/10.1002/esp.2262, 2012.

Legleiter, C. J. and Overstreet, B. T.: Mapping gravel bed river bathymetry from space, J. Geophys. Res.-Earth, 117, https://doi.org/10.1029/2012JF002539, 2012.

Legleiter, C. J. and Roberts, D. A.: Effects of channel morphology and sensor spatial resolution on image-derived depth estimates, Remote Sens. Environ., 95, 231–247, https://doi.org/10.1016/j.rse.2004.12.013, 2005.

Legleiter, C. J., Roberts, D. A., and Lawrence, R. L.: Spectrally based remote sensing of river bathymetry, Earth Surf. Proc. Land., 34, 1039–1059, https://doi.org/10.1002/esp.1787, 2009.

Lejot, J., Delacourt, C., Piégay, H., Fournier, T., Trémélo, M.-L., and Allemand, P.: Very high spatial resolution imagery for channel bathymetry and topography from an unmanned mapping controlled platform, Earth Surf. Proc. Land., 32, 1705–1725, https://doi.org/10.1002/esp.1595, 2007.

Liceaga-Correa, M. A. and Euan-Avila, J. I.: Assessment of coral reef bathymetric mapping using visible Landsat Thematic Mapper data, Int. J. Remote Sens., 23, 3–14, https://doi.org/10.1080/01431160010008573, 2002.

Lyons, M., Phinn, S., and Roelfsema, C.: Integrating Quickbird multi-spectral satellite and field data: Mapping bathymetry, seagrass cover, seagrass species and change in Moreton Bay, Australia in 2004 and 2007, Remote Sens., 3, 42–64, https://doi.org/10.3390/rs3010042, 2011.

Lyzenga, D. R.: Remote sensing of bottom reflectance and water attenuation parameters in shallow water using aircraft and Landsat data, Int. J. Remote Sens., 2, 71–82, https://doi.org/10.1080/01431168108948342, 1981.

Lyzenga, D. R., Malinas, N. P., and Tanis, F. J.: Multispectral bathymetry using a simple physically based algorithm, IEEE Trans. Geosci. Remote Sens., 44, 2251–2259, https://doi.org/10.1109/TGRS.2006.872909, 2006.

Mandlburger, G., Pfennigbauer, M., Wieser, M., Riegl, U., and Pfeifer, N.: Evaluation Of A Novel Uav-Borne Topo-Bathymetric Laser Profiler, ISPRS – Int. Arch. Photogramm. Remote Sens. Spat. Inf. Sci., XLI-B1, 933–939, https://doi.org/10.5194/isprs-archives-XLI-B1-933-2016, 2016.

Manley, P. L. and Singer, J. K.: Assessment of sedimentation processes determined from side-scan sonar surveys in the Buffalo River, New York, USA, Environ. Geol., 55, 1587–1599, https://doi.org/10.1007/s00254-007-1109-8, 2008.

Marcus, W. A., Legleiter, C. J., Aspinall, R. J., Boardman, J. W., and Crabtree, R. L.: High spatial resolution hyperspectral mapping of in-stream habitats, depths, and woody debris in mountain streams, Geomorphology, 55, 363–380, https://doi.org/10.1016/S0169-555X(03)00150-8, 2003.

Nitsche, F. O., Ryan, W. B. F., Carbotte, S. M., Bell, R. E., Slagle, A., Bertinado, C., Flood, R., Kenna, T., and McHugh, C.: Regional patterns and local variations of sediment distribution in the Hudson River Estuary, Estuar. Coast. Shelf Sci., 71, 259–277, https://doi.org/10.1016/j.ecss.2006.07.021, 2007.

Overstreet, B. T. and Legleiter, C. J.: Removing sun glint from optical remote sensing images of shallow rivers, Earth Surf. Proc. Land., 42, 318–333, 2017.

Powers, J., Brewer, S. K., Long, J. M., and Campbell, T.: Evaluating the use of side-scan sonar for detecting freshwater mussel beds in turbid river environments, Hydrobiologia, 743, 127–137, https://doi.org/10.1007/s10750-014-2017-z, 2015.

Ridolfi, E. and Manciola, P.: Water level measurements from drones: A Pilot case study at a dam site, Water (Switzerland), https://doi.org/10.3390/w10030297, 2018.

Rovira, A., Batalla, R. J., and Sala, M.: Fluvial sediment budget of a Mediterranean river: The lower Tordera (Catalan Coastal Ranges, NE Spain), Catena, 60, 19–42, https://doi.org/10.1016/j.catena.2004.11.001, 2005.

Schäppi, B., Perona, P., Schneider, P., and Burlando, P.: Integrating river cross section measurements with digital terrain models for improved flow modelling applications, Comput. Geosci., 36, 707–716, https://doi.org/10.1016/j.cageo.2009.12.004, 2010.

Smith, M., Vericat, D., and Gibbins, C.: Through-water terrestrial laser scanning of gravel beds at the patch scale, Earth Surf. Proc. Land., 37, 411–421, https://doi.org/10.1002/esp.2254, 2012.

Smith, M. W. and Vericat, D.: Evaluating shallow-water bathymetry from through-water terrestrial laser scanning under a range of hydraulic and physical water quality conditions, River Res. Appl., 30, 905–924, https://doi.org/10.1002/rra.2687, 2014.

Snellen, M., Siemes, K., and Simons, D. G.: Model-based sediment classification using single-beam echosounder signals., J. Acoust. Soc. Am., 129, 2878–2888, https://doi.org/10.1121/1.3569718, 2011.

Spicer, K. R., Costa, J. E., and Placzek, G.: Measuring flood discharge in unstable stream channels using ground-penetrating radar, Geology, 25, 423–426, https://doi.org/10.1130/0091-7613(1997)025<0423:MFDIUS>2.3.CO;2, 1997.

Strayer, D. L., Malcom, H. M., Bell, R. E., Carbotte, S. M., and Nitsche, F. O.: Using geophysical information to define benthic habitats in a large river, Freshw. Biol., 51, 25–38, https://doi.org/10.1111/j.1365-2427.2005.01472.x, 2006.

Stumpf, R. P., Holderied, K., and Sinclair, M.: Determination of water depth with high-resolution satellite imagery over variable bottom types, Limnol. Oceanogr., 48, 547–556, https://doi.org/10.4319/lo.2003.48.1_part_2.0547, 2003.

Tamminga, A., Hugenholtz, C., Eaton, B., and Lapointe, M.: Hyperspatial Remote Sensing of Channel Reach Morphology and Hydraulic Fish Habitat Using an Unmanned Aerial Vehicle (Uav): a First Assessment in the Context of River Research and Management, River Res. Appl., 31, 379–391, https://doi.org/10.1002/rra.2743, 2014.

Tauro, F., Petroselli, A., and Arcangeletti, E.: Assessment of drone-based surface flow observations, Hydrol. Process., 30, 1114–1130, https://doi.org/10.1002/hyp.10698, 2015a.

Tauro, F., Pagano, C., Phamduy, P., Grimaldi, S., and Porfiri, M.: Large-Scale Particle Image Velocimetry From an Unmanned Aerial Vehicle, IEEE/ASME Trans. Mechantronics, 20, 1–7, https://doi.org/10.1109/TMECH.2015.2408112, 2015b.

Tauro, F., Porfiri, M., and Grimaldi, S.: Surface flow measurements from drones, J. Hydrol., 540, 240–245, https://doi.org/10.1016/j.jhydrol.2016.06.012, 2016.

Virili, M., Valigi, P., Ciarfuglia, T., and Pagnottelli, S.: A prototype of radar-drone system for measuring the surface flow velocity at river sites and discharge estimation, Geophys. Res. Abstr., 17, 2015–12853, 2015.

Walker, D. J. and Alford, J. B.: Mapping Lake Sturgeon Spawning Habitat in the Upper Tennessee River using Side-Scan Sonar, North Am. J. Fish. Manag., 36, 1097–1105, https://doi.org/10.1080/02755947.2016.1198289, 2016.

Weng, J., Coher, P., and Herniou, M.: Camera Calibration with Distortion Models and Accuracy Evaluation, IEEE T. Pattern Anal., 14, 965–980, https://doi.org/10.1109/34.159901, 1992.

Westaway, R. M., Lane, S. N., and Hicks, D. M.: Remote sensing of clear-water, shallow, gravel-bed rivers using digital photogrammetry, Photogramm. Eng. Rem. S., 67, 1271–1281, 2001.

Winterbottom, S. J. and Gilvear, D. J.: Quantification of channel bed morphology in gravel-bed rivers using airborne multispectral imagery and aerial photography, Regul. Rivers-Research Manag., 13, 489–499, https://doi.org/10.1002/(SICI)1099-1646(199711/12)13:6<489::AID-RRR471>3.0.CO;2-X, 1997.

Wong, G. S. K. and Zhu, S.: Speed of sound in seawater as a function of salinity, temperature, and pressure, J. Acoust. Soc. Am., 97, 1732–1736, https://doi.org/10.1121/1.413048, 1995.

Woodget, A. S., Carbonneau, P. E., Visser, F., and Maddock, I. P.: Quantifying submerged fluvial topography using hyperspatial resolution UAS imagery and structure from motion photogrammetry, Earth Surf. Proc. Land., 40, 47–64, https://doi.org/10.1002/esp.3613, 2015.

11

Singularity-sensitive gauge-based radar rainfall adjustment methods for urban hydrological applications

L.-P. Wang[1], S. Ochoa-Rodríguez[2], C. Onof[2], and P. Willems[1]

[1]Hydraulics Laboratory, Katholieke Universiteit Leuven, 3001 Heverlee (Leuven), Belgium
[2]Department of Civil and Environmental Engineering, Imperial College London, London, SW7 2AZ, UK

Correspondence to: L.-P. Wang (lipen.wang@bwk.kuleuven.be)

Abstract. Gauge-based radar rainfall adjustment techniques have been widely used to improve the applicability of radar rainfall estimates to large-scale hydrological modelling. However, their use for urban hydrological applications is limited as they were mostly developed based upon Gaussian approximations and therefore tend to smooth off so-called "singularities" (features of a non-Gaussian field) that can be observed in the fine-scale rainfall structure. Overlooking the singularities could be critical, given that their distribution is highly consistent with that of local extreme magnitudes. This deficiency may cause large errors in the subsequent urban hydrological modelling. To address this limitation and improve the applicability of adjustment techniques at urban scales, a method is proposed herein which incorporates a local singularity analysis into existing adjustment techniques and allows the preservation of the singularity structures throughout the adjustment process. In this paper the proposed singularity analysis is incorporated into the Bayesian merging technique and the performance of the resulting singularity-sensitive method is compared with that of the original Bayesian (non singularity-sensitive) technique and the commonly used mean field bias adjustment. This test is conducted using as case study four storm events observed in the Portobello catchment (53 km^2) (Edinburgh, UK) during 2011 and for which radar estimates, dense rain gauge and sewer flow records, as well as a recently calibrated urban drainage model were available. The results suggest that, in general, the proposed singularity-sensitive method can effectively preserve the non-normality in local rainfall structure, while retaining the ability of the original adjustment techniques to generate nearly unbiased estimates. Moreover, the ability of the singularity-sensitive technique to preserve the non-normality in rainfall estimates often leads to better reproduction of the urban drainage system's dynamics, particularly of peak runoff flows.

1 Introduction

Traditionally, urban hydrological applications have relied mainly upon rain gauge data as input. While rain gauges generally provide accurate point rainfall estimates near the ground surface, they cannot properly capture the spatial variability of rainfall, which has a significant impact on the urban hydrological system and thus on the modelling of urban runoff (Gires et al., 2012; Schellart et al., 2012). Thanks to the development of radar technology, weather radar data have been playing an increasingly important role in urban hydrology (Krämer et al., 2007; Liguori et al., 2011). Radars can survey large areas and better capture the spatial variability of the rainfall, thus improving the short-term predictability of rainfall and flooding. However, the accuracy of radar measurements is in general insufficient, particularly in the case of extreme rainfall magnitudes (Einfalt et al., 2004, 2005). This is due to the fact that, instead of being a direct measurement, radar rainfall intensity is derived indirectly from measured radar reflectivity. As a result, both radar reflectivity measurements and the reflectivity–intensity conversion process are subject to multiple sources of error.

First, errors in radar reflectivity measurements may arise from blockage of the radar beam, attenuation, ground clutter, anomalous propagation of the signal, among other sources (Collier, 1996; Einfalt et al., 2004; Harrison et al., 2000). Although the radar reflectivity measurements undergo a number of corrections before they are converted into rainfall intensity, it is virtually impossible to have error-free reflectivity measurements. Second, the conversion between radar reflectivity (Z) and rainfall rate (R) uses the so-called Z–R relationship, $Z = a\,R^b$ (Marshall and Palmer, 1948), where a and b are variables generally deduced by physical approximation or empirical calibration. This can be theoretically linked to the rain drop size distribution, which varies for different rainfall types. Operationally, a number of "static" Z–R relations are usually derived to generate radar rainfall rates for different rainfall types (e.g. stratiform, convective and tropical storms), and the associated a and b variables are calibrated based upon long-term comparisons (Collier, 1986; Krajewski and Smith, 2002). However, in reality, it is almost impossible to classify a single storm purely under a specific rainfall type. Consequently, it is not entirely appropriate to use a static Z–R relation to derive rainfall intensities, even for a single storm event. This has been confirmed by several studies, which indicate that rain drop size distribution is a highly dynamic process and may significantly or suddenly change within a storm event (Smith et al., 2009; Ulbrich, 1983). Because of this, the Z–R derived rainfall intensity cannot effectively reflect the short-term dynamics of true rainfall intensities and may statistically compromise with intermediate rainfall intensities.

In order to overcome these drawbacks of radar rainfall estimates while preserving their spatial description of rainfall fields, it is possible to dynamically adjust them using rain gauge measurements. Many studies on this subject have been carried out over the last few years, though most of them focus on hydrological applications at large scales (Anagnostou and Krajewski, 1999; Fulton et al., 1998; Germann et al., 2006; Goudenhoofdt and Delobbe, 2009; Harrison et al., 2000, 2009; Seo and Smith, 1991; Thorndahl et al., 2014). Only few studies have examined the applicability of these adjustment techniques to urban-scale hydrological applications and concluded that they can effectively reduce rainfall bias, thus leading to improvements in the reproduction of hydrological outputs (Smith et al., 2007; Vieux and Bedient, 2004; Villarini et al., 2010; Wang et al., 2013). However, despite the improvements achieved with the current adjustment techniques, underestimation of storm peaks can still be seen after adjustment and this is particularly evident in the case of small drainage areas (such as those of urban catchments) and extreme rainfall magnitudes (Wang et al., 2012; Ochoa-Rodríguez et al., 2013). This may be due to the fact that the underlying adjustment techniques, mainly based upon Gaussian (first and/or second statistical moments) approximations (Goudenhoofdt and Delobbe, 2009; Krajewski, 1987; Todini, 2001), cannot properly cope with the so-called "singularities" (which imply non-normality and often correspond to local extreme magnitudes) observed at small scales (Schertzer et al., 2013; Tchiguirinskaia et al., 2012). In fact, it is often the case that the radar image captures the spatial structure of striking local extremes (albeit the actual rainfall depth/intensity may be inaccurate), but these structures are lost or smoothened throughout the merging process. This deficiency may cause large errors in the subsequent urban hydrological modelling.

To address this limitation and improve the applicability of adjustment techniques at urban scales, a method is proposed herein which incorporates a local singularity (identification) analysis (Cheng et al., 1994; Schertzer and Lovejoy, 1987; Wang et al., 2012) into existing adjustment techniques and enables the preservation of the singularity structures throughout the adjustment process. The singularity-sensitive method is particularly intended to improve geostatistical-based merging techniques (e.g. co-kriging, Krajewski, 1987; Bayesian merging, Todini, 2001; kriging with external drift, Wackernagel, 2003 and conditional merging, Sinclair and Pegram, 2005), which seek to represent the spatial covariance structure of the rainfall field or its errors by making use of the semi-variogram (Goudenhoofdt and Delobbe, 2009). Being a second-order tool, the semi-variogram cannot adequately capture higher-order features of the rainfall field, thus causing these to be lost in the merging process.

The proposed singularity-sensitive method was initially developed and preliminarily tested in the reconstruction of a storm event which led to reported flooding in the Maida Vale area, Central London, in June 2009 (Wang and Onof, 2013; Wang et al., 2014). The radar rainfall product for this event showed strong and localised singularity structures, but the accuracy of the actual estimates was poor. A dynamic gauge-based adjustment was conducted using the Bayesian data merging method (Todini, 2001), which in previous studies had been shown to outperform other adjustment methods (Mazzetti and Todini, 2004; Wang et al., 2013). Nonetheless, for this particular event records from only a few rain gauge sites were available and these were located away from the area of interest and at points where less intense radar rainfall was observed. Under these circumstances, the aforementioned shortcomings associated with existing adjustment techniques became evident. The Bayesian data merging method proved inadequate as it smoothed out the singularity structures, which had the effect of considerably reducing the peak rainfall intensities. It was then that the singularity-sensitive method was devised and effectively incorporated into the Bayesian merging technique. The resulting singularity-sensitive Bayesian merging method led to rainfall fields which better preserved the spatial structure as captured by the radar and better reproduced peak rainfall intensities.

In the present paper the formulation of the proposed singularity-sensitive method is explained in detail and new numerical strategies aimed at improving the use of singularity information are introduced. Moreover, the method is further tested using as case study four storm events observed in the Portobello catchment (53 km^2) (Edinburgh, UK) during 2011 and for which radar estimates, a spatially dense net-

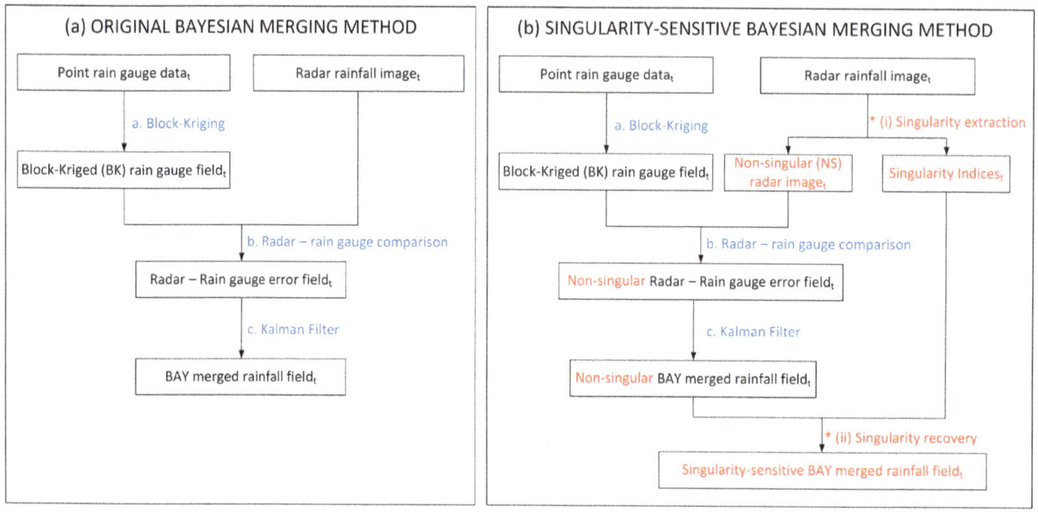

Figure 1. Schematics of **(a)** original Bayesian merging method (adapted from Fig. 2 in Mazzetti (2012)) and **(b)** singularity-sensitive Bayesian merging method.

work of rain and flow gauges, as well as a recently calibrated urban drainage model were available.

The paper is organised as follows. In Sect. 2 a detailed explanation is provided of the theoretical development of the singularity-sensitive method, including the newly implemented numerical strategies. In Sect. 3 we present the case study, including a description of the study area and data set, the performance criteria used to evaluate the proposed methodology, and the results of the testing. Lastly, in Sect. 4 the main conclusions are presented and future work is discussed.

2 Formulation of the singularity-sensitive Bayesian data merging method

Firstly, a description is provided of the two key techniques used in this paper: the Bayesian data merging method and the local singularity analysis. Afterwards, the proposed method for integrating these two techniques is explained. Intermediate results of each of the steps described in this section, which help illustrate the main features of the proposed methodology, can be found in the Supplement.

2.1 Bayesian radar–rain gauge data merging method

The Bayesian data merging method (BAY) is a dynamic adjustment method (applied independently at each time step) intended for real-time applications (Todini, 2001). The underlying idea of this method is to analyse and quantify the uncertainty of rainfall estimates (in terms of error co-variance) from multiple data sources – in this case, radar and rain gauge sensors – and then combine these estimates in such a way that the overall (estimation) uncertainty is minimised. The BAY merging method consists of the following steps (illustrated in Fig. 1a):

a. For each time step t, the point rain gauge (RG) measurements are interpolated into a synthetic rainfall field using the block kriging (BK) technique. The result of this step is an interpolated rain gauge rainfall field, with areal estimates at each radar grid location (y_t^{RG}), and which are accompanied by the associated estimation error co-variance function ($V_{\varepsilon_t}^{\mathrm{RG}}$), representing the uncertainty of rain gauge estimates.

b. The interpolated rain gauge rainfall field is compared against the radar field (y_t^{RD}), based upon which a field of errors (estimated as the bias at each radar grid location: $\varepsilon_t^{\mathrm{RD}} = y_t^{\mathrm{RD}} - y_t^{\mathrm{RG}}$) is obtained empirically. Assuming that areal rain gauge estimates are unbiased, the expectation value ($\mu_{\varepsilon_t^{\mathrm{RD}}}$) and the co-variance function ($V_{\varepsilon_t}^{\mathrm{RD}}$) of this radar–rain gauge error field at each time step is used to represent, respectively, the mean bias and the uncertainty of radar estimates.

c. Using a Kalman filter (Kalman, 1960), the two rainfall fields are optimally combined such that the overall estimation uncertainty is minimised. In the Kalman filter the radar data and the interpolated rain gauge estimates act, respectively, as "a priori estimate" and "measurement". The degree of "uncertainty" of each type of estimate constitutes a gain value (the so-called Kalman gain, K_t) at each radar grid location, and determines the proportion of each type of estimate that is used to compute the merged output. The use of this gain value ensures the minimisation of the overall estimation uncertainty and is expressed as

$$K_t = V_{\varepsilon_t^{\mathrm{RD}}}\left(V_{\varepsilon_t^{\mathrm{RD}}} + V_{\varepsilon_t^{\mathrm{RG}}}\right)^{-1}, \quad (1)$$

and the optimally merged output, BAY (i.e. the a posteriori estimates y_t'' in the Kalman filter) can be obtained from

$$y_t'' = y_t' + K_t\left(y_t^{\mathrm{RG}} - y_t'\right), \quad (2)$$

where y_t' is the "unbiased" radar rainfall estimate (i.e. $y_t' = y_t^{\mathrm{RD}} - \mu_{\varepsilon_t^{\mathrm{RD}}}$) used as the a priori estimate in the Kalman filter.

It can be seen that the Kalman gain is a function of the co-variances of radar and rain gauge estimation errors. When $V_{\varepsilon_t^{\mathrm{RD}}} \gg V_{\varepsilon_t^{\mathrm{RG}}}$ (or $K_t \approx 1$, i.e. radar estimates have significantly higher uncertainty than the rain gauge ones), the radar estimates are less trustworthy and the output estimates will be very similar to the interpolated rain gauge field. In contrast, when $V_{\varepsilon_t^{\mathrm{RG}}} \gg V_{\varepsilon_t^{\mathrm{RD}}}$ (or $K_t \approx 0$), the output will be closer to the radar estimates.

It is in steps b and c where the problems associated with the Bayesian merging technique, and geostatistical techniques in general, arise. The (second-order) co-variance function that these techniques employ to characterise radar–rain gauge errors cannot well capture local singularity structures. Instead, in second-order models singularities may be mistakenly regarded as errors in the radar data, thus leading to higher estimated radar uncertainty, $V_{\varepsilon_t^{\mathrm{RD}}}$. As a result, the radar data will be trusted less, leading to smoother merged outputs, which are closer to the interpolated rain gauge field.

2.2 Local singularity analysis

Various types of hazardous geo-processes, including precipitation, often result in anomalous amounts of energy release or mass accumulation confined to narrow intervals in time and/or space. The property of anomalous amounts of energy release or mass accumulation is termed *singularity* and it is often associated with structures depicting fractality or multifractality (Agterberg, 2007; Cheng, 1999; Lovejoy and Mandelbrot, 1985; Schertzer and Lovejoy, 1987). Several mathematical models and methodologies have been developed to respectively characterise and treat singularities. In this work, the local singularity analysis proposed by Cheng et al. (1994) has been adopted to identify and extract singularities form rainfall fields. Cheng's method, which has been widely used for estimation of geo-chemical concentrations (Agterberg, 2007; Cheng and Zhao, 2011; Cheng et al., 1994), employs the definition of coarse Hölder exponent to characterise singularities. According to this model, singularities are defined by the fact that the areal average measure (in this work, areal rainfall) centred on point x (taken as the centre of a radar pixel) varies as a power function of the area (Evertsz and

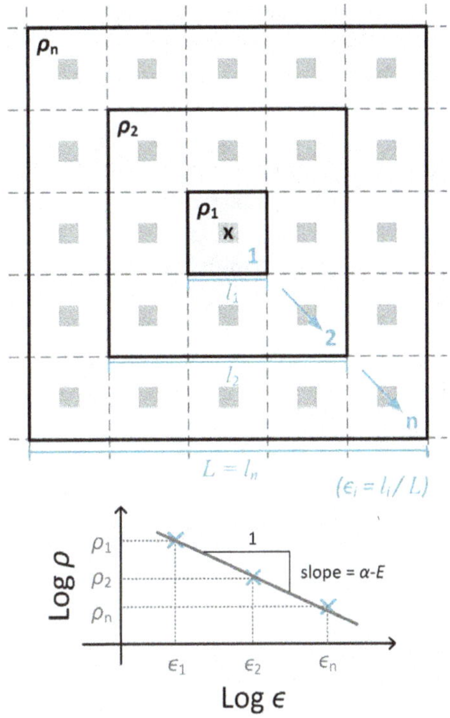

Figure 2. Schematic of the local singularity analysis (adapted from Wang et al., 2014).

Mandelbrot, 1992). This power-law relationship can be formulated as an equation (Cheng et al., 1994):

$$\rho(x, \epsilon) = c(x)\epsilon^{\alpha(x)-E}, \quad (3)$$

where $\rho(x, \epsilon)$ represents the density of measure (e.g. concentration of geo-data; in the context of this paper, rainfall intensity) over a square area with side-length l and associated scale ϵ ($\epsilon = l/L$, where L is the side-length of the largest square area under consideration) centred at a specific location x; $c(x)$ is a constant value (in the context of this paper, a constant intensity value) at x; $\alpha(x)$ is the singularity index (or the coarse Hölder exponent); and $E = 2$ is the Euclidean dimension of the plane.

A schematic of the estimation of the constant value $c(x)$ and singularity index $\alpha(x)$ from gridded data is provided in Fig. 2. For a given pixel with centre x (centre of top plot in Fig. 2), the mean rainfall intensities at different spatial scales (centred in x) can be calculated (i.e. rainfall intensities ρ_1, $\rho_2, \ldots, \rho_n$, respectively at scales $\epsilon_1, \epsilon_2, \ldots, \epsilon_n$). Then, the logarithms of these mean values and the associated spatial scales are compared (bottom plot in Fig. 2). The constant value $c(x)$ and singularity index $\alpha(x)$ of the data set can be derived by applying a simple linear regression analysis, where the slope and the y-intercept of the regression line correspond, respectively, to the terms $(\alpha(x) - E)$ and $\log c(x)$. A detailed explanation of the computation of $c(x)$ and $\alpha(x)$ can be found in previous studies (Agterberg, 2012b; Chen et al., 2007; Cheng et al., 1994). It is worth mentioning that the estimation of

$c(x)$ and $\alpha(x)$ can be trusted only if a good linear relation is observed (i.e. if the scaling behaviour is well followed).

Going back to the definition of singularity, Eq. (3) constitutes a useful tool to decompose an areal rainfall intensity at a given location x into two components (Wang et al., 2012): (1) the background (or non-singular, NS) magnitude $c(x)$, which is invariant as measuring scale ϵ changes and is more approximately normal than the original field; and (2) a local "scaling" multiplier, the magnitude of which changes as the measuring scale ϵ changes, according to the local singularity index $\alpha(x)$. When $\alpha(x) < 2$, the rainfall magnitude strikingly increases as the measuring scale ϵ decreases (namely local enrichment); this corresponds to a "peak" singularity. In contrast, when $\alpha(x) > 2$, the rainfall magnitude decreases as ϵ decreases (i.e. local depletion), and it is therefore a "trough" singularity. When $\alpha(x) = 2$, there is no singularity: the rainfall intensity $\rho(x, \epsilon)$ within a $\epsilon \times \epsilon$ area remains the same as scale changes (i.e. $\rho(x, \epsilon) = c(x)$).

In practice however, there is a drawback to this local singularity analysis. Because it carries out a "local" analysis, the singularity exponents are usually obtained from a small number of data samples. This increases the uncertainty of the estimation of $\alpha(x)$. The consequence of this drawback is that the singularity is incorrectly estimated or incompletely extracted; therefore, $c(x)$ is an unreliable or incomplete non-singular value. To circumvent this, two numerical strategies were employed in this study. The first one involves constraining the value of the estimated singularity exponents within a certain range. This can avoid obtaining unreasonably large or small singularity exponents. A number of ranges, symmetric to the non-singular condition (i.e. $\alpha(x) = 2$), were selected for testing. They are (from the widest to narrowest intervals): SIN1 = [0, 4], SIN2 = [0.5, 3.5], SIN3 = [1, 3], SIN4 = [1.5, 2.5] and SIN5 = [1.75, 2.25]. These "truncated" singularity ranges were empirically chosen according to the authors' experience and the fact that the distribution of $\alpha(x)$ is seldom largely skewed to a specific side of $\alpha(x) = 2$. It can be generally expected that, the wider the range is, the more singularity information (both local enrichment and depletion) from the radar images is taken into account in the merging process. The impact of different singularity ranges in rainfall estimation and hydraulic simulation is further discussed in Sect. 3.3.

The second numerical strategy is to decompose the rainfall field using an iterative procedure (Agterberg, 2012b; Chen et al., 2007):

$$c^{(k-1)}(x) = c^{(k)}(x)\epsilon^{\alpha^{(k)}(x)-E}, \tag{4}$$

where the iterative index $k =, 0, 1, 2, \ldots, n$. As $k = 0$, $c^{(-1)}(x) = \rho(x, \epsilon)$ (i.e. the original value) and $c^{(0)}(x)$ is the "calculated non-singular" value from the first iteration, which is equal to $c(x)$ from the non-iterative calculation above (Eq. 3). This $c^{(0)}(x)$ is then used as the left-hand-side value of Eq. (4) to calculate the "non-singular" value at the next iteration, and so on. Substituting Eq. (4) into Eq. (3), one can obtain an iterative local singularity analysis equation:

$$\rho(x, \epsilon) = c^*(x)\epsilon^{\alpha^*(x)-E}, \tag{5}$$

where

$$\begin{cases} c^*(x) = c^{(n)}(x) \\ \alpha^*(x) = \alpha^{(0)}(x) + \sum_{k=1}^{n} \left(\alpha^{(k)}(x) - E\right). \end{cases} \tag{6}$$

The criterion to terminate the iteration procedure is when $\alpha^{(k)}(x) \approx E$ (which is equivalent to $c^{(k-1)}(x) \approx c^{(k)}(x)$). That means the singularity components have been clearly removed from the data.

Moreover, in this work a spatial-scale range of 1–9 km, which results in a total of five rainfall intensity samples (at scales 1, 3, 5, 7 and 9 km), was used in the singularity analysis. This range was selected for two main reasons. Firstly, our analyses revealed that a good linear behaviour was generally observed within this scale range, while small-scale structures were still preserved in the resulting rainfall product. As such, the selected spatial scale range was deemed to represent a good balance between estimation uncertainty (which depends upon the number of samples employed in the calculations) and local feature preservation. Secondly, a scaling break at approximately 8–16 km has been reported in studies in which 1 km radar rainfall data were analysed (Gires et al., 2012; Tchiguirinskaia et al., 2012). This means that the rainfall data at spatial-scale regimes ranging from 1 to 8–16 km comply with the same or similar statistical or physical behaviour. This scaling range has also been used in other applications to represent relatively local characteristics of rainfall fields (Bowler et al., 2006).

Lastly, a 10-iteration singularity analysis was applied in order to ensure that most of the singularity exponents could be extracted. The downside of conducting many iterations is the longer computational time, which may be an issue for real-time applications. Nonetheless, in practice, approximately 4–6 iterations are sufficient for effectively removing most of the singularity.

2.3 Incorporation of the local singularity analysis into the Bayesian merging method

The underlying idea of the proposed method is to use the local singularity analysis to decompose each radar image into a non-singular image and a singularity map before applying the Bayesian merging (step (i) in Fig. 1b). The non-singular radar image (NS-RD), which has a distribution closer to normality (thus being more suitable for Gaussian-based treatments), is merged with the point rain gauge data following the Bayesian procedure. This yields a non-singular Bayesian merged field (NS-BAY). Afterwards, the singularity map is applied back and proportionally to the NS-BAY merged field (step (ii) in Fig. 1b), thus yielding a singularity-sensitive merged field (SIN). This is done by multiplying each pixel

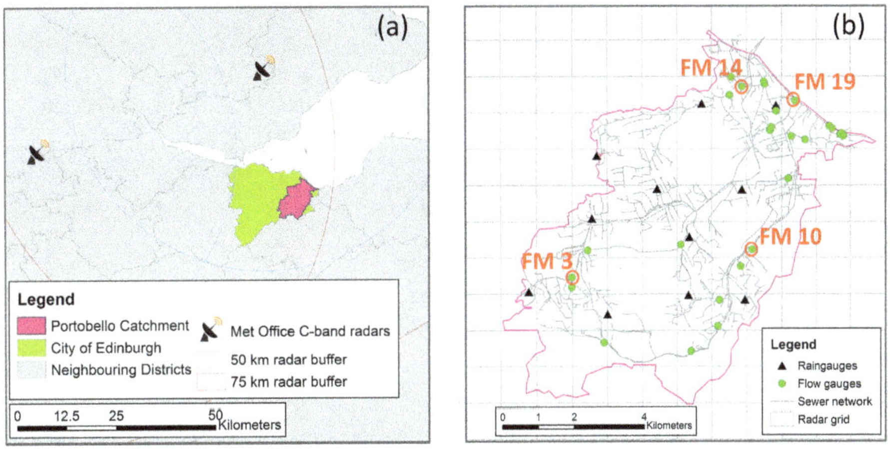

Figure 3. Portobello catchment **(a)** general location; **(b)** sensor location, sewer network and radar grid over the catchment. Flow gauges FM 3, 10, 14 and 19 (the circled round markers in **(b)**, respectively located at the up-, mid- and downstream parts of the catchment) are particularly selected for visual inspection in Sect. 3.3.2.

value of the NS-BAY field by the following ratio:

$$r(x) = \epsilon^{\alpha^*(x)-E} \tag{7}$$

which corresponds to the ratio difference between the original radar field (RD) and the non-singular radar field (NS-RD). In this way, a singularity-sensitive merged field (SIN), which better retains the local singularity structures embedded in the original RD field, is obtained.

It is worth noting that the proposed singularity-sensitive merging method does not always increase the reliability of RD estimates. Such increase only happens when the RD estimates exhibit high singularity and thus cannot be well handled using Gaussian approximations.

A particular phenomenon which may cause problems in the application of the proposed methodology and is therefore worth highlighting is the eventual presence of singularity structures in the interpolated rain gauge field (i.e. BK field) and in the resulting supposedly non-singular Bayesian (NS-BAY) merged field. While BK fields are generally highly smooth, singularity structures may appear in the special case in which a rain gauge is located within a convective cell or a local depletion. Singularity structures in the BK field may be preserved in the NS-BAY field. When this is the case, the application of the singularity map back and proportionally to the NS-BAY field may result in double-counting of singularities. This can ultimately result in a merged (SIN) product with more singularities than those originally observed in the radar image. In order avoid this, a "moving window" smoothing has been applied to the BK field before it is merged with the NS-RD field. That is, each pixel value of the BK field is replaced by the mean of the original value and neighbouring pixel values within a 9 km diameter (which is equal to the coarsest scale considered in the local singularity analysis). In this way singularity structures potentially present in the BK field are smoothed-off.

3 Case study

The proposed SIN merging method is tested using as case study four storm events observed in the Portobello catchment (Edinburgh, UK) during 2011 and for which radar estimates, dense rain gauge and flow records, as well as a recently calibrated urban drainage model were available. Portobello is a coastal town located 5 km to the east of the city centre of Edinburgh, along the coast of the Firth of Forth, in Scotland (Fig. 3a). The catchment is predominantly urban, of residential character. It stretches over an area of approximately 53 km^2, of which 27 km^2 are drained by the sewer system. Of the drained area, 46 % corresponds to impervious surfaces. This includes a small western part of Edinburgh city centre and the surrounding southwestern region. The storm water drainage system is predominantly combined with some separate sewers and drains from the southwest to the northeast (towards the sea).

3.1 Available models and data sets

3.1.1 Urban storm-water drainage model

A semi-distributed model of the storm-water drainage system of the Portobello catchment, including its sewer system (Fig. 3b), was set up by the water utility of the area in the commercial modelling package InfoWorks CS v13.0. In this model the whole catchment surface is split into sub-catchment units through which rainfall is applied (within each sub-catchment rainfall is assumed to be uniform). Each sub-catchment comprises a mix of pervious and impervious surfaces whose runoff drains to a common outlet point, which corresponds to an inlet node of the sewer system (i.e. a gully or a manhole). Each sub-catchment is characterised by a number of parameters, including total area, length, slope, proportion of each land use, amongst others. Based upon

Table 1. Selected rainfall events over the Portobello catchment.

Event	Date	Duration (h)	RG total (mm)	RG peak intensity (mm h^{-1})	RD total (mm)	RD peak intensity (mm h^{-1})
Storm 1	6–7 May 2011	7	9.25	11.21	9.67	7.93
Storm 2	23 May 2011	7	7.70	5.03	11.02	4.10
Storm 3	21–22 Jun 2011	24	32.96	8.46	26.21	3.33
Storm 4	8 Jun 2011	11	5.03	7.11	4.69	2.41

Note: the accumulation and peak intensity values shown in this table correspond to areal mean values for the entire domain under consideration (as shown in Fig. 3b).

these parameters, the runoff volume at each sub-catchment is estimated using the NewUK rainfall–runoff model (Osborne, 2001). The estimated runoff is then routed to the sub-catchment outlet using the Wallingford (double linear reservoir) model (HR Wallingford, 1983). Sewers are modelled as one-dimensional conduits and flows within them are simulated based on the full de Saint-Venant equations (i.e. fully hydrodynamic model).

The Portobello model contains a total of 1116 sub-catchments, with areas ranging between 0.02 and 24.42 ha and a mean area of 2.3 ha. Sub-catchment slopes range from 0.0 to 0.63 m m^{-1}, with a mean slope of 0.031 m m^{-1}. The model of the sewer system comprises 2917 nodes and 2907 conduits, in addition to 14 pumps. The total length of modelled sewers is 250 km. The sewer system ranges in height from 186.6 mAOD at Comiston to 3.8 mAOD along the Firth of Forth. 2 % of the modelled pipes have a gradient between 0.1 and 0.25 m m^{-1}, 55 % have a gradient between 0.01 and 0.1 m m^{-1}, and 43 % have a gradient < 0.01 m m^{-1}.

Following UK standards (WaPUG, 2002) and using solely rain gauge data as input, the model of the Portobello catchment was manually calibrated in 2011 based upon three storm events recorded during the medium-term flow survey described below. For the three storm events used for model calibration, the following mean performance statistics were obtained: Nash–Sutcliffe efficiency of 0.5782, root mean square error of 0.0373 m^3 s^{-1}, coefficient of determination (R^2) of 0.7756 and regression coefficient (β), which provides a measurement of conditional bias, of 0.8826.

3.1.2 Local monitoring data (medium term survey)

Local rainfall and flow data were collected in the Portobello catchment through a medium-term flow survey carried out between April and June 2011. The survey comprised 12 tipping bucket rain gauges and 28 flow monitoring stations (each comprising a depth and a velocity sensor, based upon which flow rates were estimated). Both rain gauge and flow records were available at a temporal resolution of 2 min. However, rain gauge records were linearly interpolated to 5 min, in order to ensure agreement with the temporal resolution at which radar estimates were available (see Sect. 3.1.3). The location of the local flow and rain gauges is shown in Fig. 3b.

3.1.3 Radar rainfall data

The Portobello catchment is within the coverage of C-band radars operated by the UK Met Office (Fig. 3b). Radar rainfall estimates for the same period as the local flow survey (i.e. April–June 2011) were obtained through the British Atmospheric Data Centre (BADC). These estimates were available at spatial and temporal resolutions of 1 km and 5 min, respectively, and correspond to a quality controlled multi-radar composite product generated with the UK Met Office Nimrod system (Golding, 1998), which incorporates corrections for the different errors inherent to radar rainfall measurements, including identification and removal of anomalous propagation (e.g. beam blockage and clutter interference), attenuation correction and vertical profile correction (for a full description of the Nimrod system, the reader may refer to Harrison et al. (2000, 2009)).

3.1.4 Storm events selected for analysis

During the monitoring period (April–June 2011), four relevant storm events were captured which comply with UK standards for calibration and verification of urban drainage models (i.e. these events have instantaneous rainfall rates > 5 mm h^{-1} and accumulation > 5 mm) (Gooch, 2009). Three of these events (referred to as Storms 1, 2 and 3) were used for calibration of the urban drainage model of the Portobello catchment (following UK standards, as mentioned above). In this study, all four storm events, including one not used in the calibration of the model (Storm 4), were used to test the proposed singularity-sensitive merging method. The dates and main statistics of the four selected events are summarised in Table 1. It is worth mentioning that, given the response time of the catchment, as well as inter-event time definition thresholds (IETD) recommended in the literature (Guo and Adams, 1998), a minimum IETD of 6 h was used as criteria to differentiate rainfall events.

As can be seen in Table 1, Storms 1, 2 and 4 are of relatively short duration, whilst Storm 3 is of much longer duration. Moreover, well-structured storm cell clusters crossing the catchment area can be found in Storms 1, 3 and 4, but not in Storm 2. This is reflected in the lower RG peak intensity observed in Storm 2.

3.2 Evaluation methodology

The performance of the proposed singularity-sensitive Bayesian method (SIN hereafter) is assessed by inter-comparison against radar (RD), rain gauge (RG) and block-kriged (BK) interpolated RG estimates, as well against adjusted estimates resulting from the original Bayesian (non singularity-sensitive) technique (BAY) and the commonly used mean field bias (MFB) adjustment method. It is important to note that, in this work, the MFB was implemented in a relatively dynamic way by computing a sample cumulative bias (B_t) at each time step as $B_t = \sum \mathrm{RG}_{\mathrm{catchment}} / \sum \mathrm{RD}_{\mathrm{co\text{-}located}}$, where $\mathrm{RG}_{\mathrm{catchment}}$ and $\mathrm{RD}_{\mathrm{co\text{-}located}}$ represent the rain gauge and the co-located radar grid rainfall estimates over the experimental catchment during the last hour. The MFB adjusted estimates are obtained by multiplying the bias (B_t) at each particular time step by the original rainfall field; that is, $\mathrm{MFB} = B_t \cdot \mathrm{RD}$.

Two evaluation strategies were applied:

1. Through analysis of the different rainfall estimates, using as main reference local rain gauge records, while also inter-comparing the behaviour of other estimates.

2. Through analysis of the hydraulic outputs obtained by feeding the different rainfall estimates as input to the hydraulic model of the Portobello catchment and comparison of these with available flow records. Note that the RG estimates were applied to the model using Thiessen polygons.

Both evaluation strategies have inherent limitations which are next described. However, they provide useful and complementary insights into the performance of the proposed merging method.

The first strategy is a natural and widespread way of assessing the performance of rainfall products. However, the fact that all precipitation estimates entail errors and that the true rainfall field is unknown, in addition to the differences in the spatial and temporal resolutions of RG and RD estimates (and the resulting merged rainfall products), renders any direct comparison of rainfall estimates imperfect (Brandes et al., 2001). In the particular case of rain gauge records, which are used as main reference in the present investigation, errors can arise from a variety of sources (Einfalt and Michaelides, 2008). In order of general importance, systematic errors common to all rain gauges include errors due to wind field deformation above the gauge orifice, errors due to wetting loss in the internal walls of the collector, errors due to evaporation from the container, and errors due to in- and out-splashing of water. In addition, tipping bucket rain gauges, such as the ones used in this investigation, are known to underestimate rainfall at higher intensities because of the rainwater amount that is lost during the tipping movement of the bucket (La Barbera et al., 2002). This is a systematic error unique to the tipping bucket rain gauge and is of the same order of importance as wind-induced losses. Besides these systematic errors, tipping bucket rain gauge records are also subject to local random errors, mainly related to their discrete sampling mechanism (Ciach, 2003; Habib et al., 2008). The order of magnitude of the two main error sources associated with tipping bucket rain gauges is 2–10 % for wind-induced losses (depending on wind speed, weight of precipitation and gauge construction parameters) (Sevruk and Nešpor, 1998) and up to 10 % for rainfall intensities of $100\,\mathrm{mm\,h^{-1}}$ and 20 % for intensities of $200\,\mathrm{mm\,h^{-1}}$, for water losses during the tipping action (Luyckx and Berlamont, 2001; Molini et al., 2005). Given that the storm events under investigation were not extremely heavy and that a basic quality-control was applied to ensure the quality of rain gauge measurements (e.g. manual comparison between neighbouring rain gauge data), the rain gauge records used in this study can be deemed acceptable as reference for the evaluation of other rainfall products. However, they are not perfect and the reader must keep in mind that the true rainfall field remains unknown.

The second strategy (i.e. hydraulic evaluation) allows some of the limitations of the rainfall evaluation strategy to be overcome, and is particularly useful when dense flow records are available, as is the case in the Portobello catchment. However, it has two main deficiencies: the fact that flow records (obtained based upon depth and velocity measurements) used in the evaluation contain errors, and the fact that the hydraulic modelling results encompass uncertainties from different sources in addition to rainfall input uncertainty (Deletic, 2012; Kavetski et al., 2006). In this regard, it is worth reminding the reader that the hydraulic model used in this study was calibrated using as input RG data. In fact, the model was calibrated based upon the data of Storms 1–3, which are used for testing in the present paper (Storm 4 is the only event not used in the calibration of the model). Since the model was "attuned" for RG inputs, it favours all RG-derived (and RG-emulating) rainfall estimates. Moreover, the relatively coarse spatial resolution of rainfall inputs (in this case $\sim 1\,\mathrm{RG}/4.4\,\mathrm{km^2}$) may have led to further biases in the model (Kavetski et al., 2006). It would be desirable to re-calibrate the hydraulic model using as input the different rainfall products analysed in this study. However, this would entail a significant amount of work which falls outside of the scope of the present study. In spite of these limitations, we believe that the hydraulic evaluation strategy using the currently available hydraulic model still provides useful insights into the performance of the proposed SIN merging method in relation to other rainfall estimates, and complements the findings of the rainfall analysis.

3.2.1 Methodology for analysis of rainfall estimates

The performance of the SIN rainfall products in relation to other rainfall estimates (including RD, RG, BAY and MFB) is evaluated in terms of accumulations and rainfall rates at the areal level (i.e. at a scale corresponding to the area over which the Portobello catchment stretches) and at individual point gauge locations. In addition, a qualitative assessment of the spatial structure of the different (gridded) rainfall products is carried out based upon visual inspection of images of the rainfall fields at the time of areal average peak intensity.

In view of the high density and coverage of the RG network over the Portobello catchment, the areal average RG estimates in the areal level analysis are assumed to be a good approximation of the "true" areal (average) rainfall over the experimental catchment (i.e. the areal reduction effect is expected to be minor – Bell, 1976) and are therefore used as reference to evaluate the performance of the different areal gridded rainfall estimates. The areal analysis includes comparison of event areal average accumulations and peak intensities, as well as comparison of intensities throughout the duration of each event (through scatterplots using RG areal average intensities as reference).

In the analysis of rainfall estimates at rain gauge point locations a cross-validation strategy was adopted and three performance statistics are used. The cross-validation strategy, also referred to as "leave-one-out", is an iterative method in which, at each iteration, data from one RG site is omitted from the calculations and the value at the "hidden" (i.e. omitted) location is estimated using the remaining data. Performance statistics are then computed from the comparison between the estimated and the known (but not used) values (Velasco-Forero et al., 2009). The following three performance statistics are used in the present study. Firstly, a sample bias ratio (B) is used to quantify the cumulative bias between gridded rainfall estimates (i.e. RD, BK, MFB, BAY and SIN) and RG estimates at each RG location over the event period under consideration. $B = 1$ means no cumulative bias between the RG and the given gridded rainfall estimates (i.e. equal rainfall accumulation recorded by RG and the gridded product at the given gauge location); $B > 1$ means that the accumulations of the gridded estimates at the point locations are greater than those recorded by RG, and $B < 1$ means the opposite. In addition to the comparison of rainfall accumulations, a simple linear regression analysis is applied to each pair of "instantaneous" (rain rate) point RG records and the co-located gridded estimates. The results of the regression analysis are presented in terms of R^2 (coefficient of determination) and β (regression coefficient, i.e. the slope or gradient of the linear regression). These two measures provide an indication of how well RG rates are replicated by the different rainfall estimates at each gauging location, both in terms of pattern and accuracy. The R^2 measure ranges from 0 to 1 and describes how much of the "observed" (RG) variability is explained by the "modelled" (RD/BK/adjusted) one. In practical terms, R^2 provides a measurement of the similarity between the patterns of the observed (i.e. RG) and "modelled" (i.e. gridded estimates) rainfall time series at a given gauging location. However, systematic bias (under- or over-estimation) of the modelled estimates cannot be detected from this measure (Krause et al., 2005). The regression coefficient, β, is therefore employed to provide this supplementary information to the R^2. $\beta \approx 1$ represents good agreement in the magnitude of the rainfall rates recorded by RG and those of the gridded estimates; $\beta > 1$ means that the rain rates associated with the gridded estimate are higher in the mean (by a factor of β) than those recorded by RG; and $\beta < 1$ means the opposite (i.e. rain rates of gridded estimates are lower in the mean than RG ones).

3.2.2 Methodology for analysis of hydraulic outputs

A qualitative analysis of the hydraulic outputs is carried out based upon visual inspection of recorded vs. simulated flow hydrographs (for the different rainfall inputs) at different points of the catchment. Furthermore, similar to the rainfall analysis, a simple linear regression analysis is applied to each pair of recorded and simulated flow time series (at each flow gauging location). The performance of the associated hydraulic simulations is evaluated using the R^2 and β statistics obtained from the linear regression analysis. In addition, the "weighted" coefficient of determination (R^2_{w}) is employed to quantify the joint performance of hydrological efficiency. This measure is defined as (Krause et al., 2005)

$$R^2_{\mathrm{w}} = \begin{cases} |\beta| \cdot R^2 & \text{for } \beta \leq 1 \\ |\beta|^{-1} \cdot R^2 & \text{for } \beta > 1, \end{cases} \tag{8}$$

where higher R^2_{w} values correspond to better hydraulic performance.

In order to minimise the influence of the errors in the flow measurements, the available flow records were quality-controlled (QC) before carrying out the statistical analysis of hydraulic outputs. The QC was carried out following UK guidelines (WaPUG, 2002). It included analysis of depth vs. flow scatterplots at each monitoring location (the shape and spread of the resulting scatterplots provides insights into the quality and consistency of depth and velocity records at each site), as well as visual inspection of the observed hydrographs at each location. Whenever a flow monitor was deemed unreliable, it was manually removed from the analysis. Likewise, with the purpose of preventing systematic hydraulic modelling errors from affecting results, whenever the model was found to be unable to replicate the recorded flows at a given location, the given flow monitor was also manually removed from the analysis. This left us with a total of 16 flow monitors for analysis. Moreover, all records associated with depth measurements below 0.1 m were left out when estimating performance statistics; this is due to the fact that at low depths, both velocity and depth records become unreliable (the 0.1 m threshold was adopted based upon UK guidelines

Table 2. Areal average rainfall accumulations and peak intensities for the different rainfall products.

	Input	Storm 1	Storm 2	Storm 3	Storm 4
AVG rainfall accumulation (mm)	RG	9.25	7.70	32.96	5.03
	RD	9.80	11.02	26.21	4.69
	BK	8.87	7.77	31.41	4.08
	MFB	8.64	7.25	31.87	4.55
	BAY	8.76	7.79	27.24	3.81
	SIN1	9.37	7.95	33.37	4.82
	SIN2	9.37	7.95	33.38	4.83
	SIN3	9.38	7.95	34.43	4.85
	SIN4	9.41	7.96	33.73	4.94
	SIN5	9.61	8.04	31.57	5.23
AVG rainfall peak intensity over 5 min ($mm\,h^{-1}$)	RG	11.21	5.03	8.46	7.11
	RD	7.93	4.10	3.33	2.41
	BK	10.66	4.54	7.59	5.09
	MFB	9.06	4.68	4.05	3.34
	BAY	10.47	3.80	6.82	4.92
	SIN1	13.17	5.08	8.00	6.97
	SIN2	13.17	5.08	8.00	6.97
	SIN3	13.17	5.08	8.00	6.98
	SIN4	13.19	5.08	8.01	7.03
	SIN5	13.53	5.09	8.27	7.25

and recommendations from studies focusing on the performance of flow gauges; Marshall and McIntyre, 2008).

3.3 Results and discussion

3.3.1 Rainfall estimates

Areal rainfall estimates

Table 2 shows the areal average (AVG) accumulations and peak intensities for the different rainfall products for the four storm events under consideration. As can be seen, the difference between RD and RG event areal average accumulations is generally small, yet it is event-varying. For Storms 1 and 2, the areal average RG totals are slightly overestimated by the RD estimates, while for Storms 3 and 4 they are underestimated, with the underestimation being largest in Storm 3. In terms of areal average peak intensities, RD estimates appear to consistently underestimate the areal average peak intensities recorded by RG. The relative difference between RG and RD areal average peak intensities is approximately 20–30 % for Storms 1 and 2, and it is as high as 60 % for Storms 3 and 4. This indicates that the RD estimates could not satisfactorily capture instantaneous rainfall rates, particularly high rainfall rate values, and corroborates the need for dynamic adjustment of RD estimates using local RG measurements.

As would be expected, the BK estimates exhibit areal average accumulations and peak intensities similar to those of the RG. Small differences are observed (in general BK values are slightly lower than RG ones) which can be generally attributed to the area-point rainfall differences (Anagnostou and Krajewski, 1999). These differences become more evident when analysing results at individual gauge locations (in next section).

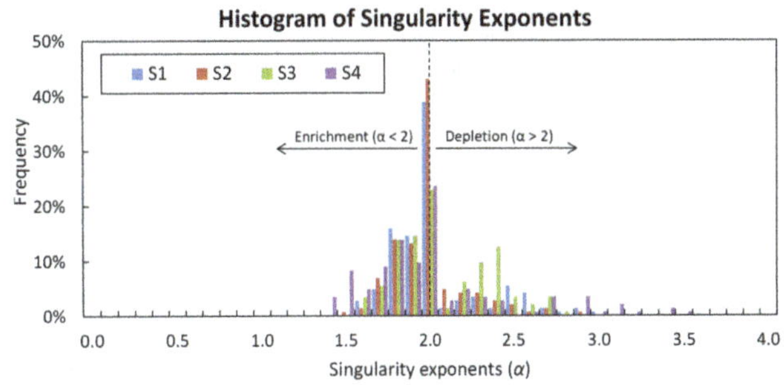

Figure 4. Histogram of singularity exponents (α) at the time of areal peak intensity for the four selected storm events.

When looking at the adjusted rainfall products (i.e. MFB, BAY and SIN), it can be seen that all of them can improve the original RD estimates, but the degree of improvement is different for each method. As expected, the MFB successfully reduces the difference in event areal average accumulations (i.e. bias), leading to areal average accumulations close to those recorded by RG. In terms of peak intensities, the MFB method leads to some improvement, but the resulting peak intensities are still significantly lower than the RG ones. Although the MFB was applied dynamically with an hourly frequency of bias correction, these results suggest that more dynamic and spatially varying (higher order) methods than the MFB are required in order to successfully adjust radar rainfall estimates for urban hydrological applications.

The BAY estimates show the least improvement in terms of event bias, with a general tendency to underestimate RG areal accumulations, which is even more marked than for BK estimates. This is particularly the case in Storms 3 and 4, in which strong singularity structures, as represented by the high frequency of α values different from 2 (Fig. 4), were observed. This tendency to underestimate can be attributed to a combined effect of the BAY method "over-trusting" the BK estimates (which show a slight underestimation tendency at the areal level), in addition to smoothing off the singularity structures (often associated to strong intensities) originally present in the RD image. With regard to peak intensities, the BAY estimates display a larger improvement than the MFB ones, which demonstrates the benefits of more dynamic and spatially varying adjustment methods. However, the areal average peak intensities of the BAY estimates still underestimate the "true" (RG) areal peak intensities. Lastly, the SIN estimates, particularly SIN ranges 1–4, exhibit very good performance: both areal average accumulations and peak intensities of SIN estimates are close to those of RG, and no systematic over- or underestimation is observed in SIN estimates. The better performance of SIN estimates suggests that the singularity analysis can in fact improve the original BAY merging method. With regard to the impact of the singularity range, it can be seen that, as the range becomes narrower

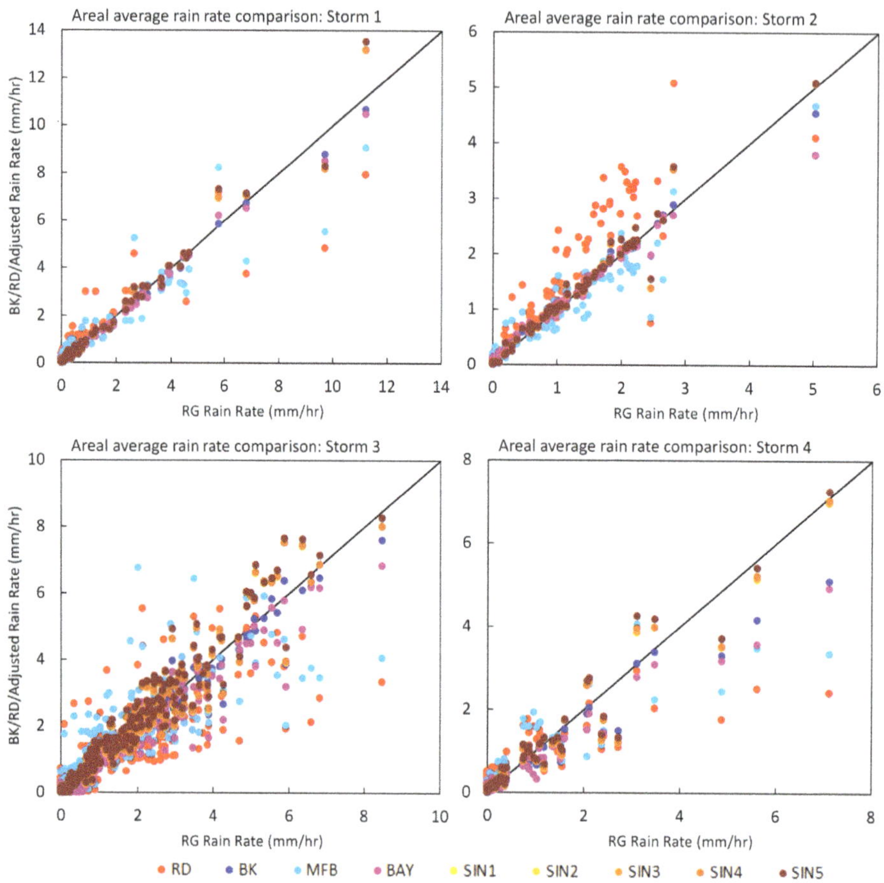

Figure 5. Scatterplots of instantaneous areal average RG vs. RD/BK/MFB/BAY/SIN rainfall rates over the Portobello catchment for the four selected events, where SIN1–SIN5 represent the SIN estimates with different "truncated" singularity ranges (from widest to narrowest).

(from SIN1 to SIN5), the areal accumulations and peak intensities tend to be higher. A particular steep increment is observed in the areal accumulations and peak intensities of SIN5, in relation to the other singularity ranges, which results in a slight overestimation of accumulations and peak intensities, as compared to the RG estimates. This is especially apparent in Storms 3 and 4. The increment in rainfall accumulation and peak intensities as the singularity range becomes narrower can be explained by the fact that at narrower ranges, only part of the singularity structures are removed before merging, and a big proportion of them remains in the radar image. This is evident from Fig. 4, where it can be seen that for Storms 3 and 4 a significant number of singularity exponents spread beyond the narrowest singularity range (i.e. SIN5: $\alpha \in [1.75, 2.25]$). The problem arises because some singularity features remain in the radar image before the BAY merging is applied, and these may be partially preserved throughout the merging process. Afterwards, when the extracted singularity component is applied back and proportionally to the merged rainfall field, it may interact (in a nonlinear fashion) with the singularity structures preserved throughout the BAY merging, thus leading to an overestimation of extremes (whether these are enrichments or depletions). These results suggest that a better approach is to be sure to remove most of the singularity structures before carrying out the merging. Therefore, very narrow ranges such as SIN5 should be avoided. In fact, it can be seen that the intermediate range of SIN3 (i.e. $\alpha \in [1, 3]$) covers most singularity exponents (see Fig. 4). Indeed, using wider singularity ranges leads to very similar results as those obtained when using the SIN3 range (notice the similarity between SIN1, SIN2 and SIN3 estimates in Table 2). This suggests that the singularity range of $[1, 3]$ is appropriate.

Figure 5 shows a further comparison of instantaneous areal average RG intensities vs. areal average BK, RD, BAY and SIN intensities throughout each of the storm events under consideration. As expected and in line with the analysis above, the areal BK estimates are generally in good agreement with areal RG estimates. Some underestimations can be observed at high RG rainfall intensities; nonetheless, most of them are fairly minor and are still within a reasonable range, which can be attributed to areal-point differences. With regard to the RD estimates, it can be seen that they tend to overestimate small rainfall intensities and underestimate the peak intensities, with underestimation being more evident in the events with relatively high intensities (i.e. Storms 1, 3 and 4). Unlike BK estimates, the difference between areal RD and RG peak intensities is too large to be entirely explained by areal-point differences. The fact that the RD estimates display relatively good performance in (long-duration) accumulations but not in (short-duration) instantaneous intensities corroborates the claim that RD estimates fail to capture the short-term dynamics of small-scale rainfall. As mentioned in Sect. 1, the main reason for this lies in the use of a static Z–R conversion function, which represents a statistical compromise for the range of rainfall rates that frequently occur (whereas the occurrence of very small and large intensities is relatively rare). Concerning the MFB method, from Fig. 5 it can be seen that it fails to satisfactorily improve RD instantaneous rainfall rates; this is particularly evident at high intensities, at which, similarly to RD estimates, MFB estimates perform poorly. An accurate representation of peak intensities is of outmost importance in the modelling and forecasting of urban pluvial flooding. This confirms that, being a first-order technique, the MFB adjustment method may be insufficient for urban-scale applications. In contrast, it can be seen that over- and underestimation errors in RD estimates can be improved by second- or higher-order adjustment techniques, such as BAY and SIN. In fact, in terms of instantaneous rainfall rates the BAY estimates display a significantly better performance than the MFB ones. However, the BAY estimates still fail to properly reproduce the highest intensities. These shortcomings of the BAY method seem to be overcome by the incorporation of the singularity analysis: indeed, the SIN estimates exhibit the best overall performance, particularly at peak intensities. In agreement with the results displayed in Table 2, in Fig. 5 it can be seen that as the singularity range becomes narrower, rainfall estimates with slightly higher intensities are generated. Moreover, it can be noticed that wider singularity ranges lead to more conservative results and appear to be a good choice.

Rainfall estimates at gauging locations

The aforementioned features of the different rainfall estimates are further highlighted through analysis at each rain gauge location; the associated statistics, including sample bias (B), regression coefficient (β) and coefficient of determination (R^2), are summarised in Fig. 6.

As expected, the RD estimates (before adjustment) display the largest differences from point RG estimates: in general, they possess the largest cumulative bias (B) and the lowest R^2, and their statistics show great variability. Moreover, the distribution of the β values indicates that RD estimates tend to largely underestimate RG instantaneous rainfall rates. This is the case for all storm events, except for Storm 2, for which rainfall intensities were low on average.

Similarly to the results of the areal (average) analysis, the individual-site BK estimates display the closest behaviour to the RG ones. This is of course expected given that the BK estimates are obtained by simple interpolation of point RG data. It can be seen that the BK estimates are nearly unbiased (B is very close to 1) and possess the highest R^2 medians (i.e. closest to 1), as well as the narrowest R^2 boxes and whiskers. However, when looking at the distribution of β values of the BK estimates, it can be seen that most of the time the whole boxes and whiskers are below 1. This reflects a systematic underestimation of rainfall rates at point gauging locations, which is discussed below.

With regard to the adjusted rainfall estimates, the MFB method is found to bring original radar estimates slightly closer to RG ones, but the improvement seems insufficient. As expected, the main improvement of MFB estimates is found in the bias (B), which is significantly reduced (thus becoming closer to 1). In terms of instantaneous rainfall rates, the improvement provided by MFB is very limited. This is reflected in the low R^2 values and in the poor β scores, which remain remarkably close to those of the original RD estimates. Similar to the results of the areal analysis, these results suggest that the MFB adjustment method is insufficient for urban-scale applications, in which small-scale rainfall dynamics are critical.

When looking at the statistics of the BAY estimates, it can be noticed that these behave similarly to the BK ones: their bias is also small, the R^2 is generally high and the β values are systematically below 1. The similarity in the behaviour of BK and BAY estimates suggests that for the selected events, the BAY method tends to trust the (smooth interpolated) BK estimates more than the RD estimates. As explained in Sects. 1 and 2, this is the main shortcoming of the BAY method. The systematic underestimation of rainfall rates observed in BK and BAY estimates (reflected by β values systematically below 1) can be partially explained by the areal reduction effect. However, considering that the individual-site comparison was conducted using instantaneous rainfall estimates for very short time intervals (i.e. 5 min), one would expect the tendency of the areal-point differences to be of higher randomness, rather than of a systematic nature. This suggests that the systematic underestimation may be a joint consequence of the areal reduction effect and of the underlying second-order approximation (which smooths off some local extreme magnitudes).

With regard to the SIN estimates, it can be seen that their bias is small (close to 1) and that the distribution of their R^2 values is somewhere between that of the BAY and RD estimates. This indicates that, as compared to the original BAY estimates, the SIN method incorporates more spatial features from the RD estimates throughout the merging process, while retaining the accuracy of the RG estimates. In terms of β, it can be seen that although the median values are usually below 1, they are generally much closer to 1 than other rainfall estimates. Moreover, their distribution is more variable than that associated with BK and BAY estimates (which display a systematic underestimation). In line with the re-

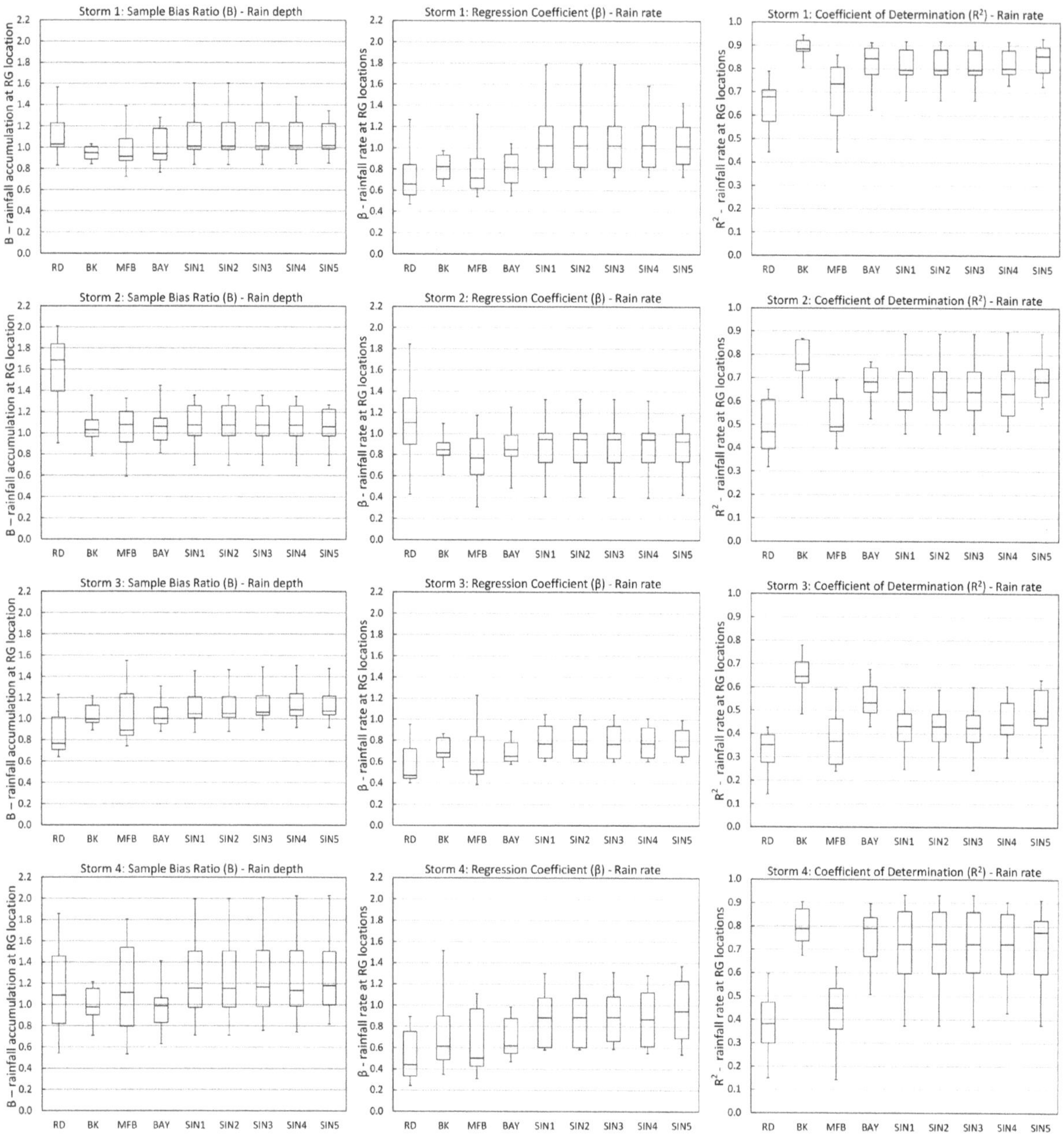

Figure 6. Boxplots displaying the distribution of sample bias ratio (B) (left column panels), regression coefficient (β) (middle column panels) and coefficient of determination (R^2) (right column panels) estimated between the different gridded rainfall estimates and RG records at individual rain gauge locations following a cross-validation approach.

sults of the areal analysis, this serves to further highlight two important features of the proposed SIN method: it generally respects the commonly observed areal reduction effect, and it integrates more small-scale randomness from RD data in the data-merging process. Regarding the singularity ranges, a similar behaviour can be observed for all of them, although a slight tendency can be observed for the bias (B) and β values to increase as the singularity range becomes narrower. This is in agreement with the results of the areal analysis and suggests that working with intermediate ranges, as opposed to very narrow ones which can truncate the singular structures, is advisable.

Spatial structure of rainfall fields

Snapshot images of the different gridded rainfall products at the time of peak areal intensity for the four storm events under consideration are shown in Fig. 7. Due to space constraints, images of only one of the SIN ranges (the intermediate one, SIN3: $\alpha \in [1, 3]$) are shown in this figure. As men-

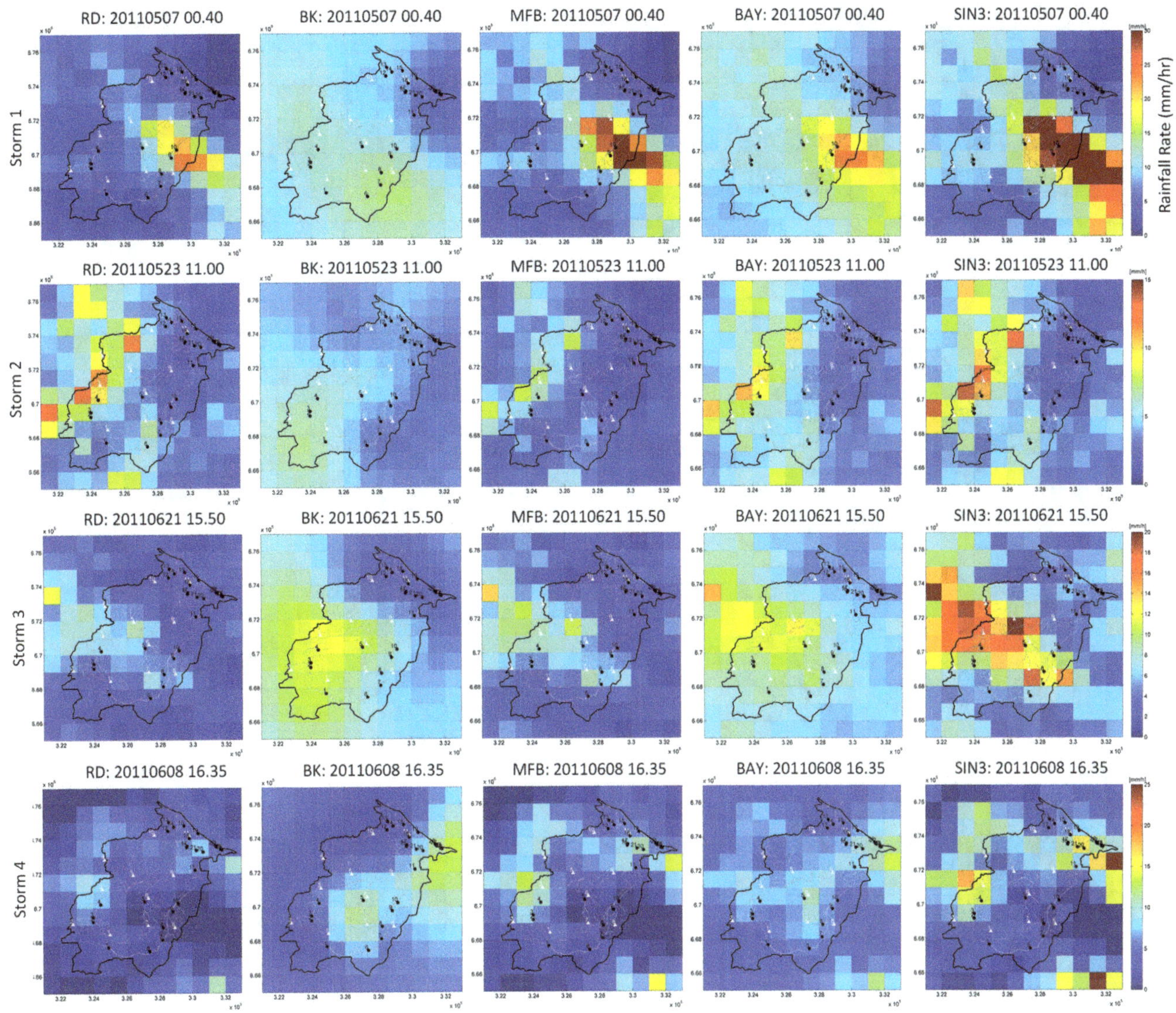

Figure 7. Snapshot images of the different spatial rainfall products at the time of peak areal intensity for Storms 1 (top panels) to 4 (bottom panels) over the Portobello catchment. From left to right panels: RD, BK, MFB, BAY and SIN3 (with singularity range [1, 3]) estimates. The black polygon indicates the boundary of the Portobello catchment, and the black and white markers respectively represent the location of flow and rain gauges.

tioned in the previous sections, the SIN3 range covers most of the singularity indices and consistently led to good results for the storms under consideration.

It can be seen that the spatial structure of the BK rainfall field (fully based upon rain gauge data) is highly symmetric and smooth, and is rather unrealistic. With regard to the adjusted rainfall products (MFB, BAY and SIN), it can be noticed that the proportion of radar (RD) and BK interpolated rain gauge features that are preserved varies according to the method. The MFB fields fully inherit the spatial structure of the RD fields; the only change is that the actual intensity values are scaled up or down by an areal ratio derived from the sample bias between mean rain gauge and radar rainfall estimates. In agreement with the quantitative results presented above, it can be seen that the structure of the BAY peak rainfall fields is often similar to that of the BK ones and is smoother than the original RD image. Singularity structures are often present in rainfall fields during peak intensity periods (such as the ones shown in Fig. 7). As explained in Sect. 2, the presence of these structures causes the RD fields to be considered highly uncertain and therefore these are less trusted in the BAY merging process. This results in BAY peak intensity merged fields closer to the BK ones, instead of to RD ones. Some spatial features from RD can be still observed in the BAY fields – for example, in the lower-right area of the BAY image of Storm 1 and in the

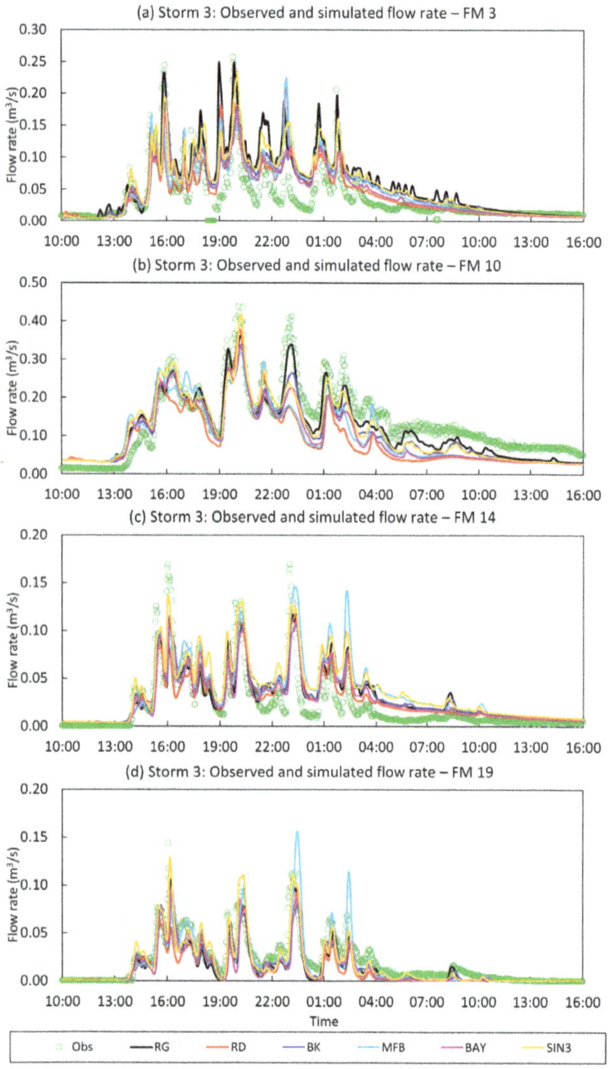

Figure 8. Observed flows vs. simulated flows with RG, RD, BK, BAY and SIN3 rainfall inputs at selected flow gauging sites of Portobello catchment during Storm 3. Selected gauging sites: FM3: upstream end of the catchment (top panels); FM10: mid-stream area (bottom panels). The location of the selected monitoring sites is shown in Fig. 3.

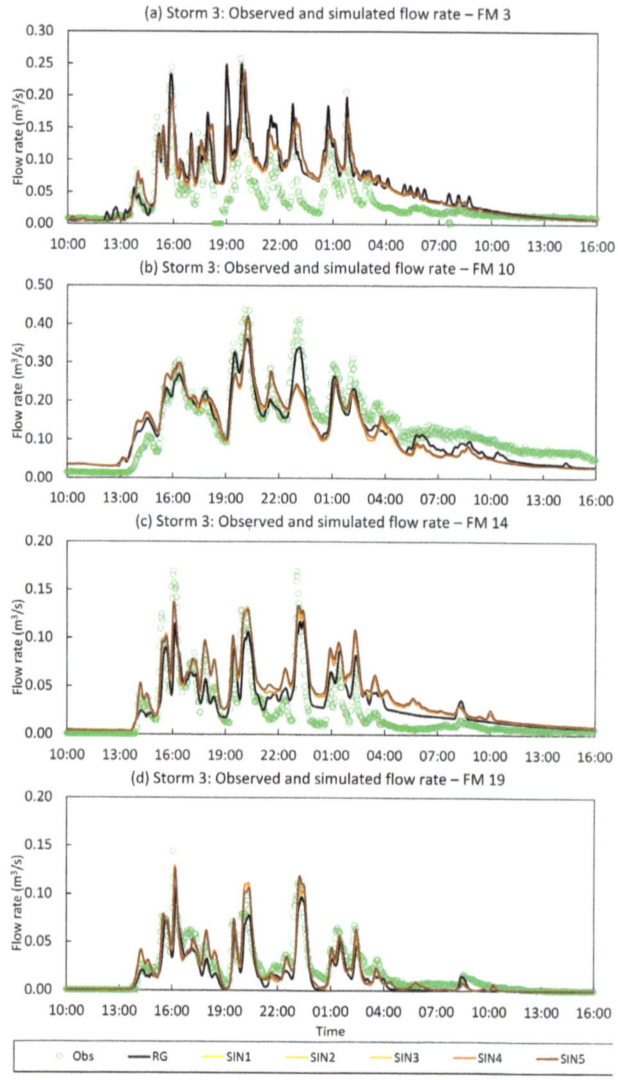

Figure 9. Observed flows vs. simulated flows with SIN1–SIN5 rainfall inputs at selected flow gauging sites of Portobello catchment during Storm 3. Selected gauging sites: FM14: upstream end of a small branch of the sewer system (top panels); FM19: downstream end of the catchment (bottom panels). The location of the selected monitoring sites is shown in Fig. 3.

middle-left area of the BAY image of Storm 3 (see Fig. 7, top). However, these features appear to be much smoother and spreading over a larger area, as compared to their structure in the original RD image. As compared to the BAY peak rainfall fields, the SIN fields display less smooth and more realistic structures which preserve more features of the original RD fields. In general, the inspection of the snapshot images of the different rainfall products confirms the findings of the areal and point gauge analyses regarding the ability of the SIN method to better preserve the singularity structures present in rainfall fields (and captured by RD) throughout the merging process, as compared to the original BAY merging method.

3.3.2 Hydraulic modelling results

Figures 8 and 9 shows example observed vs. simulated flow hydrographs for the different rainfall inputs at four gauging locations (see locations in Fig. 3) during Storm 3. Note that Storm 3 is an event with high rainfall accumulations, rainfall rates and strong singularity structures. Figure 10 summarises the performance statistics resulting from the simple linear regression analysis (i.e. β, R^2 and R^2_{w}) conducted at each flow gauging station for each storm event.

From the hydrographs in Figs. 8 and 9 it can be seen that, in terms of pattern and timing, all simulated flows (resulting from the different rainfall inputs) are generally in good

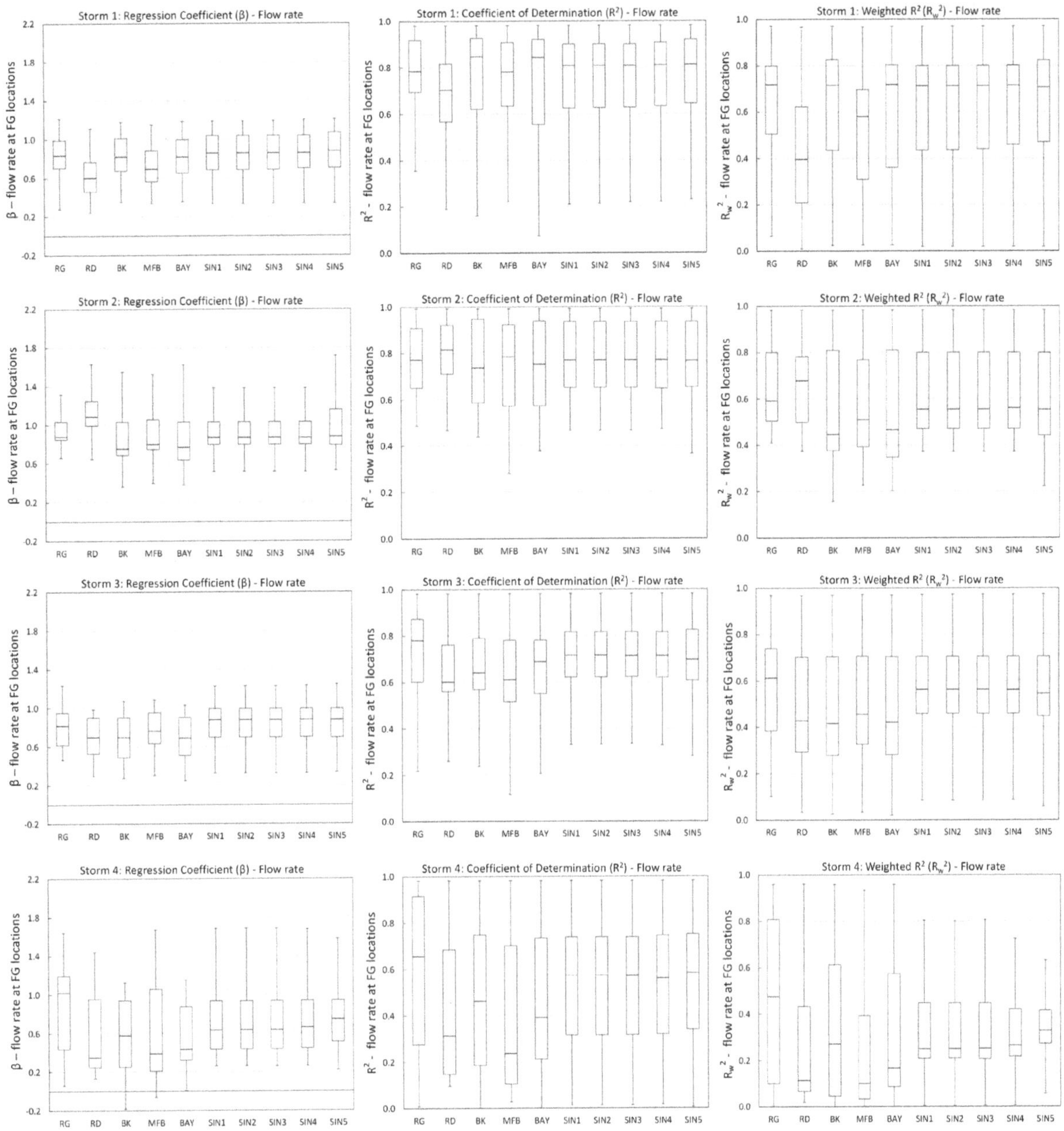

Figure 10. Boxplots displaying the distribution of regression coefficient (β) (left column panels), coefficient of determination (R^2) (middle column panels) and weighted coefficient of determination (R_w^2) (right column panels) statistics derived from the linear regression analysis conducted for each pair of recorded and simulated flow time series at each gauging location.

agreement with observations. This indicates that all rainfall products, including the RD before adjustment, can well capture the general dynamics of rainfall fields. The main difference between the simulated flows lies in their ability to reproduce flow peaks, which in turn is a function of the ability of the different rainfall estimates to reproduce peak rainfall rates in terms of magnitude, timing and spatial distribution. In line with the results of the rainfall analysis, the flow hydrographs associated with BK estimates are close to the ones associated with RG records; however, the former display a smoother behaviour and generally lead to flow peaks that are lower than the recorded ones and the ones resulting from RG inputs. The RD outputs can well match some of the observed flow peaks, but significantly underestimate others (e.g. flow peak at around 23 h in FM10); this can be attributed to the underestimation of peak intensities observed in RD estimates. The MFB associated flows show little improvement over the RD ones and in many cases they even lead to a worse performance (e.g. see overestimation of peak flows at around 02:00 UTC in FM14 and FM19). This is further confirmed by the statistics in Fig. 10 and corroborates the claim that the application of a "blanket" MFB correction over the area of interest is insufficient for urban applications. The BAY outputs, on the other hand, show a consistent improvement over the RD outputs. Nonetheless, in agreement with the results of the rainfall analysis, the BAY estimates behave similarly to the BK ones and lead to smooth flow peaks which often underestimate observations. The SIN outputs also show a consistent improvement over RD estimates, but, unlike BAY outputs, the SIN outputs do not smooth off flow peaks and instead show a better ability to reproduce these, sometimes leading to a better match of observed flow peaks than RG associated outputs (which were used as input for the calibration of the model). With regard to the singularity ranges, it can be seen that the differences observed in the SIN1–SIN5 rainfall estimates are mostly filtered out when converting rainfall to runoff. As a result, the flow outputs of all SIN estimates are very similar and their hydrographs can barely be differentiated.

The preliminary conclusions drawn from the visual inspection of the selected hydrographs are corroborated by the statistics in Fig. 10. As would be expected, the RG outputs generally show the best performance in all statistics (except for Storm 2). It is nonetheless noteworthy that the β values for RG estimates (for which the model was calibrated) are generally below 1. This reveals a slight bias of the model to underestimate flows and partially explains the fact that β values associated with all rainfall inputs are mostly below 1. With regard to the BK (i.e. interpolated RG) associated outputs, the tendency to underestimate, which is observed in the rainfall analysis, becomes even more evident in the hydraulic outputs: BK's β values are significantly lower than 1 and lower than the β values associated with RG estimates. Different from the results of the rainfall analysis, in which those products closest to RG estimates (including BK) displayed the best performance in terms of R^2, in the hydraulic analysis BK outputs generally lead to a deterioration in R^2 values (see statistics of Storms 2, 3 and 4). This suggests that the smoothing caused in the BK interpolation affects the small-scale dynamics of rainfall fields, leading to poor representation of associated flow dynamics. Contrary to RG-associated outputs, RD flow estimates generally display the worst performance. In agreement with the results of the rainfall analysis, RD outputs show a tendency to largely underestimate flows (as indicated by β values well below one and much lower than those obtained for RG outputs). Moreover, they display relatively low R^2 and associated R_w^2 values. Nonetheless, a special case is observed in Storm 2, when RD estimates, which in the rainfall analysis displayed the poorest performance, yielded the best flow simulations, thus emphasising the added value that RD estimates can provide, as well as the complementary information provided by the hydraulic evaluation strategy. In the cases in which RD outputs perform poorly (i.e. in Storms 1, 3 and 4), all adjusted rainfall estimates lead to improvements over RD hydraulic results, with the degree of improvement varying according to the adjustment technique. In Storm 2, when RD outputs displayed the best performance, the different merging methods showed to retain different degrees of RD features. Overall, it can be seen that the MFB estimates provide little improvement over the original RD estimates. In the cases in which RD led to systematic underestimation of flows, the MFB estimates managed to slightly reduce this underestimation by bringing β values closer to 1, as compared to those of the original RD estimates. However, in Storm 2, in which RD outputs performed best, MFB caused a large deterioration of β values, whereas the SIN estimates managed to keep β scores closer to 1. In terms of R^2, the MFB estimates do not provide much improvement and can actually lead to a deterioration of this statistic (e.g. Storms 2 and 4), suggesting that the application of the MFB adjustment can alter the spatial-temporal structure of the original RD fields. The BAY outputs show a greater improvement than MFB, particularly in terms of R^2. Nonetheless, in agreement with the rainfall analysis and with the visual inspection of hydrographs, the BAY outputs behave remarkably similarly to BK ones. One of the main features of the BAY outputs is that they lead to systematically lower flows than RG estimates (note β values consistently lower than those of RG outputs). This confirms the smoothing of rainfall peaks that occurs when second-order approximations are applied. Lastly, it can be seen that the SIN outputs display the greatest improvement over original RD outputs, both in terms of correcting systematic bias, as well as in terms of well reproducing rainfall and associated flow patterns. As compared to MFB and BAY outputs, the SIN outputs generally display β values closer to 1, higher R^2 values (sometimes even higher than those of RG outputs) and consequently higher R_w^2 values. The better performance of SIN hydraulic outputs over BAY ones, particularly in terms of β, provides an a posteriori confirmation that it was right, in the SIN method, to view the singularities in the radar field as actual features of the real rainfall. Similarly as was observed in the hydrographs in Figs. 8 and 9, the performance of the different SIN ranges is very similar. In line with the results of the rainfall analysis, SIN5 shows a slight tendency towards higher flows, which sometimes resulted in better hydraulic statistics. Nonetheless, the differences are small and, based upon the findings of the rainfall analysis, the adoption of a wider SIN range and removal of most singularities appears to be a more conservative option. However, this aspect must be further investigated using a wider range of storm events and pilot catchments.

Regarding the difference in hydraulic performance between the events used for model calibration (i.e. Storms 1–3) and the independent event (i.e. Storm 4), it can be seen that the statistics of the hydraulic outputs during Storm 4 are generally lower than for the other three storm events (Fig. 10). This can be expected, given that the model was "tuned" to give a good fit for the calibration events. However, as discussed above, the general features and relative performance of the hydraulic outputs associated with the different rainfall inputs was generally consistent in all storm events under consideration. The particular differences that were observed were due to the nature of a given storm and not to the fact that a given event was used for model calibration or not.

4 Conclusions and future work

In this paper, a new gauge-based radar rainfall adjustment method was proposed, which aims at better merging rainfall estimates obtained from rain gauges and radars, at the small spatial and temporal scales characteristic of urban catchments. The proposed method incorporates a local singularity analysis into the Bayesian merging technique (Cheng et al., 1994; Todini, 2001). Through this incorporation, the merging process preserves the fine-scale singularity (non-Gaussian) structures present in rainfall fields and captured by radar, which are often associated with local extremes and are generally smoothed off by currently available radar–rain gauge merging techniques, mainly based upon Gaussian approximations.

Using as case study four storm events observed in the Portobello catchment (53 km^2) (Edinburgh, UK) in 2011, the performance of the proposed singularity-sensitive Bayesian data-merging (SIN) method, in terms of adjusted rainfall estimates and the subsequent runoff estimates, was evaluated and compared against that of the original Bayesian data-merging (BAY) technique and the widely used mean field bias correction (MFB) method. This analysis clearly brought out the benefits of introducing the singularity-sensitive method. The results suggest that the proposed SIN method can effectively identify, extract and preserve the singular structures present in radar images while retaining the accuracy of rain gauge (RG) estimates. This is reflected in the better ability of the SIN method to reproduce instantaneous rainfall rates, rainfall accumulations and associated runoff flows. This method clearly outperforms the commonly used MFB adjustment, which simply fails to reproduce the dynamics of rainfall in urban areas, and the original BAY method, which shows an overall good performance but smooths off peak rainfall magnitudes, thus leading to underestimation of runoff extremes.

In this study the sensitivity of the SIN results to the "degree" of singularity that is removed from the radar image and preserved throughout the merging process was also tested. While the impact of it was found to be generally small, the results suggest that partially removing singularities could have a negative impact on the results. Therefore, removing most singularity exponents from the original radar image is advisable.

While the proposed singularity method has shown great potential to improve the merging of radar and rain gauge data for urban hydrological applications, further testing including more storm events and pilot catchments is still required in order to ensure that the results are not case specific and to draw more robust conclusions about the applicability of the proposed method. Other aspects on which further work is recommended are the following:

a. The current version of the singularity-sensitive method shows a slight tendency to overestimate rainfall rates and accumulations. This is likely to be due to one of two aspects, or a combination of them:

   - In the eventual case in which a rain gauge is located within the core of a convective cell, the resulting interpolated (block-kriged; BK) field may end up having singularity structures and, as explained in Sect. 2.3, this may ultimately lead to "double-counting" of singularities. For the time being, a moving-window smoothing has been applied on the BK field before it is merged with the non-singular radar field, so as to remove singularity structures potentially present in the BK field. While this has proven to be an acceptable solution, we believe it can be further refined. Other methods for dealing with this particular problem have been tested, including extraction of singularities from the point rain gauge records using the singularity exponents derived from the co-located radar pixels, and extraction of singularities from the BK field using local singularity analysis. However, these have proven unstable and highly uncertain. Further work to better deal with this issue is required.
   - The asymmetric distribution of singularity exponents and the numerical stability of singularity extraction from a small set of data samples. This drawback could be improved by forcing the mean of non-singular components to remain equal to the original radar estimates (Agterberg, 2012a). Alternatively, other techniques for singularity identification and extraction could be used. For example, the wavelet transformation (Kumar and Foufoula-Georgiou, 1993; Mallat and Hwang, 1992; Robertson et al., 2003; Struzik, 1999), Principal Component Analysis (Gonzalez-Audicana et al., 2004; Zheng et al., 2007) and Empirical Mode Decomposition (Nunes et al., 2003, 2005) techniques are widely recommended in the literature.

b. Given that the proposed singularity-sensitive merging method is particularly intended to improve rainfall estimates for (small-scale) urban areas, it would be interesting to test it using higher spatial-temporal resolution data (e.g. from X-band radars).

Lastly, a suggestion often made to us and therefore worth briefly discussing is to use a transformation in order to bring the distribution of the radar field closer to normality before the merging (be it with the Bayesian or other geo-statistical method) is conducted. However, doing this would somehow miss the point of the proposed method. The key point here is that for a non-Gaussian structure, moments beyond the second order are important, as each brings new information worth preserving. To create a more "normal" field is not the purpose of the singularity extraction; instead, it is the consequence after removing singularities from the rainfall fields, which can be physically associated with abnormal energy concentration, such as "convective" cells, and which in the proposed method are set aside to ensure their preservation throughout the merging process.

Acknowledgements. The authors would like to acknowledge the support of the Interreg IVB NWE RainGain project, the Research Foundation-Flanders (FWO) and the PLURISK project for the Belgian Science Policy Office of which this research is part. Special thanks go to Alex Grist and Richard Allitt, from Richard Allitt Associates, for providing the rain gauge data and the hydraulic model, and for their constant support with the hydraulic simulations. Thanks are also due to the UK Met Office and the BADC (British Atmospheric Data Centre) for providing Nimrod (radar) data, to Innovyze for providing the InfoWorks CS software, and to Cinzia Mazzetti and Ezio Todini for making freely available to us the RAINMUSIC software package for meteorological data processing. Lastly, the authors would like to thank the reviewers, Scott Sinclair and Cinzia Mazzetti, and the Editor, Uwe Ehret, for their insightful and constructive comments which helped improved the manuscript significantly.

Edited by: U. Ehret

References

Agterberg, F. P.: Mixtures of multiplicative cascade models in geochemistry, Nonlin. Processes Geophys., 14, 201–209, doi:10.5194/npg-14-201-2007, 2007.

Agterberg, F. P.: Multifractals and geostatistics, J. Geochem. Explor., 122, 113–122, doi:10.1016/j.gexplo.2012.04.001, 2012a.

Agterberg, F. P.: Sampling and analysis of chemical element concentration distribution in rock units and orebodies, Nonlin. Processes Geophys., 19, 23–44, doi:10.5194/npg-19-23-2012, 2012b.

Anagnostou, E. N. and Krajewski, W. F.: Real-time radar rainfall estimation, Part I: Algorithm formulation, J. Atmos. Ocean. Tech., 16, 189–197, 1999.

Bell, F. C.: The Areal Reduction Factor in Rainfall Frequency Estimation, Centre for Ecology & Hydrology (CEH), Wallingford, 1976.

Bowler, N. E., Pierce, C. E., and Seed, A. W.: STEPS: a probabilistic precipitation forecasting scheme which merges an extrapolation nowcast with downscaled NWP, Q. J. Roy Meteorol. Soc., 132, 2127–2155, doi:10.1256/qj.04.100, 2006.

Brandes, E. A., Ryzhkov, A. V and Zrnić, D. S.: An evaluation of radar rainfall estimates from specific differential phase, J. Atmos. Ocean. Tech., 18, 363–375, doi:10.1175/1520-0426(2001)018<0363:AEORRE>2.0.CO;2, 2001.

Chen, Z., Cheng, Q., Chen, J., and Xie, S.: A novel iterative approach for mapping local singularities from geochemical data, Nonlin. Processes Geophys., 14, 317–324, doi:10.5194/npg-14-317-2007, 2007.

Cheng, Q.: Multifractality and spatial statistics, Comput. Geosci., 25, 949–961, doi:10.1016/S0098-3004(99)00060-6, 1999.

Cheng, Q. and Zhao, P.: Singularity theories and methods for characterizing mineralization processes and mapping geo-anomalies for mineral deposit prediction, Geosci. Front., 2, 67–79, doi:10.1016/j.gsf.2010.12.003, 2011.

Cheng, Q., Agterberg, F. P., and Ballantyne, S. B.: The separation of geochemical anomalies from background by fractal methods, J. Geochem. Explor., 51, 109–130, doi:10.1016/0375-6742(94)90013-2, 1994.

Ciach, G. J.: Local random errors in tipping-bucket rain gauge measurements, J. Atmos. Ocean. Tech., 20, 752–759, doi:10.1175/1520-0426(2003)20<752:LREITB>2.0.CO;2, 2003.

Collier, C. G.: Accuracy of rainfall estimates by radar, part I: Calibration by telemetering raingauges, J. Hydrol., 83, 207–223, doi:10.1016/0022-1694(86)90152-6, 1986.

Collier, C. G.: Applications of Weather Radar Systems, 2nd Edn., Wiley, Chichester, England, 1996.

Deletic, A., Dotto, C. B. S., McCarthy, D. T., Kleidorfer, M., Freni, G., Mannina, G., Uhl, M., Henrichs, M., Fletcher, T. D., Rauch, W., Bertrand-Krajewski, J. L., and Tait, S.: Assessing uncertainties in urban drainage models, Phys. Chem. Earth Pt. A/B/C, 42–44, 3–10, doi:10.1016/j.pce.2011.04.007, 2012.

Einfalt, T. and Michaelides, S.: Quality control of precipitation data, in: Precipitation: Advances in Measurement, Estimation and Prediction SE-5, edited by: Michaelides, S., Springer, Berlin, Heidelberg, 101–126, 2008.

Einfalt, T., Arnbjerg-Nielsen, K., Golz, C., Jensen, N.-E., Quirmbach, M., Vaes, G., and Vieux, B.: Towards a roadmap for use of radar rainfall data in urban drainage, J. Hydrol., 299, 186–202, doi:10.1016/j.jhydrol.2004.08.004, 2004.

Einfalt, T., Jessen, M., and Mehlig, B.: Comparison of radar and raingauge measurements during heavy rainfall, Water Sci. Technol., 51, 195–201, 2005.

Evertsz, C. J. G. and Mandelbrot, B. B.: Multifractal measures, in: Chaos and Fractals, edited by: Peitgen, H.-O., Jurgens, H., and Saupe, D., Springer, New York, 922–953, 1992.

Fulton, R. A., Breidenbach, J. P., Dong-Jun, S., and Miller, D. A.: The WSR-88D rainfall algorithm, Weather Forecast., 13, 377–395, doi:10.1175/1520-0434(1998)013<0377:TWRA>2.0.CO;2, 1998.

Germann, U., Galli, G., Boscacci, M., and Bolliger, M.: Radar precipitation measurement in a mountainous region, Q. J. Roy Meteorol. Soc., 132, 1669–1692, doi:10.1256/qj.05.190, 2006.

Gires, A., Onof, C., Maksimović, C., Schertzer, D., Tchiguirinskaia, I., and Simões, N.: Quantifying the impact of small scale unmeasured rainfall variability on urban runoff through multifractal downscaling: a case study, J. Hydrol., 442–443, 117–128, doi:10.1016/j.jhydrol.2012.04.005, 2012.

Golding, B. W.: Nimrod: a system for generating automated very short range forecasts, Meteorol. Appl., 5, 1–16, doi:10.1017/S1350482798000577, 1998.

Gonzalez-Audicana, M., Saleta, J. L., Catalan, R. G., Garcia, R.: Fusion of multispectral and panchromatic images using improved IHS and PCA mergers based on wavelet decomposition, IEEE T. Geosci. Remote, 42, 1291–1299, doi:10.1109/TGRS.2004.825593, 2004.

Gooch, M. N.: Use of rainfall data from flow surveys, WaPUG User Note No. 6, Chartered Institute of Water and Environmental Management (CIWEM), London, 2009.

Goudenhoofdt, E. and Delobbe, L.: Evaluation of radar-gauge merging methods for quantitative precipitation estimates, Hydrol. Earth Syst. Sci., 13, 195–203, doi:10.5194/hess-13-195-2009, 2009.

Guo, Y. and Adams, B. J.: Hydrologic analysis of urban catchments with event-based probabilistic models: 1. Runoff volume, Water Resour. Res., 34, 3421–3431, doi:10.1029/98WR02449, 1998.

Habib, E., Meselhe, E. and Aduvala, A.: Effect of local errors of tipping-bucket rain gauges on rainfall–runoff simulations, J. Hydrol. Eng., 13, 488–496, doi:10.1061/(ASCE)1084-0699(2008)13:6(488), 2008.

Harrison, D. L., Driscoll, S. J., and Kitchen, M.: Improving precipitation estimates from weather radar using quality control and correction techniques, Meteorol. Appl., 7, 135–144, doi:10.1017/S1350482700001468, 2000.

Harrison, D. L., Scovell, R. W., and Kitchen, M.: High-resolution precipitation estimates for hydrological uses, P. I. Civil Eng-Wat. M., 162, 125–135, doi:10.1680/wama.2009.162.2.125, 2009.

HR Wallingford: Wallingford Procedure for Design and Analysis of Urban Storm Drainage, Wallingford, UK, 1983.

Kalman, R. E.: A new approach to linear filtering and prediction problems, J. Basic Eng.-T. ASME, 82, 35–45, doi:10.1115/1.3662552, 1960.

Kavetski, D., Kuczera, G., and Franks, S. W.: Bayesian analysis of input uncertainty in hydrological modeling: 1. Theory, Water Resour. Res., 42, W03407, doi:10.1029/2005WR004368, 2006.

Krajewski, W. F.: Cokriging radar-rainfall and rain gage data, J. Geophys. Res., 92, 9571–9580, doi:10.1029/JD092iD08p09571, 1987.

Krajewski, W. F. and Smith, J. A.: Radar hydrology: rainfall estimation, Adv. Water Resour., 25, 1387–1394, doi:10.1016/S0309-1708(02)00062-3, 2002.

Krämer, S., Fuchs, L., and Verworn, H.: Aspects of radar rainfall forecasts and their effectiveness for real time control-the example of the sewer system of the City of Vienna, Water Pract. Technol., 2, 42–49, 2007.

Krause, P., Boyle, D. P., and Bäse, F.: Comparison of different efficiency criteria for hydrological model assessment, Adv. Geosci., 5, 89–97, doi:10.5194/adgeo-5-89-2005, 2005.

Kumar, P. and Foufoula-Georgiou, E.: A multicomponent decomposition of spatial rainfall fields: 1. Segregation of large- and small-scale features using wavelet transforms, Water Resour. Res., 29, 2515–2532, doi:10.1029/93WR00548, 1993.

La Barbera, P., Lanza, L. G., and Stagi, L.: Tipping bucket mechanical errors and their influence on rainfall statistics and extremes, Water Sci. Technol., 45, 1–9, 2002.

Liguori, S., Rico-Ramirez, M. A., Schellart, A. N. A., and Saul, A. J.: Using probabilistic radar rainfall nowcasts and NWP forecasts for flow prediction in urban catchments, Atmos. Res., 103, 80–95, doi:10.1016/j.atmosres.2011.05.004, 2011.

Lovejoy, S. and Mandelbrot, B. B.: Fractal properties of rain, and a fractal model, Tellus A, 37, 209–232, doi:10.1111/j.1600-0870.1985.tb00423.x1985.

Luyckx, G. and Berlamont, J.: Simplified method to correct rainfall measurements from tipping bucket rain gauges, in: Urban Drainage Modeling: Proceedings of the Speciality Symposium of the World Water and Environmental Resource Congress, edited by: Brashear, R. W. and Maksimović, Č., American Society of Civil Engineers, Washington, D.C., 767–776, 2001.

Mallat, S. and Hwang, W.-L.: Singularity detection and processing with wavelets, IEEE T. Inform. Theory, 38, 617–643, doi:10.1109/18.119727, 1992.

Marshall, J. and Palmer, W.: The distribution of raindrops with size, J. Meteorol., 5, 165–166, doi:10.1175/1520-0469(1948)005<0165:TDORWS>2.0.CO;2, 1948.

Marshall, M. and McIntyre, N.: Field verification of bed-mounted ADV meters, Proc. ICE – Water Manage., 161, 199–206, doi:10.1680/wama.2008.161.4.199, 2008.

Mazzetti, C.: Data interpolation and multi-sensors Bayesian combinations, RainMusic User's Manual and References, PROGEA srl, Bologna, Italy, 2012.

Mazzetti, C. and Todini, E.: Combining raingauges and radar precipitation measurements using a Bayesian approach, in: geoENV IV – Geostatistics for Environmental Applications, edited by: Sanchez-Vila, X., Carrera, J., and Gómez-Hernández, J. J., Kluwer Academic Publishers, Springer, the Netherlands, 401–412, 2004.

Molini, A., Lanza, L. G., and La Barbera, P.: The impact of tipping-bucket raingauge measurement errors on design rainfall for urban-scale applications, Hydrol. Process., 19, 1073–1088, doi:10.1002/hyp.5646, 2005.

Nunes, J., Bouaoune, Y., Delechelle, E., Niang, O., and Bunel, P.: Image analysis by bidimensional empirical mode decomposition, Image Vis. Comput., 21, 1019–1026, doi:10.1016/S0262-8856(03)00094-5, 2003.

Nunes, J., Guyot, S., and Deléchelle, E.: Texture analysis based on local analysis of the bidimensional empirical mode decomposition, Mach. Vis. Appl., 16, 177–188, doi:10.1007/s00138-004-0170-5, 2005.

Ochoa-Rodríguez, S., Wang, L.-P., Grist, A., Allitt, R., Onof, C., and Maksimović, Č.: Improving the applicability of radar rainfall estimates for urban pluvial flood modelling and forecasting, in: Urban Drainage Group Autumn Conference and Exhibition 2013: Future Thinking and Challenges, 13–15 November 2013, Nottingham, UK, 19, 2013.

Osborne, M. P.: A New Runoff Volume Model, Wastewater Planning Users Group WaPUG, Chartered Institute of Water and Environmental Management (CIWEM), UK, 2001.

Robertson, A. N., Farrar, C. R., and Sohn, H.: Singularity detection for structural health monitoring using Hölder exponents, Mech. Syst. Signal Pr., 17, 1163–1184, doi:10.1006/mssp.2002.1569, 2003.

Schellart, A. N. A., Shepherd, W. J., and Saul, A. J.: Influence of rainfall estimation error and spatial variability on sewer flow prediction at a small urban scale, Adv. Water Resour., 45, 65–75, doi:10.1016/j.advwatres.2011.10.012, 2012.

Schertzer, D. and Lovejoy, S.: Physical modeling and analysis of rain and clouds by anisotropic scaling multiplicative processes, J. Geophys. Res., 92, 9693–9714, doi:10.1029/JD092iD08p09693, 1987.

Schertzer, D., Tchiguirinskaia, I., and Lovejoy, S.: Multifractality: at least three moments!, Interactive comment on "Just two moments! A cautionary note against use of high-order moments in multifractal models in hydrolog" by F. Lombardo et al., Hydrol. Earth Syst. Sci. Discuss., 10, C3103–C3109, 2013.

Seo, D. and Smith, J.: Rainfall estimation using raingages and radar – a Bayesian approach: 1. Derivation of estimators, Stoch. Hydrol. Hydraul., 5, 17–29, doi:10.1007/BF01544175, 1991.

Sevruk, B. and Nešpor, V.: Empirical and theoretical assessment of the wind induced error of rain measurement, Water Sci. Technol., 37, 171–178, doi:10.1016/S0273-1223(98)00330-8, 1998.

Sinclair, S. and Pegram, G.: Combining radar and rain gauge rainfall estimates using conditional merging, Atmos. Sci. Lett., 6, 19–22, doi:10.1002/asl.85, 2005.

Smith, J. A., Baeck, M. L., Meierdiercks, K. L., Miller, A. J., and Krajewski, W. F.: Radar rainfall estimation for flash flood forecasting in small urban watersheds, Adv. Water Resour., 30, 2087–2097, doi:10.1016/j.advwatres.2006.09.007, 2007.

Smith, J. A., Hui, E., Steiner, M., Baeck, M. L., Krajewski, W. F., and Ntelekos, A. A.: Variability of rainfall rate and raindrop size distributions in heavy rain, Water Resour. Res., 45, W04430, doi:10.1029/2008WR006840, 2009.

Struzik, Z. R.: Local effective Hölder exponent estimation on the wavelet transform maxima tree, in: Fractals: Theory and Applications in Engineering, edited by: Dekking, M., Véhel, J. L., Lutton, E., and Tricot, C., Springer, London, UK, 93–112, 1999.

Tchiguirinskaia, I., Schertzer, D., Hoang, C. T., and Lovejoy, S.: Multifractal study of three storms with different dynamics over the Paris region, in: Weather Radar and Hydrology, edited by: Moore, R. J., Cole, S. J., and Illingworth, A. J., IAHS, International Association of Hydrological Sciences (IAHS), Exeter, UK, 421–426, 2011.

Thorndahl, S., Nielsen, J. E., and Rasmussen, M. R.: Bias adjustment and advection interpolation of long-term high resolution radar rainfall series, J. Hydrol., 508, 214–226, doi:10.1016/j.jhydrol.2013.10.056, 2014.

Todini, E.: A Bayesian technique for conditioning radar precipitation estimates to rain-gauge measurements, Hydrol. Earth Syst. Sci., 5, 187–199, doi:10.5194/hess-5-187-2001, 2001.

Ulbrich, C. W.: Natural variations in the analytical form of the raindrop size distribution, J. Clim. Appl. Meteorol., 22, 1764–1775, doi:10.1175/1520-0450(1983)022<1764:NVITAF>2.0.CO;2, 1983.

Velasco-Forero, C. A., Sempere-Torres, D., Cassiraga, E. F., and Gómez-Hernández, J.: A non-parametric automatic blending methodology to estimate rainfall fields from rain gauge and radar data, Adv. Water Resour., 32, 986–1002, doi:10.1016/j.advwatres.2008.10.004, 2009.

Vieux, B. E. and Bedient, P. B.: Assessing urban hydrologic prediction accuracy through event reconstruction, J. Hydrol., 299, 217–236, doi:10.1016/j.jhydrol.2004.08.005, 2004.

Villarini, G., Smith, J. A., Lynn Baeck, M., Sturdevant-Rees, P., and Krajewski, W. F.: Radar analyses of extreme rainfall and flooding in urban drainage basins, J. Hydrol., 381, 266–286, doi:10.1016/j.jhydrol.2009.11.048, 2010.

Wackernagel, H.: Multivariate Geostatistics, An Introduction with Applications, Springer, Berlin, 2003.

Wang, L.-P. and Onof, C.: High-resolution rainfall field reconstruction based upon Kriging and local singularity analysis, in: Hydrofractals '13, HF-10, 17–19 October 2013, Kos Island, Greece, 2013.

Wang, L.-P., Onof, C., Ochoa-Rodríguez, S., and Simões, N.: Analysis of kriged rainfields using multifractals, in: 9th International Workshop on Precipitation in Urban Areas: Urban Challenges in Rainfall Analysis, 6–9 December 2012, St. Moritz, Switzerland, 138–142, 2012.

Wang, L.-P., Ochoa-Rodríguez, S., Simões, N. E., Onof, C., and Maksimović, Č.: Radar-raingauge data combination techniques: a revision and analysis of their suitability for urban hydrology, Water Sci. Technol., 68, 737–747, doi:10.2166/wst.2013.300, 2013.

Wang, L.-P., Ochoa-Rodríguez, S., Willems, P., and Onof, C.: Improving the applicability of gauge-based radar rainfall adjustment methods to urban pluvial flood modelling and forecasting using local singularity analysis, in: International Symposium on Weather Radar and Hydrology (WRaH), 7–9April 2014, Washington, D.C., 10 pp., 2014.

WaPUG: Code of Practice for the Hydraulic Modelling of Sewer Systems, Wastewater Planning Users Group WaPUG, Chartered Institute of Water and Environmental Management (CIWEM), UK, 2002.

Zheng, Y., Hou, X., Bian, T., and Qin, Z.: Effective image fusion rules Of multi-scale image decomposition. in: 5th International Symposium on Image and Signal Processing and Analysis, 27–29 September 2007, Istanbul, Turkey, 362–366, doi:10.1109/ISPA.2007.4383720, 2007.

12

Discharge hydrograph estimation at upstream-ungauged sections by coupling a Bayesian methodology and a 2-D GPU shallow water model

Alessia Ferrari[1], Marco D'Oria[1], Renato Vacondio[1], Alessandro Dal Palù[2], Paolo Mignosa[1], and Maria Giovanna Tanda[1]

[1]Department of Engineering and Architecture, University of Parma, Parma, Italy
[2]Department of Mathematical, Physical and Computer Sciences, University of Parma, Parma, Italy

Correspondence: Alessia Ferrari (alessia.ferrari@unipr.it)

Abstract. This paper presents a novel methodology for estimating the unknown discharge hydrograph at the entrance of a river reach when no information is available. The methodology couples an optimization procedure based on the Bayesian geostatistical approach (BGA) with a forward self-developed 2-D hydraulic model. In order to accurately describe the flow propagation in real rivers characterized by large floodable areas, the forward model solves the 2-D shallow water equations (SWEs) by means of a finite volume explicit shock-capturing algorithm. The two-dimensional SWE code exploits the computational power of graphics processing units (GPUs), achieving a ratio of physical to computational time of up to 1000. With the aim of enhancing the computational efficiency of the inverse estimation, the Bayesian technique is parallelized, developing a procedure based on the Secure Shell (SSH) protocol that allows one to take advantage of remote high-performance computing clusters (including those available on the Cloud) equipped with GPUs. The capability of the methodology is assessed by estimating irregular and synthetic inflow hydrographs in real river reaches, also taking into account the presence of downstream corrupted observations. Finally, the procedure is applied to reconstruct a real flood wave in a river reach located in northern Italy.

1 Introduction

The definition of discharge hydrographs in specific river sections is still a relevant hydraulic problem not only for flood modelling purposes but also for more practical issues related to flood-protection measures, hydropower plants, water resource management, the design of new structures, etc. Flood-routing techniques, either hydrological or hydraulic, are extensively studied and are widely used to estimate discharge hydrographs in downstream ungauged sites based on data available at upstream gauged stations (forward propagation). However, the flow hydrograph is often required in a river section that is completely ungauged and does not have useful upstream information for its definition. In these cases, discharge hydrographs at specific sites can be estimated by coupling rainfall-runoff and forward flood-propagation models. However, rainfall-runoff models (Beven, 2011) present several uncertainties associated, for example, with the choice of the model for the basin schematization, the evaluation of the effective rainfall, and the calibration procedure. An alternative approach is to assess the upstream unknown flow hydrograph using only the information in terms of the discharge values or water levels available downstream from the selected site and possibly the characteristics of the river reach. In the literature, this approach is known as reverse flow routing (D'Oria and Tanda, 2012), an ill-posed inverse problem that presents two main challenges; the solution may be non-unique, and instabilities may arise during the inversion. The traditional attempts of solving the reverse flow routing problem are based on two main approaches: the solution of a reverse form of the Saint Venant equations (e.g. Eli et al., 1974; Szymkiewicz, 1993; Dooge and Bruen, 2005; Bruen and Dooge, 2007) and the back-oriented application of hydrological routing schemes (e.g. Das, 2009; Koussis et al., 2012; Koussis and Mazi, 2016). Beyond the approximations introduced by the hydrological routing schemes, the aforementioned procedures were applied to simplified reach geometries and flow conditions. In almost all cases, especially considering downstream information affected by errors, instabilities and spurious oscillations appeared; low-pass filters with subjective parameters were sometimes used to dampen the estimated inflow fluctuations. D'Oria and Tanda (2012) and Zucco et al. (2015) provide additional references and details on the reverse flow routing problem.

In addition to the above procedures, the estimation of an unknown upstream flow hydrograph based only on downstream information (observations) can be performed via optimization methods. These techniques aim at finding the upstream flow hydrograph that, routed downstream, best matches the available observations. D'Oria and Tanda (2012) solved the reverse flow routing problem by adopting a novel Bayesian geostatistical approach (BGA) as an optimization procedure that considers the flow hydrograph as a continuous random function that presents autocorrelation. The authors showed the capability of the BGA methodology, in combination with a forward hydraulic model, to estimate the discharges in an upstream-ungauged section based only on an available downstream flow hydrograph: the solution was stable also in the presence of corrupted downstream flow values. The forward model, which solves the 1-D Saint Venant equations, was considered already implemented and calibrated and was able to describe the hydraulic routing process with sufficient accuracy. The BGA method was further extended in order to adopt stage hydrographs instead of discharge ones as downstream observations (D'Oria et al., 2014). Saghafian et al. (2015) identified the upstream hydrograph of a river reach given the downstream one by using a genetic algorithm coupled with a forward hydraulic model that solves the 1-D Saint-Venant equations under the kinematic wave approximations. Only some minor oscillations and instabilities occurred during the inversion, but the authors applied the procedure to a rectangular prismatic channel, and no errors were added to the downstream observations. Zucco et al. (2015) investigated the reverse flow routing process in natural channels and estimated the discharge hydrograph in ungauged sections by means of a genetic algorithm coupled with a simplified routing model. The parametric forward model was based on the continuity equation written in a characteristic form, lumped over the entire river reach, and on simplified rating curves at the channel ends. In addition, the unknown inflow hydrograph was assumed to be distributed in time as a Pearson type III function with three parameters, thus preventing the possibility of estimating real flood waves with irregular shapes (e.g. multi-peak hydrographs).

All the previously cited works adopted 1-D hydraulic models or simplified hydrological routing schemes in combination with different optimization procedures. Nevertheless, in many real cases, the complex hydrodynamic field generated by the flood propagation cannot be accurately described under 1-D assumptions, and it is necessary to adopt schemes based on the 2-D shallow water equations, even if this poses the drawback of the computational burden and requires a detailed terrain survey. However, nowadays, bathymetric data can be easily obtained from high-resolution digital terrain models (DTM), and fast 2-D numerical models have been developed. With the purpose of estimating the discharge hydrograph in an upstream-ungauged river section, having water level information only in a downstream observation site, this paper extends the BGA methodology for reverse flow routing from D'Oria and Tanda (2012) and D'Oria et al. (2014) to a 2-D forward algorithm in order to model natural rivers with complex geometry, including flood plains and floodable areas. With this aim, the stable, accurate and fast PARFLOOD graphics processing unit (GPU) code (Vacondio et al., 2014, 2016, 2017), which solves the conservative form of the 2-D shallow water equations on a finite volume scheme, is adopted as forward model and is coupled with the inverse estimation procedure. In order to reduce the computational time, the Jacobian matrix estimation procedure, which is the key point of the BGA method, has been parallelized. Additionally, a host–server data management procedure has been implemented to exploit the computational power of remote large modern supercomputer and/or cloud HPC resources. The capability of the optimization procedure has been tested by estimating real or pseudo-real inflow hydrographs in natural river reaches, where 1-D models cannot accurately describe the flood propagation. Moreover, during the discharge estimation, the presence of downstream corrupted observations has also been taken into account, since registered data at gauging stations are quite often affected by instrumental errors.

The paper is organized as follows; in Sect. 2 the theory of the Bayesian geostatistical approach is illustrated. A step-by-step description of the inverse procedure is provided in Sect. 3: the parallel implemented scheme, the forward model optimization for reducing the run times, and the iteration management between the local host and the remote server are described in detail. Section 4 is dedicated to the application of the procedure to synthetic test cases concerning the estimation of inflow hydrographs with different shapes in two rivers in northern Italy. The practicability of the inverse procedure for reconstructing a historical flooding event is presented in Sect. 5. Some concluding remarks are finally outlined in Sect. 6.

2 Theory of the Bayesian geostatistical approach

The optimization software adopted to solve the reverse flow routing problem is the bgaPEST (Fienen et al., 2013), which implements the Bayesian geostatistical approach of Kitanidis (1995), and it is developed according to the PEST (model independent parameter estimation) software (Doherty, 2016). The bgaPEST is appropriate for solving inverse problems (in a context of a highly parameterized inversion), which are characterized by unknown parameters that are correlated to one another in space or time, for example, the discharge values of a flow hydrograph. The first applications of the inverse methodology were related to the estimation of spatial parameter fields in a groundwater context (Kitanidis and Vomvoris, 1983; Hoeksema and Kitanidis, 1984, among others), but later the method was adopted to evaluate unknown time functions in different areas (e.g. Snodgrass and Kitanidis, 1997; Michalak et al., 2004; Butera et al., 2013; D'Oria and Tanda, 2012; D'Oria et al., 2015; Leonhardt et al., 2014).

2.1 Bayes' theorem

The crux of the adopted bgaPEST, as well as other methods based on the Bayesian approach, is Bayes' theorem, which reads

$$p(\boldsymbol{s}|\boldsymbol{y}) \propto L(\boldsymbol{y}|\boldsymbol{s})\, p(\boldsymbol{s}), \tag{1}$$

where $\boldsymbol{s}$ is the vector of the unknown parameters, $\boldsymbol{y}$ is the vector of the measured data, $p(\boldsymbol{s}|\boldsymbol{y})$ is the posterior probability density function (pdf) of $\boldsymbol{s}$ given $\boldsymbol{y}$, $L(\boldsymbol{y}|\boldsymbol{s})$ is the likelihood function, and $p(\boldsymbol{s})$ is the prior probability density function of $\boldsymbol{s}$. Since the present work aims at estimating an upstream hydrograph in an ungauged section, assuming the knowledge of downstream water levels, $\boldsymbol{s}$ represents the discharge values over time of the unknown inflow hydrograph, whereas $\boldsymbol{y}$ denotes the downstream water level observations. Following Eq. (1), the posterior pdf can be seen as a combination between a priori knowledge on the parameters (prior pdf), where a priori means that the observed data are still not considered, and information about parameters contained in the measured data (likelihood function) (Glickman and Van Dyk, 2007). In the BGA method proposed by Kitanidis (1995), the prior pdf and the likelihood function are described by means of Gaussian distributions, and the best set of parameter $\boldsymbol{s}$ is obtained by maximizing the posterior pdf.

2.1.1 The likelihood function

The likelihood function $L(\boldsymbol{y}|\boldsymbol{s})$ in Eq. (1) characterizes the deviation between observed data and model results (Fienen et al., 2013). Starting from the results of the forward model, $L(\boldsymbol{y}|\boldsymbol{s})$ delineates how a particular set of parameters $\boldsymbol{s}$ is able to reproduce the observations $\boldsymbol{y}$ in space and/or time, thus accounting for the epistemic errors. The investigated inverse problem presents different sources of errors that are related to the conceptual schematization of the inverse procedure, the numerical forward model, and the data measurement. In the likelihood function, the errors are assumed to be independent and identically distributed, with the zero mean and covariance matrix expressed as follows;

$$\mathbf{R} = \sigma_R^2 \mathbf{I}, \tag{2}$$

where σ_R^2 denotes the variance that expresses the misfit between observed and modelled data, and $\mathbf{I}$ is the identity matrix.

2.1.2 The prior probability density function

The prior knowledge about $\boldsymbol{s}$ ($p(\boldsymbol{s})$ in Eq. 1) is limited to the definition of a mean value (unknown and estimated during the procedure) and a characteristic about the continuity and/or smoothness of the parameter field described through a covariance function. It furnishes a soft knowledge about the structure/shape of the unknowns and provides a regularization of the solution; the prior pdf can also be used to enforce non-negativity to the parameters (D'Oria and Tanda, 2012). The prior mean is defined as:

$$E[\boldsymbol{s}] = \mathbf{X}\boldsymbol{\beta}, \tag{3}$$

where E is the expected value, $\boldsymbol{\beta}$ is the vector of drift coefficients, and $\mathbf{X}$ is a known matrix of basis functions. In our case, $\boldsymbol{\beta}$ is a single unknown scalar, but different drift coefficients can be used to introduce discontinuities in the stochastic function to be estimated (e.g. when the unknown parameters are likely to form distinct populations). For example, in the context of reverse flow routing problems, multiple values of $\boldsymbol{\beta}$ are adopted if more than one inflow hydrograph must be estimated at the same time (e.g. the inflow on both the upstream branches of a river confluence). The matrix of the basis function, $\mathbf{X}$, links each unknown parameter with the corresponding element of $\boldsymbol{\beta}$ and, at the same time, specifies the model of the mean (e.g. constant mean, mean with a trend, etc.); in our case the mean is constant and therefore $\mathbf{X}$ is a single vector of one (Fienen et al., 2008).

The prior covariance matrix of the unknown parameters $\mathbf{Q}_{\mathrm{ss}}$ is then defined as

$$\mathbf{Q}_{\mathrm{ss}} = E\left[(\boldsymbol{s} - \mathbf{X}\boldsymbol{\beta})(\boldsymbol{s} - \mathbf{X}\boldsymbol{\beta})^T\right]. \tag{4}$$

In the context of geostatistics, the covariance matrix $\mathbf{Q}_{\mathrm{ss}}$ is a function of the separation distance (in time in this case) between the parameters and describes their deviations from the mean behaviour. Different functions can be adopted to describe the covariance. For example, it can be assumed as a linear function, represented through a limiting case of the exponential covariance function (Fienen et al., 2008) according to the following relation;

$$\mathbf{Q}_{\mathrm{ss}}(\theta) = \theta l \exp\left(-\frac{|\boldsymbol{d}|}{l}\right), \tag{5}$$

where $\boldsymbol{d}$ represents the vector of the separation times between all the parameter pairs ($d_{i,j} = t_i - t_j$, with i, $j = 1, \ldots, N_\mathrm{p}$, t denoting the time associated with each parameter and N_p the total number of unknowns), l a fixed integral scale ($l = 10\max d$), and θ the slope (structural parameter) that governs the correlation between the discharge values of the unknown hydrograph. A different formulation (D'Oria et al., 2014) defines the prior covariance matrix $\mathbf{Q}_\mathrm{ss}$ by means of a Gaussian function characterized by two structural parameters (σ_s^2 and l);

$$\mathbf{Q}_\mathrm{ss}(\sigma_\mathrm{s}^2, l) = \sigma_\mathrm{s}^2 \exp\left(-\frac{\left|\boldsymbol{d}^2\right|}{l^2}\right), \tag{6}$$

where σ_s^2 denotes the variance. The linear function (Eq. 5) enforces only continuity to the solution, whereas the Gaussian function (Eq. 6) also adds a degree of smoothness, but the final solution is still driven by the observations.

2.1.3 The posterior probability density function

With the assumptions made, the likelihood and prior terms that compose the posterior pdf of Eq. (1) can be rewritten as follows (Fienen et al., 2009; D'Oria and Tanda, 2012; D'Oria et al., 2014);

$$L(\boldsymbol{y}|\boldsymbol{s}) = \exp\left(-\frac{1}{2}(\boldsymbol{y} - \boldsymbol{h}(\boldsymbol{s}))^T \mathbf{R}^{-1} (\boldsymbol{y} - \boldsymbol{h}(\boldsymbol{s}))\right), \tag{7}$$

$$p(\boldsymbol{s}) = \exp\left(-\frac{1}{2}(\boldsymbol{s} - \mathbf{X}\boldsymbol{\beta})^T \mathbf{Q}_\mathrm{ss}^{-1} (\boldsymbol{s} - \mathbf{X}\boldsymbol{\beta})\right). \tag{8}$$

In the likelihood function, the term $\boldsymbol{h}(\boldsymbol{s})$ represents the modelled values in the same place and time as the available observations $\boldsymbol{y}$. Therefore, to evaluate $\boldsymbol{h}(\boldsymbol{s})$, a forward model of the considered river reach that is able to describe the hydraulic routing process is required in order to provide the corresponding downstream water levels for a given set of parameter $\boldsymbol{s}$.

Recalling that the aim of the inverse procedure is to obtain the vector of the unknown parameters $\boldsymbol{s}$, as well as to quantify the uncertainty in the estimation, the solution is found by maximizing the posterior pdf or, more conveniently, minimizing its negative logarithm (objective function) (Fienen et al., 2013).

In case a linear relationship between parameters and observations (linear forward model) holds, a computationally efficient method to find the best estimate $\hat{\boldsymbol{s}}$ of vector $\boldsymbol{s}$ (and $\hat{\boldsymbol{\beta}}$ of $\boldsymbol{\beta}$) is obtained by introducing the vector $\boldsymbol{\xi} = (\mathbf{H}\mathbf{Q}_\mathrm{ss}\mathbf{H}^T + \mathbf{R})^{-1}(\boldsymbol{y} - \mathbf{H}\mathbf{X}\hat{\boldsymbol{\beta}})$ and solving the following linear system of equations (Fienen et al., 2009);

$$\begin{cases} \hat{\boldsymbol{s}} = \mathbf{X}\hat{\boldsymbol{\beta}} + \mathbf{Q}_\mathrm{ss}\mathbf{H}^T\boldsymbol{\xi} \\ \begin{bmatrix} \mathbf{H}\mathbf{Q}_\mathrm{ss}\mathbf{H}^T + \mathbf{R} & \mathbf{H}\mathbf{X} \\ \mathbf{X}^T\mathbf{H}^T & 0 \end{bmatrix} \begin{bmatrix} \boldsymbol{\xi} \\ \hat{\boldsymbol{\beta}} \end{bmatrix} = \begin{bmatrix} \boldsymbol{y} \\ 0 \end{bmatrix}, \end{cases} \tag{9}$$

where $\mathbf{H}$ is the sensitivity (Jacobian) matrix, representing how the observations $\boldsymbol{y}$ are influenced by each unknown parameter s_i (D'Oria et al., 2015). However, for the particular problem under investigation, $\boldsymbol{h}(\boldsymbol{s})$ is non-linear and matrix $\mathbf{H}$ therefore depends on $\boldsymbol{s}$. Following the quasi-linear geostatistical approach (Kitanidis, 1995), the relationship between observations and parameters can be successively linearized about a candidate solution $\boldsymbol{s}_k$, where k represents each iteration;

$$\boldsymbol{h}(\boldsymbol{s}) \approx \boldsymbol{h}(\boldsymbol{s}_k) + \widetilde{\mathbf{H}}_k (\boldsymbol{s} - \boldsymbol{s}_k). \tag{10}$$

Then, a correction to the measurements is applied according to the following relation;

$$\boldsymbol{y}_k = \boldsymbol{y} - \boldsymbol{h}(\boldsymbol{s}_k) + \widetilde{\mathbf{H}}_k \boldsymbol{s}_k. \tag{11}$$

Therefore, the sensitivity matrix is evaluated at each iteration as follows (D'Oria et al., 2014);

$$\widetilde{\mathbf{H}}_k = \left.\frac{\partial \boldsymbol{h}(\boldsymbol{s})}{\partial \boldsymbol{s}}\right|_{\boldsymbol{s}_k}. \tag{12}$$

Analogously to the linear system in Eq. (9), the linearized system is solved according to

$$\begin{bmatrix} \widetilde{\mathbf{H}}_k\mathbf{Q}_\mathrm{ss}\widetilde{\mathbf{H}}_k^T + \mathbf{R} & \widetilde{\mathbf{H}}_k\mathbf{X} \\ \mathbf{X}^T\widetilde{\mathbf{H}}_k^T & 0 \end{bmatrix} \begin{bmatrix} \boldsymbol{\xi}_{k+1} \\ \hat{\boldsymbol{\beta}}_{k+1} \end{bmatrix} = \begin{bmatrix} \boldsymbol{y}_k \\ 0 \end{bmatrix}, \tag{13}$$

and the next estimate of the parameters is evaluated by means of

$$\widetilde{\boldsymbol{s}}_{k+1} = \mathbf{X}\hat{\boldsymbol{\beta}}_{k+1} + \mathbf{Q}_\mathrm{ss}\widetilde{\mathbf{H}}_k^T \boldsymbol{\xi}_{k+1}. \tag{14}$$

A proper selection of the covariance function structural parameters (θ, σ_s^2 and l) and optionally of the epistemic error variance σ_R^2 is important in order to reach a good solution. The structural parameters are estimated from the data using a Bayesian adaptation of the restricted maximum likelihood (RML) method of Kitanidis (1995) that allows one to reach the best compromise between the fitting of the modelled data with the observations and the prior information (Fienen et al., 2013). Dealing with non-linear problems, unknowns ($\boldsymbol{s}$) and structural parameters must be iteratively estimated in successive steps. The linearization process ends if the improvement (absolute difference between two successive iterations) in the objective function is below a user-defined value or if the maximum number of iterations N_i is reached. The structural parameter iteration loop (outer loop) progresses until the L_2-norm of the differences between structural parameter values at consecutive iterations is below a user-defined value or if the maximum number of iterations N_o is reached (Fienen et al., 2013). Finally, at the end of the estimation, the linearized uncertainties of the unknowns can be evaluated in terms of the posterior covariance matrix of the estimated parameters (Fienen et al., 2013). The diagonal elements of this matrix represent the posterior variance (σ^2) of the estimated parameters; thus, the 95 % credibility interval of the solution is evaluated as $\pm 2\sigma^2$.

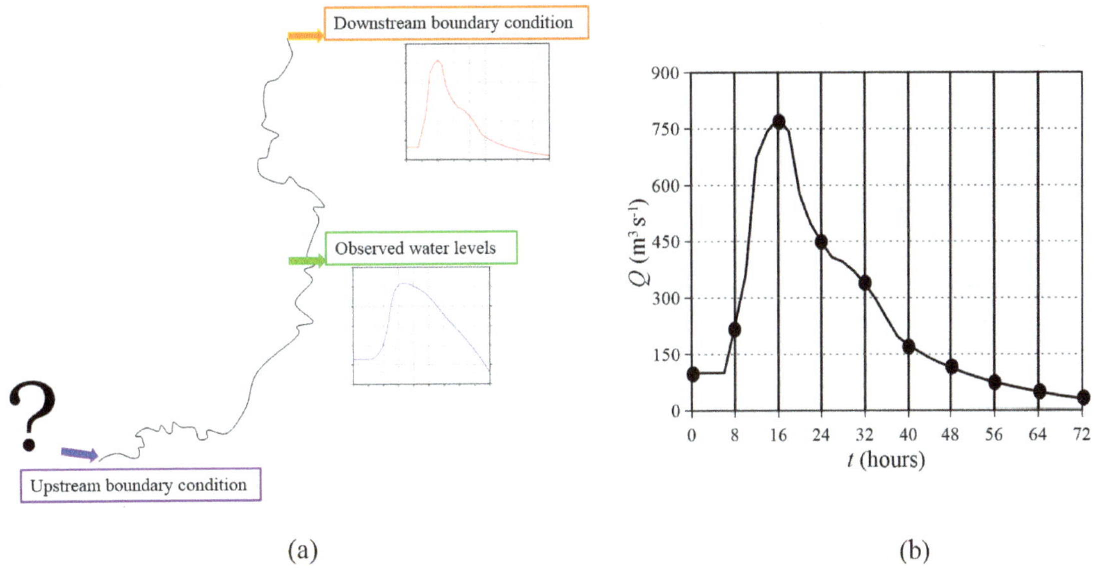

Figure 1. Definition of the reverse flood routing problem **(a)** and of the unknown parameters **(b)**.

3 Description of the Bayesian estimation procedure

After having described the theory of the Bayesian geostatistical approach in Sect. 2, some operational information about the BGA inverse procedure is now illustrated. As mentioned in the introduction and illustrated in Fig. 1a, the goal of the adopted BGA methodology is the estimation of the discharge hydrograph in an upstream-ungauged river section (identified by a question mark in Fig. 1a), with information about water levels being observed in a downstream section (intermediate site in Fig. 1a). A boundary condition downstream of the observation site must also be specified; this can be based on observed data or can be approximated extending the computational domain far away from the intermediate section. The inverse method estimates the N_{p} parameters (the vector of the unknown parameters $\mathbf{s}$ in Eq. 1) that originate from the discretization of the upstream discharge hydrograph by means of time intervals, which are regular in this case (Fig. 1b).

The BGA algorithm solves the inverse problem by means of the following steps.

First, the unknown parameters and the structural ones are initialized. The first ones may all be assumed equal to a constant discharge value coherent with the considered river, whereas the starting values for the structural parameters are assigned so that the variability between contiguous parameters is small (flat solution, with a high degree of correlation); complexity is then introduced during the optimization process if supported by the data. The variance of the epistemic errors is assumed as being close to the expected one.

Assuming that the first guess for the unknown parameters is the upstream boundary condition, the hydraulic forward model is run, and the resulting water levels are extracted at the observation site. The simulation of a base run once a particular set of parameters has been assumed (deriving from the initialization or from previous estimation steps) represents a mandatory step for the Jacobian matrix evaluation, which is performed at this point in the procedure in order to quantify how each observation is influenced by the variation of each estimable parameter. Particularly, Eq. (12) is approximated using a finite difference method; hence each element of the matrix is evaluated as the ratio between the variation of each observation (numerator) for given variation of each parameter (denominator) with respect to the base run. Therefore, in addition to the base run, the hydraulic forward model is further run as many times as the number of parameters to estimate N_{p}. With each run, a single value of the upstream boundary condition is modified by a known quantity with respect to the previous value, and the hydraulic forward model is run again. As a consequence, each simulation tests the sensitivity of the resulted water levels (all the observations at once) to the variation of a single parameter i.

In order to exemplify this step, Fig. 2a shows the discharge imposed as upstream boundary condition for a base run of an intermediate set of parameters; after the propagation, the resulting water levels extracted at the observation site are shown in Fig. 2c. To test the sensitivity to parameter i, in Fig. 2b, the considered parameter is changed by a known quantity, and a new upstream boundary condition is defined (solid line); it is worth noting that the solid and the dotted lines differ only for the parameter i. The water levels resulting from this single parameter variation are shown in Fig. 2d (solid line); they are identical to ones of the base run until time $i-1$, whereas after that time, they differ from those of the base run (dotted line). The computation of the differences between the resulting water levels of the simulation i and the base run (solid and dotted lines) and the variation of parame-

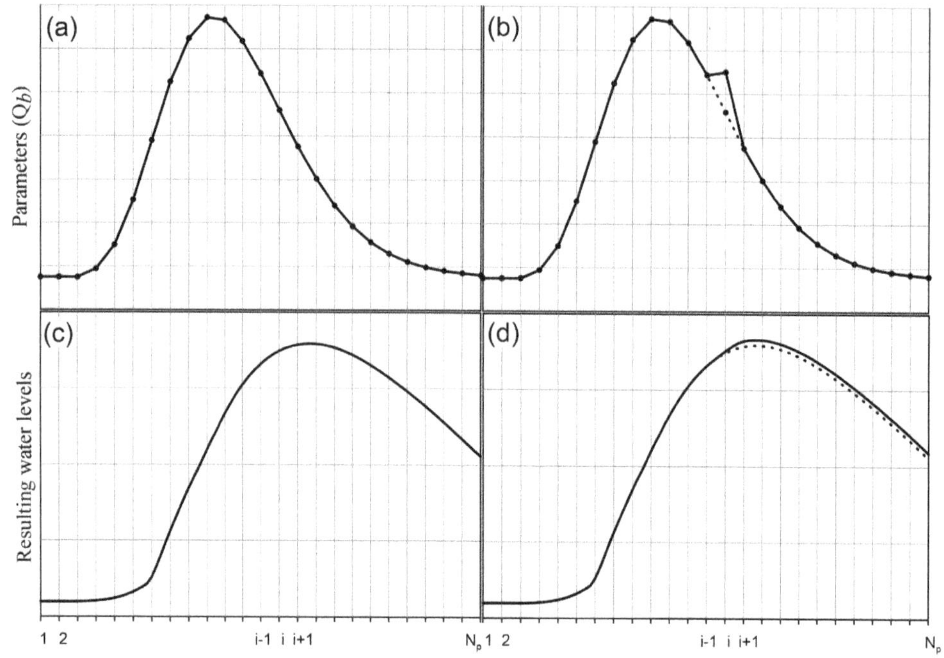

Figure 2. Example of the base run **(a)** and of the run i for the Jacobian matrix evaluation **(b)**.

ter i allows for the computing of the column i of the Jacobian matrix, which is a $N_{\mathrm{obs}} \times N_{\mathrm{p}}$ matrix where N_{obs} represents the number of the observations. After N_{p} runs, the Jacobian (sensitivity) matrix is evaluated and a new set of parameters $\boldsymbol{s}$ is estimated (Eq. 14).

This procedure is repeated until convergence or the maximum number of iteration N_{i} is reached. Then, the structural parameters are estimated using the last set of parameters $\boldsymbol{s}$. Due to the non-linearity of the forward problem, the model and the structural parameter estimation is repeated until convergence or the maximum number of iterations N_{o} is reached. Therefore, the BGA implementation requires running the forward model N_{t} times according to the following relation (Fienen et al., 2013);

$$N_{\mathrm{t}} = \left(N_{\mathrm{p}} + 1\right) N_{\mathrm{o}} N_{\mathrm{i}} + 1. \tag{15}$$

The whole BGA procedure previously described is illustrated in Fig. 3a.

3.1 Parallelization of the Jacobian matrix evaluation

The most relevant contribution to the total computational time required by the inverse procedure is ascribed to the forward model runs (i.e. the computation of each element in the Jacobian matrix) rather than to the bgaPEST operations. However, since each of the N_{p} runs in Eq. (15) checks the sensitivity of the observations to the variation of a single parameter, the solution of a run does not affect the other ones. Therefore, in order to reduce the computational burden, the independent N_{p} runs can be potentially performed in parallel.

In this work, the PARFLOOD two-dimensional-GPU numerical model presented in Vacondio et al. (2014) and Vacondio et al. (2017) has been adopted for routing the inflow hydrograph. Therefore, the bgaPEST routine to evaluate the Jacobian matrix has been parallelized in order to take advantage of the computational capability of modern high-performance computing (HPC) clusters, which are usually equipped with many GPUs. The implemented parallel procedure, which is illustrated in the flow chart of Fig. 3b, handles the parallelism among host and GPUs by means of the Secure Shell network protocol (SSH) and manages the most operative parts of the parallelism (login, run, etc.) outside of the bgaPEST code. In the serial version (Fig. 3a), the crucial part of the Jacobian matrix evaluation consists of a DO-loop over the parameters. Considering the parameter i, the input file that will be read by the forward model is first written, then the model is run, and the resulting values are finally read. In the modified version (Fig. 3b), this main loop is split in three parts: first, the input files (equal to N_{p}) in which a particular parameter is modified are written, then the forward model is run (N_{p} times), and a second loop is finally performed to read all the resulted values.

3.2 The forward model

In the parallel bgaPEST (Fig. 3b), the "run forward model" instruction actually runs a shell script that controls the file transfer between the host (a standard PC or a single node of a cluster) and the HPC platform, the creation of the N_{p} simulations for the Jacobian matrix evaluation, and the run of the

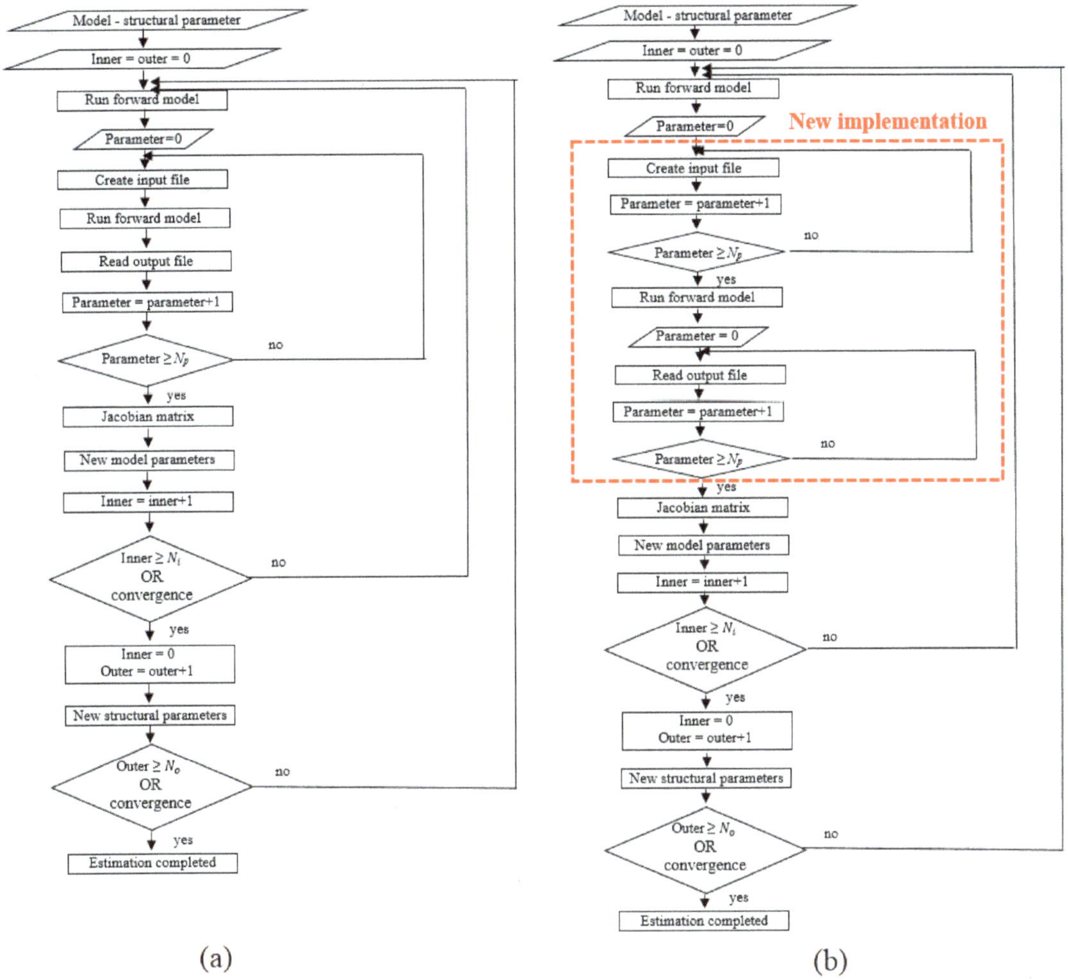

Figure 3. Illustration of BGA algorithm in the serial **(a)** and parallel **(b)** version.

two-dimensional SWE-GPU code on the device (GPU). In the present work, a cluster with 10 NVIDIA® Tesla® P100 GPUs hosted by the University of Parma was adopted. As shown in Fig. 4, the bgaPEST algorithm runs on the CPU of a computer, where the N_p simulations (assumed equal to 3 for the sake of simplicity in Fig. 4) are first created and then sent to the server user partition by means of the SSH protocol. Here, the cluster access node schedules all the jobs submitted by the users, using the HPC scheduler Portable Batch System (PBS). Then, each simulation is assigned to a specific GPU node. At the end of the computation, the observations are extracted and the output files remain on the cluster partition until the CPU verifies via SSH the end of the simulation and copies the results back. The procedure illustrated in Fig. 4 and later described represents the parallelization of the Jacobian matrix computation.

Listing 1 provides a detailed description of the "run forward model" shell file. In order to use the algorithm for different test cases and potentially on different HPC clusters, all the paths are first declared together with the involved variables (number of parameters to estimate, time interval among parameters, and start and end of the simulation; line 2). Then, the algorithm (line 3) checks if the considered run is one useful for the Jacobian matrix evaluation where a given parameter varies, or if it is the base run. Considering the first *if* condition as true (line 3), the script generates and copies the input files for all the N_p simulations to the server (lines 5–7). These files contain the same bathymetrical initial conditions (water level and velocity) and roughness configuration but have a different upstream boundary condition; each simulation tests the sensitivity of the observations to the variation of a given model parameter. Moreover, all the simulations adopt the same grid (Cartesian or multi-resolution), which is generated only once at the beginning of the procedure. It is relevant to note that all the N_p simulations do not have to be run from time t_{start} to time t_{end}; in fact, the variation of parameter i affects the observations only after time t_{i-1} (see Fig. 2). The PARFLOOD model guarantees the possibility of using the results of the base run and starting simulations from time t_{i-1}. The theoretical physical time T required to

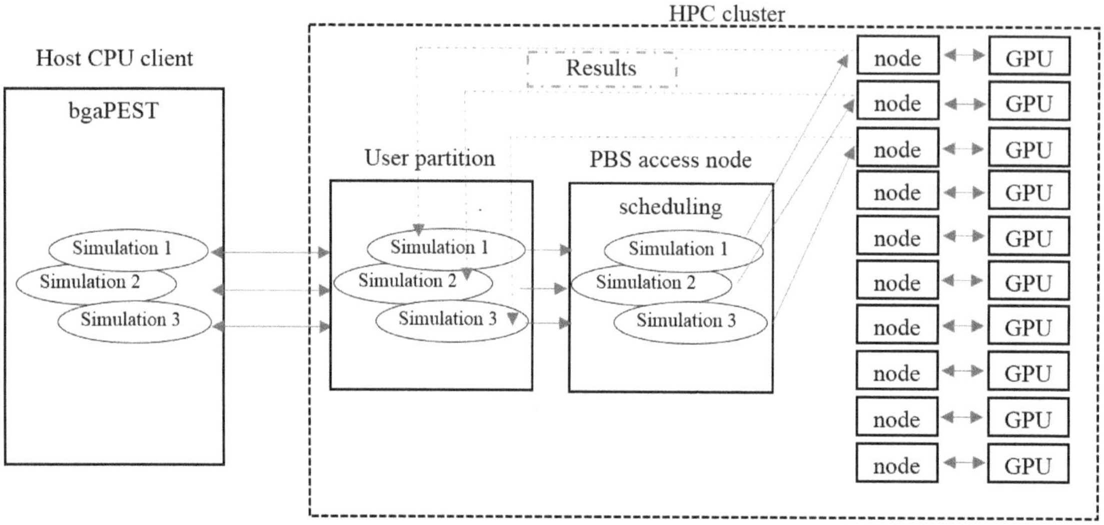

Figure 4. Schematization of the data transfer assuming three parameters and thus three parallel simulations.

evaluate the Jacobian matrix simulating each of the N_p runs from t_{start} to t_{end} is equal to

$$T = N_p (N_p - 1) \Delta t, \tag{16}$$

where Δt denotes the constant time interval between two consecutive parameters.

Conversely, the physical time T^* required to simulate all the N_p runs by restarting the ith simulation from time t_{i-1} instead of t_{start} is equal to

$$T^* = (N_p - 1)\Delta t + \sum_{i=2}^{N_p} \left[N_p - (i-1)\right] \Delta t. \tag{17}$$

As pointed out by Eqs. (16)–(17) and exemplified in Fig. 5, this simple operation allows for the reaching of a relevant decrease of the total computational time. Therefore, at line 8, the algorithm computes the time needed to restart the simulation.

In order to perform the simulation, the host logs in to the HPC cluster by means of the SSH protocol (line 9) and a sleep condition ensures the login procedure (line 10). Then the job is submitted to the queue of the cluster using external parameters for passing the name of the simulation folder and the time for restart (line 11); the submitted job contains the reference to the PBS queue and the link to the executable two-dimensional SWE-GPU code. At the end of the simulation, the water levels at the observation site are automatically extracted. Once the job is submitted, the SSH login is closed (line 12). After having submitted all the simulations, for each parameter (line 15), the code regularly (line 18) tests the presence of the end_file via SSH, which states the end of the simulation (line 20) and waits in case it is missing (line 25). Once the simulation is finished, the resulting observations are copied back to the host client (line 28), and the folder is removed from the server (line 29).

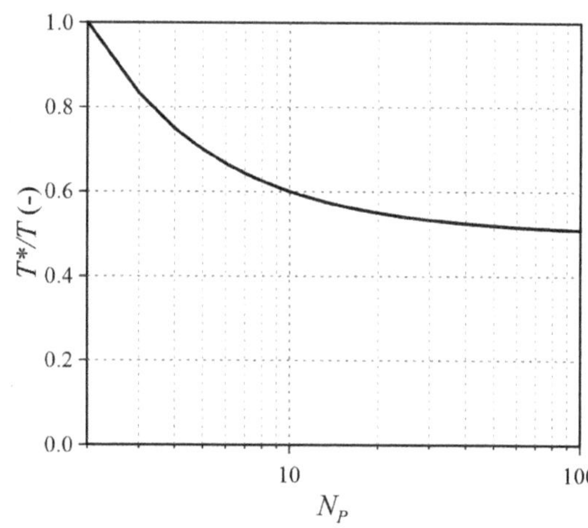

Figure 5. Time reduction T^*/T as a function of the number of estimable parameters (the x-axis is in logarithmic scale).

Conversely, the *else* condition (line 30) is true for the base run. The simulation folder with the input files is copied to the server (line 31), and the job is submitted (line 34). Then, the algorithm periodically verifies the end of the simulation and copies the results back to the host client (lines 39–49). It is relevant to note that the base run is performed first, whereas the other N_p runs can be performed in parallel.

4 Application of the inverse methodology to synthetic test cases

In the context of applying the BGA method described above, it is worth noting that reference solutions for inverse problems are by definition unavailable, since the goal of the methodology is the estimation of an upstream inflow hydro-

```
#!/bin/bash
Variable and path declaration
if [parameter run];
    then
    for (( i=1; i<=Np; i++ ))
        do
          Create and copy the simulation folder to the server (simulation_i)
          Compute the time from which restarting the simulation
          ssh server name << EOF
          sleep 15
          ssh submission: frame number for restart , name of the test case, job to submit
          exit
          EOF
    done
    for (( i=1; i<=Np; i++ ))
        do
          end_file=0
          while [ end_file -eq 0 ];
          do
          ssh server name  find server_path -iname end_file.txt | wc -l > end_file
          if [end_file];
          then
               continue
          else
               sleep 10
          fi
    done
    Copy the file with observation from server to CPU
    Remove the simulation folder on the cluster
else
    Create and copy the simulation folder to server
    ssh server name << EOF
    sleep 15
    ssh submission: frame number for restart , name of the test case, job to submit
    exit
    EOF
    end_file=0
    while [ end_file -eq 0 ];
    do
          ssh server name  find server_path -iname end_file.txt | wc -l > end_file
          if [end_file];
          then
               continue
          else
               sleep 10
          fi
    done
    Copy the file with observation from server to CPU
fi
```

Listing 1. Run forward model for the parallel bgaPEST scheme.

graph that is unknown at the beginning of the process. Therefore, in this section the inflow hydrographs in two natural rivers in northern Italy are estimated, and the reference solutions, which are necessary in order to validate the inverse procedure, are obtained as follows (D'Oria et al., 2014). Considering the domain in Fig. 6, a selected inflow discharge Q^{ref} is routed from the upstream section A to the downstream boundary D, where a rating curve is imposed far away from C. The resulting water level hydrographs are extracted at sites B and C. The inverse procedure is then applied to the sub-domain illustrated with solid line in Fig. 6 by assuming the water levels in sites B and C (derived in step 1) as the observations and downstream boundary condition, respectively. The information in sub-reach C–D is only preparatory for setting up the synthetic cases, and it is not used in the inverse procedure. Imposing a rating curve in D allows one to obtain water levels with a non-unique stage–discharge relationship in section C, which is more close to the real situations when applying the inverse procedure. The methodology estimates the inflow Q^{est} assuming that no information is available on the discharge (or water stage) at the inflow section A.

Quantitative information about the accuracy of the inverse methodology is provided by evaluating the differences between the reference Q^{ref} and the estimated Q^{est} hydrographs by means of three different indicators. First, the Nash–Sutcliffe efficiency criterion (Nash and Sutcliffe, 1970) E_{h} was adopted according to the following relation;

$$E_{\text{h}} = \left[1 - \frac{\sum_{i=1}^{N_{\text{p}}} (Q_i{}^{\text{ref}} - Q_i{}^{\text{est}})^2}{\sum_{i=1}^{N_{\text{p}}} (Q_i{}^{\text{ref}} - \overline{Q}^{\text{ref}})^2}\right] \cdot 100, \tag{18}$$

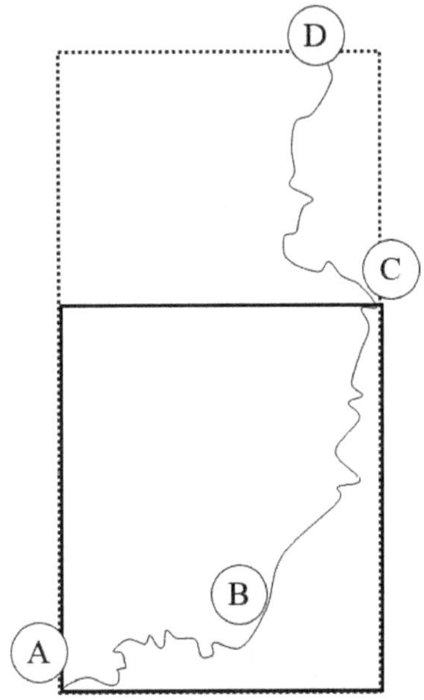

Figure 6. Exemplification of a test case definition.

where N_{p} is the number of parameters, Q_i^{ref} and Q_i^{est} are the ith reference and estimated inflow values, respectively, and $\overline{Q}^{\text{ref}}$ is the mean value of the reference hydrograph. Then, the root-mean-square error (RMSE) was evaluated as follows:

$$\text{RMSE} = \sqrt{\frac{\sum_{i=1}^{N} (Q_i{}^{\text{ref}} - Q_i{}^{\text{est}})^2}{N_{\text{p}}}}. \tag{19}$$

Finally, the estimation error in the peak discharge E_{p} was assessed as

$$E_{\text{p}} = \left[\frac{Q_{\text{p}}^{\text{est}}}{Q_{\text{p}}^{\text{ref}}} - 1\right] \cdot 100, \tag{20}$$

where $Q_{\text{p}}^{\text{est}}$ and $Q_{\text{p}}^{\text{ref}}$ denote the peak discharge value of the estimated and reference hydrographs, respectively.

4.1 Inflow hydrograph estimation on the Parma River

The first test concerns the estimation of a hypothetical discharge hydrograph at the entrance of the Parma River (northern Italy). Figure 7a illustrates the studied domain and the locations of the upstream boundary condition A, the observation site B, and the downstream boundary section C. The domain includes a 20 km long embanked reach that is characterized by several meanders and flood plains. As shown in Fig. 7, the flow field significantly varies at low and high discharge values due to the river morphology. At the beginning of the flood wave, the flow is characterized by both low water depths (Fig. 7b) and velocity (Fig. 7d). Conversely, at the arrival of the flood peak, most of the meanders are cut by the flow, as shown in Fig. 7c and e for water depths and velocity, respectively. This makes the adoption of 1-D numerical

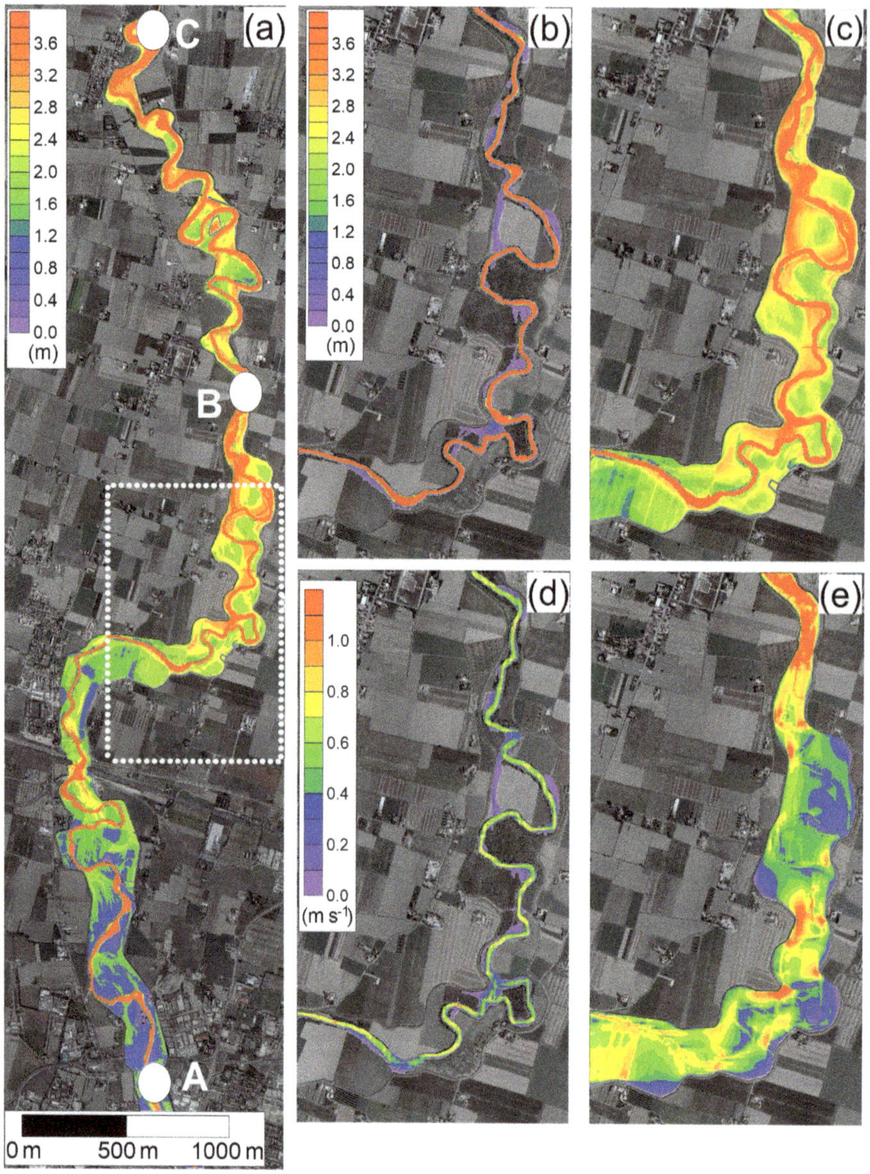

Figure 7. Map of the maximum simulated water depths for the Parma River. **(a)** the upstream (A) and downstream (C) boundary conditions and the intermediate observation site (B) are indicated. With reference to the area marked with dotted white line in **(a)**, **(b)** and **(c)** represent the water depths and **(d)** and **(e)** the velocity field at low and high discharge values, respectively.

schemes not suitable to accurately describe the flood propagation.

The bathymetry was derived from a 1 m resolution DTM obtained through a LiDAR survey carried out in drought condition. The domain was discretized by means of a Cartesian grid with cell sizes $\Delta x = \Delta y = 4$ m, and about 275×10^3 computing cells were adopted. The Manning roughness coefficient was assumed equal to $0.05\,\mathrm{s\,m^{-1/3}}$. The steady-state values of water depth and velocity fields, obtained considering the initial discharge value of the hydrograph, were adopted as initial conditions.

The inflow condition to be estimated was assumed as follows (D'Oria et al., 2015);

$$Q(t) = A + B \cdot f(t, b, k), \tag{21}$$

where t denotes the time, A the base flow (constant value), B the volume above the base flow (constant value), and f the gamma distribution, which states that

$$f(t, b, k) = \frac{1}{k^b \Gamma(b)} t^{b-1} e^{-\frac{t}{k}}, \tag{22}$$

where $\Gamma(b)$ represents the gamma function defined through the parameters b and k that denote the shape and the scale parameter, respectively. The parameters of the gamma distribution were set as follows: $A = 100\,\mathrm{m^3\,s^{-1}}$, $B = 3 \times 10^7\,\mathrm{m^3}$, $b = 6$, and $k = 10\,000\,\mathrm{s}$. The resulted flood wave presented a peak value of about $630\,\mathrm{m^3\,s^{-1}}$ at time $(b-1)k \approx 14\,\mathrm{h}$ (Fig. 8a).

During the estimation, when the sensitivity to the first parameter p_1 is investigated, the steady-state flow for the initial discharge is also recomputed. This means that parameter p_1 determines not only the first value of the estimated flood wave but also governs the initial condition of the river reach.

The inflow hydrograph duration was limited to 40 h, and it was discretized using 2 h time interval ($N_\mathrm{p} = 21$), whereas the observation stage hydrographs were discretized every 0.5 h. The prior pdf was defined by means of a Gaussian covariance function, and the initial structural parameters were set as reported in Table 1. In order to avoid non-physical discharge values during the computations, non-negativity was enforced to the unknown parameters by performing the estimation in a logarithmic space. The initial model parameter values were defined by applying the Linesearch tool of the bgaPEST, which dampens the solution between successive iterations (Fienen et al., 2013) and avoids numerical instabilities that may occur starting from a first choice of the parameters too far from the true one.

The inflow hydrograph was estimated first considering true observations (the variance was set equal to $10^{-8}\,\mathrm{m^2}$ to take into account the truncation error). Then, the same discharge hydrograph was defined corrupting the observed water levels with random errors uniformly distributed with maximum deviations of ± 0.05 m and variance $10^{-3}\,\mathrm{m^2}$ (Fig. 8b).

Qualitative assessment of the inverse methodology is achieved by comparing the reference with the estimated inflow hydrograph, as well as the observed with the modelled water levels in the observation site. Considering the simulation without errors in the observations, Fig. 9 shows that the estimated flood wave overlaps the reference one (a), and the modelled water levels agree almost perfectly with the measured ones (b).

The results of the simulation with random errors corrupting the observations are depicted in Fig. 10. The estimated flood wave matches well with the reference one again, presenting a misfit relative to the peak value lower than 5 %, and the modelled water levels similarly reproduce the reference ones with a residual of less than 1 %. Only the last value of the reconstructed flood wave is slightly overestimated, since the more the tested parameter nears the end of the wave, the fewer observations contain information about the related effects, as illustrated by the increasing range of the 95 % credibility interval. However, the "true" discharge values are inside the 95 % credibility interval, thus confirming the high accuracy of the solution. In addition to this behaviour at the end of the discharge hydrograph (that can be postponed extending the hydrograph total duration), very small differences between the observed and modelled variables appear when abrupt changes in the inflow function are present (e.g. the initial transition from the steady state to the flood wave). This behaviour is due to the regularization introduced into the solution by the prior information that imposes some degree of continuity and/or smoothness to the estimated hydrograph. However, the residuals are practically negligible, and abrupt discontinuities in the inflow hydrographs are not common in natural floods.

Table 1. Parma River: initial and estimated structural parameters and epistemic error variance.

		No random errors	Random errors
σ_R^2 (m^2)	Initial	–	1.00E-4
	Estimated	–	1.09E-3
σ_s^2 (m^6 s^{-2})	Initial	5.00E+2	5.00E+2
	Estimated	1.07E+3	5.36E+1
l (s)	Initial	6.48E+4	6.48E+4
	Estimated	2.90E+4	5.28E+4

Table 2. Parma River: Nash–Sutcliffe E_h, root-mean-square error (RMSE), and error in the peak discharge E_p values.

	E_h (–)	RMSE (m^3 s^{-1})	E_p (%)
No random errors	99.99	0.49	−0.04
Random errors	99.88	6.65	0.15

The structural parameters and the epistemic error variance estimated in the presence and absence of corrupted observations are reported in Table 1.

An assessment of the methodology accuracy has been quantified by means of the Nash–Sutcliffe E_h, root-mean-square error (RMSE), and error in the peak discharge E_p values reported in Table 2. The E_h values are greater than 99 %, the E_p values are almost negligible and the RMSE is less than $0.5\,\mathrm{m^3\,s^{-1}}$ without random errors and reaches the maximum value of $6\,\mathrm{m^3\,s^{-1}}$ with corrupted observations.

4.2 Inflow hydrograph estimation on the Secchia River

The second test case concerns both a different river reach and shape of the inflow hydrograph. The studied domain includes a 25 km long reach of the Secchia River (northern Italy) between the outflow of the flood control reservoir of Rubiera-Campogalliano, located west of the town of Modena (point A) and the gauging station of Ponte Bacchello (point C), referring the water level observations to the gauging station of Ponte Alto (point B) (Fig. 11). The modelled river reach is characterized by the presence of many flood plains and floodable areas that influence the flood propagation. The bathymetry was derived from a 1 m resolution DTM obtained through a LiDAR survey carried out in drought conditions.

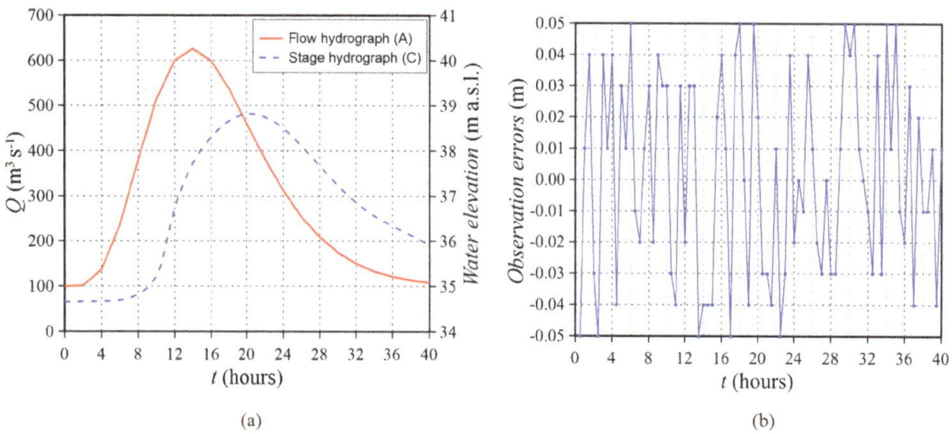

Figure 8. Parma River: flow and stage hydrographs at sections A and C, respectively, **(a)** and observation error distribution **(b)**.

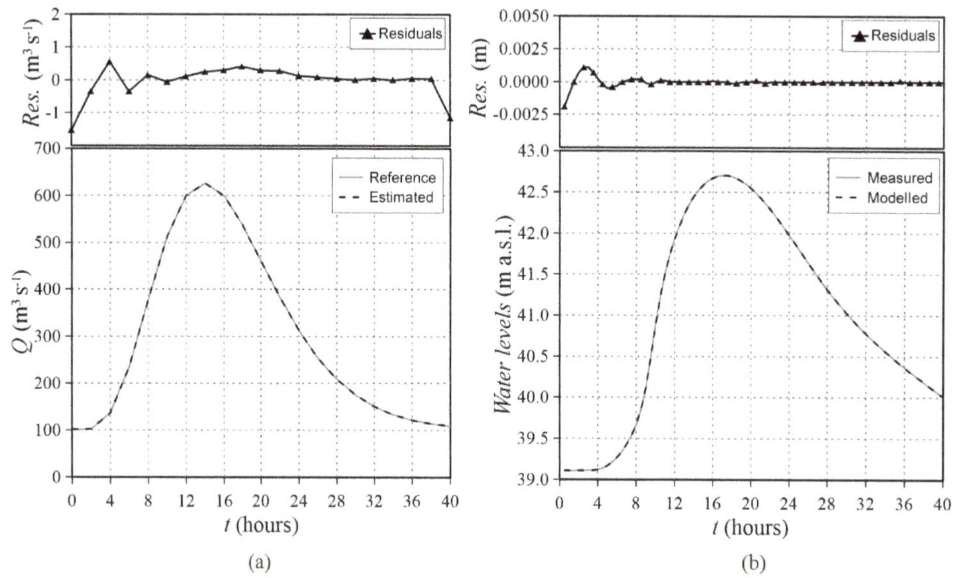

Figure 9. Parma River: reference and estimated inflow hydrograph **(a)** and observed (uncorrupted) and modelled water levels **(b)**. The residuals between reference and estimated values are also reported.

The domain was discretized by means of a non-uniform Block-Uniform Quadtree (BUQ) grid (Vacondio et al., 2017), resulting in 77×10^3 computing cells. The Manning roughness coefficient in the riverbed was assumed equal to $0.05\,\mathrm{s\,m^{-1/3}}$ (Vacondio et al., 2016).

The discharge hydrograph to be estimated is the flood wave of a 20-year return period of the Secchia River, with a peak value of about $780\,\mathrm{m^3\,s^{-1}}$. In order to increase the non-smoothness of the wave, a quite abrupt increment that separates the initial steady-state condition ($100\,\mathrm{m^3\,s^{-1}}$) from the rising limb was introduced (Fig. 12a). It is noteworthy that this flow hydrograph is characterized by a pseudo-real irregular shape, which cannot be properly approximated by an analytical parametric function (e.g. Gamma distribution, Pearson function). The inflow hydrograph ended in 72 h and was discretized using 2 h time interval ($N_\mathrm{p} = 37$), whereas the observed stage hydrograph was discretized every 0.5 h. The inflow hydrograph was first estimated assuming that the true water levels extracted at section B only had a truncation error resulting in a variance of $10^{-8}\,\mathrm{m^2}$, and then considering corrupted observations with random errors uniformly distributed with maximum deviations of $\pm 0.05\,\mathrm{m}$ and variance of $10^{-3}\,\mathrm{m^2}$ (Fig. 12b). Figure 12a also depicts the discharge hydrograph at the downstream boundary condition section C in order to highlight the attenuation effect exerted by the flood plains and floodable areas.

As before, the parameters were estimated in a logarithmic space, and their initial values were calculated by adopting the Linesearch tool of the bgaPEST (Fienen et al., 2013). The prior pdf was described by means of a linear and Gaus-

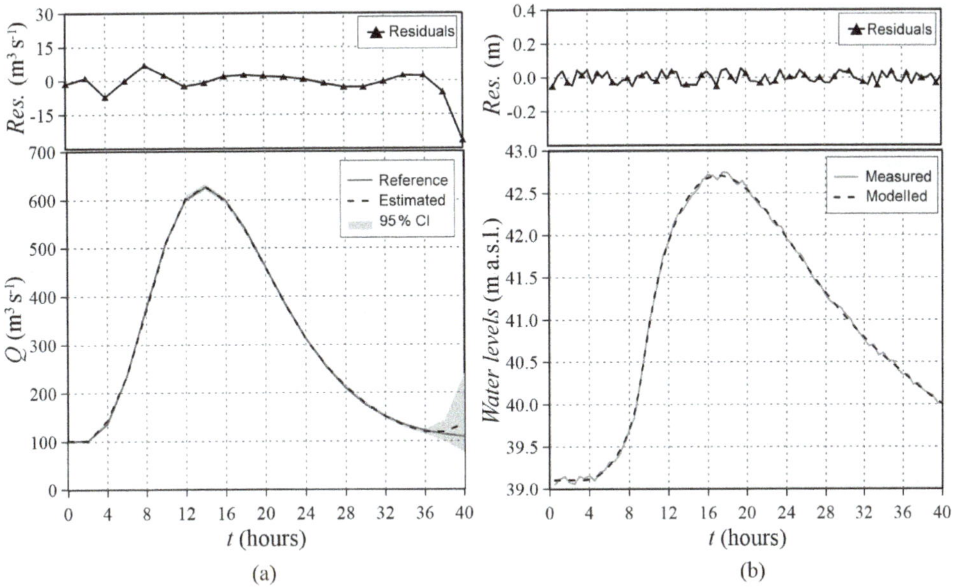

Figure 10. Parma River: reference and estimated (with a 95 % credibility interval) inflow hydrographs **(a)** and observed (corrupted) and modelled water levels **(b)**. The residuals between reference and estimated values are also reported.

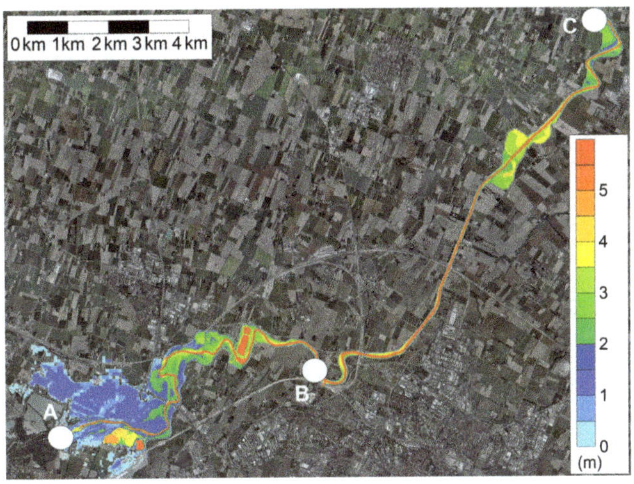

Figure 11. Map of the water depths at the flood peak occurrence on the Secchia River, with indication of the upstream (A) and downstream (C) boundary conditions and the intermediate observation site (B).

sian covariance function in the configuration with and without corrupted observations, respectively (Table 3).

As shown in Fig. 13 for the simulation without corrupted observations, the estimated flood wave matches almost perfectly with the reference one, and the modelled water levels agree with the measured ones.

The results of the simulation with corrupted observations depicted in Fig. 14 highlight that both the shape and the peak value are well captured. The small discrepancies between the estimated peak flood wave and the reference one are essentially caused by the fact that the portion with the peak is discretized with only a few parameters and that the adopted covariance function smooths the solution.

Table 3. Secchia River: initial and estimated structural parameters and epistemic error variance.

		No random errors	Random errors
θ ($m^6\,s^{-3}$)	Initial	1.00E-10	–
	Estimated	3.97E-6	–
σ_R^2 (m^2)	Initial	–	1.00E-4
	Estimated	–	1.11E-3
σ_s^2 ($m^6\,s^{-2}$)	Initial	–	5.00E+2
	Estimated	–	1.38E+1
l(s)	Initial	–	4.32E+4
	Estimated	–	3.88E+4

The structural parameters and the epistemic error variance estimated in the presence and absence of corrupted observations are reported in Table 3.

The indicators used for evaluating the accuracy of the methodology are reported in Table 4. The Nash–Sutcliffe efficiency $E_{\rm h}$ values exceed 99 %, the errors in the peak flow $E_{\rm p}$ are almost negligible, and the RMSE is less than 1 $m^3\,s^{-1}$ without random errors and reaches the maximum value of 16 $m^3\,s^{-1}$ with corrupted observations; these values highlight the accuracy of the procedure in estimating the overall shape and peak of the inflow hydrograph.

For this case, some details about the computational characteristics are reported in Table 5.

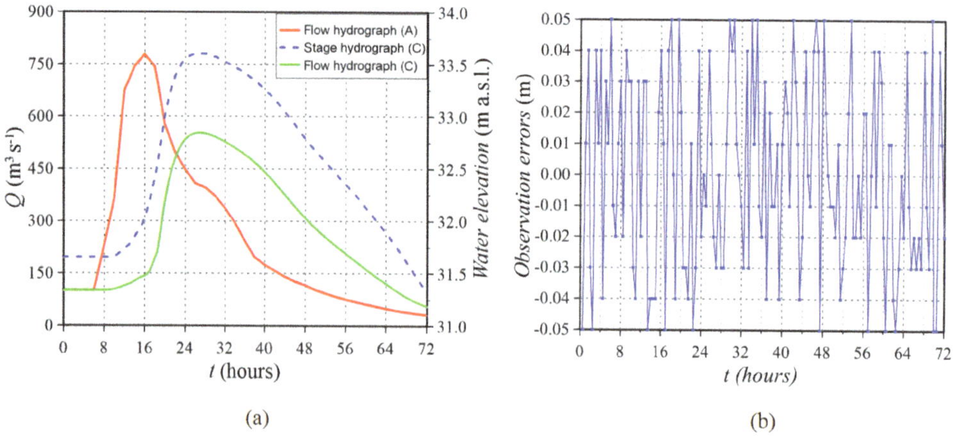

Figure 12. Secchia River: flow hydrograph in section A and flow and stage hydrographs in section C **(a)** and observation error distribution **(b)**.

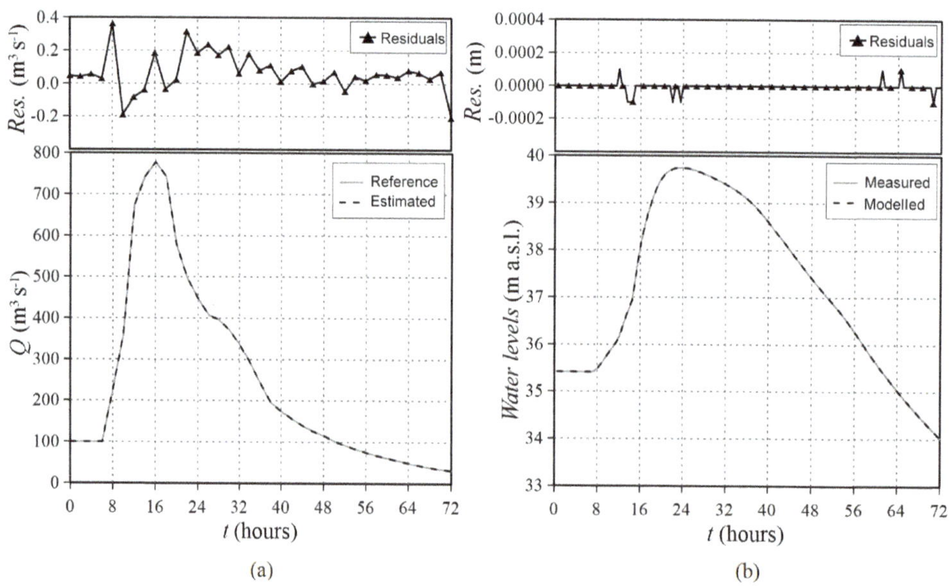

Figure 13. Secchia River: reference and estimated inflow hydrograph **(a)** and observed (uncorrupted) and modelled water levels **(b)**. The residuals between reference and estimated values are also reported.

Table 4. Secchia River: Nash–Sutcliffe E_h, root-mean-square error (RMSE), and error in the peak discharge E_p values.

	E_h (–)	RMSE (m^3 s^{-1})	E_p (%)
No random errors	99.99	0.13	−0.02
Random errors	99.44	16.57	2.89

The computational time of the whole inflow hydrograph simulation (72 h) is 9.62 min, whereas the simulations for evaluating the Jacobian matrix and testing parameters 2–37 required a computational time progressively lower than 9.62 min, thanks to the restart option illustrated in the Sect. 3. In order to evaluate the total time required by the inverse procedure, it is noteworthy that dealing with an HPC cluster the global run time depends on the number of the available GPUs. However, this test was performed using 10 GPUs, and the computational cost of the 609 runs was about 13 h. Since the implemented procedure that manages the interaction between host and server can be used for different HPC clusters, the availability of a cluster equipped with N_p GPUs would have allowed the estimation of the flood wave in about 8 h. On the other side, the adoption of the serial bgaPEST procedure and the PARFLOOD code as routing model would have required about 4 days of computations, which means about 8 times higher than the parallel procedure proposed here. Particularly interesting is the hypothetical evaluation of the computational time for a serial BGA procedure and the adoption of a serial CPU code as the forward hydraulic model. Vacon-

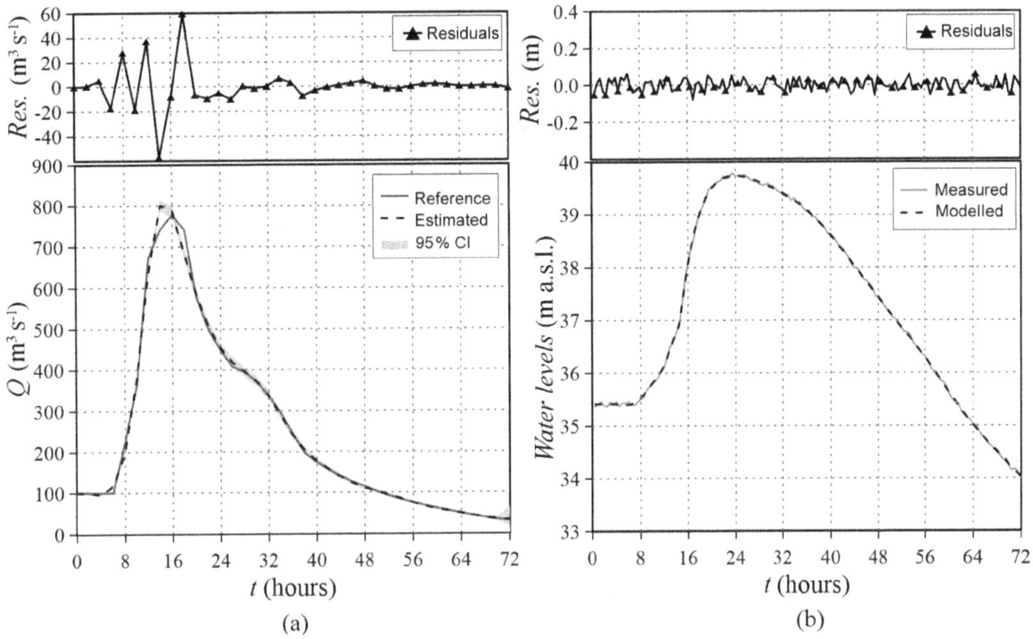

Figure 14. Secchia River: reference and estimated (with a 95 % credibility interval) inflow hydrograph **(a)** and observed (corrupted) and modelled water levels **(b)**. The residuals between reference and estimated values are also reported.

Table 5. Secchia River: characteristics of the simulation.

Number of parameters N_p	37
Physical total time of the inflow hydrograph	72 h
Physical total time of the run testing the 1st parameter p_1, assuming 100 h for reaching the steady state condition	172 h
Computational time of the whole inflow hydrograph simulation (72 h)	9.62 min
Computational time of the run testing the 1st parameter (172 h)	19.38 min
Number of the BGA iterations N_i for the model parameter estimation	4
Number of the BGA iterations N_o for the structural parameter estimation	4
Total number of simulations N_t (Eq. 15)	609

dio et al. (2014) pointed out that the PARFLOOD code led to a speed-up of up to 2 orders of magnitude if compared to a serial CPU code. Therefore, if a serial BGA procedure and the GPU forward model would have required about 4 computational days, the inverse problem solution with a serial forward code would have ended in 400 computational days, making the use of the inverse procedure practically infeasible.

5 Reconstruction of a historical event: the December 2009 flood wave on the Secchia River

The inverse procedure is now validated by investigating the December 2009 flooding event on the Secchia River, which is one of the most significant events that occurred in the last 10 years in this river. The Interregional Agency for the Po River (AIPo) monitored the river and provided the water stage hydrographs recorded in the two gauging stations indicated in Fig. 11 with letters B and C, respectively. As shown in Fig. 15, the recorded water levels present more than one rising and recession limb; thus, besides the challenges related to a real field application, this test also aims at addressing the estimation of an inflow with multiple peaks. In order to estimate the discharge in section A (Fig. 11), the water levels recorded at points B and C were assumed as observations and the downstream boundary condition, respectively. The event was simulated from 21:00 LT on 22 December 2009 to 12:00 LT on 26 December, with a total duration of 87 h. The water levels were recorded every 0.5 h, whereas the unknown inflow hydrograph was discretized into 88 parameters (one per hour, $N_p = 88$).

The studied domain is the same as the one previously adopted for a synthetic inflow; thus, the reader is kindly referred to Sect. 4.2 for the information about bathymetry, initial conditions, and the roughness configuration.

As before, the parameters were estimated in a logarithmic space and their initial values were calculated by adopting the Linesearch tool of the bgaPEST (Fienen et al., 2013). The

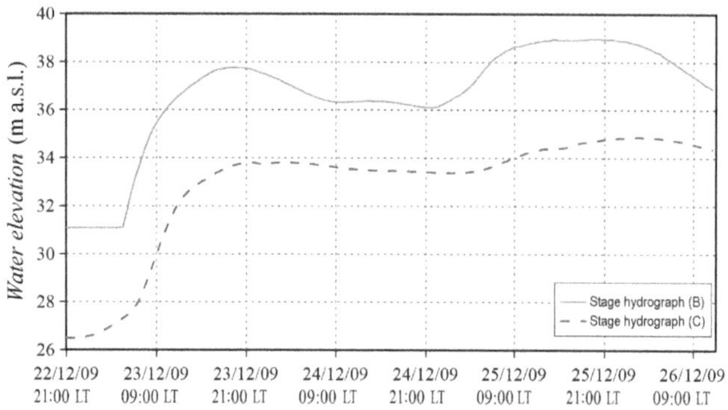

Figure 15. December 2009 recorded stage hydrographs on the Secchia River at sections B and C.

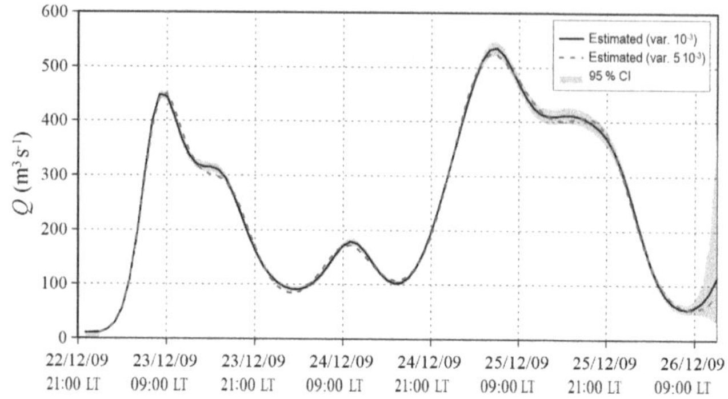

Figure 16. Secchia 2009 event: estimated inflow hydrographs assuming the epistemic error variance equal to 10^{-3} m^2 and 5×10^{-3} m^2. The 95 % credibility interval applies to the simulation with the epistemic variance equal to 10^{-3} m^2.

Table 6. Secchia 2009 event: initial and estimated structural parameters.

	σ_s^2 (m^6 s^{-2})	l (s)
Initial	5.00E+2	6.48E+4
Estimated	1.49E+1	3.36E+4

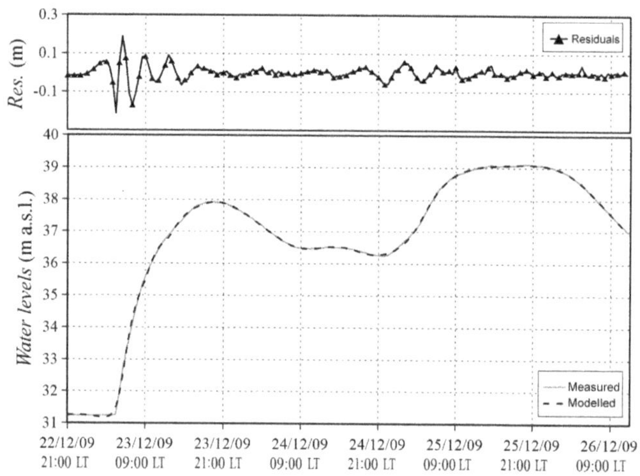

Figure 17. Secchia 2009 event: observed and modelled water levels in section B. The residuals between recorded and estimated values are also reported.

prior pdf was described by means of a Gaussian covariance function; the initial and estimated structural parameters are reported in Table 6.

Figure 16 shows the estimated flood wave (and the 95 % credibility interval), which presents an irregular shape and two main peaks, as it could be expected from the observed stage hydrograph. Moreover, an additional small intermediate peak is captured that was not as evident in the registered water levels at section B (Fig. 15), even if a little pronounced local maximum can be seen around 15:00 LT on 24 December 2009. The resulting flood wave presents neither instabilities nor oscillations. During the computation, the variance of the epistemic error was assumed equal to 10^{-3} m^2; as shown in Sect. 4, this means considering the observed water levels corrupted with random errors with maximum deviations of ± 0.05 m. In Fig. 16, the flood wave estimated by increasing the variance by half an order of magnitude is also depicted (dotted line); the solution appears slightly smoothed in a few points but substantially similar when compared to the inflow resulting in the smaller variance, which is thus considered the estimated inflow of the studied event. The comparison between modelled and measured water levels at section B is presented in Fig. 17; it is relevant to note that the residuals between the two hydrographs are mostly less than 2 cm, and only in a few points of the first rising limb do they reach the highest value of 18 cm.

With the aim of validating the methodology for this real application, it is noteworthy that the upstream section of the river is located immediately downstream from a flood control reservoir equipped with water level sensors. Therefore, the "reference" discharge hydrograph has been obtained from the dam's geometrical data (i.e. number and dimension of the bottom openings, crest length of the spillway, etc.) and the recorded water levels adopting the classic hydraulic theory of sluice gates and spillways.

Due to the uncertainty in evaluating the discharge coefficients and the fact that during flood events a large amount of wood debris reduces the outflow discharge from the bottom openings (especially during the depletion phase) and interferes with the overflow spillway, the discharge hydrograph has been calculated by adopting equally likely coefficients (Fig. 18). The flood wave estimated by the inverse procedure is in good agreement with the one calculated using the flood reservoir data; the main differences are after the highest peak, which is well reproduced, although the inverse methodology provides a smoother solution. For this real application, even though the river roughness coefficient was already calibrated in previous studies (Vacondio et al., 2016), an additional inverse Bayesian estimation was performed with a dif-

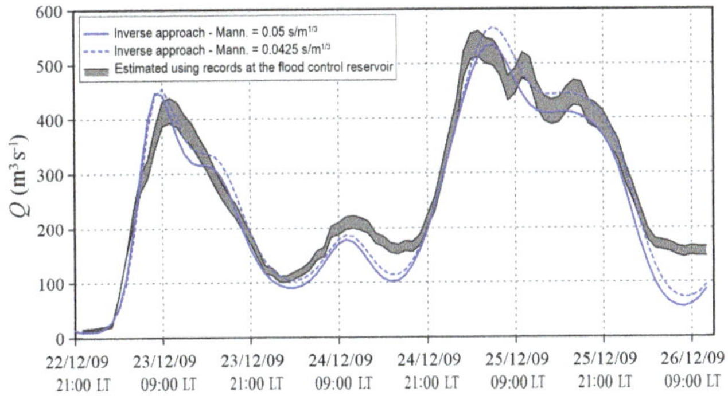

Figure 18. Secchia 2009 event: comparison among the inflow hydrographs obtained from the inverse procedure using two different Manning coefficients and the envelope of different solutions obtained using the records at the flood control reservoir.

ferent value in order to assess the effect of this coefficient on the solution. Particularly, the Manning coefficient, originally set to $0.05\,\mathrm{s\,m^{-1/3}}$, was decreased by 15 % ($0.0425\,\mathrm{s\,m^{-1/3}}$), which, for example, can happen due to seasonal changes in vegetation. As shown in Fig. 18, the estimated flood waves are similar, and the highest difference, which is in correspondence with the main peak, is less than 6 %. Therefore, the influence of assuming a "wrong" roughness coefficient is less than linear in the discharge estimation. Despite all the involved approximations, this comparison confirms that the proposed inverse procedure is capable of estimating inflow hydrographs with multiple peaks and irregular shapes in real rivers.

6 Conclusions

In this work the inverse problem of estimating the unknown inflow hydrograph in an upstream-ungauged section, having water level information only in downstream sites, has been solved by means of a Bayesian methodology. The key aspects in the solution of this problem have been the adoption of a parallel two-dimensional SWE code running on GPUs and the performance of the simulations over a HPC cluster. The parallelization of the runs useful for the Jacobian matrix computation and the implementation of an ad hoc procedure, which allows one to take advantage of any HPC cluster with GPUs, have provided a remarkable reduction of the computational costs. In a test case, this parallel procedure reduced the computational time by a factor of 8 against running the two-dimensional SWE code on a single GPU. Furthermore, the analysis of the runtimes has highlighted that the use of a parallel hydraulic forward routing model is the *conditio sine qua non* for solving this type of inverse problem, whereas the adoption of a serial code would lead to inadmissible computational times. The inverse procedure has been validated considering two different natural rivers; in both tests, no instabilities due to the adopted inverse procedure or to the availability of a stable, fast, and accurate forward hydraulic model arose. Moreover, the obtained results have highlighted that the implemented procedure estimates the unknown inflow hydrographs with different and irregular shapes and in the presence of corrupted observations well; quantitative indicators have proved the accuracy of the methodology. In all the presented tests, the resulting Nash–Sutcliffe efficiency criterion exceeded 99 %, the error in the peak discharge was less than 3 %, and the RMSE was less than 2 %. Finally, the proposed inverse procedure allowed for the estimation of a historical flood wave characterized by the presence of multiple peaks, without causing instabilities in the solution. The test cases were simulated while taking advantage of the HPC cluster of the University of Parma. However, since the implemented procedure is general, it is possible to adopt clouds of GPUs or online mini clusters, which are now common and accessible to everyone. The adopted Bayesian software (bgaPEST) is open access, and two-dimensional SWE models are a quite common tools for practitioners, even if few of them were fast enough to perform the necessary simulations with a reasonable computing time until now. Therefore, the 2-D coupled methodology proposed here can be adopted in the near future by professional hydrologists, too, who are involved, for example, in the design of hydraulic infrastructures, as well as for engineers working on water resource management (i.e. irrigation systems, hydroelectric power stations, etc.) or forensic activities. The future development of the methodology will also focus on the possibility of reconstructing the flood waves in the presence of levee breaches and flooding outside the river region, where the adoption of a two-dimensional SWE model is mandatory.

Competing interests. The authors declare that they have no conflict of interest.

Acknowledgements. This work was partially supported by Ministry of Education, Universities and Research under the Scientific Independence of young Researchers project, grant number RBSI14R1GP, CUP code D92I15000190001. This research benefits from the HPC (High Performance Computing) facility of the University of Parma, Italy. The Interregional Agency for the Po River (AIPo) is also gratefully acknowledged for providing data. The authors are grateful to the editor, the anonymous reviewer, and Antonis D. Koussis for the valuable suggestions on the early version of this manuscript.

Edited by: Roberto Greco

References

Beven, K. J.: Rainfall-runoff modelling: the primer, John Wiley & Sons, 2011.

Bruen, M. and Dooge, J. C. I.: Harmonic analysis of the stability of reverse routing in channels, Hydrol. Earth Syst. Sci., 11, 559–568, https://doi.org/10.5194/hess-11-559-2007, 2007.

Butera, I., Tanda, M. G., and Zanini, A.: Simultaneous identification of the pollutant release history and the source location in groundwater by means of a geostatistical approach, Stoch. Env. Res. Risk A., 27, 1269–1280, 2013.

Das, A.: Reverse stream flow routing by using Muskingum models, Sadhana, 34, 483–499, 2009.

Doherty, J. E.: PEST, Model-Independent Parameter Estimation – User Manual, sixth ed., Tech. rep., Watermark Numerical Computing, Brisbane, Australia, 2016.

Dooge, J. and Bruen, M.: Problems in reverse routing, Acta Geophysica Polonica, 53, 357–371, 2005.

D'Oria, M. and Tanda, M. G.: Reverse flow routing in open channels: A Bayesian Geostatistical Approach, J. Hydrol., 460, 130–135, 2012.

D'Oria, M., Mignosa, P., and Tanda, M. G.: Bayesian estimation of inflow hydrographs in ungauged sites of multiple reach systems, Advances in Water Resources, 63, 143–151, 2014.

D'Oria, M., Mignosa, P., and Tanda, M. G.: An inverse method to estimate the flow through a levee breach, Adv. Water Resour., 82, 166–175, 2015.

Eli, R., Wiggert, J., and Contractor, D.: Reverse flow routing by the implicit method, Water Resour. Res., 10, 597–600, 1974.

Fienen, M., Hunt, R., Krabbenhoft, D., and Clemo, T.: Obtaining parsimonious hydraulic conductivity fields using head and transport observations: A Bayesian geostatistical parameter estimation approach, Water Resour. Res., 45, W08405, https://doi.org/10.1029/2008WR007431, 2009.

Fienen, M. N., Clemo, T., and Kitanidis, P. K.: An interactive Bayesian geostatistical inverse protocol for hydraulic tomography, Water Resour. Res., 44, W00B01, https://doi.org/10.1029/2007WR006730, 2008.

Fienen, M. N., D'Oria, M., Doherty, J. E., and Hunt, R. J.: Approaches in highly parameterized inversion: bgaPEST, a Bayesian geostatistical approach implementation with PEST: documentation and instructions, Tech. rep., US Geological Survey, available at: https://pubs.usgs.gov/tm/07/c09/ (last access: 12 October 2018), 2013.

Glickman, M. E. and Van Dyk, D. A.: Basic bayesian methods, Topics in Biostatistics, 319–338, 2007.

Hoeksema, R. J. and Kitanidis, P. K.: An application of the geostatistical approach to the inverse problem in two-dimensional groundwater modeling, Water Resour. Res., 20, 1003–1020, 1984.

Kitanidis, P. K.: Quasi-linear geostatistical theory for inversing, Water Resour. Res., 31, 2411–2419, 1995.

Kitanidis, P. K. and Vomvoris, E. G.: A geostatistical approach to the inverse problem in groundwater modeling (steady state) and one-dimensional simulations, Water Resour. Res., 19, 677–690, 1983.

Koussis, A. D. and Mazi, K.: Reverse flood and pollution routing with the lag-and-route model, Hydrolog. Sci. J., 61, 1952–1966, 2016.

Koussis, A. D., Mazi, K., Lykoudis, S., and Argiriou, A. A.: Reverse flood routing with the inverted Muskingum storage routing scheme, Nat. Hazards Earth Syst. Sci., 12, 217–227, https://doi.org/10.5194/nhess-12-217-2012, 2012.

Leonhardt, G., D'Oria, M., Kleidorfer, M., and Rauch, W.: Estimating inflow to a combined sewer overflow structure with storage tank in real time: evaluation of different approaches, Water Sci. Technol., 70, 1143–1151, 2014.

Michalak, A. M., Bruhwiler, L., and Tans, P. P.: A geostatistical approach to surface flux estimation of atmospheric trace gases, J. Geophys. Res.-Atmos., 109, D14109, https://doi.org/10.1029/2003JD004422, 2004.

Nash, J. E. and Sutcliffe, J. V.: River flow forecasting through conceptual models part I – A discussion of principles, J. Hydrol., 10, 282–290, 1970.

Saghafian, B., Jannaty, M., and Ezami, N.: Inverse hydrograph routing optimization model based on the kinematic wave approach, Eng. Optimiz., 47, 1031–1042, 2015.

Snodgrass, M. F. and Kitanidis, P. K.: A geostatistical approach to contaminant source identification, Water Resour. Res., 33, 537–546, 1997.

Szymkiewicz, R.: Solution of the inverse problem for the Saint Venant equations, J. Hydrol., 147, 105–120, 1993.

Vacondio, R., Dal Palù, A., and Mignosa, P.: GPU-enhanced finite volume shallow water solver for fast flood simulations, Environ. Modell. Softw., 57, 60–75, 2014.

Vacondio, R., Aureli, F., Ferrari, A., Mignosa, P., and Dal Palù, A.: Simulation of the January 2014 flood on the Secchia River using a fast and high-resolution 2D parallel shallow-water numerical scheme, Nat. Hazards, 80, 103–125, 2016.

Vacondio, R., Dal Palù, A., Ferrari, A., Mignosa, P., Aureli, F., and Dazzi, S.: A non-uniform efficient grid type for GPU-parallel Shallow Water Equations models, Environ. Modell. Softw., 88, 119–137, 2017.

Zucco, G., Tayfur, G., and Moramarco, T.: Reverse flood routing in natural channels using genetic algorithm, Water Resour. Manag., 29, 4241–4267, 2015.

13

Remote sensing algorithm for surface evapotranspiration considering landscape and statistical effects on mixed pixels

Zhi Qing Peng[1,2], Xiaozhou Xin[1], Jin Jun Jiao[1,2], Ti Zhou[1,2], and Qinhuo Liu[1,3]

[1]State Key Laboratory of Remote Sensing Science, Institute of Remote Sensing and Digital Earth, Chinese Academy of Sciences, Beijing, 100101, China
[2]University of Chinese Academy of Sciences, Beijing, 100049, China
[3]Joint Center for Global Change Studies (JCGCS), Beijing, 100875, China

Correspondence to: Xiaozhou Xin (xin_xzh@163.com)

Abstract. Evapotranspiration (ET) plays an important role in surface–atmosphere interactions and can be monitored using remote sensing data. However, surface heterogeneity, including the inhomogeneity of landscapes and surface variables, significantly affects the accuracy of ET estimated from satellite data. The objective of this study is to assess and reduce the uncertainties resulting from surface heterogeneity in remotely sensed ET using Chinese HJ-1B satellite data, which is of 30 m spatial resolution in VIS/NIR bands and 300 m spatial resolution in the thermal-infrared (TIR) band. A temperature-sharpening and flux aggregation scheme (TSFA) was developed to obtain accurate heat fluxes from the HJ-1B satellite data. The IPUS (input parameter upscaling) and TRFA (temperature resampling and flux aggregation) methods were used to compare with the TSFA in this study. The three methods represent three typical schemes used to handle mixed pixels from the simplest to the most complex. IPUS handles all surface variables at coarse resolution of 300 m in this study, TSFA handles them at 30 m resolution, and TRFA handles them at 30 and 300 m resolution, which depends on the actual spatial resolution. Analyzing and comparing the three methods can help us to get a better understanding of spatial-scale errors in remote sensing of surface heat fluxes. In situ data collected during HiWATER-MUSOEXE (Multi-Scale Observation Experiment on Evapotranspiration over heterogeneous land surfaces of the Heihe Watershed Allied Telemetry Experimental Research) were used to validate and analyze the methods. ET estimated by TSFA exhibited the best agreement with in situ observations, and the footprint validation results showed that the R^2, MBE, and RMSE values of the sensible heat flux (H) were 0.61, 0.90, and 50.99 W m^{-2}, respectively, and those for the latent heat flux (LE) were 0.82, −20.54, and 71.24 W m^{-2}, respectively. IPUS yielded the largest errors in ET estimation. The RMSE of LE between the TSFA and IPUS methods was 51.30 W m^{-2}, and the RMSE of LE between the TSFA and TRFA methods was 16.48 W m^{-2}. Furthermore, additional analysis showed that the TSFA method can capture the subpixel variations of land surface temperature and the influences of various landscapes within mixed pixels.

1 Introduction

Five types of methods have been developed to estimate evapotranspiration (ET) or latent heat flux (LE) via remote sensing.

1. Surface energy balance models calculate LE as a residual term. According to the partitioning of the sources and sinks of the soil–plant–atmosphere continuum (SPAC), surface energy balance models can be classified as one-source (Bastiaanssen et al., 1998; Su, 2002; Allen et al., 2007; Long and Singh, 2012a) or two-source models (Shuttleworth and Wallace, 1985; Norman et al., 1995; Xin and Liu, 2010; Zhu et al., 2013).

2. Penman–Monteith models are used to calculate LE by using the Penman–Monteith equation and numerous surface resistance parameterization schemes that

control the diffusion of evaporation from soil surfaces and transpiration from plant canopies. These two-source Penman–Monteith models separate soil evaporation from plant transpiration (Cleugh et al., 2007; Mu et al., 2011; Leuning et al., 2008; Chen et al., 2013; Sun et al., 2013; Mallick et al., 2015).

3. Land surface temperature–vegetation index (LST-VI) space methods assign the dry and wet edges of the LST-VI feature space as minimum and maximum ET, respectively. These methods interpolate the media, and use the Penman–Monteith or Priestley–Taylor equation to calculate the LE (Jiang and Islam, 1999, 2001; Sun et al., 2011; Long and Singh, 2012b; Yang and Shang, 2013; Fan et al., 2015; Zhang et al., 2005).
4. Priestley–Taylor models expand the range of the Priestley–Taylor coefficient in the Priestley–Taylor equation (Jiang and Islam, 2003; Jin et al., 2011) or combine the physiological force factors with the energy component of ET (Fisher et al., 2008; Yao et al., 2013).
5. Additional methods include empirical/statistical methods (Wang and Liang, 2008; Yebra et al., 2013) and the use of complementary-based models (Venturini et al., 2008) and land-process models with data assimilation schemes (Bateni and Liang, 2012; Xu et al., 2015).

If the operational algorithm can be described as a linear combination of inputs, or if the surface variables and landscapes are homogeneous at the pixel scale, scale error does not exist (Hu and Islam, 1997). However, it is difficult to develop linear operational models due to the complexity of mass and heat transfer processes between the atmosphere and land surface. ET estimation models have been generally developed for simple and homogeneous surface conditions. However, heterogeneity is a natural attribute of the surface of the Earth. Therefore, larger spatial-scale errors occur when these remotely sensed models are applied to calculate the regional ET using satellite data.

In previous studies, researchers have coupled high- and low-resolution satellite data and statistically quantified the inhomogeneity of mixed pixels to correct the scale error in ET estimations using (1) temperature downscaling, which converts images from a lower (coarser) to higher (finer) spatial resolution using statistical-based models with regression or stochastic relationships among parameters (Kustas et al., 2003; Norman et al., 2003; Cammalleri et al., 2013; Ha et al., 2013); (2) the correction-factor method, which uses subpixel landscapes information to determine the correction factor of scale bias (Maayar and Chen, 2006); and (3) the area-weighting method, which calculates roughness length and sensible heat flux based on subpixel landscapes (Xin et al., 2012). These correction methods mainly focus on two problems: inhomogeneity of landscapes and inhomogeneity of surface variables.

Studies have shown that different landscapes (Blyth and Harding, 1995; Moran et al., 1997; Bonan et al., 2002; McCabe and Wood, 2006) and the subpixel variations of surface variables, such as stomatal conductance (Bin and Roni, 1994), or leaf area index (Bonan et al., 1993; Maayar and Chen, 2006), can cause errors in turbulent heat flux estimations. Surface variables' inhomogeneity is rather difficult to evaluate, as the subpixel variation of surface variables can be large, even in the pure pixels. For example, generally, temperatures over land surfaces vary strongly in space and time, and it is common for the LST to vary by more than 10 K over just a few centimeters of distance or by more than 1 K in less than a minute over certain cover types (Z. L. Li et al., 2013). However, in case of mixed pixels, surface variables such as land surface temperature are commonly considered as a single value to represent the entire pixel area in ET estimation models, which results in large errors.

The focus of this study is on the effects of surface heterogeneity when estimating ET. Based on the satellite products that are currently available, three methods were used to analyze the uncertainties produced by surface heterogeneity: (1) input parameter upscaling (IPUS) does not consider the surface heterogeneities at all. It was designed to simulate the satellites that have identical spatial resolutions in both the visible near-infrared (VNIR) and thermal infrared (TIR) bands; (2) temperature resampling and flux aggregation (TRFA) does not consider the heterogeneity of LST; and (3) temperature sharpening and flux aggregation (TSFA) considers all the surface heterogeneities. These methods were designed for use with the majority of satellite data or products that have inconsistent spatial resolutions between the VNIR and TIR bands, such as the Landsat and HJ-1B satellites.

The surface variables in this paper were mainly derived from HJ-1B satellite data. The Chinese HJ-1A/B satellites were launched on 6 September 2008, and were designed for disaster and environmental monitoring, as well as other applications. The HJ-1B satellites are equipped with two charge-coupled device (CCD) cameras and one infrared scanner (IRS) with spatial resolutions of 30 and 300 m, respectively. Compared to high-temporal-resolution satellite data, such as the MODIS satellite data, or high-spatial-resolution satellite data, such as the Landsat 7 or 8 satellites data, HJ-1B data have the advantage of a high spatiotemporal resolution. Since the satellites were launched, the HJ-1/CCD time series data have been widely used in China to accurately classify land cover (Zhong et al., 2014a) and monitor various environmental disasters (Wang et al., 2010). Land-based variables, such as leaf area index (LAI), land surface temperature (LST), and downward longwave radiation (L_d), have been retrieved by the HJ-1B satellites using algorithms developed by Chen et al. (2010), H. Li et al. (2010, 2011), and Yu et al. (2013), respectively. These variables lay the foundation for ET research.

Although the HJ-1B satellites provide CCD data with a high spatial resolution of 30 m, the spatial resolution of the TIR band is only 300 m. Thus, surface heterogeneity effects must be considered when estimating the heat flux.

2 Methodology

2.1 Temperature-sharpening method based on statistical relationships

Surface thermal dynamics affect ET. The spatial resolution of TIR images is usually not as high as the spatial resolution of visible near-infrared (VNIR) bands because the energy of VNIR photons is higher than the energy of thermal photons. Thus, the inhomogeneity of TIR images would be larger than the inhomogeneity of VNIR images. Since the land surface temperature is calculated from the TIR band, the uncertainty of the variables becomes unpredictable when the inhomogeneity of TIR images is enhanced. Therefore, land surface temperature data should be derived with a high spatial resolution.

The land surface temperature can be reconstructed at the spatial resolution of the VNIR images by using a statistical temperature-sharpening strategy proposed by Kustas et al. (2003). This method assumes that the negative correlation between the normalized difference vegetation index (NDVI) and LST is invariant. The NDVI reflects vegetation growth and coverage, and the LST reflects surface thermal dynamics. The LST decreases with increasing vegetation cover. The scatter plot between the LST and NDVI values forms a feature space that is applicable at different scales when a sufficient number of pixels exist.

HJ-1B satellite images can provide vegetation and thermal information at spatial resolutions of 30 and 300 m, respectively. The 300 m resolution thermal data cannot sufficiently distinguish the surface temperatures of small targets within pixels. However, this issue can be addressed by temperature sharpening based on the functional relationship between NDVI and LST. A flowchart of temperature sharpening is shown in Fig. 1, and LST at the NDVI pixel resolution can be derived based on the following steps (Kustas et al., 2003):

1. The NDVI_{30} is aggregated to 300 m NDVI (NDVI_{300}). Then, the NDVI_{300} is divided into three classes ($0 \le \mathrm{NDVI}_{300} < 0.2$, $0.2 \le \mathrm{NDVI}_{300} < 0.5$, and $0.5 \le \mathrm{NDVI}_{300}$).

2. A subset of pixels is selected from the scene where the NDVI is as homogeneous as possible at a pixel resolution of 300 m based on the coefficient of variation (CV). The CVs are calculated using the original 30 m NDVI data (NDVI_{30}) as follows:

$$\mathrm{CV} = \frac{\mathrm{SD}}{\mathrm{mean}}, \tag{1}$$

where SD and mean are the standard deviation and the average values of 10×10 pixels of NDVI_{30}, respectively. The CVs are sorted from smallest to largest. Lower CVs correspond to more homogeneous land surface values, and a threshold should be determined to guarantee that a sufficient number of pixels is available for least squares fitting between NDVI_{300} and T_{300}. Therefore, the fractions of 25 % of the lowest CVs are selected from each class.

3. A least squares expression is established between NDVI_{300} and T_{300} using the selected pixels.

$$\hat{T}_{300}(\mathrm{NDVI}_{300}) = a + b \times \mathrm{NDVI}_{300} + c \times \mathrm{NDVI}_{300}^2 \tag{2}$$

4. For each 30 m pixel within a 300 m pixel, $\hat{T}_{30}$ can be calculated according to Eq. (2) as follows:

$$\hat{T}_{30}(\mathrm{NDVI}_{30}) = a + b \times \mathrm{NDVI}_{30} + c \times \mathrm{NDVI}_{30}^2 + \Delta\hat{T}_{300}, \tag{3}$$

where $\Delta\hat{T}_{300} = T_{300} - \hat{T}_{300}$ is the deviation between the regressed temperature and the temperature that was observed by the satellite at 300 m.

2.2 Area-weighting method based on landscape information

Coarse pixels are inhomogeneous because various types of land use may be included. Using a dominant type to represent such a large landscape is irrational. The spatial resolution of LST is significantly increased by temperature sharpening in Sect. 2.1. Consequently, all inputs of ET algorithms can be obtained at high spatial resolutions. Then, inhomogeneity issues can be greatly diminished by dividing the landscape into finer pixels.

Combined with a high-resolution classification map, subpixel-scale parameters can be used in the ET algorithm, which is more rational than using a dominant class type because different landscapes may require different ET algorithms. The surface energy fluxes can be averaged linearly due to the conservation of energy (Kustas et al., 2003), and a simple average that calculates the arithmetic mean over subpixels is the best choice for flux upscaling (Ershadi et al., 2013). Thus, the aggregated flux at a low resolution $F(x,y)$ is the arithmetic mean of all the $n \times n$ subpixel fluxes that constitute the contributing flux $F(x_i, y_j)$ at coordinate (x_i, y_j):

$$F(x,y) = \frac{1}{n \times n} \sum_{i=1}^{n} \sum_{j=1}^{n} F\left(x_i, y_j\right). \tag{4}$$

Because the average of the subpixels fluxes is equal to the area-weighted sum of each land-type result, the final coarse result can be derived from the area-weighted sum of each land type result within the landscape. The main steps in the area-weighting process are shown below (Xin et al., 2012):

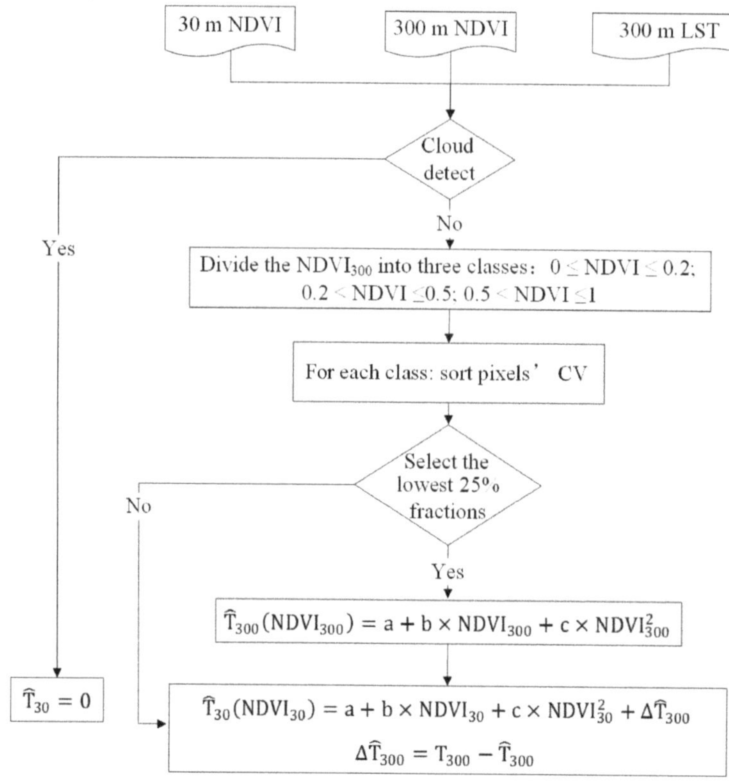

Figure 1. Flowchart of temperature sharpening.

1. The VNIR and TIR input data sets are geometrically corrected and registered.
2. The area ratios of different land cover types within each pixel of a low-spatial-resolution classification image are counted.
3. According to the fine-classification data, different parameterization schemes can be used in the ET algorithm to calculate the subpixel flux, such as the net radiation (R_n), soil heat flux (G), and sensible heat flux (H).
4. To calculate the regional flux, the flux of the large pixel is calculated by the area-weighting method as follows:

$$F = \sum_{i=1}^{n} w_i \cdot F_i, \tag{5}$$

where w_i is the fractional area contributing flux F_i of class type i and F is the aggregated flux at the coarse resolution. The LE is computed as a residual of the surface energy balance in the TSFA (see Sect. 2.3) process, in which a high-spatial-resolution image is used to reduce the number of mixed pixels.

2.3 Pixel ET algorithm

The surface energy balance describes the energy between the land surface and atmosphere. The energy budget is commonly expressed as follows:

$$R_n = \mathrm{LE} + H + G, \tag{6}$$

where R_n is the net radiation, G is the soil heat flux, H is the sensible heat flux, and LE is the latent heat flux absorbed by water vapor when it evaporates from the soil surface and transpires from plants through stomata. The widely used one-source energy balance model considers a homogeneous SPAC medium and ignores the inhomogeneity and structure. In this case, LE can be expressed as follows:

$$\mathrm{LE} = \frac{\rho c_p}{\gamma} \cdot \frac{e_s - e_a}{r_a + r_s}, \tag{7}$$

where γ is the psychometric constant; e_s and e_a are the aerodynamic saturation vapor pressure and atmospheric water vapor pressure, respectively; and r_a and r_s are the water vapor transfer aerodynamic resistance and surface resistance, respectively. Surface resistance includes soil resistance and canopy resistance. The surface resistance is influenced by the physiological characteristics of the vegetation and the water supply of roots. Thus, it is difficult to obtain surface resistance via remote sensing, and surface resistance is highly uncertain, particularly over heterogeneous surfaces. To avoid error introduced by the uncertainty of the surface resistance, LE is computed as a residual of the surface energy balance equation.

R_n is the difference between incoming and outgoing radiation and is calculated as follows:

$$R_n = S_d(1-\alpha) + \varepsilon_s L_d - \varepsilon_s \sigma T_{rad}^4, \quad (8)$$

where S_d is the downward shortwave radiation, α is the surface broadband albedo, ε_s is the emissivity of the land surface, L_d is the downward atmospheric longwave radiation, $\sigma = 5.67 \times 10^{-8}\,\mathrm{W\,m^{-2}\,K^{-4}}$ is the Stefan–Boltzmann constant, and T_{rad} is the surface radiation temperature.

G is commonly estimated using the empirical relationship with R_n. Because the canopy exerts a significant influence on G, the fractional canopy coverage FVC is used to determine the ratio of G to R_n as follows:

$$G = R_n \times [\Gamma_c + (1-\mathrm{FVC}) \times (\Gamma_s - \Gamma_c)], \quad (9)$$

where Γ_s is 0.315 for bare soil and Γ_c is 0.05 for a full vegetation canopy (Su, 2002). H is the transfer of turbulent heat between the surface and atmosphere, which is driven by a temperature difference and controlled by resistances that depend on local atmospheric conditions and land cover properties (Kalma et al., 2008). According to gradient diffusion theory, the equation for H is as follows:

$$H = \rho c_p \frac{T_{aero} - T_a}{r_a}, \quad (10)$$

where ρ is the density of the air; c_p is the specific heat of the air at a constant pressure; T_{aero} is the aerodynamic surface temperature obtained by extrapolating the logarithmic air temperature profile to the roughness length for heat transport; T_a is the air temperature at a reference height; and r_a is the aerodynamic resistance, which influences the heat transfer between the source of the turbulent heat flux and the reference height. Aerodynamic resistance was calculated based on the Monin–Obukhov similarity theory (MOST) using a stability correction function (Paulson, 1970; Ambast et al., 2002). The zero-plane displacement height, d, and roughness length, $z_{0\,m}$, were parameterized by the schemes proposed by Choudhury and Monteith (1988).

In this approach, H must be accurately estimated. However, calculating H using Eq. (10) is difficult. Because remote sensing cannot obtain T_{aero}, the value of T_{aero} is generally replaced with the radiative surface temperature T_{rad}, which is not always equal to T_{aero}. The difference between these terms for homogeneous and full-coverage vegetation is approximately 1–2° (Choudhury et al., 1986), and it can reach 10° in sparsely vegetative areas (Kustas, 1990). The method that corrects for this discrepancy adds excess resistance r_{ex} to r_a. We used the brief method proposed by Chen (1988) to calculate r_{ex}: $r_{ex} = 4/u_*$.

Figure 2 shows the flowchart for merging ET retrieval and temperature sharpening based on HJ-1B satellites.

The spatial-scale effect is generally revealed by a discrepancy between different upscaling methods. In one method, parameters are upscaled to a large scale before calculating the heat flux. In the other method, heat flux is calculated at a small scale, and the results are then upscaled. In this study, the resolution of the final output result is 300 m. To evaluate the heterogeneity-reducing effect of TSFA, two other upscaling methods, called IPUS and TRFA, were implemented (see Fig. 3). In the case of IPUS, the inputs of the energy balance model are first retrieved at 30 m resolution (see information of HJ-1B satellite data in Sect. 3.2.1) and then aggregated to 300 m resolution. Subsequently, these 300 m inputs are used in the one-source energy balance model to obtain the four energy balance components at 300 m resolution. In TRFA, the LST at 300 m is first resampled to 30 m using the nearest neighbor method and the 30 m resolution inputs are used for estimating ET. The outputs of the four energy-balance components of the TRFA are obtained using the area-weighting method shown in Sect. 2.2.

3 Study area and data set

3.1 Study area

Our study was conducted in the middle stream of the Heihe River basin (HRB), which is located near the city of Zhangye in the arid region of Gansu province in northwestern China (100.11–100.16° E, 39.10–39.15° N). The middle reach of the HRB is a typical desert–oasis agriculture ecosystem dominated by maize and wheat. Areas of the Gobi Desert and the alpine vegetation in the Qilian Mountains are located near the study area (see Fig. 4). The artificial oasis is highly heterogeneous, which impacts the thermal dynamics and hydraulic features. Consequently, the water use efficiency and ET are variable. The Heihe River basin has long served as a test bed for integrated watershed studies, as well as land surface and hydrological experiments. Comprehensive experiments, such as the Watershed Allied Telemetry Experimental Research (WATER) project (Li et al., 2009), and an international experiment – the Heihe Basin Field Experiment (HEIFE) as part of the World Climate Research Programme (WCRP), have been conducted in this basin. One major objective of HiWATER was to capture the strong land surface heterogeneities and associated uncertainties within a watershed (X. Li et al., 2013).

3.2 Data set

In this study, data are mainly derived from the HJ-1B satellite. We combined these data with ancillary data and in situ Multi-Scale Observation Experiment on Evapotranspiration over heterogeneous land surfaces of The Heihe Watershed Allied Telemetry Experimental Research (HiWATER-MUSOEXE) data to estimate and validate the HJ-1B land surface variables and heat fluxes.

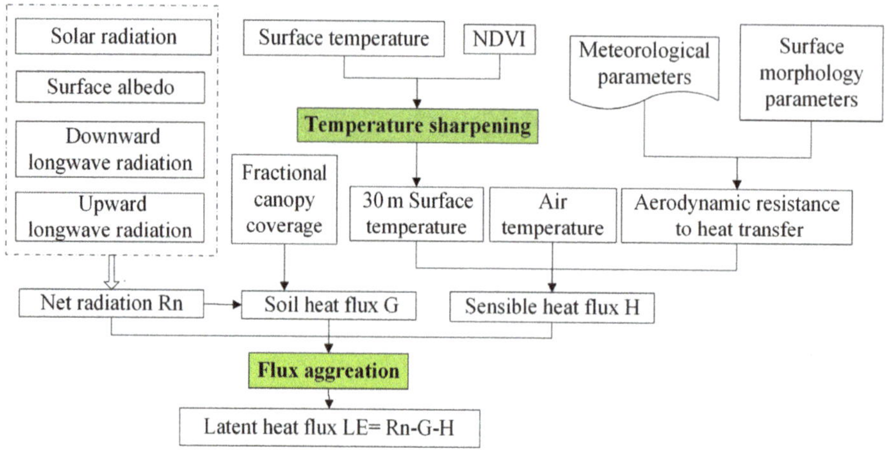

Figure 2. Flowchart of ET retrieval using the temperature-sharpening and flux aggregation method.

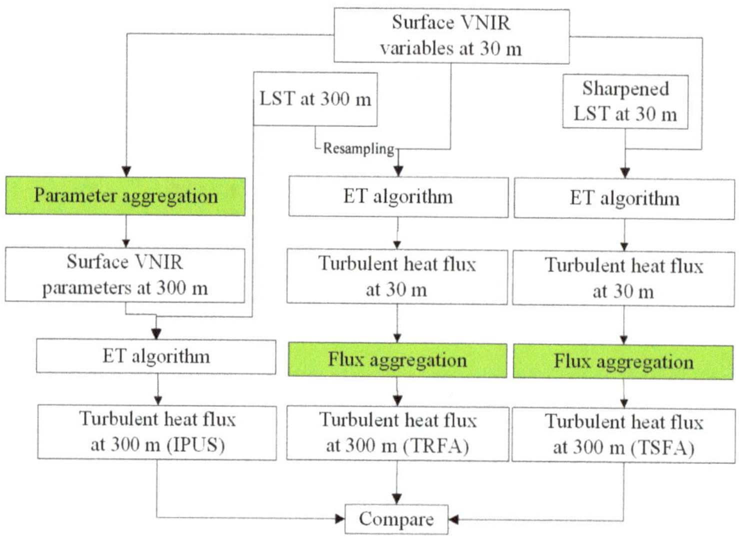

Figure 3. Flowchart of the three upscaling methods for retrieving evapotranspiration.

3.2.1 Remote sensing data

HJ-1B satellite data

The specifications of HJ-1B are shown in Table 1. The satellite has quasi-sun-synchronous orbits at an altitude of 650 km, a swath width of 700 km and a revisit period of 4 days. Combined, the revisit period of the satellites is 48 h. Because HJ-1A/B CCDs lack an onboard calibration system, cross-calibration methods were proposed to calibrate the CCD instruments (Zhang et al., 2013; Zhong et al., 2014b). The image quality of the HJ-1A/B CCD is stable, the performances of each band are balanced (Zhang et al., 2013) and the radiometric performance of the HJ-1A/B CCD sensors is similar to the performances of the Landsat-5 TM, Observer-1 (EO-1) Advanced Land Imager, and Terra ASTER. The image quality of the HJ-1A/B CCD is very similar to the image quality of Landsat-5 TM (Jiang et al., 2013). In addition, the accuracy of the TIR band's onboard calibration meets the land surface temperature retrieval requirements but not the sea surface temperature retrieval requirements (J. Li et al., 2011). The Center for Resources Satellite Data and Application (CRESDA) in China releases calibration coefficients annually on its website (http://www.cresda.com). These data are freely available from the CRESDA website (http://218.247.138.121/DSSPlatform/index.html).

We used the HJ-1B satellite data from the HRB region in 2012. Because many variable-retrieving algorithms require clear-sky conditions when calculating ET, we combined data quality information with visual interpretation to select satellite images without clouds. Considering the period of ground observations discussed in Sect. 3.2.2, we obtained data for 11 days: 19 and 30 June; 8 and 27 July; 2, 15, 22, and 29 August; 2, 13, and 14 September.

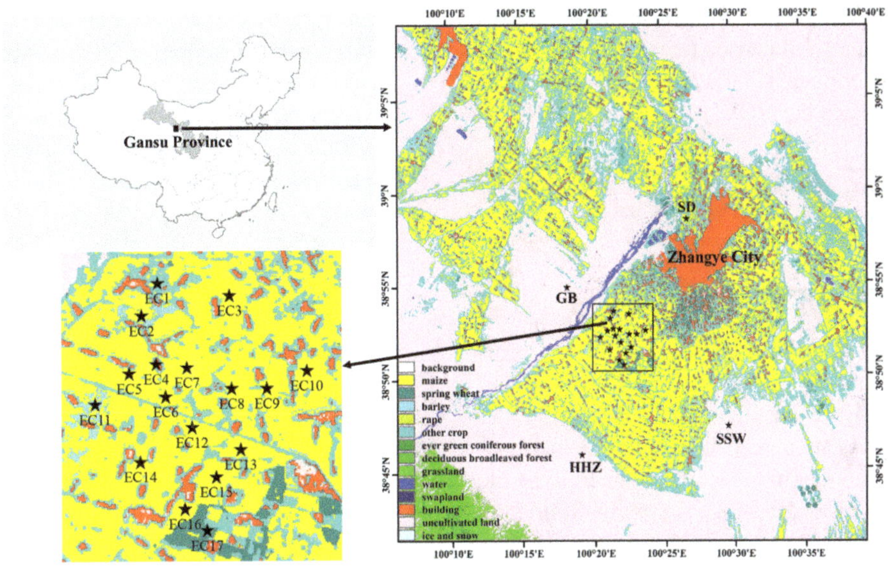

Figure 4. Study area and distribution of EC towers in HiWATER-MUSOEXE.

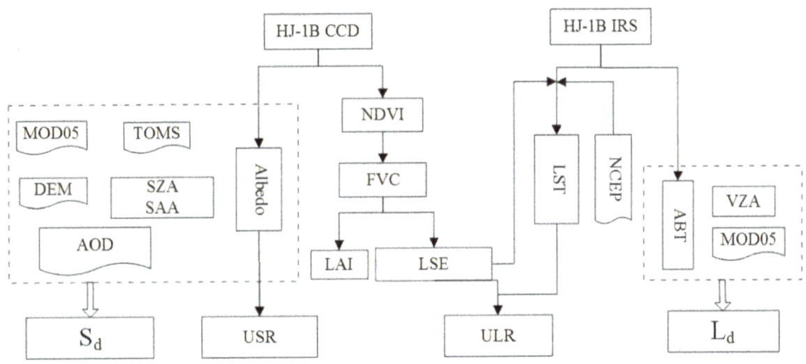

Figure 5. Flowchart of land surface variable retrieval. The abbreviations are defined as follows: SZA: solar zenith angle; SAA: solar azimuth angle; VZA: view zenith angle; AOD: aerosol optical depth; ABT: at-nadir brightness temperature; S_d: downward shortwave radiation; USR: upward shortwave radiation; ULR: upward longwave radiation; and L_d: downward longwave radiation.

Table 1. Specifications of the HJ-1B main payloads.

Sensor	Band	Spectral range (μm)	Spatial resolution (m)	Swath width (km)	Revisit time (days)
CCD	1	0.43–0.52	30	360 (single) 700 (two)	4
	2	0.52–0.60			
	3	0.63–0.69			
	4	0.76–0.90			
IRS	5	0.75–1.10	150	720	4
	6	1.55–1.75			
	7	3.50–3.90			
	8	10.5–12.5	300		

The HJ-1B satellite data of the HRB were preprocessed, including geometric correction, radiometric calibration, and atmospheric correction. The following surface variables are needed in Eqs. (1) to (10): downward shortwave radiation, downward longwave radiation, emissivity, albedo, fractional vegetation coverage (FVC), cloud mask data, meteorological data, LAI, and LST. Figure 5 illustrates a flowchart of the retrieval of these variables.

1. Surface albedo was obtained from the top of the atmosphere (TOA) reflectance by the HJ-1A/B satellite using a lookup table based on an angular bin regression relationship according to the algorithm proposed by Liang et al. (2005) and Q. Liu et al. (2011). The surface albedo and bidirectional reflectance distribution function (BRDF) of the HJ-1A/B satellite in the regression procedure were monitored using POLDER-3/PARASOL BRDF data sets, and BRDF was used to obtain the TOA reflectance in the 6S (second simulation of a satellite signal in the solar spectrum) radiation transfer mode.

2. The NDVI is the regression kernel of the temperature-sharpening procedure and is used to calculate the FVC. Atmospherically corrected surface reflectance values were used to calculate the NDVI as follows:

$$\mathrm{NDVI} = \frac{\rho_{\mathrm{nir}} - \rho_{\mathrm{red}}}{\rho_{\mathrm{nir}} + \rho_{\mathrm{red}}} \tag{11}$$

and

$$\mathrm{FVC} = \frac{\mathrm{NDVI} - \mathrm{NDVI_s}}{\mathrm{NDVI_v} + \mathrm{NDVI_s}}, \tag{12}$$

where ρ_{nir} and ρ_{red} are the reflectances in the near-infrared and red band, respectively, and $\mathrm{NDVI_v}$ and $\mathrm{NDVI_s}$ are the fully vegetated and bare soil NDVI values, respectively. As an important input for the parameterization of surface roughness length and aerodynamic resistance, the LAI was determined using the following equation (Nilson, 1971):

$$P(\theta) = e^{-G(\theta)\cdot\Omega\cdot\mathrm{LAI}/\cos(\theta)} \tag{13}$$

$$P(\theta) = 1 - \mathrm{FVC}, \tag{14}$$

where θ is the zenith angle, $P(\theta)$ is the angular distribution of the canopy gap fraction, $G(\theta)$ is the projection coefficient (0.5) and Ω is the total foliage clumping index, which can be obtained from the GLC global clumping index database according to the land use type (He et al., 2012).

3. Land surface emissivity (LSE) is needed to calculate the R_n and is extremely important for retrieving LST. In this paper, LSE was calculated using the FVC as follows (Valor and Caselles, 1996):

$$\varepsilon = \varepsilon_v \cdot \mathrm{FVC} + \varepsilon_g(1 - \mathrm{FVC}) + 4 < d\varepsilon > \cdot \mathrm{FVC} \cdot (1 - \mathrm{FVC}) \tag{15}$$

where ε is the LSE, $< d\varepsilon >$ is an effective value of the cavity effect of emissivity, the mean $d\varepsilon$ of all vegetation species in this study is $< d\varepsilon > = 0.015$, ε_v and ε_g are the vegetation and ground emissivity, respectively.

4. Land surface temperature is a single-channel parametric model for retrieving LST based on HJ-1B/IRS TIR data developed by H. Li et al. (2010) was employed to obtain the LST. This model was developed from a parametric model based on MODTRAN4 using NCEP atmospheric profile data.

5. Downward shortwave radiation was calculated in this study by applying the algorithm proposed by L. Li et al. (2010). MOD05, TOMS, aerosol and solar angle data were used to estimate the direct light flux and diffuse light flux using a lookup table that was generated via the 6S radiation transfer mode (Vermote et al., 2006). This method considered the influences of complex terrain, and a topographic correction was performed by using products of the ASTER digital elevation model (DEM).

6. Downward longwave radiation (L_d) was calculated by the algorithm proposed by Yu et al. (2013). The TOA brightness temperature of the HJ-1B thermal channel was used to substitute the atmospheric effective temperature. Effective atmospheric emissivity was parameterized as an empirical function of the water vapor content. These values were substituted for atmospheric temperature and atmospheric emissivity to estimate the value of L_d. Because this L_d retrieval method was only valid for clear-sky conditions, cloud masking information was used to determine clear skies. When cloud contamination existed in the image, the brightness temperature was relatively low, causing the L_d to be lower than that in the cloudless images.

Ancillary data

Ancillary data were used because the bands of the satellite could not invert all of the variables needed to retrieve ET.

1. MODIS provides atmospheric water vapor data (MOD05), including a 1 km near-infrared product and a 5 km thermal-infrared product, every day. The 1 km near-infrared water vapor product was used to retrieve L_d in this study.

2. For surface elevation data, we used the 30 m resolution global digital elevation model (GDEM) based on ASTER, which covers 83° N–83° S, to derive S_d.

3. For atmosphere ozone data, a total ozone mapping spectrometer (TOMS), which was carried on an Earth Probe (EP) satellite, was used to derive S_d. The TOMS-EP provided daily global atmospheric ozone data at a resolution of $1° \times 1.25°$ (L. Li et al., 2010).

4. For atmosphere profile data, global reanalysis data from the National Centers for Environmental Prediction (NCEP) were used to derive LST. These data were generated globally every 6 h (0:00, 06:00, 12:00, 18:00 UTC) for every 1° of latitude and longitude (H. Li et al., 2010).

3.2.2 HiWATER experiment data set

The in situ HRB observation data were provided by HiWATER. From June to September 2012, HiWATER designed nested observation matrices over 30 km × 30 km and 5.5 km × 5.5 km within the middle stream oasis in Zhangye to focus on the heterogeneity of the scale effect in HiWATER-MUSOEXE.

In the larger observation matrix, four eddy covariance (EC) systems and one superstation were installed in the oasis–desert ecosystem. Each station was supplemented with an automatic meteorological station (AMS) to record meteorological and soil variables and monitor the spatial–temporal variations of ET and its associated factors (X. Li et al., 2013).

Table 2. The in situ HiWATER-MUSOEXE station information.

Station	Longitude	Latitude	Tower height (m)	Altitude (m)	Land cover
EC1	100.36° E	38.89° N	3.8	1552.75	vegetation
EC2	100.35° E	38.89° N	3.7	1559.09	maize
EC3	100.38° E	38.89° N	3.8	1543.05	maize
EC4	100.36° E	38.88° N	4.2	1561.87	building
EC5	100.35° E	38.88° N	3	1567.65	maize
EC6	100.36° E	38.87° N	4.6	1562.97	maize
EC7	100.37° E	38.88° N	3.8	1556.39	maize
EC8	100.38° E	38.87° N	3.2	1550.06	maize
EC9	100.39° E	38.87° N	3.9	1543.34	maize
EC10	100.40° E	38.88° N	4.8	1534.73	maize
EC11	100.34° E	38.87° N	3.5	1575.65	maize
EC12	100.37° E	38.87° N	3.5	1559.25	maize
EC13	100.38° E	38.86° N	5	1550.73	maize
EC14	100.35° E	38.86° N	4.6	1570.23	maize
EC15	100.37° E	38.86° N	4.5	1556.06	maize
EC17	100.37° E	38.85° N	7	1559.63	orchard
GB	100.30° E	38.91° N	4.6	1562	uncultivated land – Gobi
SSW	100.49° E	38.79° N	4.6	1594	uncultivated land – desert
SD	100.45° E	38.98° N	5.2	1460	swamp land

The station information is shown in Table 2, and the distribution of the stations is shown in Fig. 4. Within the artificial oasis, an observation matrix composed of 17 EC towers and ordinary AMSs exists where the superstation was located. The land surface was heterogeneous and dominated by maize, maize intercropped with spring wheat, vegetables, orchards, and residential areas (X. Li et al., 2013). Because the EC16 and HHZ stations lacked R_n and G observation data, they were excluded from this study.

The ground observation data included the H and LE. Reliable methods were used to ensure the quality of the turbulent heat flux data. Before the main campaign, an intercomparison of all instruments was conducted in the Gobi Desert (Xu et al., 2013). After basic processing, including spike removal and corrections for density fluctuations (WPL correction), a four-step procedure was performed to control the quality of the EC data. In this procedure, data were rejected when (1) the sensor had been malfunctioning, (2) precipitation occurred within 1 h before or after collection, (3) the ratio of missing data was greater than 3 % in the 30 min raw record and, (4) the friction velocity was below 0.1 m s^{-1} at night (for more details, see S. M. Liu et al., 2011; Xu et al., 2013; Liu et al., 2016). EC outputs are available every 30 min. G was measured by using three soil heat plates at a depth of 6 cm at each site, and the surface G was calculated using the method proposed by Yang and Wang (2008) based on the soil temperature and moisture above the plates. Surface meteorological variables, such as wind speed, wind direction, relative humidity, and air pressure, were used to interpolate images using the inverse distance weighting. Researchers can obtain these data from the websites of the Cold and Arid Regions Science Data Center at Lanzhou (http://card.westgis.ac.cn/) or the Heihe Plan Data Management Center (http://www.heihedata.org/).

Table 3. The station validation results of land surface temperature.

Station	R^2	MBE (K)	RMSE (K)	Station	R^2	MBE (K)	RMSE (K)
EC1	0.82	0.18	1.74	EC11	0.42	1.59	2.98
EC2	0.82	0.59	1.54	EC12	0.87	0.62	1.51
EC3	0.69	0.38	1.90	EC13	0.83	0.44	1.48
EC4	0.83	−9.87	10.04	EC14	0.73	1.43	2.44
EC5	0.83	1.71	2.34	EC15	0.74	1.53	2.41
EC6	0.61	0.30	2.44	EC17	0.78	1.20	2.32
EC7	0.82	0.39	1.40	GB	0.69	0.12	2.33
EC8	0.83	0.45	1.55	SSW	0.59	1.38	3.82
EC9	0.63	2.31	3.15	SD	0.76	−3.83	4.84
EC10	0.68	1.32	2.45				

Energy imbalances are common in ground flux observations. The conserving Bowen ratio (H / LE) and residual closure technique are often used to force the energy balance. Computing the LE as a residual variable may be a better method for energy balance closure under conditions with large LEs (small or negative Bowen ratios due to strong advection) (Kustas et al., 2012). Thus, the residual closure method was applied because the oasis effect was distinctly observed in the desert–oasis system on clear days during the summer (S. M. Liu et al., 2011).

4 Results and analysis

4.1 Evaluation of surface variables

To control model inputs and analyze error sources, the coarse-resolution land surface temperature, downward shortwave radiation, downward longwave radiation, R_n, and G were evaluated using in situ data.

The ground-based land surface temperature, T_s, was calculated using the Stefan–Boltzmann law from the AMS measurements of the longwave radiation fluxes (Li et al., 2014) as follows:

$$T_s = \left[\frac{L^{\uparrow} - (1-\varepsilon_s)\cdot L^{\downarrow}}{\varepsilon_s \cdot \sigma}\right]^{\frac{1}{4}}, \quad (16)$$

in which $L^{\uparrow}$ and $L^{\downarrow}$ are in situ surface upwelling and atmospheric downwelling longwave radiation, respectively, and ε_s is the surface broadband emissivity, which is regarded as the pixel value of the HJ-1B at the AMS. The coefficient of determination R^2, mean bias error (MBE) and root mean square error (RMSE) of the LST are 0.71, −0.14 K, and 3.37 K, respectively. As shown in Table 3, the accuracy of EC4 is low. The main causes of the large errors are as follows: (1) buildings and soil/vegetation are distinct materials, the LSE algorithm may not be suitable for buildings and

Table 4. The station validation results of downward shortwave radiation.

Station	R^2	MBE ($W m^{-2}$)	RMSE ($W m^{-2}$)	Station	R^2	MBE ($W m^{-2}$)	RMSE ($W m^{-2}$)
EC1	0.97	25.23	27.73	EC11	0.90	30.11	33.76
EC2	0.84	28.29	33.57	EC12	0.96	24.35	26.43
EC3	0.97	17.56	19.25	EC13	0.93	12.41	17.92
EC4	0.98	6.07	9.34	EC14	0.98	32.40	33.49
EC5	0.98	10.60	12.29	EC15	0.94	26.71	29.71
EC6	0.93	27.68	30.71	EC17	0.94	−20.25	24.54
EC7	0.89	−17.69	27.59	GB	0.89	25.34	30.63
EC8	0.83	15.63	25.50	SSW	0.63	18.51	34.93
EC9	0.96	−2.27	9.96	SD	0.98	5.70	13.82
EC10	0.94	−3.50	11.97				

Table 5. The station validation results of downward longwave radiation.

Station	R^2	MBE ($W m^{-2}$)	RMSE ($W m^{-2}$)	Station	R^2	MBE ($W m^{-2}$)	RMSE ($W m^{-2}$)
EC1	0.85	4.16	17.21	EC11	0.93	−2.72	10.55
EC2	0.88	0.11	14.23	EC12	0.87	−0.84	14.80
EC3	0.91	−35.65	37.88	EC13	0.86	−7.28	15.98
EC4	0.88	3.36	16.38	EC14	0.82	4.07	16.42
EC5	0.88	−0.79	15.02	EC15	0.85	17.67	23.06
EC6	0.84	2.55	15.43	EC17	0.90	−1.11	12.87
EC7	0.75	−5.90	19.72	GB	0.88	9.50	27.82
EC8	0.80	−1.35	17.49	SSW	0.85	25.33	34.50
EC9	0.86	10.44	17.99	SD	0.85	−26.54	34.08
EC10	0.87	7.98	16.05				

(2) the EC4 foundation is non-uniform and is not suitable for validation. After removing the EC4 data, the R^2, MBE, and RMSE values of the LSTs were 0.83, 0.69 K, and 2.51 K, respectively. The LST errors of SSW and SD were large due to large errors on particular days. For example, although it was briefly cloudy above station SSW on 27 July, this area was not identified as cloudy in the cloud detection process.

The R^2, MBE, and RMSE values of S_d were 0.81, 13.80 $W m^{-2}$, and 25.35 $W m^{-2}$, respectively. The station validation results are shown in Table 4. The accuracy of SSW is low. Because cloudy conditions occurred briefly on 27 July, few ground observations were obtained, and S_d was significantly overestimated. After removing these data, the R^2, MBE, and RMSE values of S_d at SSW were 0.87, 10.90 $W m^{-2}$, and 21.13 $W m^{-2}$, respectively.

The R^2, MBE, and RMSE values of the HRB L_d were 0.73, 0.28 $W m^{-2}$, and 21.24 $W m^{-2}$, respectively. As seen in Table 5, the accuracies at EC3, SD, and SSW were low. The low accuracies at EC3 and SD potentially resulted from (1) high humidity, which resulted in low at-nadir brightness temperatures and low retrieved L_d, or (2) instrument error, which occurred because the EC3 ground observations were always greater than those of the other stations during the same period. Although SSW was located in a desert, the ground–air temperature difference was large. The L_d retrieval may have a large error because the models use surface temperature when estimating L_d to approximate or substitute the near-surface temperature (Yu et al., 2013). The corrected error of our L_d retrieving algorithm resulted from the ground–air temperature difference in non-vegetated areas. The inaccuracy of the SSW LST may influence the L_d results.

The R^2, MBE, and RMSE values of the HRB R_n were 0.70, −9.64 $W m^{-2}$, and 42.77 $W m^{-2}$, respectively. The station-validated results of R_n are shown in Table 6, which indicates that the accuracies of EC4, EC7, EC17, and SSW were relatively low. According to the sensitivity analysis of Eq. (8), L_d and S_d are highly sensitive variables when calculating R_n, while the albedo, LSE, and LST are not as sensitive. Although LST was not a sensitive variable, the MBE and RMSE values of LST at EC4 reached −9.87 and 10.04 K because the land cover of EC4 was maize at 300 m resolution. However, the observation tower was located in a built-up area, which potentially caused errors when estimating R_n. The accuracies of S_d and L_d at EC7 were low on several days, and MBE = −43.40 $W m^{-2}$ and RMSE = 50.50 $W m^{-2}$ after removing these data. EC17 was located in an orchard, and the signal that was received by the sensors at EC17 was affected

Table 6. The station net radiation validation results.

Station	R^2	MBE ($W m^{-2}$)	RMSE ($W m^{-2}$)	Station	R^2	MBE ($W m^{-2}$)	RMSE ($W m^{-2}$)
EC1	0.76	−2.55	30.61	EC11	0.86	−15.13	28.05
EC2	0.79	2.52	25.24	EC12	0.90	−8.46	19.38
EC3	0.86	−35.84	42.97	EC13	0.88	−25.73	32.34
EC4	0.84	76.64	80.25	EC14	0.90	4.23	18.18
EC5	0.85	−24.41	32.34	EC15	0.84	8.33	23.01
EC6	0.82	4.35	23.44	EC17	0.89	−62.62	68.11
EC7	0.61	−58.66	67.83	GB	0.77	−10.40	38.86
EC8	0.83	−20.62	32.45	SSW	0.44	23.05	62.93
EC9	0.87	−29.60	36.27	SD	0.75	19.98	35.24
EC10	0.83	−24.35	33.51				

Table 7. The station validation results of the soil heat flux.

Station	R^2	MBE ($W m^{-2}$)	RMSE ($W m^{-2}$)	Station	R^2	MBE ($W m^{-2}$)	RMSE ($W m^{-2}$)
EC1	0.50	19.73	31.53	EC11	0.71	4.23	19.23
EC2	0.24	20.78	28.72	EC12	0.53	20.29	24.79
EC3	0.03	−1.15	36.28	EC13	0.91	−0.89	17.27
EC4	0.45	18.50	22.29	EC14	0.82	−1.89	18.72
EC5	0.38	41.87	60.19	EC15	0.78	6.68	15.80
EC6	0.83	−5.91	14.57	EC17	0.49	8.26	33.59
EC7	0.28	7.50	24.65	GB	0.29	−17.86	26.81
EC8	0.68	−5.73	20.15	SSW	0.01	30.41	51.87
EC9	0.61	6.83	26.96	SD	0.71	−4.79	13.71
EC10	0.41	7.68	28.67				

by the complex vertical structure of the orchard ecosystem. The information of substrate plants may be ignored, leading to albedo retrieval errors. An albedo bias of 0.03 can lead to an R_n error of approximately 20 $W m^{-2}$ when the solar incoming radiation is large. As previously discussed, it was briefly cloudy on 27 July, and after removing those data, the R^2, MBE, and RMSE values of the R_n obtained at station SSW were 0.72, 8.20 $W m^{-2}$, and 37.60 $W m^{-2}$, respectively.

The R^2, MBE, and RMSE values of the G in the HRB were 0.57, 8.51 $W m^{-2}$, and 29.73 $W m^{-2}$, respectively. The station-validated G results are shown in Table 7. At EC5, the soil temperature and moisture were the same at different depths after 19 July, which resulted in a surface G that was equal to the G at a depth of 6 cm. G below the surface was usually less than the G at the soil surface; thus, the validation results of G at EC5 indicate that G was overestimated. At SSW, the brief cloudy period decreased the observed soil surface temperature, which decreased the calculated surface G. However, the remotely sensed G did not reflect this situation. In this case, the G was overestimated because the R_n was overestimated. After removing the data on 27 July, the R^2, MBE, and RMSE values of the G at SSW were 0.17, 19.34 $W m^{-2}$, and 33.30 $W m^{-2}$, respectively.

4.2 Validation of heat fluxes by TSFA

Figure 6 provides the turbulent heat flux results calculated by TSFA on 13 September 2012. The spatial distribution of the turbulent heat flux is obvious. The H values of buildings and uncultivated land, including land patches in the Gobi region, barren areas, and desert areas, were high, in addition to the LEs of water and agricultural areas in the oasis. The southern areas of the images show uncultivated barren land bordering the Qilian Mountains that resulted from snowmelt and the downward movement of water. In these areas, the groundwater levels are high and the soil moisture content is approximately 30 % based on in situ measurements at a depth of 2 cm. Therefore, the LE values of barren areas in the south are higher than the LE values of desert areas in the southeast, although both areas were classified as uncultivated land.

Studies have shown that validation methods that consider the source area are more appropriate for evaluating ET models than traditional validation methods based on a single pixel (Jia et al., 2012; Song et al., 2012). In this study, a user-friendly tool presented by Neftel et al. (2008), which is based on the Eulerian analytic flux footprint model proposed by Kormann and Meixner (2001) was used to calculate the footprints of the function parameters. The continuous footprint

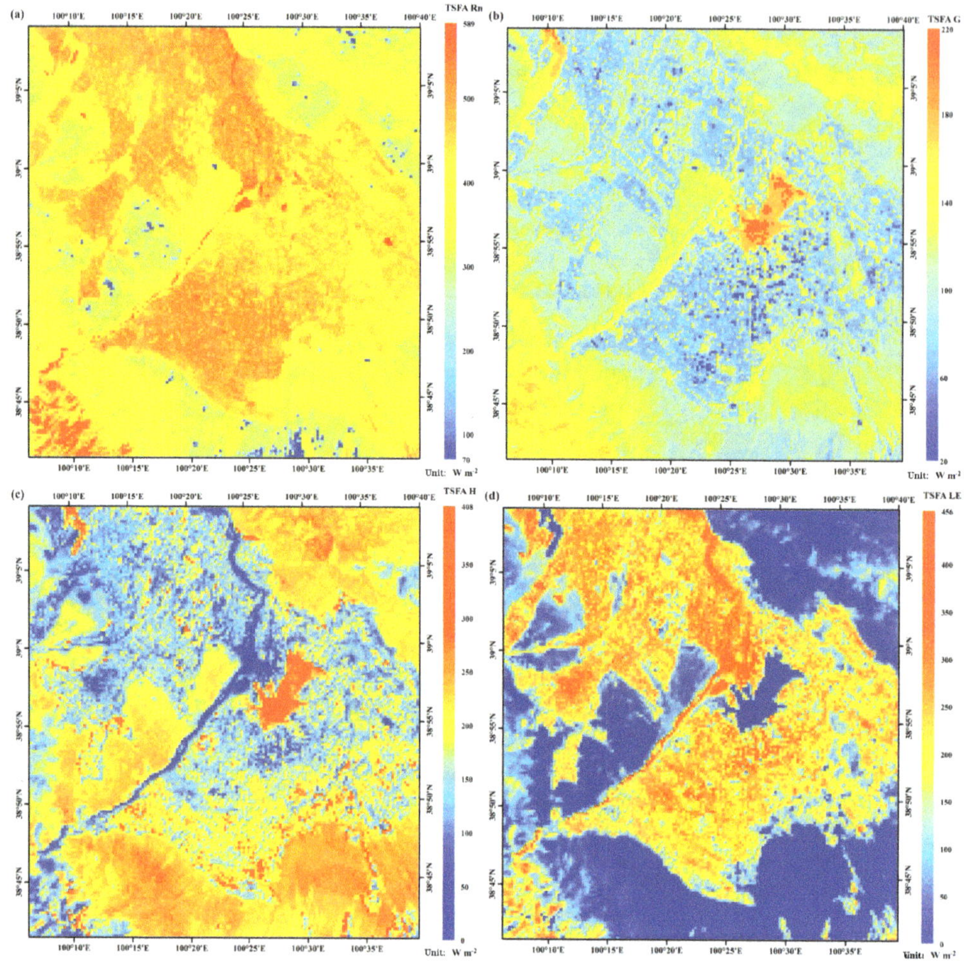

Figure 6. Maps of the four energy components, **(a)** R_{n}, **(b)** G, **(c)** H, and **(d)** LE, calculated by TSFA on 13 September 2012.

function was dispersed based on the relative weights of the pixels in which the source area was located.

The footprint validation results of the TSFA turbulent heat fluxes are shown in Fig. 7 and Table 8. The R^2, MBE, and RMSE of H were 0.61, 0.90 W m^{-2}, and 50.99 W m^{-2}, respectively, and those of LE were 0.82, −20.54 W m^{-2}, and 71.24 W m^{-2}, respectively. Because LE was calculated as a residual term, it was impacted by R_{n}, surface G, and H. The errors of all inputs may contribute to the LE, which complicates the error sources of the LE. These errors are discussed in detail in Sect. 4.3.2 and 4.4.

As shown in Fig. 7, the majority of the H values are small because June, July, August, and September constitute the growing season when ET greatly cools the air. The temperature difference between the land surface and air is small, leading to a low H. The points with large H values are influenced by uncultivated land. In our study area, the Gobi region, barren area and desert area comprise the uncultivated land. The points in the scatter plot with large H values represent desert, where the H values reach approximately 300 W m^{-2}. Some points in the H scatter plot are less than zero due to inversion from the oasis effect or irrigation. For example, the HiWATER soil moisture data show that irrigation occurred on 22 August 2012. Irrigation is the main source of water within the oasis and cools the land surface to temperatures below the air temperature. In addition, irrigation leads to errors in LST retrieval because it increases the atmospheric water vapor content, as discussed in Sect. 4.1. The model error is further analyzed in Sect. 4.4.

4.3 Comparison between TSFA, TRFA, and IPUS

To verify whether the TSFA method can simulate the heterogeneity of the land surface, the TRFA and IPUS methods were also implemented for comparison purposes. These three methods were evaluated using (1) validation of TRFA and IPUS based on in situ measurements and (2) quantitative analysis based on the spatial distribution and scatter plots of the four energy balance components.

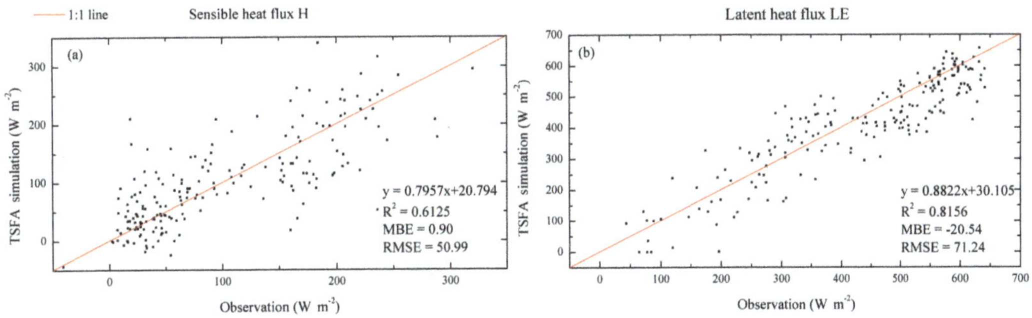

Figure 7. Scatter plot of the TSFA turbulent heat flux results.

Table 8. In situ validation results of heat flux using the TSFA.

	TSFA-H ($\mathrm{W\,m^{-2}}$)			TSFA-LE ($\mathrm{W\,m^{-2}}$)		
Date	R^2	MBE	RMSE	R^2	MBE	RMSE
19 Jun	0.39	44.73	66.38	0.69	−44.15	80.60
30 Jun	0.73	23.71	38.96	0.88	−63.81	77.83
8 Jul	0.55	32.70	58.72	0.85	−43.02	72.32
27 Jul	0.90	−34.34	43.59	0.92	26.74	57.60
3 Aug	0.80	−4.77	18.92	0.78	−4.58	47.86
15 Aug	0.74	−18.37	38.82	0.93	4.75	35.41
22 Aug	0.40	31.64	66.21	0.65	−44.44	93.81
29 Aug	0.79	23.01	38.36	0.79	−50.45	77.99
2 Sep	0.21	−45.10	74.81	0.54	24.39	69.31
13 Sep	0.25	−9.64	41.01	0.59	−59.36	82.77
14 Sep	0.31	−34.11	50.88	0.47	27.99	67.50

4.3.1 Validation of the TRFA and IPUS heat fluxes

Table 9 provides the in situ validation results of H and LE calculated using the IPUS and TRFA methods. Compared to validation results of TSFA in Table 8, the TSFA produced a better retrieval accuracy than the TRFA, and the TRFA was better than the IPUS on all days and the MBE and RMSE values of TSFA decreased and the R^2 of TSFA increased on most days. Table 9 shows that the improvements in accuracy between TRFA and IPUS were relatively larger than those between TSFA and TRFA. Compared to the IPUS results, the TRFA results were similar to the TSFA results because subpixel landscapes and subpixel variations of most variables were considered. Thus, TRFA effectively decreased the scale error that resulted from heterogeneity because the 30 m VNIR data were fully used. However, the performance of the TRFA method is unstable. For example, on 3 and 29 August, the TRFA results were slightly worse than the IPUS results. This situation occurred because the different subpixel landscape temperatures were considered as equal to the values estimated at the 300 m resolution. Thus, when values of LST at 300 m scale have large retrieval errors, the turbulent heat flux retrieval error may be amplified by the subpixel landscapes.

Variations in landscape characteristics systematically trigger variations in surface variables. Landscape inhomogeneity can be classified using two conditions: nonlinear vegetation density variations between subpixels (e.g., different types of vegetation mixed with each other or with bare soil) and coarse pixels containing different landscapes (e.g., vegetation or bare soil mixed with buildings or water). To evaluate the effects of TSFA, stations with a typical severe heterogeneous surface at EC4, a weak heterogeneous surface at EC11, a typical pixel (called "TP" hereafter) at the boundary of the oasis and bare soil (sample 62, line 102 in the image of study area), and a uniform surface at EC15, were selected to analyze the temperature-sharpening results.

EC4 is used as an example because its land cover and subpixel variations of temperature were complicated. Table 10 compares the turbulent heat fluxes calculated using the IPUS, TRFA, and TSFA methods. Significant differences were observed between the TSFA and IPUS results and between the TRFA and IPUS results due to the heterogeneity of the surface. The LE calculated using the TSFA method was more consistent with in situ measurements than the LE calculated using the IPUS method because the MBE and RMSE decreased considerably, the R^2 increased, and the accuracy was improved by approximately 40 $\mathrm{W\,m^{-2}}$. However, the LE calculated by the TRFA was more accurate than the LE calculated by the TSFA, as discussed below.

The H calculated by using the TSFA method was more accurate than the H calculated by using the TRFA and IPUS methods. The RMSE of the results from the TRFA method was relatively close to the RMSE of the results from the TSFA method because the TRFA method also considers the effects of the heterogeneity of landscapes. In addition, the H values obtained from the TRFA method were always greater than those obtained from the TSFA method. Because the TSFA turbulent heat flux results are the same as the TRFA turbulent heat flux results for buildings and water bodies in our pixel ET algorithm, the differences between the TSFA and TRFA results depend on the vegetation and bare soil. Additionally, the 300 m resolution LST is larger than the LST of the subpixels, such as pixels containing vegetation or bare soil. This relationship occurs for two reasons: (1) the coarse

Table 9. In situ validation results of the turbulent heat fluxes of IPUS and TRFA.

	IPUS-H ($W\,m^{-2}$)			IPUS-LE ($W\,m^{-2}$)			TRFA-H ($W\,m^{-2}$)			TRFA-LE ($W\,m^{-2}$)		
Date	R^2	MBE	RMSE	R^2	MBE	RMSE	R^2	MBE	RMSE	R^2	MBE	RMSE
19 Jun	0.32	48.53	71.70	0.66	−47.68	86.02	0.39	52.28	70.98	0.65	−46.71	85.93
30 Jun	0.50	41.45	67.30	0.80	−81.75	102.33	0.69	42.64	60.85	0.86	−78.50	93.98
8 Jul	0.34	44.17	77.45	0.63	−66.75	118.63	0.44	54.20	76.00	0.82	−63.82	89.11
27 Jul	0.81	−33.14	50.01	0.83	25.61	74.26	0.84	−23.53	41.76	0.86	14.82	65.21
3 Aug	0.84	−5.23	33.50	0.74	−3.98	60.49	0.80	7.76	37.51	0.76	−18.23	62.71
15 Aug	0.64	−23.28	47.89	0.85	10.32	54.98	0.70	−14.77	39.99	0.89	0.59	45.22
22 Aug	0.31	41.50	74.81	0.61	−53.60	102.12	0.40	40.63	69.94	0.65	−54.17	98.97
29 Aug	0.72	27.15	44.16	0.76	−54.76	83.20	0.75	30.79	44.97	0.77	−59.43	86.22
2 Sep	0.28	−52.44	83.25	0.51	32.89	76.48	0.21	−45.77	75.84	0.52	24.37	71.69
13 Sep	0.08	−11.45	57.50	0.61	−57.38	81.83	0.06	−11.89	49.63	0.54	−57.78	84.58
14 Sep	0.12	−36.52	67.38	0.28	19.46	89.30	0.03	−34.34	64.85	0.38	25.41	75.96

Table 10. Comparison of the turbulent heat flux results at EC4.

EC4	H ($W\,m^{-2}$)				LE ($W\,m^{-2}$)			
Date	EC	IPUS	TRFA	TSFA	EC	IPUS	TRFA	TSFA
19 Jun	150.65	105.86	154.71	142.13	278.55	402.60	344.05	357.79
30 Jun	138.32	99.91	153.53	126.88	341.98	419.83	358.12	386.07
8 Jul	117.04	63.47	131.79	112.16	361.16	502.60	424.85	444.01
27 Jul	136.41	4.87	85.99	72.33	306.53	543.48	452.01	467.96
3 Aug	68.97	36.51	111.73	74.76	389.63	498.21	414.67	454.23
15 Aug	104.60	12.69	88.26	82.56	357.34	522.31	436.43	441.95
22 Aug	125.34	85.93	120.68	93.18	318.08	415.15	370.76	400.99
29 Aug	82.93	73.06	103.84	74.76	317.68	362.04	322.77	355.16
2 Sep	162.05	93.74	144.49	132.60	280.41	375.42	315.16	326.29
13 Sep	119.42	151.44	157.07	130.85	263.18	234.93	222.62	249.59
14 Sep	110.02	88.24	128.37	99.33	262.33	333.82	285.04	314.91

	IPUS			TRFA			TSFA		
Variable	R^2	MBE	RMSE	R^2	MBE	RMSE	R^2	MBE	RMSE
EC4-H	0.11	−44.65	61.73	0.25	5.88	26.33	0.51	−16.93	26.54
EC4-LE	0.49	99.21	119.55	0.56	42.69	62.40	0.60	63.92	76.78

pixels contain buildings and result in a larger 300 m resolution LST, and (2) the LSTs were underestimated at EC4 (as shown in Table 3), which would underestimate the value of $\Delta\hat{T}_{300}$ in Eq. (3) and, consequently, the sharpening temperature at 30 m and H. Because the LE was calculated as a residual item in the energy balance equation, the errors of the other three energy balance components accumulate in the LE term. At EC4, the R_n was overestimated by approximately $80\,W\,m^{-2}$, as discussed in Sect. 4.1, but the scale effect of R_n was not obvious (see Sect. 4.3.2), and the G was overestimated by approximately $20\,W\,m^{-2}$. These results decreased the accuracy of the available energy and overestimated the error by $60\,W\,m^{-2}$. Because the TRFA overestimates H, the underestimation of H in the TSFA would result in larger overestimation of LE than that estimated by the TRFA. Consequently, the LE calculated by using the TSFA method is less accurate than the LE calculated by the TRFA method.

Figure 8 shows that the classes and temperatures of 10×10 subpixels at 30 m correspond to the pixels with a resolution of 300 m at the EC tower. In the IPUS upscaling scheme, the 300 m pixels included buildings, maize, and vegetable crops at the 30 m resolution and were identified as maize. The canopy height gap between maize and vegetables was large during our study period, resulting in the overestimation of the canopy height. For additional details, see the error analysis in Sect. 4.4. However, because buildings corresponded with $H = 0.6R_n$ in this study, ignoring the contributions of buildings would result in the underestimation of H. Figure 8a shows the temperature-sharpening results in the EC4 pixel on 29 August. The temperature retrieved at 300 m scale was 303.49 K. Compared with the in situ measurement

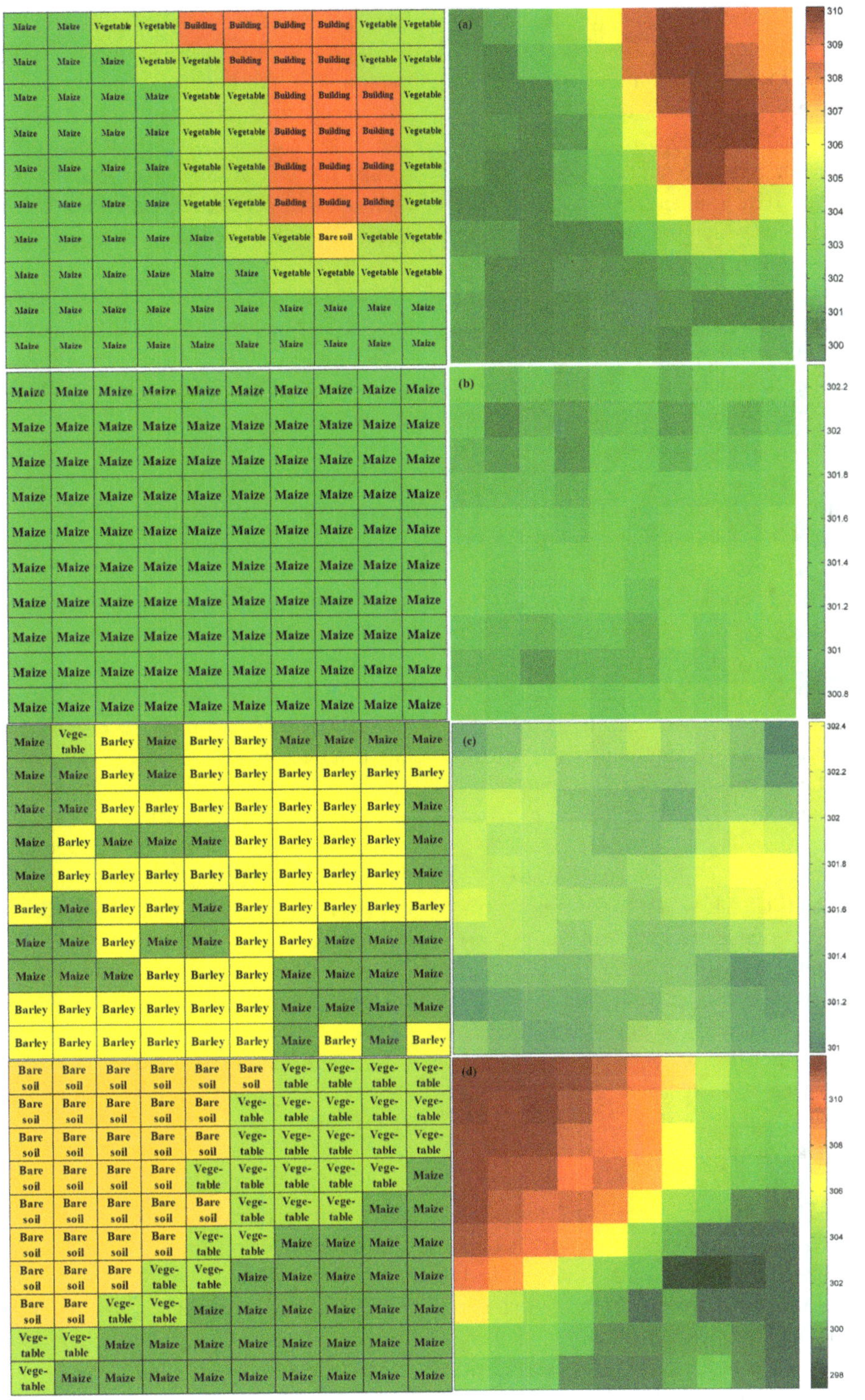

Figure 8. Distribution of classes and temperatures over **(a)** EC4, **(b)** EC15, **(c)** EC11, and **(d)** TP on 29 August 2012.

of 313.24 K, the temperature was underestimated at a resolution of 300 m. Even when substituting the in situ temperature into the ET model, the value of H reached 399.60 W m^{-2} and the LE became 0 W m^{-2}. When substituting the in situ temperature into the TRFA method, H was 396.49 W m^{-2} and LE was 18.7 W m^{-2}, indicating that the LE was underestimated and the H was overestimated with large errors. After processing by temperature sharpening, the distribution of the temperature at the 30 m resolution agreed with the classification. Temperature sharpening improved the description of heterogeneity based on the thermodynamic-driven force of the turbulent heat flux. These results apply to the ET model based on the classification map and high-resolution inputs and correspond to more accurate sensible heat flux estimates.

The land surface of EC15 was uniform and consisted of pure pixels covering maize fields. Consequently, the temperature distribution at 30 m resolution was very homogeneous, and the surface temperature variations were comprised within a range of 1.6 K. Table 11 shows the in situ validation results at EC15, for which the overall accuracy is not high due to the low LST retrieval accuracy on 8 July, which is discussed in Sect. 4.4.1. For homogeneous surfaces, the gaps between IPUS, TRFA, and TSFA were not large (within 10 W m^{-2}), and the accuracy did not improve (MBE and RMSE did not exhibit obvious variations). Statistically sharpening the temperature may increase the uncertainty of the model results for homogeneous surfaces; however, this influence could be omitted.

The weak heterogeneous land surface at EC11 contained barley, maize, and vegetables in a 300 m pixel resolution with a fractional area of 58 : 41 : 1 and was classified as barley at the 300 m resolution. The distributions of the classes and temperatures are shown in Fig. 8c. The pixel belongs to the first conditions of heterogeneity (nonlinear vegetation density variation between subpixels). Table 12 shows the in situ validation results of EC11. The improvements in the accuracies of H and LE by temperature resampling or sharpening were not as obvious as the improvements at EC4, which contained totally different landscapes (the other inhomogeneous condition).

Theoretically, the LE from the TSFA and TRFA at EC11 should be smaller than the IPUS LE values in the energy balance system. The height of maize (ranges from 0.3 to 2 m) was generally higher than the height of barley (ranges from 0.9 to 1.1 m) in the study area from June to August. Taller vegetation resulted in larger surface roughness and smaller aerodynamic resistance, which led to larger H values and smaller LE values, and vice versa (e.g., vegetables with a canopy height of 0.2 m). When using the TSFA and TRFA methods, patch landscapes consisting of different crops, such as maize and vegetables, were considered. Thus, the LE was smaller than the IPUS LE. On 19 June, the canopy height of maize was 0.74 m, which was lower than the canopy height of barley (1 m) and indicated that the H values that resulted from the TRFA and TSFA methods were less than H that resulted from the IPUS method. Because our validation method considered the influence of source area, the in situ turbulent heat flux validation results included the effects of neighboring pixels (i.e., on 3 August, the turbulent heat flux values of the pixel corresponding to the location of EC11 was only assigned a weight of 37 % in the source area).

The differences between the TSFA and TRFA methods were small and resulted from the LST differences between the 30 m resolution temperature-sharpening results and the LST retrieved at the 300 m resolution, but these differences were not evident at EC11. For example, on 29 August, the temperature range was 1.4 K, as shown in Fig. 8c. This temperature was even less than the temperature range at EC15 because the observation system at EC15 was a superstation with a 40 m tall tower that may cause large shadow effects and result in a relatively large temperature range. Hence, the temperature-sharpening effect is not obvious after aggregating the flux at the 300 m resolution under dense vegetation canopies. However, temperature sharpening can still decrease the heterogeneity that resulted from thermal dynamics.

The excess errors at EC11 was caused by the relatively low LST accuracy, with R^2, MBE, and RMSE values of 0.42, 1.59 K, and 2.98 K, respectively. On August 29, the temperature retrieved at 300 m scale was 301.6 K, and the observed ground temperature was 300.20 K. The LST at the 300 m resolution was slightly overestimated. When the in situ temperature was substituted into the IPUS algorithm, the value of H decreased to 16.06 W m^{-2} and LE became 467.43 W m^{-2}. When the in situ temperature was substituted into the TRFA scheme, the value of H was 22.43 W m^{-2} and the LE was 461.58 W m^{-2}, which were similar to the ground observations.

Another typical pixel located at the boundary of the bare soil and the oasis with no flux measurements was used to evaluate the correction effects of landscapes and temperature sharpening. The land surface of the TP contained maize, vegetables, and bare soil at a fraction of 35 : 31 : 34. Table 13 shows that when neither the heterogeneity of the landscape nor the LST are considered, the relative error of LE reached 180 W m^{-2}. In addition, if only the LST heterogeneity is not considered, the LE relative error reached 48 W m^{-2}. This result also reveals that the influences of landscape inhomogeneity are greater than the influences of inhomogeneity on the LST in mixed pixels.

4.3.2 Comparison of the TRFA and IPUS methods

Using data of 13 September as an example, the spatial distributions of the four components of the energy balance calculated by the IPUS and TRFA methods are shown in Figs. 9 and 10, respectively. TSFA minus IPUS and TSFA minus TRFA, which display the spatial distributions of the scale effect, are shown in Fig. 11. Scatterplots of TSFA vs. IPUS and TRFA are shown in Fig. 12.

A comparison of Fig. 6 with Fig. 9 shows that the spatial distributions of the fluxes greatly change, except for R_n. The TSFA results are synoptically smoother than the IPUS results because the land cover types and temperature distributions in mixed pixels are not considered in IPUS. For example, the boundary between the oasis and uncultivated land becomes a belt of intermediate G, H, and LE because these mixed pixels include uncultivated land and vegetation. However, mixed pixels are classified as the dominant land use type in the parameterization process of IPUS. This result overlooks the contributions of heat flux from complex land use types and overestimates or underestimates the turbulent

Table 11. Comparison of the turbulent heat fluxes results at EC15.

EC15	H (W m^{-2})				LE (W m^{-2})			
Date	EC	IPUS	TRFA	TSFA	EC	IPUS	TRFA	TSFA
19 Jun	92.55	106.60	109.25	99.81	419.47	427.19	419.99	429.98
30 Jun	42.37	43.99	45.51	44.67	551.73	527.12	525.17	526.09
8 Jul	18.34	217.53	235.48	209.90	620.95	425.71	397.49	424.86
27 Jul	27.68	21.22	31.11	24.30	597.76	589.58	579.43	586.47
3 Aug	2.33	33.32	−0.07	0.01	592.37	565.20	601.33	601.33
15 Aug	48.81	32.31	46.28	44.62	553.74	561.92	547.48	549.11
22 Aug	54.59	154.34	151.77	158.60	473.68	408.37	410.80	405.07
29 Aug	9.80	94.97	95.01	90.91	473.54	399.25	398.52	402.93
13 Sep	176.96	265.62	209.65	257.81	307.72	165.40	221.68	173.58
14 Sep	188.34	198.15	197.04	196.60	274.98	275.07	276.05	276.56

	IPUS			TRFA			TSFA		
Variable	R^2	MBE	RMSE	R^2	MBE	RMSE	R^2	MBE	RMSE
EC15-H	0.40	40.64	74.64	0.33	45.93	80.81	0.40	40.36	72.88
EC15-LE	0.74	−52.11	83.48	0.71	−48.80	82.51	0.74	−49.00	81.94

Table 12. Comparison of the turbulent heat flux results at EC11.

EC11	H (W m^{-2})				LE (W m^{-2})			
Date	EC	IPUS	TRFA	TSFA	EC	IPUS	TRFA	TSFA
19 Jun	33.94	173.69	158.12	158.18	531.46	391.60	407.42	407.40
30 Jun	25.03	3.29	23.12	21.37	635.22	586.37	566.48	568.28
8 Jul	32.29	68.17	97.16	96.13	601.98	567.73	538.77	539.81
27 Jul	21.42	−1.17	−1.58	−3.77	587.70	618.80	619.19	621.46
3 Aug	7.01	24.85	20.34	19.52	614.28	575.03	585.29	586.16
15 Aug	38.94	12.51	15.52	16.02	567.07	584.31	581.31	580.82
22 Aug	69.25	73.45	83.11	84.38	516.07	483.23	473.60	472.40
29 Aug	29.77	48.21	60.9	60.81	473.22	427.92	415.32	415.45
2 Sep	193.97	154.58	197.01	197.49	306.62	361.96	319.54	319.03
13 Sep	288.37	168.42	176.4	177.71	160.29	216.53	208.49	207.19
14 Sep	240.33	268.91	256.29	256.40	199.52	156.00	168.63	168.55

	IPUS			TRFA			TSFA		
Variable	R^2	MBE	RMSE	R^2	MBE	RMSE	R^2	MBE	RMSE
EC11-H	0.61	−1.07	61.31	0.57	−0.36	63.24	0.67	−0.21	55.50
EC11-LE	0.88	−19.83	63.16	0.89	−18.12	60.02	0.90	−21.29	58.11

heat flux by approximately 50 W m^{-2}. Since the TSFA can integrate the effects of these land areas and reveals the relative actual surface conditions, the heat flux results of TSFA vary less dramatically than those of IPUS, as shown in the figures. The results are similar in the oasis area.

Based on Figs. 6 and 10, the TRFA and TSFA methods are similar. Because the TRFA method considers the subpixel landscapes that could be a significant source of error in the ET models, the difference between the TSFA and TRFA methods mainly resulted from the differences between the sharpened LST and retrieved, resampled LST of subpixels at the 30 m resolution. In addition, the bias between the TSFA and TRFA is not as obvious as the bias between the TSFA and IPUS methods, as shown in Fig. 11c–f. Furthermore, Fig. 11f shows that the LEs calculated by using the TSFA method in most oasis areas were slightly greater than the LEs calculated by using the TRFA method, which yielded values of approximately 20 W m^{-2}.

The quadrangular shape with a relatively unstable bias shown in Fig. 11a is caused by the L_d that was calculated from the MOD05 water vapor product which exists as a quadrangular shape even after preprocessing the instrument mal-

Table 13. Comparison of the turbulent heat flux results at TP.

	H (W m^{-2})			LE (W m^{-2})		
Date	IPUS	TRFA	TSFA	IPUS	TRFA	TSFA
19 Jun	186.31	149.73	143.98	321.04	358.22	364.79
30 Jun	383.65	191.59	158.79	67.03	259.36	292.89
8 Jul	498.36	240.20	204.18	0.29	259.25	293.41
27 Jul	276.79	136.06	84.01	206.52	347.64	402.23
3 Aug	214.14	75.45	53.72	252.37	392.08	416.41
15 Aug	214.14	98.24	72.05	252.37	368.64	393.68
22 Aug	436.48	369.28	276.70	0.00	67.79	162.80
29 Aug	235.29	117.16	67.21	183.62	302.41	356.75
2 Sep	423.61	212.15	180.92	0.00	211.77	241.36
13 Sep	338.00	285.04	216.26	0.00	53.62	122.58
14 Sep	270.44	148.20	100.19	115.19	238.43	286.51

	IPUS			TRFA		
Variable	R^2	MBE	RMSE	R^2	MBE	RMSE
TP-H	0.62	174.47	185.49	0.95	42.28	48.01
TP-LE	0.71	−175.91	186.63	0.97	−43.11	49.04

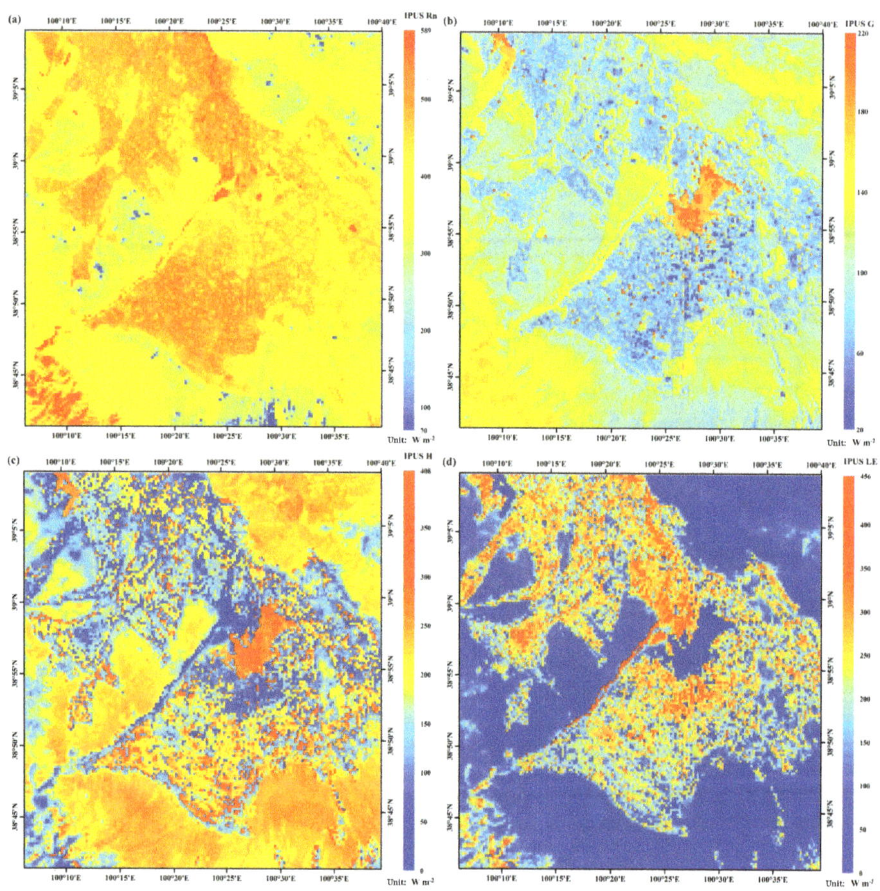

Figure 9. Maps of the four energy components, **(a)** R_n, **(b)** G, **(c)** H, and **(d)** LE, calculated using the IPUS method on 13 September 2012.

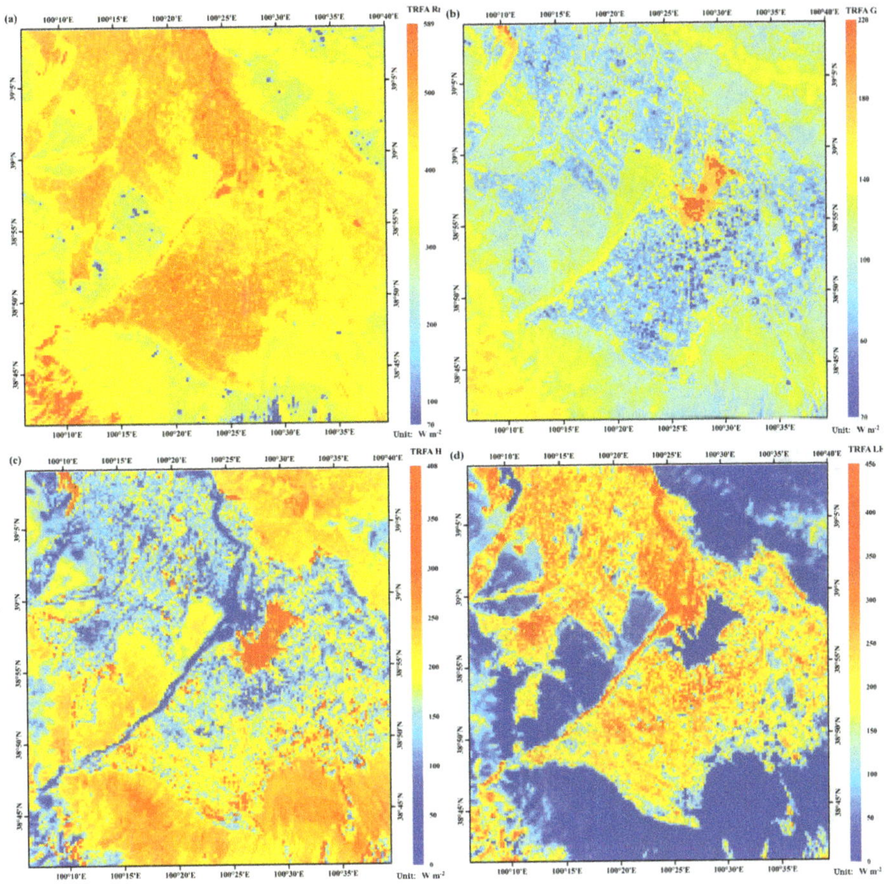

Figure 10. Maps of the four energy components, **(a)** R_n, **(b)** G, **(c)** H and **(d)** LE, calculated using the TRFA method on 13 September 2012.

function gap. In Fig. 11, the differences of the four energy components of the pure pixels between these three methods are within 5 W m^{-2}, and the mixed pixels have different ranges.

Figure 12 shows the scatter plots between the results of the TSFA method and the other two methods for all four energy balance components. Figure 12a and e show that R_n does not vary much between the three methods and the scatter is centralized around the 1 : 1 line. However, regarding the spatial-scale effect, the differences in G, H, and LE calculated by using the IPUS and TSFA methods are obvious. The scatter plots disperse at the mixed pixels, and the differences between the TRFA and TSFA results are relatively smaller. When using the TSFA method, the temperature-sharpening results can be divided into results that are higher and lower than the LST retrieved at 300 m. Compared to the LST retrieved at 300 m when using the TRFA method, a higher LST is counterbalanced by a lower LST when calculating H using the TSFA. Thus, the effect of temperature heterogeneity is neutralized in this case. This observation is potentially resulted from the temperature-sharpening algorithms because they tend to overestimate the subpixel LST for cooler landscapes and underestimate the subpixel LST for warmer areas in the image (Kustas et al., 2003).

However, LE is calculated as a residual; thus, the difference of LE is resulted from G and H. When the 300 m mixed pixels contained various types of landscapes, they were categorized as one type of landscape in the IPUS method and a single temperature value was used to evaluate the thermal dynamic effects when using the TRFA method. Pixels with highly different G, H, and LE values are mainly distributed near the mixed pixels, as shown in Fig. 11. An explanation for these deviations is provided below.

The parameterization of G and H are based on the land cover type. For example, for buildings, $G = 0.4R_n$ (Kato and Yamaguchi, 2005) (which is usually greater than the G of vegetation and bare soil deduced from Eq. 9) and $H = 0.6R_n$, and for water, $G = 0.226R_n$ and LE $= R_n - G$. From the land cover map shown in Fig. 4, four major classes exist in the study area: buildings with a high H, uncultivated land with a relatively high H, cropland with a relatively low H, and water with $H = 0$.

1. If a pixel contains cropland and buildings and is categorized as cropland, the building area within the pixel is ignored in the IPUS method. In this case, G and H

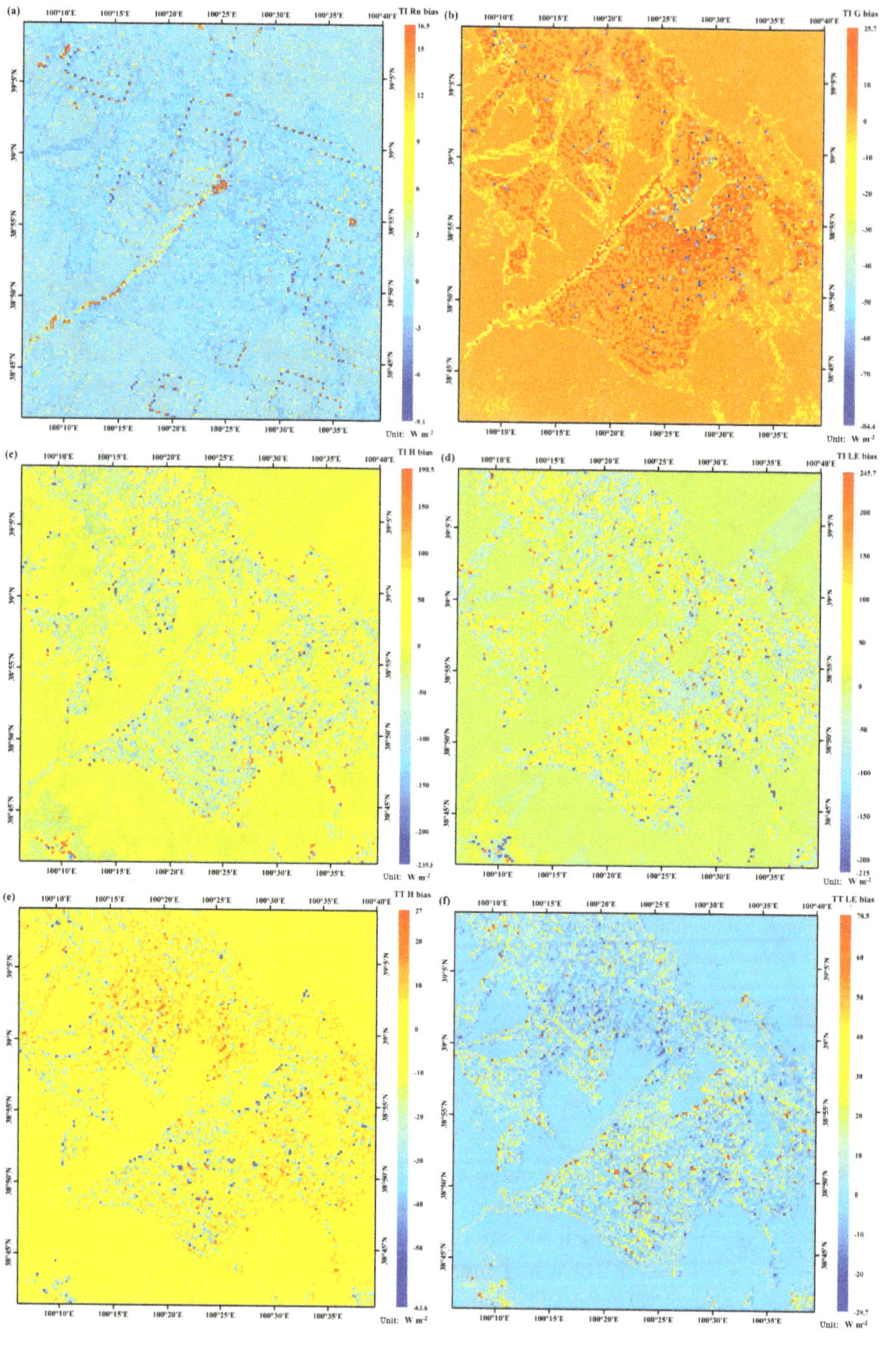

Figure 11. Maps of the bias of the energy balance components calculated using the TSFA method minus the IPUS method: **(a)** R_n, **(b)** G, **(c)** H, **(d)** LE; and TSFA minus TRFA: **(e)** H and **(f)** LE.

are underestimated and LE is overestimated. In addition, after considering the landscapes using the TRFA method, the LE is underestimated and H is overestimated because the pixels contain buildings that are still reflected indistinctly by LST at 300 m as the detailed temperature heterogeneity cannot be represented by the TRFA method. These points are shown in green in Fig. 12. However, if the pixel is categorized as buildup, the building area within a pixel is exaggerated, which causes G and H to be overestimated and LE to be underestimated when using the IPUS method. This situation is similar to that illustrated by the green points associated with the TRFA results and is shown by the red points in Fig. 12.

2. At the boundary of the oasis and uncultivated land, the mixed pixels are divided into cropland, LE is over-

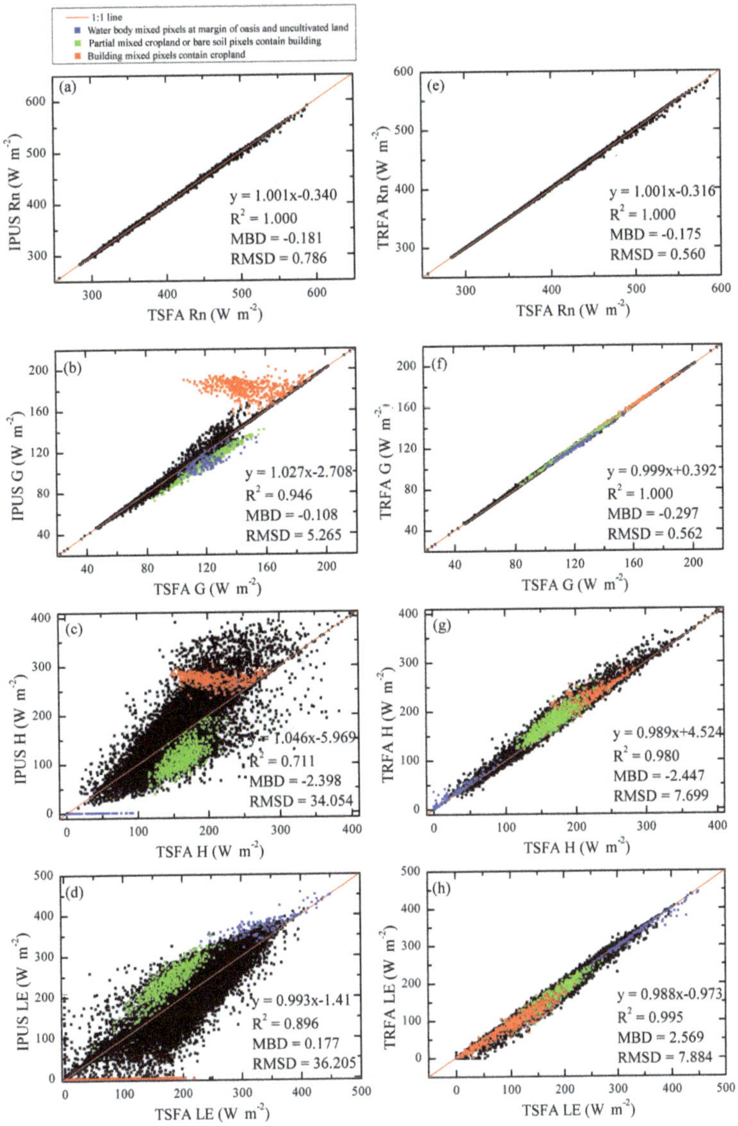

Figure 12. Scatter plots between the TSFA and IPUS results: **(a)** R_n, **(b)** G, **(c)** H, and **(d)** LE; and TSFA and TRFA results: **(e)** R_n, **(f)** G, **(g)** H, and **(h)** LE. MBD and RMSD are the mean bias deviation and root mean square deviation between the TSFA and IPUS results, respectively.

estimated, G and H are underestimated in the IPUS method, and vice versa. LE is also overestimated in the pixels containing water and other types of land cover (generally bare soil in our study area). These pixels are categorized as water and are shown as blue points in Fig. 12. Some of the blue LE points calculated by using the TSFA method are slightly smaller than those calculated using the TRFA method for pixels containing vegetation. At noon, the temperature of vegetation in those pixels is lower than that of water bodies.

3. In mixed pixels that contain various crops, such as maize and vegetables, LE is underestimated if the area of maize within the pixel is overestimated because the canopy height of the maize is taller than that of vegetables. This relationship results in the overestimation of H when using the IPUS and TRFA methods. In addition, G depends on the FVC values of the crops when using the IPUS method. Moreover, G depends on R_n when using the TRFA method and is nearly identical to the values of G obtained using the TSFA method.

At the study area, we compared the TRFA and IPUS methods to quantify the ability of the TSFA method to simulate the heterogeneities of the land surface on 13 September (see Table 14). In pure pixels, the LE biases among the IPUS, TRFA, and TSFA methods were small. In mixed pixels, the LE bias between the TSFA and IPUS methods varied from 35.36 to

65.66 W m^{-2}, and the bias between the TSFA and TRFA methods varied from 4.41 to 22.53 W m^{-2}. More class types in mixed pixels correspond to larger biases. Table 15 shows the bias of the mixed pixels that contain buildings and bare soil between the three methods. In mixed pixels with buildings, the IPUS and TRFA methods generally underestimated LE and had large bias values compared to those of the TSFA method. In mixed pixels without buildings and bare soil, the bias between TRFA (or IPUS) and TSFA was relatively small, which indicates that the landscape and temperature inhomogeneity were accounted for by the TSFA method. The aforementioned analyses demonstrate that the TSFA method can consider the heterogeneous effects of mixed pixels.

Considering the landscapes and inhomogeneous distribution of LST, the TSFA method ensures that none of the end members (30 m pixels) are ignored or exaggerated. Thus, the distribution of LE calculated using the TSFA method is smoother and more rational than the distributions of LE calculated using the other methods. At the regional scale, the TSFA method describes the heterogeneity of the land surface more precisely. The degree of achievable estimation accuracy is discussed hereafter.

4.4 Error analysis

Since LE is calculated as a residual term in the energy balance equations, the sensitivity of H was analyzed first. Land surface variables (including LST, LAI, canopy height, and FVC) and meteorological variables including wind speed, air temperature, air pressure, and relative humidity are the major factors for H sensitive analysis. Figure 13 presents a case of sensitivity analysis results for H. In this case, LST is 303.9 K, and it ranges from 298.4–309.4 K with a step size of 0.5 K, LAI is set to 1.4 and it ranges from 0.14–2.66 with a step size of 0.14. Canopy height is 1 m and it ranges from 0.1–1.9 m with a step size of 0.1 m. Additionally, FVC = 0.5, wind speed $u = 2.48$ m s^{-1}, air temperature $T_a = 297.9$ K, air pressure = 97.2 kPa, and RH = 40.29 %. In addition, the land cover is maize, and the reference H is 230.2 W m^{-2}.

The air pressure is stable over a short period and has little effect on the ET results. Although excess resistance was calculated from the friction velocity, the meteorological data were provided by ground observations; thus, the meteorological data are relatively accurate. As shown in Fig. 13, LAI, canopy height, and LST are sensitive variables.

The parameterization of the momentum roughness length indicates that H is sensitive to LAI, with decreasing sensitivity when the LAI is greater than 1. When the LAI is less than 1, the momentum roughness length increases as the LAI increases, and turbulent exchange is enhanced. However, when the LAI is greater than 1, the plant canopy is regarded as a continuum that is not a sensitive variable. Because our study area is dominated by agriculture and the study period extended from July to September, the crops in the HRB middle stream grew quickly, thus, the LAI was usually greater than 1. Thus, LST and canopy height are the main sources of error.

4.4.1 Errors in LST

As shown in Fig. 13, 1 K LST bias would result in 21 % error of H while H is 230.2 W m^{-2}. However, the sensitivity of the LST is unstable and depends on the strength of the turbulence. The strength of the turbulence determines the mass and energy transport and the resistance of heat transfer, which influences the sensitivity to the LST. A weaker turbulence corresponds to a weaker LST sensitivity, and vice versa.

A sensitivity analysis of LE induced by LST was also performed. In order to exclude the influence of other factors, stations were chosen with homogeneous landscapes within coarse pixels. These results are shown in Table 16. The LE results obtained from the observed LST are consistent with the in situ observations and have less bias. LE was overestimated when the LST was underestimated, and vice versa. Because the magnitude of LE was greater than that of H, the relative error of LE was less than the relative error of H. However, 1 K of LST bias resulted in an average LE error of 30 W m^{-2}, which is consistent with the sensitivity analysis of H shown in Fig. 13. Specifically, 1 K of LST bias would result in an LE biases of 8.7 W m^{-2} (in desert, SSW) to 84.4 W m^{-2} (in oasis, EC8), which indicates that the sensitivity of LST is unstable.

4.4.2 Errors in canopy height

In this paper, canopy height was obtained from a phenophase and classification map. Thus, the accuracy of the canopy height was mainly dependent on the classification accuracy and plant growth state. Even within the same region, the canopy height of a crop can differ due to differences in seeding times and soil attributes, such as soil moisture and fertilization.

The land use type was orchard at EC17. However, in our land classification map, the land use at EC17 was other crops, which includes vegetables and orchards. Thus, it was difficult to set the canopy height. In our study area, most of the other crops were vegetables (canopy height of 0.2 m), and the height of the orchard was approximately 4 m; thus, a value of 0.2 m would overestimate the LE. The LE estimations with incorrect canopy heights and correct orchard canopy height at EC17 are shown in Table 17. The days of large LST bias were removed, and the bias between the model and ground observations decreased. The excess errors were caused by errors in the LST and land use, such as buildings and maize in the mixed pixels.

Except for the error source discussed previously, the following sources of error were unavoidable:

1. Although the remotely sensed turbulent heat flux is instantaneous, the EC data are averaged over time. Thus, the timescales do not match in the validation.

Table 14. Comparison of the latent heat flux in pixels containing different numbers of class types.

Number of class types in pixels	IPUS (W m^{-2})			TRFA (W m^{-2})			Pixel number
	R^2	MBD	RMSD	R^2	MBD	RMSD	
1	1.00	0.21	0.21	1.00	0.05	0.61	11 398
2	0.85	−7.18	35.36	1.00	−0.35	4.41	8212
3	0.66	−2.32	52.55	0.98	−7.33	12.56	4762
4	0.49	1.88	65.66	0.96	−11.56	16.55	2824
5	0.98	−30.92	62.69	0.96	−16.90	22.53	4

Notes: the number of class types in mixed pixels represents the number of classification types that were contained in the pixels. For example, 1 represents the pure pixels and 2 represents mixed pixels containing two land use types. MBD and RMSD are the mean bias deviation and root mean square deviation, respectively, between the TSFA results and the TRFA and IPUS results.

Table 15. Comparison of the latent heat fluxes of typical mixed pixels.

Types of mixed pixels	IPUS (W m^{-2})			TRFA (W m^{-2})			Pixel number
	R^2	MBD	RMSD	R^2	MBD	RMSD	
Mixed pixels containing buildings	0.58	−1.02	61.94	0.97	−9.64	14.66	4918
Mixed pixels not containing buildings	0.81	−5.49	39.21	0.99	−2.12	7.60	10 884
Mixed pixels containing bare soil	0.73	−1.52	49.04	0.98	−5.96	11.86	9049
Mixed pixels not containing bare soil	0.65	−7.55	45.28	0.98	−2.46	7.83	6753

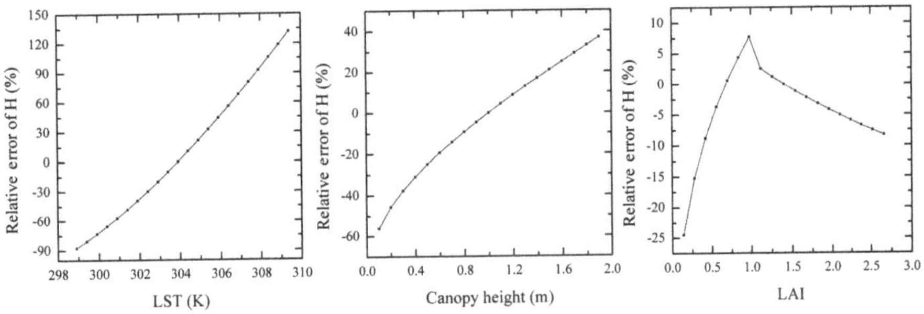

Figure 13. Sensitivity analysis of the surface variables for sensible heat flux.

2. The calibration coefficients of the HJ-1B satellite's CCD and IRS drifts because of instruments aging.

3. Geometric correction causes half-pixel bias equal to or less than the deviation of the artificially subjective interpretation.

A one-source model and simplified parameterization schemes were used in this paper to determine surface roughness lengths and heat transfer coefficients. The one-source model combines soil evaporation and plant transpiration and assumes that SPAC is a one-source continuum. This assumption is reasonable when the surface is densely covered by vegetation but relies on the accuracy of the difference between the LST and air temperature, as previously mentioned. When a one-source model is applied to an area covered by sparse vegetation, such as semiarid or arid areas, this assumption is irrational.

5 Discussions

The TSFA describes the surface heterogeneity much better than the IPUS and TRFA. The IPUS aggregates the land surface variables from 30 to 300 m, which results in the loss of land surface details and leads to the scale effects. Although the TRFA uses 30 m information from VNIR bands and partially decreases the heterogeneity, it treats the pivotal variable LST as homogeneous at 300 m resolution, which results in considerable error. In summary, the advantages of the TSFA method are described as follows:

Table 16. The results of the LST error analyses at the stations with homogeneous landscapes.

Station	Date	Retrieved LST (K)	Observed LST (K)	LST bias (K)	EC-LE ($W m^{-2}$)	LE from retrieved LST ($W m^{-2}$)	LE from observed LST ($W m^{-2}$)	LE relative error (%)	H relative error (%)
EC8	19 Jun	304.92	301.74	3.18	415.89	321.80	399.78	−22.62	68.58
EC7	30 Jun	302.5	299.35	3.15	611.22	453.59	557.97	−25.79	886.08
EC10	8 Jul	303.58	300.5	3.08	617.83	504.44	549.53	−18.35	390.24
EC15	8 Jul	303.55	300.13	3.42	620.95	425.71	603.73	−31.44	450.57
EC7	27 Jul	298.87	300.55	−1.68	577.59	643.56	566.62	11.42	−132.47
SSW	27 Jul	307.86	316.82	−8.96	119.35	238.07	78.43	99.48	−60.36
EC8	22 Aug	299.58	297.77	1.81	543.56	416.23	467.42	−23.42	88.59
EC10	22 Aug	301.61	298.04	3.57	503.82	398.82	513.67	−20.84	138.61
EC15	22 Aug	300.59	297.69	2.9	473.68	408.37	495.49	−13.79	129.60
EC8	29 Aug	301.54	300.44	1.1	514.31	402.93	428.78	−21.66	63.91
EC15	29 Aug	301.41	299.84	1.57	473.54	399.25	459.66	−15.69	182.34
SSW	2 Sep	304.9	303.42	1.48	226.88	127.96	149.83	−43.60	11.36

Notes: LST bias is calculated as the retrieved LST minus the observed LST; EC-LE is the in situ latent heat flux; LE relative error is the relative error between the retrieved and observed LST and is expressed as ((LE from retrieved LST) − (LE from observed LST))/(LE from observed LST) × 100 %, and H relative error is calculated in the same way.

Table 17. The results of the canopy height error analyses at EC17.

Date	EC-LE ($W m^{-2}$)	LE from incorrect canopy height ($W m^{-2}$)	LE-I relative error (%)	LE from correct canopy height ($W m^{-2}$)	LE-C relative error (%)
15 Aug 2012	499.62	562.06	12.50	521.83	4.45
2 Aug 2012	366.27	519.01	41.70	396.54	8.26
2 Sep 2012	377.96	471.68	24.80	336.52	−10.96
14 Sep 2012	465.38	352.78	−24.20	258.07	−44.55

Notes: LE-I relative error is the relative error between the LE from incorrect canopy height and observed LE and is expressed as ((LE from incorrect canopy height) − (EC-LE))/(EC-LE) × 100 %; LE-C relative error is the relative error between the LE from correct canopy height and observed LE and is expressed as ((LE from correct canopy height) − (EC-LE))/(EC-LE) × 100 %.

1. The temperature-sharpening algorithm in TSFA is capable of decreasing the influences of the heterogeneity of the LST, which is consistent with results from previous studies (Kustas et al., 2003; Bayala and Rivas, 2014; Mukherjee et al., 2014). As analyzed in Sect. 4.3, the non-consideration of the heterogeneity of LST in mixed pixels is ill-founded and causes errors when estimating ET.

2. In the one-source energy balance model, different parameterization schemes were employed for different landscapes. In the IPUS, a single land cover is assigned to a mixed pixel, which results in a large error. However, the TSFA method is used to calculate the surface flux at 30 m and is aggregated to 300 m using the area-weighting method, which considers all of the subpixel landscapes and improves the accuracy.

Some problems exist in the temperature-sharpening algorithms. The temperature-downscaling method used in this paper caused "boxy" anomalies in parts of the sharpened temperature fields in large pixels because of the constant residual term, $\Delta\hat{T}_{300}$, in Eq. (3) within large pixels. This situation also occurred in the temperature-sharpening algorithm proposed by Agam et al. (2007). In addition, our temperature-sharpening algorithm tends to overestimate the subpixel LST for cooler landscapes and underestimate the subpixel LST for warmer areas (Kustas et al., 2003). This inaccurate estimation causes errors that are difficult to evaluate when estimating the turbulent heat flux. For example, the small turbulent heat flux bias between TSFA and TRFA was caused by a counterbalancing effect as analyzed in Sect. 4.3.1. Additional temperature-sharpening algorithms under heterogeneous surfaces should be evaluated using real data sets when applied in ET models (Ha et al., 2011).

The retrieval methods of land surface variables were validated in other areas. For example, the albedo algorithm was previously applied to retrieve Global Land Surface Satellite (GLASS) products (Liang et al., 2014), the LST retrieval algorithm was validated in the Haihe River basin in northern China (H. Li et al., 2011), and the soil heat flux correction algorithm was validated in the GAME-Tibet campaign (Yang and Wang, 2008). Since the surface of the Heihe River basin is very heterogeneous, additional comparisons of our algorithm in other areas would be helpful.

In addition, to correct the discrepancy between remotely sensed radiative surface temperature and aerodynamic temperature at the source of heat transport, a brief and well-performed parameterization scheme (under uniformly flat plant surface) of excess resistance was used to calculate the aerodynamic resistance of heat transfer (Jiao et al., 2014).

Since our study is based on mixed pixels, multiple parameterization methods should be compared to select the optimum method.

Because of the sensitive variables of the one-source energy balance model used in this paper, the accuracies of the LST and canopy height greatly influenced the turbulent heat flux. HJ-1B IRS has a single thermal channel, and the single-channel LST-retrieving algorithm may be unstable under wet atmospheric conditions (water vapor contents higher than $3\,\mathrm{g\,cm^{-2}}$) (H. Li et al., 2010), which may create a bottleneck for ET estimations by HJ-1B. The canopy height is a priori knowledge based on phenophase classifications and would influence the accuracy of the surface roughness calculation. Multi-source remote sensing data could be used to improve the accuracy of calibrations and land surface variable estimation. Active microwave and lidar data (Colin and Faivre, 2010) could be used to obtain the canopy height, which would decrease the dependence on the accuracy of the classification.

The energy balance closure has a significant influence on the evaluation of the model calculated heat flux results. In our study area, the EC energy balance closure ratio was greater than 0.75 (S. M. Liu et al., 2011). Studies have shown that uncaptured low-frequency eddies (Von Randow et al., 2008), extension of averaging time (Charuchittipan et al., 2014), and lack of an accurate accounting of heat storage terms (Meyers and Hollinger, 2004) are potential reasons for the energy imbalance and so forth. The conserving Bowen ratio and residual closure technique are often used to force energy balance. We chose the residual closure method at last because the conserving Bowen ratio method yields irrational sensible heat flux due to small or negative Bowen ratios (large LEs due to the oasis effect) in the oasis–desert system. Energy balance closure was problematic at times for turbulent flux system and tended to be associated with significant discrepancies in LE (Prueger et al., 2005).

Since a footprint model was used in the validation, the footprints' discrepancies between in situ measurements and remote sensing pixel may cause biases. For example, model validation results were calculated based on the relative weights of the footprint model and multiplied by the heat flux results of the coarse pixels in the source area from the upwind direction. However, the heat fluxes of coarse pixels included the contributions of non-overlapped subpixels within the coarse pixel. These pixels are influenced by the heterogeneity of underlying surface, and it would cause uncertainties in the validation.

6 Conclusion

The effects of surface heterogeneity in ET estimation have been studied here by employing the IPUS, TRFA, and TSFA methods over heterogeneous surface. Compared to the IPUS and TRFA methods, the TSFA method exhibits more consistent agreement with in situ measurements (energy balance forced by the residual closure method) based on the footprint validation results. The IPUS approach does not consider surface heterogeneity at all, which causes significant error in the heat fluxes (i.e., $186\,\mathrm{W\,m^{-2}}$). The TRFA considers heterogeneity of landscapes besides LST heterogeneity, with a heat flux error (i.e., $49\,\mathrm{W\,m^{-2}}$) that is less than that of IPUS. However, this error is non-negligible. As a sensitive variable of the ET model, canopy height is mainly determined by classification, and the application of classification at a 30 m resolution can improve the accuracy of the canopy height. Additionally, the sharpened surface temperature at a resolution of 30 m decreases the thermodynamic uncertainty caused by the land surface. The TSFA method can capture the heterogeneities of the land surface and integrate the effects of landscapes in mixed pixels that are neglected at coarse spatial resolutions.

HJ-1B satellite data have advantages because of their high spatiotemporal resolution and free access. Because the satellites are still in operation, the long-term data are promising for applications of monitoring energy budgets.

7 Data availability

For this study we used the HJ-1B data in 2012, which are accessible at CRESDA via http://218.247.138.121/DSSPlatform/productSearch.html. The land cover maps of Heihe River basin are available on the website of CARD via http://card.westgis.ac.cn/hiwater/view/uuid/6bbf9a3f-e7d8-4255-9ecb-131e1543316d. MOD05 data are available at LP DAAC of NASA via https://lpdaac.usgs.gov/data_access/data_pool. GDEM data are available from Global Data Explorer via http://gdex.cr.usgs.gov/gdex/. TOMS Level-3 data are available at the official TOMS website via https://ozoneaq.gsfc.nasa.gov/data/ozone/. Atmosphere profile data come from NECP at http://rda.ucar.edu/datasets/ds083.2/. In situ observations are accessible from CARD via http://card.westgis.ac.cn/hiwater/mso.

Acknowledgements. The authors would like to thank Franz Meixner of the Max Planck Institute for Biogeochemistry in Mainz and Albrecht Neftel of the Agroscope Research Station in Zurich for providing the footprint calculation tool. We are grateful to the research team of Shao Min Liu at Beijing Normal University for providing eddy covariance and automatic meteorological station data. This study was supported by the Special Fund from the Chinese Academy of Sciences (KZZD-EW-TZ-18) and the Chinese Natural Science Foundation Project (grant no. 41371360). The authors thank Bo Zhong at Institute of Remote Sensing and Digital Earth, Chinese Academy of Sciences for his very helpful comments and many corrections. Generous help for revising the paper was provided by the editors and reviewers.

Edited by: L. Pfister

References

Agam, N., Kustas, W. P., Anderson, M. C., Li, F., and Colaizzi, P. D.: Utility of thermal sharpening over Texas high plains irrigated agricultural fields, J. Geophys. Res.-Atmos., 112, D19110, doi:10.1029/2007JD008407, 2007.

Allen, R., Tasumi, M., and Trezza, R.: Satellite-Based Energy Balance for Mapping Evapotranspiration with Internalized Calibration (METRIC) – Model, J. Irrig. Drain. E.-ASCE, 133, 380–394, doi:10.1061/(ASCE)0733-9437(2007)133:4(380), 2007.

Ambast, S. K., Keshari, A. K., and Gosain, A. K.: An operational model for estimating Regional Evapotranspiration through Surface Energy Partitioning (RESEP), Int. J. Remote Sens., 23, 4917–4930, doi:10.1080/01431160110114501, 2002.

Bastiaanssen, W. G. M., Menenti, M., Feddes, R. A., and Holtslag, A. A. M.: A remote sensing surface energy balance algorithm for land (SEBAL). 1. Formulation, J. Hydrol., 212–213, 198–212, doi:10.1016/S0022-1694(98)00253-4, 1998.

Bateni, S. M. and Liang, S.: Estimating surface energy fluxes using a dual-source data assimilation approach adjoined to the heat diffusion equation, J. Geophys. Res.-Atmos. 117, D17118, doi:10.1029/2012JD017618, 2012.

Bayala, M. I. and Rivas, R. E.: Enhanced sharpening procedures on edge difference and water stress index basis over heterogeneous landscape of sub-humid region, Egypt. J. Remote Sens. Space Sci., 17, 17–27, doi:10.1016/j.ejrs.2014.05.002, 2014.

Bin, L. and Roni, A.: The Impact of Spatial Variability of Land-Surface Characteristics on Land-Surface Heat Fluxes, J. Climate, 7, 527–537, doi:10.1175/1520-0442(1994)007<0527:TIOSVO>2.0.CO;2, 1994.

Blyth, E. M. and Harding, R. J.: Application of aggregation models to surface heat flux from the Sahelian tiger bush, Agr. Forest Meteorol., 72, 213–235, doi:10.1016/0168-1923(94)02164-F, 1995.

Bonan, G. B., Pollard, D., and Thompson, S. L.: Influence of Subgrid-Scale Heterogeneity in Leaf Area Index, Stomatal Resistance, and Soil Moisture on Grid-Scale Land–Atmosphere Interactions, J. Climate, 6, 1882–1897, doi:10.1175/1520-0442(1993)006<1882:IOSSHI>2.0.CO;2, 1993.

Bonan, G. B., Levis, S., Kergoat, L., and Oleson, K. W.: Landscapes as patches of plant functional types: An integrating concept for climate and ecosystem models, Global Biogeochem. Cy., 16, 5-1–5-23, doi:10.1029/2000GB001360, 2002.

Cammalleri, C., Anderson, M. C., Gao, F., Hain, C. R., and Kustas, W. P.: A data fusion approach for mapping daily evapotranspiration at field scale, Water Resour. Res., 49, 4672–4686, doi:10.1002/wrcr.20349, 2013.

Charuchittipan, D., Babel, W., Mauder, M., Leps, J.-P., and Foken, T.: Extension of the Averaging Time in Eddy-Covariance Measurements and Its Effect on the Energy Balance Closure, Bound.-Lay. Meteorol., 152, 303–327, doi:10.1007/s10546-014-9922-6, 2014.

Chen, J.: An important drawback and the improvement of the evapotranspiration model with remote sensing, Chinese Sci. Bull., 6, 454–457, 1988.

Chen, J., Chen, B. Z., Black, T. A., Innes, J. L., Wang, G. Y., Kiely, G., Hirano, T., and Wohlfahrt, G.: Comparison of terrestrial evapotranspiration estimates using the mass transfer and Penman–Monteith equations in land surface models, J. Geophys. Res.-Biogeo., 118, 1715–1731, doi:10.1002/2013jg002446, 2013.

Chen, W., Cao, C., He, Q., Guo, H., Zhang, H., Li, R., Zheng, S., Xu, M., Gao, M., Zhao, J., Li, S., Ni, X., Jia, H., Ji, W., Tian, R., Liu, c., Zhao, Y., and Li, J.: Quantitative estimation of the shrub canopy LAI from atmosphere-corrected HJ-1 CCD data in Mu Us Sandland, Sci. China Earth Sci., 53, 26–33, 2010.

Choudhury, B. J. and Monteith, J. L.: A four-layer model for the heat budget of homogeneous land surfaces, Q. J. Roy. Meteorol. Soc., 114, 373–398, doi:10.1002/qj.49711448006, 1988.

Choudhury, B. J., Reginato, R. J., and Idso, S. B.: An analysis of infrared temperature observations over wheat and calculation of latent heat flux, Agr. Forest Meteorol., 37, 75–88, doi:10.1016/0168-1923(86)90029-8, 1986.

Cleugh, H. A., Leuning, R., Mu, Q. Z., and Running, S. W.: Regional evaporation estimates from flux tower and MODIS satellite data, Remote Sens. Environ., 106, 285–304, doi:10.1016/j.rse.2006.07.007, 2007.

Colin, J. and Faivre, R.: Aerodynamic roughness length estimation from very high-resolution imaging LIDAR observations over the Heihe basin in China, Hydrol. Earth Syst. Sci., 14, 2661–2669, doi:10.5194/hess-14-2661-2010, 2010.

Ershadi, A., McCabe, M. F., Evans, J. P., and Walker, J. P.: Effects of spatial aggregation on the multi-scale estimation of evapotranspiration, Remote Sens. Environ., 131, 51–62, doi:10.1016/j.rse.2012.12.007, 2013.

Fan, L., Xiao, Q., Wen, J., Liu, Q., Tang, Y., You, D., Wang, H., Gong, Z., and Li, X.: Evaluation of the Airborne CASI/TASI Ts-VI Space Method for Estimating Near-Surface Soil Moisture, Remote Sensing, 7, 3114–3137, doi:10.3390/rs70303114, 2015.

Fisher, J. B., Tu, K. P., and Baldocchi, D. D.: Global estimates of the land–atmosphere water flux based on monthly AVHRR and ISLSCP-II data, validated at 16 FLUXNET sites, Remote Sens. Environ., 112, 901–919, doi:10.1016/j.rse.2007.06.025, 2008.

Ha, W., Gowda, P. H., and Howell, T. A.: Downscaling of Land Surface Temperature Maps in the Texas High Plains with the TsHARP Method, Gisci. Remote Sens., 48, 583–599, doi:10.2747/1548-1603.48.4.583, 2011.

Ha, W., Gowda, P. H., and Howell, T. A.: A review of downscaling methods for remote sensing-based irrigation management: part I, Irrigation Science, 31, 831–850, doi:10.1007/s00271-012-0331-7, 2013.

He, L., Chen, J. M., Pisek, J., Schaaf, C. B., and Strahler, A. H.: Global clumping index map derived from the MODIS BRDF product, Remote Sens. Environ., 119, 118–130, doi:10.1016/j.rse.2011.12.008, 2012.

Hu, Z. L. and Islam, S.: A framework for analyzing and designing scale invariant remote sensing algorithms, IEEE T. Geosci. Remote, 35, 747–755, doi:10.1109/36.581996, 1997.

Jia, Z. Z., Liu, S. M., Xu, Z. W., Chen, Y. J., and Zhu, M. J.: Validation of remotely sensed evapotranspiration over the Hai River Basin, China, J. Geophys. Res.-Atmos., 117, D13113, doi:10.1029/2011JD017037, 2012.

Jiang, B., Liang, S., Townshend, J. R., and Dodson, Z. M.: Assessment of the Radiometric Performance of Chinese HJ-1 Satellite CCD Instruments, IEEE J. Sel. Top. Appl., 6, 840–850, doi:10.1109/JSTARS.2012.2212236, 2013.

Jiang, L. and Islam, S.: A methodology for estimation of surface evapotranspiration over large areas using remote sensing observations, Geophys. Res. Lett., 26, 2773–2776, doi:10.1029/1999GL006049, 1999.

Jiang, L. and Islam, S.: Estimation of surface evaporation map over Southern Great Plains using remote sensing data, Water Resour. Res., 37, 329–340, doi:10.1029/2000WR900255, 2001.

Jiang, L. and Islam, S.: An intercomparison of regional latent heat flux estimation using remote sensing data, Int. J. Remote Sens., 24, 2221–2236, doi:10.1080/01431160210154821, 2003.

Jiao, J. J., Xin, X. Z., Yu, S. S., Zhou, T., and Peng, Z. Q.: Estimation of surface energy balance from HJ-1 satellite data, J. Remote Sens., 18, 1048–1058, doi:10.11834/jrs.20143322, 2014.

Jin, Y. F., Randerson, J. T., and Goulden, M. L.: Continental-scale net radiation and evapotranspiration estimated using MODIS satellite observations, Remote Sens. Environ., 115, 2302–2319, doi:10.1016/j.rse.2011.04.031, 2011.

Kalma, J., McVicar, T., and McCabe, M.: Estimating Land Surface Evaporation: A Review of Methods Using Remotely Sensed Surface Temperature Data, Surv. Geophys., 29, 421–469, doi:10.1007/s10712-008-9037-z, 2008.

Kato, S. and Yamaguchi, Y.: Analysis of urban heat-island effect using ASTER and ETM+ Data: Separation of anthropogenic heat discharge and natural heat radiation from sensible heat flux, Remote Sens. Environ., 99, 44–54, doi:10.1016/j.rse.2005.04.026, 2005.

Kormann, R. and Meixner, F.: An Analytical Footprint Model For Non-Neutral Stratification, Bound.-Lay. Meteorol., 99, 207–224, doi:10.1023/A:1018991015119, 2001.

Kustas, W. P.: Estimates of Evapotranspiration with a One- and Two-Layer Model of Heat Transfer over Partial Canopy Cover, J. Appl. Meteorol., 29, 704–715, doi:10.1175/1520-0450(1990)029<0704:EOEWAO>2.0.CO;2, 1990.

Kustas, W. P., Norman, J. M., Anderson, M. C., and French, A. N.: Estimating subpixel surface temperatures and energy fluxes from the vegetation index – radiometric temperature relationship, Remote Sens. Environ., 85, 429–440, doi:10.1016/S0034-4257(03)00036-1, 2003.

Kustas, W. P., Moran, M. S., and Meyers, T. P.: The Bushland Evapotranspiration and Agricultural Remote Sensing Experiment 2008 (BEAREX08) Special Issue, Adv. Water Resour., 50, 1–3, doi:10.1016/j.advwatres.2012.11.006, 2012.

Leuning, R., Zhang, Y. Q., Rajaud, A., Cleugh, H., and Tu, K.: A simple surface conductance model to estimate regional evaporation using MODIS leaf area index and the Penman–Monteith equation, Water Resour. Res., 44, W10419, doi:10.1029/2007WR006562, 2008.

Li, H., Liu, Q. H., Zhong, B., Du, Y. M., Wang, H. S., and Wang, Q.: A single-channel algorithm for land surface temperature retrieval from HJ-1B/IRS data based on a parametric model, 2010 IEEE International Geoscience and Remote Sensing Symposium (IGARSS), Honolulu, Hawaii, USA, 2448–2451, 2010.

Li, H., Liu, Q., Jiang, J., Wang, H., and Sun, L.: Validation of the land surface temperature derived from HJ-1B/IRS data with ground measurements, 2011 IEEE International Geoscience and Remote Sensing Symposium (IGARSS), Vancouver, Canada, 293–296, 2011.

Li, H., Sun, D., Yu, Y., Wang, H., Liu, Y., Liu, Q., Du, Y., Wang, H., and Cao, B.: Evaluation of the VIIRS and MODIS LST products in an arid area of Northwest China, Remote Sens. Environ., 142, 111–121, doi:10.1016/j.rse.2013.11.014, 2014.

Li, J., Gu, X., Li, X., Yu, T., Chen, H., and Long, M.: Validation of HJ-1B Thermal Band On-board Calibration and Its Sensitivity Analysis, Remote Sens. Inform., 1, 3–9, doi:10.3969/j.issn.1000-3177.2011.01.001, 2011.

Li, L., Xin, X. Z., Su, G. L., and Liu, Q. H.: Photosynthetically active radiation retrieval based on HJ-1A/B satellite data, Sci. China Earth Sci., 53, 81–91, doi:10.1007/s11430-010-4142-5, 2010.

Li, X., Li, X. W., Li, Z. Y., Ma, M. G., Wang, J., Xiao, Q., Liu, Q., Che, T., Chen, E. X., Yan, G. J., Hu, Z. Y., Zhang, L. X., Chu, R. Z., Su, P. X., Liu, Q. H., Liu, S. M., Wang, J. D., Niu, Z., Chen, Y., Jin, R., Wang, W. Z., Ran, Y. H., Xin, X. Z., and Ren, H. Z.: Watershed Allied Telemetry Experimental Research, J. Geophys. Res.-Atmos., 114, D22103, doi:10.1029/2008JD011590, 2009.

Li, X., Cheng, G. D., Liu, S. M., Xiao, Q., Ma, M. G., Jin, R., Che, T., Liu, Q. H., Wang, W. Z., Qi, Y., Wen, J. G., Li, H. Y., Zhu, G. F., Guo, J. W., Ran, Y. H., Wang, S. G., Zhu, Z. L., Zhou, J., Hu, X. L., and Xu, Z. W.: Heihe Watershed Allied Telemetry Experimental Research (HiWATER): Scientific Objectives and Experimental Design, B. Am. Meteorol. Soc., 94, 1145–1160, doi:10.1175/BAMS-D-12-00154.1, 2013.

Li, Z. L., Tang, B. H., Wu, H., Ren, H., Yan, G., Wan, Z., Trigo, I. F., and Sobrino, J. A.: Satellite-derived land surface temperature: Current status and perspectives, Remote Sens. Environ., 131, 14–37, doi:10.1016/j.rse.2012.12.008, 2013.

Liang, S., Stroeve, J., and Box, J. E.: Mapping daily snow/ice shortwave broadband albedo from Moderate Resolution Imaging Spectroradiometer (MODIS): The improved direct retrieval algorithm and validation with Greenland *in situ* measurement, J. Geophys. Res.-Atmos., 110, D10109, doi:10.1029/2004JD005493, 2005.

Liang, S. L., Zhang, X. T., Xiao, Z. Q., Cheng, J., Liu, Q., and Zhao, X.: Global LAnd Surface Satellite (GLASS) Products: Algorithms, Validation and Analysis, in: Springer Briefs in Earth Sciences, 1st Edn., Springer International Publishing, Beijing, doi:10.1007/978-3-319-02588-9, 2014.

Liu, Q., Qu, Y., Wang, L. Z., Liu, N. F., and Liang, S. L.: Glass-Global Land Surface Broadband Albedo Product: Algorithm Theoretical Basis Document, Version 1, College of Global Change and Earth System Science, Beijing Norman University, Beijing, 1–50, 2011.

Liu, S. M., Xu, Z. W., Wang, W. Z., Jia, Z. Z., Zhu, M. J., Bai, J., and Wang, J. M.: A comparison of eddy-covariance and large aperture scintillometer measurements with respect to the energy balance closure problem, Hydrol. Earth Syst. Sci., 15, 1291–1306, doi:10.5194/hess-15-1291-2011, 2011.

Liu, S., Xu, Z., Song, L., Zhao, Q., Ge, Y., Xu, T., Ma, Y., Zhu, Z., Jia, Z., and Zhang, F.: Upscaling evapotranspiration measurements from multi-site to the satellite pixel scale over heterogeneous land surfaces, Agr. Forest Meteorol., doi:10.1016/j.agrformet.2016.04.008, in press, 2016.

Long, D. and Singh, V. P.: A modified surface energy balance algorithm for land (M-SEBAL) based on a trapezoidal framework, Water Resour. Res., 48, W02528, doi:10.1029/2011WR010607, 2012a.

Long, D. and Singh, V. P.: A Two-source Trapezoid Model for Evapotranspiration (TTME) from satellite imagery, Remote Sens. Environ., 121, 370–388, doi:10.1016/j.rse.2012.02.015, 2012b.

Maayar, E. M. and Chen, J. M.: Spatial scaling of evapotranspiration as affected by heterogeneities in vegetation, topography, and soil texture, Remote Sens. Environ., 102, 33–51, doi:10.1016/j.rse.2006.01.017, 2006.

Mallick, K., Boegh, E., Trebs, I., Alfieri, J. G., Kustas, W. P., Prueger, J. H., Niyogi, D., Das, N., Drewry, D. T., Hoffmann, L., and Jarvis, A. J.: Reintroducing radiometric surface temperature into the Penman–Monteith formulation, Water Resour. Res., 51, 6214–6243, doi:10.1002/2014WR016106, 2015.

McCabe, M. F. and Wood, E. F.: Scale influences on the remote estimation of evapotranspiration using multiple satellite sensors, Remote Sens. Environ., 105, 271–285, doi:10.1016/j.rse.2006.07.006, 2006.

Meyers, T. P. and Hollinger, S. E.: An assessment of storage terms in the surface energy balance of maize and soybean, Agr. Forest Meteorol., 125, 105–115, doi:10.1016/j.agrformet.2004.03.001, 2004.

Moran, M. S., Humes, K. S., and Pinter Jr., P. J.: The scaling characteristics of remotely-sensed variables for sparsely-vegetated heterogeneous landscapes, J. Hydrol., 190, 337–362, doi:10.1016/S0022-1694(96)03133-2, 1997.

Mu, Q. Z., Zhao, M. S., and Running, S. W.: Improvements to a MODIS global terrestrial evapotranspiration algorithm, Remote Sens. Environ., 115, 1781–1800, doi:10.1016/j.rse.2011.02.019, 2011.

Mukherjee, S., Joshi, P. K., and Garg, R. D.: A comparison of different regression models for downscaling Landsat and MODIS land surface temperature images over heterogeneous landscape, Adv. Space Res., 54, 655–669, doi:10.1016/j.asr.2014.04.013, 2014.

Neftel, A., Spirig, C., and Ammann, C.: Application and test of a simple tool for operational footprint evaluations, Environ. Pollut., 152, 644–652, doi:10.1016/j.envpol.2007.06.062, 2008.

Nilson, T.: A theoretical analysis of the frequency of gaps in plant stands, Agr. Forest Meteorol., 8, 25–38, 1971.

Norman, J. M., Kustas, W. P., and Humes, K. S.: Source approach for estimating soil and vegetation energy fluxes in observations of directional radiometric surface temperature, Agr. Forest Meteorol., 77, 263–293, doi:10.1016/0168-1923(95)02265-Y, 1995.

Norman, J. M., Anderson, M. C., Kustas, W. P., French, A. N., Mecikalski, J., Torn, R., Diak, G. R., Schmugge, T. J., and Tanner, B. C. W.: Remote sensing of surface energy fluxes at 101-m pixel resolutions, Water Resour. Res., 39, 1221, doi:10.1029/2002WR001775, 2003.

Paulson, C. A.: The mathematical representation of wind speed and temperature profiles in the unstable atmospheric surface layer, J. Appl. Meteorol., 9, 857–861, 1970.

Prueger, J. H., Hatfield, J. L., Parkin, T. B., Kustas, W. P., Hipps, L. E., Neale, C. M. U., MacPherson, J. I., Eichinger, W. E., and Cooper, D. I.: Tower and Aircraft Eddy Covariance Measurements of Water Vapor, Energy, and Carbon Dioxide Fluxes during SMACEX, J. Hydrometeorol., 6, 954–960, doi:10.1175/JHM457.1, 2005.

Shuttleworth, W. J. and Wallace, J.: Evaporation from sparse crops – an energy combination theory, Q. J. Roy. Meteorol. Soc., 111, 839–855, 1985.

Song, Y., Wang, J. M., Yang, K., Ma, M. G., Li, X., Zhang, Z. H., and Wang, X. F.: A revised surface resistance parameterisation for estimating latent heat flux from remotely sensed data, Int. J. Appl. Earth Obs., 17, 76–84, doi:10.1016/j.jag.2011.10.011, 2012.

Su, Z.: The Surface Energy Balance System (SEBS) for estimation of turbulent heat fluxes, Hydrol. Earth Syst. Sci., 6, 85–100, doi:10.5194/hess-6-85-2002, 2002.

Sun, L., Sun, R., Li, X. W., Chen, H. L., and Zhang, X. F.: Estimating Evapotranspiration using Improved Fractional Vegetation Cover and Land Surface Temperature Space, J. Resour. Ecol., 2, 225–231, doi:10.3969/j.issn.1674-764x.2011.03.005, 2011.

Sun, L., Liang, S. L., Yuan, W. P., and Chen, Z. X.: Improving a Penman–Monteith evapotranspiration model by incorporating soil moisture control on soil evaporation in semiarid areas, Int. J. Digit. Earth, 6, 134–156, doi:10.1080/17538947.2013.783635, 2013.

Valor, E. and Caselles, V.: Mapping land surface emissivity from NDVI: Application to European, African, and South American areas, Remote Sens. Environ., 57, 167–184, doi:10.1016/0034-4257(96)00039-9, 1996.

Venturini, V., Islam, S., and Rodriguez, L.: Estimation of evaporative fraction and evapotranspiration from MODIS products using a complementary based model, Remote Sens. Environ., 112, 132–141, doi:10.1016/j.rse.2007.04.014, 2008.

Vermote, E. F., Tanre, D., Deuze, J. L., Herman, M., Morcrette, J. J., and Kotchenova, S. Y.: Second Simulation of a Satellite Signal in the Solar Spectrum-Vector, 6S User Guide Version 3, http://6s.ltdri.org/pages/manual.html (last access: October 2016), 2006.

Von Randow, C., Kruijt, B., Holtslag, A. A. M., and de Oliveira, M. B. L.: Exploring eddy-covariance and large-aperture scintillometer measurements in an Amazonian rain forest, Agr. Forest Meteorol., 148, 680–690, doi:10.1016/j.agrformet.2007.11.011, 2008.

Wang, K. C. and Liang, S. L.: An improved method for estimating global evapotranspiration based on satellite determination of surface net radiation, vegetation index, temperature, and soil moisture, IEEE International Geoscience and Remote Sensing Symposium, IGARSS 2008, Boston, Massachusetts, USA, III-875–III-878, 2008.

Wang, Q., Wu, C., Li, Q., and Li, J.: Chinese HJ-1A/B satellites and data characteristics, Sci. China Earth Sci., 53, 51–57, doi:10.1007/s11430-010-4139-0, 2010.

Xin, X. and Liu, Q.: The Two-layer Surface Energy Balance Parameterization Scheme (TSEBPS) for estimation of land surface heat fluxes, Hydrol. Earth Syst. Sci., 14, 491–504, doi:10.5194/hess-14-491-2010, 2010.

Xin, X., Liu, Y. N., Liu, Q., and Tang, Y.: Spatial-scale error correction methods for regional fluxes retrieval using MODIS data, J. Remote Sens., 16, 207–231, 2012.

Xu, T. R., Bateni, S. M., and Liang, S. L.: Estimating Turbulent Heat Fluxes With a Weak-Constraint Data Assimilation Scheme: A Case Study (HiWATER-MUSOEXE), IEEE T. Geosci. Remote, 12, 68–72, doi:10.1109/LGRS.2014.2326180, 2015.

Xu, Z. W., Liu, S. M., Li, X., Shi, S. J., Wang, J. M., Zhu, Z. L., Xu, T. R., Wang, W. Z., and Ma, M. G.: Intercomparison of surface energy flux measurement systems used during the HiWATER–MUSOEXE, J. Geophys. Res.-Atmos., 118, 13140–13157, doi:10.1002/2013JD020260, 2013.

Yang, K. and Wang, J.: A temperature prediction-correction method for estimating surface soil heat flux from soil temperature and moisture data, Sci. China Ser. D, 51, 721–729, doi:10.1007/s11430-008-0036-1, 2008.

Yang, Y. T. and Shang, S. H.: A hybrid dual-source scheme and trapezoid framework-based evapotranspiration model (HTEM) using satellite images: Algorithm and model test, J. Geophys. Res.-Atmos., 118, 2284–2300, doi:10.1002/jgrd.50259, 2013.

Permissions

All chapters in this book were first published in HESS, by Copernicus Publications; hereby published with permission under the Creative Commons Attribution License or equivalent. Every chapter published in this book has been scrutinized by our experts. Their significance has been extensively debated. The topics covered herein carry significant findings which will fuel the growth of the discipline. They may even be implemented as practical applications or may be referred to as a beginning point for another development.

The contributors of this book come from diverse backgrounds, making this book a truly international effort. This book will bring forth new frontiers with its revolutionizing research information and detailed analysis of the nascent developments around the world.

We would like to thank all the contributing authors for lending their expertise to make the book truly unique. They have played a crucial role in the development of this book. Without their invaluable contributions this book wouldn't have been possible. They have made vital efforts to compile up to date information on the varied aspects of this subject to make this book a valuable addition to the collection of many professionals and students.

This book was conceptualized with the vision of imparting up-to-date information and advanced data in this field. To ensure the same, a matchless editorial board was set up. Every individual on the board went through rigorous rounds of assessment to prove their worth. After which they invested a large part of their time researching and compiling the most relevant data for our readers.

The editorial board has been involved in producing this book since its inception. They have spent rigorous hours researching and exploring the diverse topics which have resulted in the successful publishing of this book. They have passed on their knowledge of decades through this book. To expedite this challenging task, the publisher supported the team at every step. A small team of assistant editors was also appointed to further simplify the editing procedure and attain best results for the readers.

Apart from the editorial board, the designing team has also invested a significant amount of their time in understanding the subject and creating the most relevant covers. They scrutinized every image to scout for the most suitable representation of the subject and create an appropriate cover for the book.

The publishing team has been an ardent support to the editorial, designing and production team. Their endless efforts to recruit the best for this project, has resulted in the accomplishment of this book. They are a veteran in the field of academics and their pool of knowledge is as vast as their experience in printing. Their expertise and guidance has proved useful at every step. Their uncompromising quality standards have made this book an exceptional effort. Their encouragement from time to time has been an inspiration for everyone.

The publisher and the editorial board hope that this book will prove to be a valuable piece of knowledge for researchers, students, practitioners and scholars across the globe.

Yao, Y. J., Liang, S. L., Cheng, J., Liu, S. M., Fisher, J. B., Zhang, X. D., Jia, K., Zhao, X., Qin, Q. M., Zhao, B., Han, S., Jie, Zhou, G. S., Zhou, G. Y., Li, Y. L., and Zhao, S. H.: MODIS-driven estimation of terrestrial latent heat flux in China based on a modified Priestley–Taylor algorithm, Agr. Forest Meteorol., 171–172, 187–202, doi:10.1016/j.agrformet.2012.11.016, 2013.

Yebra, M., Van Dijk, A., Leuning, R., Huete, A., and Guerschman, J. P.: Evaluation of optical remote sensing to estimate actual evapotranspiration and canopy conductance, Remote Sens. Environ., 129, 250–261, doi:10.1016/j.rse.2012.11.004, 2013.

Yu, S., Xin, X., and Liu, Q.: Estimation of clear-sky longwave downward radiation from HJ-1B thermal data, Sci. China Earth Sci., 56, 829–842, doi:10.1007/s11430-012-4507-z, 2013.

Zhang, R., Sun, X., Wang, W., Xu, J., Zhu, Z., and Tian, J.: An operational two-layer remote sensing model to estimate surface flux in regional scale: physical background, Sci. China Ser. D, 34, 200–216, 2005.

Zhang, X., Zhao, X., Liu, G., Kang, Q., and Wu, D.: Radioactive Quality Evaluation and Cross Validation of Data from the HJ-1A/B Satellites' CCD Sensors, Sensors, 13, 8564, doi:10.3390/s130708564, 2013.

Zhong, B., Ma, P., Nie, A., Yang, A., Yao, Y., Lü, W., Zhang, H., and Liu, Q.: Land cover mapping using time series HJ-1/CCD data, Sci. China Earth Sci., 57, 1790–1799, doi:10.1007/s11430-014-4877-5, 2014a.

Zhong, B., Zhang, Y., Du, T., Yang, A., Lv, W., and Liu, Q.: Cross-Calibration of HJ-1/CCD Over a Desert Site Using Landsat ETM+ Imagery and ASTER GDEM Product, IEEE T. Geosci. Remote, 52, 7247–7263, doi:10.1109/TGRS.2014.2310233, 2014b.

Zhu, G. F., Su, Y. H., Li, X., Zhang, K., and Li, C. B.: Estimating actual evapotranspiration from an alpine grassland on Qinghai-Tibetan plateau using a two-source model and parameter uncertainty analysis by Bayesian approach, J. Hydrol., 476, 42–51, doi:10.1016/j.jhydrol.2012.10.006, 2013.

List of Contributors

E. Beckers, N. Gigot and A. Degré
Univ. Liège, GxABT, Soil–Water Systems, 2 Passage des Déportés, 5030 Gembloux, Belgium

E. Plougonven and A. Léonard
Univ. Liège, Department of Applied Chemistry, Laboratory of Chemical Engineering, Sart-Tilman, 4000 Liège, Belgium

C. Roisin
Walloon Agricultural Research Centre of Gembloux (CRA-W), Department of Agriculture and Natural Environment, Soil Fertility and Water Protection Unit, 4 Rue du Bordia, 5030 Gembloux, Belgium

Y. Brostaux
Univ. Liège, GxABT, Applied Statistics, Computer Science and Mathematics Unit, 2 Passage des Déportés, 5030 Gembloux, Belgium

Alain Dib and M. Levent Kavvas
Department of Civil and Environmental Engineering, University of California, Davis, 95616, USA

Valentijn R. N. Pauwels and Edoardo Daly
Monash University, Department of Civil Engineering, Clayton, Victoria, Australia

Rafael García-Bartual and Ignacio Andrés-Doménech
Universitat Politècnica de València, Instituto Universitario de Investigación de Ingeniería del Agua y Medio Ambiente (IIAMA), Camí de Vera s/n, 46022 Valencia, Spain

Murugesu Sivapalan
Department of Civil and Environmental Engineering, Department of Geography and Geographic Information Science, University of Illinois at Urbana-Champaign, Urbana, Illinois 61801, USA

L. Foresti, M. Reyniers and L. Delobbe
Royal Meteorological Institute of Belgium, Brussels, Belgium

A. Seed
Bureau of Meteorology, Melbourne, Australia

Micheal J. Simpson and Neil I. Fox
University of Missouri, School of Natural Resources, Water Resources Program, Department of Soil, Environmental, and Atmospheric Sciences, 332 ABNR Building, Columbia, Missouri 65211, USA

Zexuan Xu
Climate and Ecosystem Sciences Division, Lawrence Berkeley National Laboratory, Berkeley, California 94720, USA

Bill X. Hu
Institute of Groundwater and Earth Sciences, Jinan University, Guangzhou, Guangdong, 510632, China

Ming Ye
Department of Earth, Ocean and Atmospheric Science, Florida State University, Tallahassee, Florida 32306, USA

Filippo Bandini, Cecile Marie Margaretha Kittel, Sheng Wang, Monica Garcia and Peter Bauer-Gottwein
Department of Environmental Engineering, Technical University of Denmark, Kgs. Lyngby, Denmark

Daniel Olesen and Jakob Jakobsen
National Space Institute, Technical University of Denmark, Kgs. Lyngby, 2800, Denmark

L. P. Wang and P. Willems
Hydraulics Laboratory, Katholieke Universiteit Leuven, 3001 Heverlee (Leuven), Belgium

S. Ochoa-Rodríguez and C. Onof
Department of Civil and Environmental Engineering, Imperial College London, London, SW7 2AZ, UK

Alessia Ferrari, Marco D'Oria, Renato Vacondio, Paolo Mignosa and Maria Giovanna Tanda
Department of Engineering and Architecture, University of Parma, Parma, Italy

Alessandro Dal Palù
Department of Mathematical, Physical and Computer Sciences, University of Parma, Parma, Italy

Index

www.ingramcontent.com/pod-product-compliance
Lightning Source LLC
LaVergne TN
LVHW081026210926
848485LV00004B/957